PERSPECTIVES ON COGNITIVE PSYCHOLOGY

CONSCIOUSNESS

ITS NATURE AND FUNCTIONS

PERSPECTIVES ON COGNITIVE PSYCHOLOGY

Additional books in this series can be found on Nova's website
under the Series tab.

Additional E-books in this series can be found on Nova's website
under the E-book tab.

PSYCHOLOGY RESEARCH PROGRESS

Additional books in this series can be found on Nova's website
under the Series tab.

Additional E-books in this series can be found on Nova's website
under the E-book tab.

PERSPECTIVES ON COGNITIVE PSYCHOLOGY

CONSCIOUSNESS

ITS NATURE AND FUNCTIONS

SHULAMITH KREITLER
AND
ODED MAIMON
EDITORS

New York

For permission to use material from this book please contact us:
Telephone 631-231-7269; Fax 631-231-8175
Web Site: http://www.novapublishers.com

Library of Congress Cataloging-in-Publication Data

Consciousness : its nature and functions / editors, Shulamith Kreitle, Oded Maimon.
p. cm.
Includes index.
ISBN 978-1-62081-096-5 (hardcover)
1. Consciousness. I. Kreitle, Shulamith. II. Maimon, Oded.
BF311.C6544 2012
153--dc23

2012003617

Published by Nova Science Publishers, Inc. † New York

CONTENTS

In: Consciousness: Its Nature and Functions
Editors: Shulamith Kreitler and Oded Maimon
ISBN 978-1-62081-096-5

Chapter 1

Introduction: Consciousness is the Next Stage of Evolution

Oded Maimon and Shulamith Kreitler

The nature of consciousness has been for centuries one of the most intriguing questions. The term was used in different contexts and with totally different meanings and interpretations, but nevertheless it has preserved its identity – it seems still to be the same issue that is being discussed.

So what is consciousness? We would have liked to start with a scientific definition, and then logically follow the different disciplines that treat its various aspects. However, reality is different. There is no one definition to which we, or for that matter, the authors who represent the different disciplines, could subscribe. Therefore, since many disciplines maintain their own unique definition of consciousness, it is necessary, until proven otherwise, to adopt another attitude. So the most advanced and open minded view dictated by the present circumstances of consciousness studies is to illuminate it from a variety of disciplines, scientific as well as spiritual, and in between. However, to get started, we will first present several possible definitions and show some of their consequences, sometime even contradictory ones. Despite the difficulty of definition, we believe that there is a broadly shared underlying intuition about what consciousness is.

Consciousness can be used as a noun, as an adjective, as a fact, as an object, as a state, as a location, as a phenomenon, epiphenomenon, real or imaginary, a concept or a construction. Yet even this long list does not exhaust the possibilities. Consciousness is often referred to as the totality of an individual's state; as including the variety of aspects of the relationship between the mind, the body and the outside world; as the individual or the collective to which the individual belongs; as the state of understanding and of realizing something; as the complex of moral principles that control or regulate the actions or thoughts of an individual, a kind of an inner sense of what is right or wrong in one's conduct or motives.

It appears that the conceptual arena is strewn with many more signs of the struggle for the definition of consciousness. The view would be distorted if we did not mention some further familiar attempts at defining consciousness, for example, as subjectivity; awareness at a given moment in time or in general; feeling oneself as a self (permanent or changing); the ability to experience emotions; or the executive control system of the mind.

Less familiar definitions of consciousness have been developed in the framework of other philosophical traditions. For example, in Chinese philosophy consciousness is defined as the "Way of Being", as the interplay between the Ying and the Yang, as that from which everything emerges and to which everything returns, but is indefinable. We can try to define it but it stays undefined.

In Indian philosophy we find definitions of a different kind, such as the claim that consciousness is a biological quality of the body (Carvaka/Materialism and NyayaVaisheshika). In other words, the self is just a body, which is illuminated by consciousness, but when life ends, nothing remains. Further definitions, which highlight other, possibly contradictory aspects of consciousness, emphasize that only consciousness exists, and out of it the true existence of the world and the body arise as ongoing transformations (the Tantric approach). However, others claim that what we see and experience by means of regular consciousness are only illusions – the so-called Maya (the Advaita-Vedanta approach), until we reach the truth. These briefly presented approaches are designed only to illustrate the variations in conceptualizing consciousness that exist in different philosophy schools even within the same culture.

The above examples illustrate how deep, broad and intractable the issue of consciousness is. It has been identified with matter, energy, spirit, with existence and non-existence. Yet it remained undefined and refuses to go away. Despite all valiant attempts to frame consciousness within any of the familiar and scientific setups, it has remained as a familiar yet elusive quality.

To our mind, the persistent attempts to crack the mystery of consciousness support two important conclusions. One conclusion is that the multiple constructs from different domains with which consciousness is somehow related have been identified, although consciousness is identical with none of them. These domains include constructs from biology (body, life), philosophy (knowledge, understanding), religion (God, spirituality), mathematics (neural network and machine learning in general, complex systems) physics (energy, phase changes), physiology (brain), psychology (emotions, cognition, awareness, personality, self, conscience) and even the paranormal (psi-phenomena), and different kinds of artistic and non-verbal experiences (natural or induced). Consciousness is none of these and similar constructs. But while investigating consciousness we should beware of overlooking the fact that consciousness is in fact related to all of these.

When we get to understand consciousness – and we hope this book will contribute to the attainment of this objective - we may be able also to clarify what the relations of consciousness to these constructs are, and to deepen our understanding of these constructs and their domains too.

Another important conclusion is that exploring and eventually understanding consciousness is very much related to our existence as human beings in the full sense of the word. Consciousness has been with us in some form all along but not as something we understood or could control. In recent decades the situation seems to have changed. There is a growing interest in consciousness on the part of all disciplines (as our book also shows) and there is an increasing concern with ways and means to understand and control consciousness, and to expand it - scientifically and artistically, experientially and cognitively, personally and collectively.

Ornstein, one of the major early explorers of consciousness in the last century claimed that as human beings we are bound to turn evolution into conscious evolution (see Chapter 2). We venture to go one step further and claim that the next stage of evolution would be the evolution of consciousness. Evolution of human beings will most probably involve not better teeth (these may be replaced if necessary), increased height (we may not need it), stronger muscles (we may use machines instead), better memory (we may complement it through various computerized devices), or sharper senses (we may improve them through different electronic and other instruments). The next phase of evolution of human beings will consist in the evolution of consciousness – its expansion, development, organization, and reinforced ability on our part to control it, shape it and put it to our use for improving our chances of survival and especially enhancing our happiness.

This book aims at contributing toward this goal. It is meant to constitute one step toward better understanding of "consciousness: its nature and functions".It discusses the different aspects of consciousness, by means of the most advanced theories and practices.

The uniqueness of this book is that it presents the current variety of approaches to consciousness ranging from the modern scientific conceptions to the various philosophical and cultural answers to the oldest problem of consciousness. The book is expected to provide the readers a kind of fertilized substratum, from which their personal concepts can grow, based on science, heritage and personal experience.

So what does consciousness really mean, what is it, what can be said about it that would do justice to this multifacetted concept or phenomenon and would render it meaningful for the readers? ? Answers to these and similar questions can be obtained in the framework of the different disciplinary approaches to the exploration of consciousness. We chose ten schools of thoughts to illuminate consciousness (not claiming for completeness). The book starts with a philosophical exposure of the topic, followed by a historical-cultural approach to the psychological evolution of consciousness preceding a chapter that deals with the biological evolution of consciousness. The following chapters focus on the anthropological, sociological, experiential and spiritual approaches. Several chapters are devoted to the psychological studies of consciousness, with an emphasis on cognition, and to the neuropsychological studies, with an emphasis on the brain. The book concludes with physical studies of the phenomena.

We cover in this book the major fields of studying consciousness. Thus the reader can appreciate this fascinating topic from many important facets that exist in today's state-of-the-art studies. We may only pray that it would contribute to the enrichment of consciousness on all levels of human life.

THE PHILOSOPHICAL APPROACH

In: Consciousness: Its Nature and Functions
Editors: Shulamith Kreitler and Oded Maimon
ISBN 978-1-62081-096-5

Chapter 2

Towards a Theory of Awareness and of Consciousness

Joseph Agassi
Tel Aviv University and York University, Toronto

Abstract

Consciousness is repeatedly observed, thus deserving efforts to explain it scientifically. Early studies of it came to prove it illusory. These are irrelevant, as persistent illusions invite explanation: why do they persist? And how do we distinguish between conscious and subconscious decisions, and between decisions and self-deception? Efforts to overcome Hume's criticism of the theory of knowledge by limiting it to conscious thinking raise interest in consciousness.

Nevertheless, the view is popular that consciousness is scientifically irrelevant. This, too, invites explanation.

Traditional psychology of learning rightly begins with the study of proper research. It demands that research should rest on no presuppositions. This is impossible. When in trouble, researchers try to articulate some of their tacit presuppositions so as to improve them. This is vital for the progress of science. The most significant supposition is that perception is passive. It is amply refuted yet it stays popular. And prevents the understanding that consciousness is a dimension of perception that is irreducible to other qualities of perception and so it requires special attention. For that criticism of received theories of perception and of consciousness are vital, especially the view that scientific controversy is objectionable. For, properly run controversy always boosts science.

Removing the obstacles on the way to progress in research on consciousness is not enough but is all that is attempted here. This paper ends with a refutation of the idea that Turing's test vindicates his dismissal of consciousness as inviting research.

Preface

Consciousness is a common experience. Awareness is either consciousness or the disposition to be conscious. Thus, studies of awareness usually come under studies of consciousness. My example for these studies is the whole field of speed reading that is almost entirely devoted to aspects of the consciousness of reading and of its outcome. The demand to ignore consciousness (as subjective and thus allegedly as not objective and thus inferior) rests on the erroneous disregard for the fact that the subjective world is a part of the world that we wish to understand. The central scientific convention regarding experience, the sole convention that the scientific community endorses unanimously ever since its establishment during the scientific revolution, is this: every repeated and repeatable observation is scientific and invites scientific explanation. Repeated observations concerning consciousness thus invite scientific explanations. These invite the construction of some framework that should guide research into consciousness. Towards this let me draw attention to errors popular in the research world that impede it.

Explanations of observations, obviously, may deny existence to their objects. For example, science denies existence to the objects of mirages and of hallucinations. Explanations of mirages present them as manifestations of an optical illusion plus the understandable self-delusion of their very thirsty observers. Hence, to explain mirage is to refer to consciousness. This makes current explanations of human conduct different from, say, the current explanation of the attraction of moth to fire that is presumably automatic. This difference annoys some thinkers and delights others. We may try to study of this fact too.

Case Study: Speed Reading

Speed reading developed as soon as it the tachistoscope showed that reading is very fast: images of words thrown on a screen for a duration order-of-magnitude far shorter than that of the perception threshold (one-tenth of a second) are read with ease. This raised problems. The theoretical problem was neurological: what makes different threshold phenomena so different in magnitude?[1] The practical problem is, how can this help improve the speed of reading? Various impediments to reading thus became apparent. The biggest impediment turned out to be readers' use of muscles other than those that move eyeballs, especially of the neck and of the lips: they get tired fast. Surprisingly many people waste energy this way *unawares*. They must become *aware* of this before they can *try* to improve. It is far from easy to help them do that. The easiest way to make people *aware* of this is to photograph them and show them their *unconscious* waste of effort. The second impediment to improvement is the *reluctance* to skim that educators instill in *effort* to enhance the *care* of *careful* reading in order to make the reader fully *conscious* of what they *read*. To overcome this, readers have to become *aware* of their skimming habit, and to learn that doing so *aware* is to raise *control* over action. Generally, we possess gearboxes that habitually regulate action — including the speed of reading: becoming *aware* of gearboxes and learning to *control* them is immensely beneficial. (Good learning is not of high speed but of variable speed.) The increased *awareness* of our *control* mechanisms improves performances. Next to speed is learning to retain the information accrued in reading. While reading it is advisable to sum up information accrued in short statements. The enhancement of speed, comprehension and memory go together and

[1] The puzzle is general for all thresholds. The paradigm case is holding a stick horizontally with both hands, sensing a blow on its middle: the slightest deviation from the middle leads to the proper sensation of the hand near it as first.

separately. The information mentioned thus far suffices for the present discussion of *awareness*.

This is not to deny that for some habits awareness is an impediment. This was discussed in great detail and with much added philosophical fluff in Eugen Herrigel's *Zen in the Art of Archery* (Yamada Shōji, 2001). This shows that we can control the degree of our awareness.

EXPLAINING *VERSUS* EXPLAINING AWAY

The first item on the agenda of any scientific discussion of consciousness is the question, does it exist at all, and does it merit scientific attention? We better ignore this question. Even if consciousness is but an illusion (an epiphenomenon of the brain, to use the philosophical jargon, Searle, 1997; see also Crick, 1994), it invites explanation, of course. The explanation of shadows is the paradigm (of epiphenomena); they do not exist. But geometrical optics explains them to full satisfaction. The explanation of mirages and hallucinations similarly assume their objects non-existent. So we may want to explain consciousness too. Nevertheless, this dual question persists: does consciousness exist and does it merit scientific attention? The best known discussion of it is that of Alan Turing: science can ignore consciousness, he said, since it is irrelevant to the problem whether machines can think (Turing, 1950, page 447). He tried to prove that science can ignore consciousness without loss; received opinion has it (erroneously) that his proof (the Turing test) is conclusive. And so his recommendation that science should ignore consciousness (for a while) was influential: a volume on the topic (Sayre, 1969) and an essay on an adjacent topic (Shaffer, 1965) display the paucity of literature on it then. Sayre suggests conscious computers are possible, although of a kind we still do not possess (Sayre, 1969, p. 170). Shaffer says, a machine that emulates human conduct perfectly is constructible, yet whether perfect simulation is good enough is in doubt (Sayre, 1969, p. 90). It would then be hard to dismiss Turing's assertion, Sayre adds, that science will dismiss this doubt as pointless. Artificial intelligence doyen Marvin Minsky went further: he took consciousness to be no more problematic than the construction of self-monitoring computers, which is obviously unproblematic (Minsky, 1988, Ch. 28 §8, see also Damásio, 1010).

This sounds odd, but it is legitimate. Thus, the explanation of sound sensation as the resonance[2] of bio-acoustic-sensors to waves in the medium keeps the subjective feeling of sound outside the discussion. Similarly, explaining a color sensation as neural photoelectric response does not explain the subjective feeling that accompanies the sensation of color. (We have no idea how to do that.) Hence, the objection that Minsky's explanation is not intuitive is invalid. Still, the question remains: is his explanation true?

Popper and Eccles said, no: consciousness is not passive monitoring but a locus of brain activity. "Our present hypothesis regards the neuronal machinery as a multiplex of ... structures: the experienced unity comes, not from a neurophysiological synthesis, but from the proposed integrating character of the self-conscious mind. ... [The] self-conscious mind is developed in order to give this unity of the self in all its conscious experience and actions." (Popper and Eccles, 1984, p. 362) There is no need here to decide between Minsky's view and

---

[2] The idea that consciousness is some sort of resonance — reverberation — within the brain is reverberating in the literature ever since Donald O. Hebb has suggested it in the1940's (Hebb, 1949, p. 73).

that of Popper and Eccles: even on his view his conclusion that science can ignore consciousness is invalid. The reason is simply that observations of consciousness, and these are many, are repeatedly reported and so they merit explanation. Let me elaborate on this point.

The loss of naïve realism is painful. Some philosophers make great efforts to retrieve it (G. E. Moore, 1924, and Wittgenstein, 1953, §306), whereas others find the loss unavoidable (Russell, 1965, opening). Sir Arthur Stanley Eddington noted (Eddington, 1928, Preface) that quantum mechanics deems illusory the desk in his study: the naïve image of it as a smooth brown surface is false: it comprises an almost empty space in which minute particles swirl at immense speed. Eddington was in two minds about that desk. At times he took its presence for granted and declared physical theory a *façon de parler*; at times he deemed theory a description of the facts more accurate than naïve realism. This duality is resolvable: scientific theories comprise a series of description of reality, where the better theory explains its predecessor as less accurate and less comprehensive. And then all past theories are series of approximations to the truth, not the final truth. Naïve realism and commonsense are then older theories, be they scientific or proto-scientific.

Nevertheless, not all corrections of past views are of the same ilk. Consider this: Aristotle said, both gravity and its opposite, levity or buoyancy, are basic. He thereby dissented from Plato, who denied that levity is basic, considering bodies that buoy up heavy, though less heavy than their environment: low gravity looks like levity. Aristotle rejected this idea and considered it muddled. Archimedes proved him the muddled party. Here is a newer example. Naïve realism and some old theories of physics take as basic both heat and cold. Modern thermodynamics takes them to be mere degrees of concentration of motion-energy of molecules. This is not to deny that heat is there, but to deny the older, important idea that heat is a thing (phlogiston and its heir caloric) as well as the idea that cold is a thing: cold is low degree of heat. Similarly, a debate lasted for about a century as to whether negative electricity exists or whether it is but the low level of positive electricity (as compared to the surroundings). Both positive and negative magnetism were deemed real, until Ampère and Faraday proved magnetic poles illusory. Ampère declared electricity a thing; Faraday and Maxwell did not. J. J. Thomson discovered the electron and thus disproved them all.

Is Eddington’s desk the same thing as the magnet that is there but not as a thing? Or is it better ignored as a thing, together with Aristotle’s fire? We want science to take account of that desk, and we would not reject it as totally as we reject the phlogiston. Wherever the difference between Eddington's desk and phlogiston lies, it is very important. It is the difference between explaining and explaining away. We thus explain self-delusion, and as we do that we also explain away the ghosts that the deluded observe. Now in many cases it does not matter overmuch whether we choose explaining or explaining-away. Thus, it is well known that Einstein first declared the pervasive ether as non-existent and then spoke freely about the ether, meaning roughly-empty space. i.e., space with forces that it embeds but with no matter. Some people found this confusing, but it is not. Yet not all cases are like that. When we explain away ghosts stories and such we do not leave much room for them just because they are not explanations so that doing away with them is no loss. In the very early days of the scientific revolution the tendency to deny the existence of colors as illusory was popular. They were soon declared proper subject for inquiry as their appearance is repeatable. Later, they were explained as wavelengths. The heat of pepper and the cold of opium were deemed thermal mattes but this was soon explained away once and for all.

AWARENESS OBSERVED

As awareness is the disposition to be conscious, it is naturally as repeatedly observable as consciousness. Unable to offer an explanation for it, I hope that my presentation will help future efforts to explain it: I present it (or rather degrees of it) as a dimension of perception that is irreducible to other qualities of perception, so that it requires special attention. Classical perception theory prevented that, and awareness of this fact enhances the comprehension of perception — in experience and in theories alike. This claim is not new. Already in the seventeenth century Leibniz expressed it explicitly: he distinguished between perception and awareness by observing the awareness of the very absence of some perception (Leibniz, 1996, pp. 112-118). The striking fact he drew attention to is that the miller wakes up when the sound of the windmill's engine stops as the wind slacks. This shows that the sleeping miller is not totally unaware of the surrounding, that in sleep awareness is reduced but not shut down. This raises intriguing questions that today engage quite a few researchers, regarding lower animals and regarding patients in comma. Do other animals possess some degree of awareness?[3] Leibniz claimed that they do, but his answer was a part of a metaphysical view that the scientific community never took seriously: all objects, he said, possess awareness to some degree, even sticks and stones. The problem regarding the state of deep comma is more interesting: is it a total shutdown of consciousness? Some evidence suggests that it is not always so. Discussion on this point illustrates that consciousness (includes expectations) regarding sense experience differs from sense experience. This refutes the perception theory that was generally received until the early twentieth century, as this theory applies equally to humans and other animals, as David Hume noted approvingly (Hume, 1986, pp. 176-9).

The observations that Leibniz made were lost on his contemporaries because he had no perception theory to replace the received one that took its building blocks to be sensations of units of perception (allegedly given to the senses) and their associations. John Locke, George Berkeley and David Hume stressed the fact that they took for granted the sensationalist theory of perception (as applicable equally to all higher animals, Hume added). This plays down the place of awareness in perception. Since that theory was the only one extant, it is understandable that they took it for granted. It was a *tour de force* that Hume stated it explicitly in stark clarity. He did so in order to show that it has no room for causality. This renders all causal assertions theoretical, which is very surprising. Worse, it presents all assertions about causality as not founded on facts, indeed as incapable of finding empirical foundations. The same holds, if less clearly, for Berkeley, who argued — rightly — that what is a thing and what not is beyond the domain of the empirical as construed at the time. At times it is empirical, for sure, but by a newer view of the empirical. What this shows is incontestable; hence, all arguments that support the claim that causality or "thinginess" is perceived are arguments against the sensationalist theory of perception. Strangely, whereas psychologists are increasingly aware of the defects of the sensationalist theory of perception, philosophers of science adamantly insists on clinging to it. Their reason is that they wish to validate the empirical foundations of science. For, in the wish to prevent circularity in their arguments, they try to present pure perceptions, namely, perceptions free of any theory. These

[3] Today the hypothesis that other animals are aware is empirically testable by reference to the spindles in their sleep (Dang-Vu *et al.*, 2010).

should be observations for science to endorse unhesitatingly, regardless of any controversy within science.

Notoriously, observation-reports are theory-laden. This leads philosophers of science to efforts to strip them of their theoretical bias, namely, of their theoretical components. This effort rests on the hypothesis that some perceptions precede theorizing. Notoriously, empirical information refutes this claim: most empirical researchers of perception (unlike most philosophers of science) consider it false. This does not deter most philosophers of science: to evade it they view the precedence of the empirical over the theoretical not a fact but a logical characteristic: the empirical component of ordinary, theory-laden observations is independent of its theoretical component. What these philosophers of science claim need not be questioned, at least not here. What we should here notice is that their theory leaves no need for any discussion of consciousness and no room for it. The casualty of this exclusion is the empirical study of consciousness. Once this enters the equation, the hypothesis that ordinary observation-reports can be divided to the purely empirical and the purely theoretical has been given up completely.

This became the cornerstone of the most influential non-sensationalist theory of knowledge, that of Immanuel Kant. This theory, however, rests on Hume's observation that no theory can rest on empirical foundations. Most philosophers of science therefore ignore it so that his perception theory influenced scientists (Gregory, 1989), whereas philosophers miss it in their eagerness to provide solid empirical foundations for scientific knowledge. To that end they try to pinpoint the weakness in Hume's argument against the view that this exercise can be successfully concluded. For, they say, we know that Hume was in error, since we know that scientific knowledge does exist and that it is empirical. This knowledge is awareness of sorts and so reference to it leads them straight to inconsistency: they ignore the fact that sensationalism is the error of Hume that they seek.

Hume tried to cope with the situation differently. Finding our theoretical apparatus defective, he tried to find out how we apply our knowledge in practice. He argued that we do so *unawares*: he observed that learning to do something very well amounts to doing it unawares. (The best examples today are things like driving a car or riding a bicycle, but walking is as good an example and to the sensitive just as impressive, perhaps even more so, seeing that we learn to walk before we develop our awareness sufficiently to remember the fact, and seeing that people who spend long periods in bed have to re-learn to walk. More impressively, the ability to stop a car at a gas station to refill its nearly empty tank is improved when it happens regularly utterly unawares.)

The problem that Hume has raised persists: how is theoretical learning from experience possible? Throughout the history of learning theory since Hume raised that problem, his proposed solution to it occurred repeatedly to students of that problem, although possibly with increased clarity. The most recent ones were those of Gilbert Ryle (Ryle 1949, p. 45) and Michael Polanyi (Polanyi, 1967, p. 4). Polanyi observed that people who apply their knowledge are not always able to describe it in words. It is clearly there, as it is operative, but efforts to articulate it may fail all the same. Polanyi called this "tacit knowledge" and noted that the knowledge that artists possess and transmit to their apprentices is such, and he declared researchers artists, experts in the art of scientific research. Ryle was a member of the school of philosophy that advocates clinging to ordinary language and as such he preferred to present his view as a verbal distinction (although this forced him to use stilted language that is far from the ordinary[4]): he distinguished knowing that [a given assertion is true] from knowing how [to perform a given act properly], tacitly suggesting that the second kind of knowledge is not hit by Hume's critique, as already Hume had explicitly suggested. Hume had added to this, we remember, that the operative kind of knowledge need not involve awareness, that, indeed, the performance of an act improves with the learning to do it unawares and thus unconsciously.

No need to discuss any specific case of awareness or consciousness here: the general case suffices. Consciousness is the general disposition to be aware. It is hard to decide how to begin discussing consciousness: the very choice of a gambit already rests on many presuppositions, and these may be highly problematic and deserve preparatory studies. This trouble is not quite specific to the study of consciousness: it is general within philosophy(Wettersten and Agassi, 1978). It seldom receives the airing it deserves. In science things are obvious: most scientific fields of study take their traditions for granted, and these include traditional presuppositions and the problems that these give rise to and that contributors to the field are invited to take as read. Scientists thus often skip the discussion or even the posing of the problems whose solutions, old or new, they try to present and to discuss; from the very start they consider obvious to their readers the problems that are at issue and their backgrounds. Many papers in philosophy follow this pattern and they are almost never right as their background suppositions and the problems that they discuss are not obvious even to experts.

Thus, most texts that discuss consciousness are troublesome: their presuppositions include the sketch of a view on consciousness that their studies develop. This way they take too much for granted, and, as often happens, the tacit assumptions behind the discussion render their studies good for home consumption only: their authors preach to their parishes.

Can we begin with statements of these assumptions? Clearly it is not always easy to know the assumptions implicit in a given discussion, as the example of the axiom that every geometrician makes and that Moritz Pasch stated only in the nineteenth century (Davis and Hersh, 1981, page 160). Can we at least in principle always make all of our presuppositions explicit? This question was raised only in the twentieth century and only as Ludwig Wittgenstein gave it his affirmative answer (Wittgenstein, 1922, Preface). He stated it emphatically and as a matter of principle, yet at once he qualified it: we can articulate only what is given to articulation, he added (*loc. cit.*). This way he adumbrated an important idea that later authors elaborated on: we have tacit ideas: we may be aware of them but we cannot articulate them. First R. G. Collingwood said so (Collingwood, 1940, pp 33-9: "people are not ordinarily aware of their absolute presuppositions") and then Polanyi did (Polanyi, 1966, "We know more than we can tell"). This is a limitation on reason, of course, as Michael Oakeshott has noted (Oakeshott, 1962, p. 61, "The ... situations of normal life are met, not by consciously applying to ourselves a rule ... but by acting in accordance with certain habits"). Thomas S. Kuhn went further: when the feeling is overwhelming that the literature is stuck, there is a feeling that a change in the tacit suppositions is in demand. He described this feeling and called the time of its prevalence a revolutionary period and he declared it brief. After it the new tacit suppositions have to be broadcast, yet without being articulated: "universally recognized scientific achievements that for a time provide model problems and solutions" are

[4] Ordinary language permits talk about knowing that an act is proper just as knowing how to confirm a truth.

the vehicles of this broadcast. He aptly called these vehicles paradigms (Kuhn, 1970, page viii).

All this is questionable: the basic presuppositions of a scientific field need not be tacit: at times they are given to explicit statement and to critical discussions, propaedeutic (pertaining to teaching) or heuristic (pertaining to discovery). Often they belong not to science proper but to the adjacent field of the philosophy of science – in general or of a specific science. Most scientific fields of study take their traditions for granted, and these include traditional problems that contributors are invited to take for granted or to criticize (on the presupposition that it is recognized as sufficiently important to discuss critically). Scientific researchers thus often skip the discussion or even the posing of the problems on their agendas, namely, those whose solutions, old or new, they try to present and discuss critically; from the very start they consider obvious to their readers what problems are at issue. The problem of consciousness is an example for this: first it was largely neglected, since psychology had little room for it. On the contrary, learning theory presented humans and animals as possessing the same learning mechanisms, we remember, and so both language and awareness were given little or no role in learning. Only with the study of hysteria — of Charcot and more so of Freud — that brought to the limelight subconscious thinking and decisions (Freud, 1896)[5], did conscious thinking and decisions win better attention. Later this led to the study of consciousness as such. It then became the central issue in the studies of some current groups of students of the human mind. This seems to have brought about a consensus of sorts: it is hoped that efforts to find out what determines the constructions of the self and that this will alter social and psychological studies. This is why consciousness has become fashionable in different disciplines. Yet this seeming consensus is misleading. It looks as if the diverse studies of consciousness are complementary; in a sense they are. Yet they come from opposing viewpoints: they bespeak competing philosophies. Their integration toward the emergence of a new encompassing approach to consciousness is perhaps impossible and perhaps it requires a new integrative philosophy, but there is no guarantee for it and there is no argument for the idea that this integrative philosophy is at all possible.

To Unify Diverse Studies

The situation is odd. The theory of knowledge is supposed to promote objectivity and thus detract from subjectivity and thus allegedly detract also from the study of subjectivity, including consciousness. Yet consciousness is central to this field: scientific research is the activity which is performed as consciously as possible: it is an integral part of the Ego in Freud's theory[6]; authors repeatedly stress the need of researchers to be alert. In particular, in the second half of the twentieth century much attention was given to the question, what

[5] To be precise, Freud does not discuss decision here. In a footnote he says, "I purposely leave out of this discussion the question of what the category is to which the association between the two memories belongs (whether it is an association by simultaneity, or by causal connection, or by similarity of content), and of what psychological character is to be attributed to the various 'memories' (conscious or unconscious)." This shows he was clear about the difficulty to square psychoanalysis with associationism; yet he kept his faith in associations all his life.

[6] In a very famous book (Hadamard, 1945, 52) Jacques Hadamard discusses attempts to control the unconscious ; he also refers to both Poincaré and Einstein as aware of the role of unconscious thinking in their work.

presuppositions do researchers rely upon in their researches? Can they be made explicit? Can they be defended? Need they be defended? Etc.

There is one famous presupposition that is taken to be common to all researchers: nature is law-abiding. This presupposition interests those who wish to justify science. Others stress presuppositions in research that are not universally endorsed. This diversity seems too obvious. Yet when historian of science Kuhn observed it, he won world fame overnight. His catch-word, "paradigm", became a keyword that is here to stay despite his having admitted that it is too vague for comfort and despite his decision to cease using it. Since his contribution is so very influential, it deserves a careful exposition. "Paradigm" was initially his word for an example of a case of scientific success that researchers allegedly emulate, we remember. These exist only in some fields of research, and he declared scientific all of them and only them. Why do some fields have paradigms and other not? Why does astronomy always command theoretical consensus — is allegedly run by paradigms — and psychology never does? No answer. No matter. The main thing according to Kuhn is, paradigms change, but not the unanimity with which they rule. What then is unanimity good for? Kuhn hardly noted this question. When he did, he said, unanimity is about the question what paradigm is the most suitable? Unanimity is therefore the most efficient way to go about a paradigm once it is there. This seems reasonable. It is a serious error nonetheless: given two complementary paradigms, it may be useful to have some researchers follow this and others follow the other. Kuhn even agreed to this: there may be two paradigms, he admitted, and even more. Thus, whereas traditional lore takes every theory that commands unanimity within science to keep this quality and never lose it, Kuhn admitted that the consensus may be temporary, yet when a paradigm shift occurs, it occurs with unanimity: disagreement about a paradigm is short-lived. This cannot be the case for dual or multi paradigms. This he ignored. Thus, his theory is severely incomplete, not to say plainly inconsistent.

How does a discipline attain a paradigm? By applying discipline: Kuhn deemed unanimity as due to regimentation. We find such regimentation in other places, of course, such as in totalitarian states. Yet leaders there are not interested in research and so their motives are alien to science. What happens if the leaders of science become tyrants? Then the scientific public will dismiss them or else they cause research to atrophy. How do researchers know when their field is vibrant and when it atrophies? Kuhn made the goat the guardian of the garden: he declared fighting deterioration is the task of the scientific leadership: when field of scientific research begins to deteriorate, they declare a scientific revolution within that field. What if they become tyrants and block all change? Then the field atrophies. All this is somewhat mysterious, but this cannot be helped: just as the activities of research scientists are above the heads of the common public, so the activities of the scientific leaders are above the heads of their peers. (The inequality of intellects, incidentally, is central to most contemporary theories of consciousness.)

Kuhn's view has the merit that, viewed from a certain angle, it seems minimal deviation from traditional lore or presupposition that science commands consensus. The lore, then, modified by Kuhn or not, is amply exemplified: scientific researchers often take for granted and thus do not mention their background knowledge and specific background situations, including their specific background problem-situations, problems, and problem-presentations. This makes it necessary to add some background information for novices or for readers from another discipline or a later era. And these are available and often make for quite intriguing reading. Indeed, Kuhn took this as his starting point, which explains his move from the discipline of the history of science to that of its philosophy.

All this becomes especially intriguing when the study of the background of a piece of research refers to controversy. (They invariably do.) For, in full contrast to traditional lore, what in science enjoys full agreement is utterly unclear. Science does command a tremendously broad consensus. Amazingly, although science applies no authority – religious, political, or any other – it commands much broader a consensus than any other fields of human culture. Philosophers take it for granted that within science what is agreed upon is rational and what is rational is unique and so not given to dispute. Yet just as scientific consensus is an observed fact, so is the prevalence of scientific dispute: contrary to the traditional view that the consensus pertains to scientific theories (the claim, that is, that all and only those theories are scientific that reason commands assent to), some but not all scientific theories command universal assent. Scientific information it is that is uncontroversial, whereas theories under scientific research are controversial. Advocates of the traditional view say, scientific progress is the rendering some controversial theories uncontroversial knowledge. This puts a barrier between scientific knowledge and scientific research!

A very significant reason for the adoption of this view comes from the adoption of the idea that action requires accord and rational action requires rational accord. Traditional theory justifies much action as science-based and explains the consensus that went into it as rational, as science-based. All this is an example of a prejudice: a strongly advocated theory despite familiar contrary information. The consensus that leads to rational group decisions is usually due to compromise, as John Watkins has observed (Watkins, 1957-8).

Scientific Justification Versus Scientific Disputes

The received view of science as consensus seems very comfortable, yet it is very frustrating, as it declares all dissent due to some serious defect, to some deviation from scientific norms. The word for this deviation is prejudice. The theory of prejudice was discovered repeatedly in diverse variants and it is therefore attributed to diverse thinkers, the latest of whom are Gordon Allport (Allport, 1954) and Leon Festinger (Festinger, 1954). Earlier, Karl Marx spoke of class prejudice; Sigmund Freud ascribed it to childhood emotional scars (traumas).

What all these versions of the theory share is the idea of the source of the trouble: observations are theory-laden. This was the discovery of Galileo and of Bacon, already four centuries ago. Galileo said, but for our theories, we could just as well say that as you stroll down the street of Florence on a moonlit night the moon jumps from rooftop to rooftop like a cat (Galileo, 1633, 1953, p. 256). Bacon went further (Bacon, 1620, Bk., I, Aph. LXXXVII and Aph. CXV). He said, anyone who entertains any hypothesis is bound to fall in love with it and then be blinded by it, so that the precondition for the ability to be a researcher is the readiness to give up all of one's opinions and entertain no idea unless proven. Galileo said, we cannot help having prejudices and this is why we should subject our views to repeated examinations. Bacon disagreed: he said we all examine our prejudices repeatedly, but to no avail. Most writers on prejudices agree with Bacon here. And so the question is, how do we get rid of prejudices? Or do we?

Despite centuries of frustration, most philosophers still continue to discuss the question, what is the rational consensus about a theory? They still resolutely ignore the much more fruitful question, what is rational disagreement? They may suggest, or at least imply, that dissent rests on ignorance. But then, how come they fail to take cognizance of the prevalence of scientific dissent? They do not discuss rational action as they take it for granted that rational action rests on rationally endorsed theory, despite the observed fact that rational action rests on compromise.

There is much truth to the traditional lore: there is a lot of consensus within science, and not only about observation reports and the desire to explain them, but also as to problems and their backgrounds — the problem-situations so-called. This is best shown when science is compared with the best of pseudo-science. It is common to compare today's physics with today's psychology, and the comparisons always come to answer the question, is psychology a science? The answer, affirmative or negative as it may be, naturally rests on a criterion, and the criterion is usually controversial, as is the answer to the question, is the criterion properly applied? Oddly, when we consider a specific controversy, these problems hardly arise, since we wish to approach the controversy with an open mind and this often suffices. We may comfortably take this as the default option. Texts whose authors express contempt for some opinion are thus rightly suspected. Of course, some opinions are contemptible, yet the default option is not to refer to them and definitely not honor them with serious arguments. Even respectable but weak opinions are often rightly ignored, at least as the default option. All this seems commonsensical. Yet it leads to the disregard of most traditional writings on consciousness since in this field contempt is alas very common.

The disregard of texts infected with expressions of contempt, let me repeat, is the default option. Some texts of this kind are too valuable to ignore. It is then advisable to cleanse their valuable parts, to apply to them a background that makes sense of the controversy to which they contribute and to a contrast of them with cleansed texts that present contrary opinion. This goes a long way to divide the sheep from the goats with no reference to the criterion of demarcation of scientific theories from unscientific ones. The criterion is important for many ends, but for the study of problem-situations it is inessential: what was thus far said here should practically suffice.

Consider texts of metaphysics proper. Theses seldom present their backgrounds; usually they refer to controversial matters while ignoring their being controversial. Consider the famous controversy on the nature of Man: many options exist; Man is a rational animal, a laughing animal, or a working animal; Man is fear and trembling, or fear and boredom, or fear and nausea. One might expect some debate about all this. There is practically none.

Consider individual psychology then. The behaviorist and the psychoanalytic literatures contemptuously ignore each other. Within the psychoanalytic tradition the different strands also ignore each other: very little genuine dialogue (namely, respectful criticism) takes place between them. Many debates concern Freud's division of the mind to conscious (super-ego-plus-ego), front-conscious, sub-conscious, and unconscious (id). Clearly, these pertain to the study of consciousness. It is doubtful whether it is worthwhile to undertake the Herculean task of clearing this vast literature. It is better to use it as a mere prompt for listing some salient repeatable empirical information on consciousness. The opening paragraph of this essay lists some such facts on consciousness; this is totally absent from that literature. But there are better examples. Perhaps the best is Hume's observation that a task repeatedly performed gets performed unawares. Perhaps here is the place to invite the role of consciousness and of awareness in common thinking. Perhaps scientific thinking is better, as there the thinking process is simpler and more conspicuous.

What is the route from a list of repeatable items of factual information to a theory that explains them? This is the classical problem of induction: how do observations convey theoretical information? Students of this problem are seldom interested in this route — possibly unawares. They hardly ever mention it: their concern is to justify induction, not to understand it. This justification they discuss — endlessly. This way they take much for granted, and, as often happens, the tacit assumptions behind their discussion prevent success (Popper, 1963, pp. 57-8).

TRADITIONAL STUDIES OF CONSCIOUSNESS

Why not take consciousness as a part of nature and its study a part of the natural sciences proper? This question is central to philosophy, and advocates of the study of the human sciences as a universal science insist on the view that we should study consciousness as a part of nature and as essential to human nature and thus as indifferent to the different cultures and traditions and to their impact on consciousness. Repeatedly philosophers call this view "naturalism" and they call themselves "naturalists". If there is a dispute here, then it has to do with the detailed specification of naturalism. The chief influence here is that of Descartes; he divided the world into the internal and the external, and he said, whatever humans share with other animals belongs to mechanics proper but the rest belongs to psychology. There are only two sciences, Descartes said, physics and psychology.

The alternative idea is that there is only one science. It belongs to Descartes' disciples, chiefly Julien Offray de La Mettrie, author of *Man a Machine*. Briefly, cognitive science is not a science but a philosophy. It comprises effort to update La Mettrie in the name of current science. It includes the part of philosophy that is traditional learning theory: the theory that we learn from experience — by induction. This theory was alternately taken to be philosophical and empirical. It was therefore refuted doubly. As a part of philosophy it was meant to show that science is rational as it rests on inductive proof. Therefore, since the days of David Hume, mock-proofs repeatedly appeared that show induction to be in no need for proof. Hume himself admitted induction on empirical grounds of sorts. (This is Hume's naturalism.) Therefore, the theory earned also empirical refutations in psychology, especially in the hands of Oswald Külpe and his disciples, of the Wurzburg and the Gestalt schools (Wettersten, 1985, p. 489; Singh, 1991, Chapters 4 and 11).

Does cognitive science go beyond traditional cognitive philosophy? Is it empirical? If not, then there still is the hope that research will render it empirical. How? What makes a research project scientific or empirical? What makes a research program scientific or empirical? Grants committees repeatedly face these questions. Are they conscious of it? Do they possess some criterion or do they employ sheer gut feelings? How does this apply to current studies of consciousness?

Views on cognitive science vary. Its interdisciplinary character, however, is not contested: cognitive science borrows ideas from philosophy, physics, and neurophysiology; from psychology, linguistics and anthropology; and from computer theory and technology. The status of neurophysiology as an empirical science is not challenged. Whether neurophysiology contributes anything substantial to cognitive science, however, is not clear.

Whatever it may contribute to the understanding of cognition, such as the fascinating matter of reverse correlation in neurophysiology (the technique for studying the way sensory neurons combine signals from different locations and generate responses), relates to animal cognition in general, not specifically to human cognition; at least not as yet. Nevertheless, the wedding of neurophysiology with computer technology, such as the pioneering study of David Marr on vision (Marr, 1982), throw light on physiological processes (not on cognition) by its use of computer simulations. Paul and Patricia Churchland go a bit further as they use neurophysiological information to improve upon the commonsense view of consciousness (Churchland, 1991). But, of course, traditionally the onus of cognitive science is to present the human mind as a machine, with the simplest task of formalizing language acquisition and learning processes so as to have computers simulate humans properly. Unfortunately, the project is often blocked by the endorsement of associationism or of some of its consequences, or, worse, of vague versions of anti-associationist linguistic theories, such as those of Noam Chomsky.

Two camps of students of consciousness are exceptionally hostile to each other: those who have swallowed whole the traditional view of association and induction, and those who reject it offhand. Consider for example the new studies that employ experiments in which high powered, sophisticated machines record electromagnetic activities in brains — human or not. They incontestably belong to natural science. Do they divulge any information on consciousness? Why do people deem observations of brain activities more revealing of consciousness than the observation of the notorious effect of alcohol on consciousness? This is no rhetoric question. On the contrary: it is intriguing. It is also pivotal: those who do not agree on this at a glance will probably dismiss this essay upon reading this claim, if they did not do so already.

Computer brain imaging does show activity associated with decisions. So decisions are in. As it happens, the brain-activity in question indicates, we are told, that awareness or consciousness of a decision appears a little while *after* the decision was made. Hence, we are triumphantly told, consciousness is illusory. Of course, opponents will say that the brain activity in question signifies that consciousness takes time to receive expression. It is clear that as long as there is no knowledge of how to read brain activity, the matter is controversial.

The interesting fact is, the dispute shows how much the current views are prejudices. And so, here the computer serves a traditional dispute but is so of little help. A strange event took place here: a recent empirical observation seems to support the philosophical disregard of awareness: a recent brain neurology study suggests that awareness of a decision comes a brief period past its occurrence (Gregori-Grgič R *et al.*, 2011). The evidence cannot possibly be decisive as too little is known about the meanings of electric activities in the brain and indeed Gregori-Grgič and his co-authors are wary. Yet preference for associationism makes many brain researchers very glad to see awareness go and the tendency to read this into the experiment is marked.

CONCLUSION: THE METAPHYSICAL DISPUTE IS UNAVOIDABLE

The discussion of consciousness is often conducted with little preliminary presentation of the problem and its background. It is even difficult to find in the literature the question explicitly worded. (To speak of *the* problem of consciousness will not do; there are too many problems regarding it and there is no consensus as to which of them is dominant.) To start the discussion of consciousness more in line with the rules of the game, it must develop much more slowly. Emotions stand in the way: the discussion usually gets heated because of religion. The religious insist that the soul survives the body and the materialists deny this. The dispute is as metaphysical as they come. Efforts to try to decide it by empirical means are foolish. It is futile to try and to settle empirically the dispute as to whether the soul leaves the body or dies with it. And yet this is the motive beyond much of the research about the mind, and more specifically about consciousness.

One major development in this direction is due to Alan Turing; it is less than a century old. He did not wish to settle the dispute; he wished to render it irrelevant. At least that is what he declared in the opening of his celebrated essay on the famous test that bears his name. He failed: many students of consciousness treat it as emotionally as before. Now the Turing test is taken more often than not as anti-religious rather than as keeping religion out of the research concerns.

Turing meant his test to eliminate the discussion of the mind (or soul) from science. Three or four traditional philosophical views of it survive. One is, matter is real but the mind is not: it does not really exist. The other is, the mind is real and matter does not really exist. The third is that both are real, although the mind is ephemeral. A possible fourth is, the two run in parallel, in coordination: one being a reflection of the other. Turing rightly ignored (as not serious) the idea that matter is unreal. He also ignored parallelism as echoing materialism. His centered on dualism: is it possible that both matter and spirit are real? Yes. And Turing argued that the assumption that the spirit is real is of no empirical import.

Turing envisaged computers that are as powerful as possible. He assumed that no one will ascribe a soul to a computer, no matter how powerful. He imagined a computer that can emulate human conduct. If this is possible, then the materialist way of explaining the conduct of a computer is the way of explaining human conduct.

This is the whole of the idea of Turing. He presented also a pictorial variant of it: imagine an investigator studying a human and a computer in only one manner; the investigator asks questions and both subjects answer them. At the end of the process the investigator has to decide which of the two is human and which not. If the investigator is unable to decide, then the ascription of a soul to one party but not to the other is of no empirical import.

This is a powerful argument. Since we cannot as yet explain even the conduct of a computer, it is clear that the discussion is in matters of principle, not of fact: no one suggest that the ascription of some spiritual powers to a computer is necessary. Since there is no reason to assume that a computer cannot emulate any human intellectual conduct, there is likewise no reason to assume that humans are more than computers of sorts.

Two arguments against Turing are known. One is due to John Searle, and is known as the Chinese Room Argument. It is, briefly, that computers have no feelings and humans do. Computers may be able to emulate human feelings but not to feel. This argument is open to critical discussion, and the literature is full of it. Yet its force is moral: we should reduce suffering, human and animal, perhaps also of machine if machines really suffer, not however if they only emulate suffering. The second argument is stronger. It is this: if the machine fails to pass Turing's test, we may improve it; if it passes the test, then we may improve the test. This is incontestable. Hence, the Turing test is not conclusive. Yet it was valued only because of its claim for conclusiveness. And so we are not rid of consciousness, not yet and possibly not ever. The long and the short of it is this: there are repeatable empirical observations that want explanations. As long as we can explain them only by the assumption of goal-directed behavior, this must stay (Mises, 1957, page 1), and as long as some of these observations refer to consciousness, the same holds for consciousness.

The long and the short of it is this. The magical philosophy that most people still hold ascribes meanings — intentions — to every thing and behind every event. The mechanistic view denies that intentions exist, conscious or not. And then it had to deny the existence of consciousness too. Is it too strange to suggest that only living things have aims and only the more developed ones are aware of them? More cannot be said in general. We know that we have conflicting unknown ends. Moreover, our ends are hewn out of our images of the world, and these are usually obviously false and in propitious moments only possibly true but not very likely so. And then it is just about impossible for us to know what we want and how we should go about to meet our ends.

BIBLIOGRAPHY

Allport, G. W. (1954). *The nature of prejudice*. Cambridge, MA: Addison-Wesley.

Churchland, Paul and Patricia (1991). *On the Contrary: Critical Essays, 1987-1997*. Chapter 11: Recent Work on Consciousness: Philosophical, Theoretical and Empirical. Cambridge, MA: MIT Press.

Collingwood, Robin George (1940). *An Essay on Metaphysics*. Oxford: Clarendon.

Crick, Francis (1994). *The Astonishing Hypothesis: The Scientific Search for the Soul*. New York: Charles Scribner's Sons.

Damásio, António (2010). Self Comes to Mind: Constructing the Conscious Brain, Pantheon.

Dang-Vu T. T. et al., (2010), "Spontaneous brain rhythms predict sleep stability in the face of noise". *Current Biology*, 20, No 15, R626-7.

Davis, Philip J. and Reuben Hersh, (1981). *The Mathematical Experience*. Boston: Birkhäuser.

Festinger, Leon (1957). *A theory of cognitive dissonance*. Evanston, IL: Row.

Freud, Sigmund, (1896). "Aetiology of Hysteria", in *The Standard Edition of the Complete Psychological Works of Sigmund Freud*, Volume 3 *(1893-1899): Early Psycho-Analytic Publications*, 187-221.

Galilei, Galileo (1953 [1633]). *Dialogue Concerning the Two Chief World Systems*. Translated by Stillman Drake. Berkeley: University of California Press.

Gregori-Grgič R, Balderi M, de'Sperati C (2011) "Delayed Perceptual Awareness in Rapid Perceptual Decisions". *PLoS ONE 6(2): e17079*; doi: 10.1371.

Gregory, Frederick (1989). "Kant's Influence on Natural Science in the German Romantic Period," in *New Trends in the History of Science*, ed. R. P. W. Visser *et al*. Amsterdam: Rodopi, pp. 53-66.

Hadamard, Jacques (1945). *An Essay On The Psychology Of Invention In The Mathematical Field.* Princeton NJ: Princeton University press.

Hebb, Donald O, (1949). *The Organization of Behavior*. New York: Wiley.

Hume, David, (1896). *A Treatise of Human Nature*. Oxford: oxford University Press.

Kuhn Thomas S. (1970). *The Structure of Scientific Revolutions*, second edition. Chicago: Chicago University Press.

Leibniz, Gottfried, 1996. *New Essays on Human Understanding*, translated by Peter Remnant and Jonathan Bennett, Cambridge: Cambridge University Press.

Marr, David (1982). *Vision*, New York: Freeman.

Minsky, Marvin. *The Society of Mind*. New York: Simon and Schuster, 1988.

Mises, Ludwig von (1957). *Theory and History*. New Haven CT: Yale University Press.

Moore, G. E. (1924). "A Defence of Common Sense", in J. H. Muirhead, ed., *Contemporary British Philosophy*. London: Allen and Unwin.

Oakeshott, Michael (1962). *Rationalism in Politics and Other Essays*. London: Methuen.

Polanyi, Michael (1967). *The Tacit Dimension*, New York: Anchor.

Polanyi, Michael. 1958. *Personal Knowledge: Towards a Post-Critical Philosophy*. Chicago: University of Chicago Press.

Popper, Karl R. (1963). *Conjectures and Refutations*. London: Routledge.

Popper, Karl R. and John Eccles (1984). *The Self and Its Brain*. London: Routledge.

Russell, Bertrand (1965). *An Inquiry Into Meaning And Truth*. Penguin.

Ryle, Gilbert (19949). *The Concept of Mind*. London: Hutchinson.

Sayre, Kenneth M. (1969). *Consciousness: A Philosophical Study of Minds and Machines*. New York: Random House.

Searle, John R. (1997). The Mystery of Consciousness. New York: The New York Review of Books, Inc.

Shaffer, Jerome A. (1965). "Recent Work on the Mind-Body Problem". *American Philosophical Quarterly, 2,* 81-104.

Shōji, Yamada (2001). "The Myth of Zen in the Art of Archery", *Japanese Journal of Religious Studies*, 28:1–2, 1-30.

Singh, Arun Kumar (1991). *The Comprehensive History of Psychology*. Delhi: Motilal Banarsidass.

Turing, A.M. (1950). "Computing Machinery and Intelligence". *Mind*, 59, 433-460.

Watkins, John, (1957-8). "Epistemology and Politics", *Proceedings of the Aristotelian Society*, 58, 79-102.

Wettersten, John (1985). "The Road through Würzburg, Vienna and Göttingen", *Philosophy of the Social Sciences*, 15, 487-505.

Wettersten, John and Joseph Agassi (1978). "Rationality, Problems, Choice", *Philosophica*, 22, 5-22.

Wittgenstein, Ludwig, 1922. *Tractatus Logico-Philosophicus*. London, Routledge.

THE HISTORICAL-CULTURAL APPROACH

In: Consciousness: Its Nature and Functions
Editors: Shulamith Kreitler and Oded Maimon

ISBN 978-1-62081-096-5

Chapter 3

PSYCHOLOGICAL EVOLUTION OF CONSCIOUSNESS

Elfriede Maria Bonet
Department of Philosophy, University of Vienna, Austria

ABSTRACT

The chapter deals with the evolution of consciousness from the psychological point of view. The question about the psychological evolution of consciousness is a relatively new one, and has emerged as a separate issue in the wake of the exploration of the biological evolution of consciousness. The answers given to the questions about how and why consciousness developed differed in each period, in line with the dominant conceptions at the time. Four different approaches to psychological evolution are presented. The approach of conscious evolution, developed by Robert Ornstein, is based on contrasting the standard, uncontrolled evolution with evolution which is controlled by human beings and contributes to safeguarding the chances of survival of humankind. The Jungian approach, represented by Neumann, claims that the development of consciousness of humanity at large parallels the stages of development in the individual. It consists in elaborating experientially the sequence of archetypes which leads systematically towards the expansion of consciousness and results in enhanced freedom, maturity, and creativity. The concept of meaning-based evolution, developed by Kreitler and Kreitler, describes the evolution of consciousness as dependent on two factors: expansion of the range of contents, processes and patterns of meaning variables available to individuals, which increases the potentialities for cognition and experiencing; and the emergence of novel organizational transformations of the cognitive system, which increase the potentialities for different states of consciousness. Jaynes' approach consists in delineating two major stages in the development of consciousness. The first stage, called the "bicameral mind", was characterized by adissociation between the two brain hemispheres. In this state humans had no reflection or deliberation and could not provide explanations for their acts other than through hallucinatory voices representing authority figures. The second stage, which started only about 3000 years ago, is characterized by the emergence of consciousness proper that was made possible by the breakdown of the bicameral mind. All four approaches emphasize the continuous expansion of consciousness as involving advances in cognition, personality and behavior, based on increased possibilities of knowing, feeling, and shaping creatively oneself and the environment.

The Question about the Origins of Consciousness

No sooner had human beings become aware of being conscious than they started to ask: How did it come about? How did it develop? Jaynes (1976, p. 2) proposed the ingenious idea that the answers to the questions about the nature and origin of consciousness depend on the specific historical period and its conceptual tools. Thus, the ancient Greeks described consciousness as an enormous space that could not be discovered even by traveling every path in it. The geological discoveries in the first half of the 19th century gave rise to the idea that consciousness is constructed like the planet, whereby states from the past survive as unconscious layers deep down in the core while the more recently acquired components are placed on the top, constituting the external layers on the surface. In the middle of the 19th century, when chemistry became the fashionable science, consciousness was considered as a compound entity that could be analyzed in the laboratory into its constituents, mainly sensations and feelings. At the end of the 19th century, when the locomotive technology flourished, consciousness was conceptualized as a machine resembling a boiler that holds down the steaming energy, striving for manifest outlets. With the establishment of the theory of biological evolution the approach to consciousness became more scientific and focused on the issue of its origin. One of the major proposed theories about the origins of consciousness was that consciousness is a property of the protoplasm which has developed phylogenetically, as an intrinsic characteristic of all living beings. According to another approach, consciousness was described as an epiphenomenon of the brain, whose development has been ruled by the principles of natural selection. It was only natural that the spectacular development of psychology in the 20th century would produce attempts to answer the question about the origin of consciousness in psychological terms. Indeed, well beyond the expectations of Jaynes, it gave rise to a new question that spurred the development of new approaches to the origins of consciousness. This was the question about the psychological evolution of consciousness. In the present chapter some of the more prominent approaches to the psychological evolution of consciousness will be reviewed.

Approaches to Psychological Evolution of Consciousness

Conscious Evolution

The importance of dealing with the evolution of consciousness has been emphasized most clearly by Robert Ornstein (1992). The title of the fourth part in his well-known book "The evolution of consciousness" presents the message: "Why there will be no further evolution without conscious evolution?" His major thesis is that biologically-based evolution, which has provided the means for adaptation up to now, is at an end. At the same time the need for further adaptations to a rapidly changing planet and culture has been increasing. Human beings are called upon to take evolution into their own hands and implement a program of massive changes in thinking, behaving, and managing life so as to ensure the adaptation and survival of humanity. In order to be successful the suggested program should fulfill several requirements: first, it should consider needs and problems in various domains, such as social, medical and ecological; second, it should be based on the newest findings in the human sciences, including psychology, education, cognitive sciences, medicine, and sociology; third, it should be both creative and adaptation-oriented; and fourth, it should be consciously regulated and controlled by human beings. In this context, conscious control implies that both the goals and the means should be selected, weighed and evaluated on the whole and in the concrete details of application. The implementation of a program of this kind requires changes on the individual level in education and thinking skills from fairly young ages as well as social changes that will probably be politically-steered. "There will be no further biological evolution without conscious evolution" (ibid, p. 267). Thus, conscious evolution is the next phase of evolution, following the former phases of biological, agricultural, industrial and technological phases of evolution.

Archetypally-Conditioned Evolution

While Ornstein focuses on the evolution of consciousness on the broad scale of socio-cultural developments, Newmann (1954) unfolds the development of human consciousness on the level of the internal dynamics of the individual, as reflected in world mythology. His approach is grounded in the Jungian psychology which deals with analyzing and promoting the relationship between conscious and unconscious processes, that is necessary for the functioning of a healthy and creative personality. The individual's development is assumed to be based on the process of individuation, which involves being exposed experientially and cognitively to a series of specific archetypes in a predetermined sequence. The archetypes are particular paradigmatic structures, such as the 'shadow' (which includes qualities opposite to those manifested by the individual), 'the anima' (the feminine image), 'the animus' (the masculine image), 'the great mother' (the motherly figure that includes both the good and the bad aspects), and 'the sage' or wise old man which represents human wisdom (Jung, 1981). Neumann's thesis is that the development of individual consciousness parallels the development of the consciousness of the human species as a whole. Both the individual and the human species undergo the same stages of archetypal elaboration, resulting in a gradually increasing clarity and depth of consciousness. The stages are described in terms of archetypal images, which start with the symbol of the Uroboros, or tail-eating serpent, and include themes represented in world mythology, such as the creation of the world, the Great Mother, the Hero, the Dragon, and Rebirth. The transformation brought about by the experiential encounters with themes of this kind and their cognitive elaborations results in the evolution of the individual's consciousness. In this manner it comes to integrate the originally unconscious components with the conscious ones, in the service of an expanded human consciousness.

Meaning-Based Evolution

A different approach to the psychological evolution of consciousness is based on the theory of meaning (Kreitler and Kreitler, 1990). Meaning is cognitive contents centered on a referent, such as a stimulus, an object or a word, whose function is to express or communicate the manner in which that referent is comprehended. Meaning provides the contents and processes necessary for cognitive acts. Therefore it is a crucial and indispensable component of cognition. From infancy onward the individual acquires contents that reflect different aspects of meaning, such as functions, causes, consequences, sensory qualities and emotions, and learns to use them in an increasing number of cognitive acts and other tasks and situations. When the number of content items increases, they get organized into groupings that represent categories, such as emotions, actions, materials, functions, etc. Some categories refer to contents while others refer to tendencies, such as formation of examples or metaphors, or using nonverbal means of expression. Categories of this kind are called meaning variables. The use of a meaning variable indicates that the individual is able to apply the contents and processes represented by that meaning variable in a variety of settings (e.g., Kreitler and Kreitler, 1994).

At any given time the cognitive system is dominated by specific meaning variables, which are responsible for a particular organization of the system. As long as the organization persists, the dominant meaning variables determine the kinds of cognitive contents and processes that will be accessible to the individual for use in any act that concerns external or internal reality.

Any meaning variable that has been activated for a longer period may affect the cognitive system. However, the effect is small and transitory. The effect is stronger as well as more pervasive and durable if the dominant meaning variables increase in number beyond a given critical mass or if they form a pattern supporting a certain kind of cognitive approach, such as personal-subjective meaning, interpersonally-shared meaning, and the abstract or concrete attitudes. A change in the cognitive system that results in a reorganisation of the system around clusters of meaning variables with particular effects may be considered as organizational changes. Some of the organizational changes affect the whole of the cognitive system and indirectly also other systems in the organism, such as emotions and personality traits (Kreitler, 2003; Kreitler and Kreitler, 1997). Extensive organizational states of this kind, that are often accompanied also by experiential manifestations, are called states of consciousness (Kreitler, 1999).

An attempt to reconstruct the track of psychological evolution leads to the assumption that in the initial stage the cognitive system underwent only cognition-prompted changes due to cognitive tasks imposed externally or internally. Changes of this kind presuppose a minimally developed cognitive system, which is capable of at least simple cognitive acts, and probably exists already at the level of the lower mammals.

It is likely that organizational transformations in the cognitive system could have arisen in one of the following ways: (a) as an extension of a particular state produced by a prolonged or difficult cognitive task (e.g., involving the need for integrating different kinds of information); (b) in response to recurrent cognitive tasks of a certain type (e.g., planning); and (c) by non-cognitive recurrent stimuli, such as intense emotions, sleep or food deprivation, intoxication or sickness, that are known to affect the cognitive system.

It was both useful and adaptive to store such changes in the organization of the cognitive system so that they would be more readily available for further use, when needed. One advantage of storing the changes is that they could be applied in order to facilitate creating optimal context conditions for particular cognitive acts. Another advantage is that they could be used in order to promote gaining information and undergoing specific experiences, for example, of mystical, revelatory, psychotic and parapsychological nature that were more likely to happen when particular organizational transformations were dominant (Bentall, 1990; Wulff, 2000).

The fact that the organizational changes were stored and could be retrieved from memory rendered them gradually independent of the original task or situation in which they had been formed. Thus, they underwent generalization in regard to usages – from one task to similar tasks and later to apparently non-similar tasks too.

Several evolutionary lines in regard to the psychological development of consciousness may be noted. One line concerns the extent of the organizational transformations in the cognitive system. In the initial phase, they may have been limited, partial and non-integrated. It is only later that the organizational transformations developed into schemata affecting the whole of the cognitive system (or brain, in line with Jaynes' claim, 1976). Some of these organizational transformations may have evoked particular attention on the part of individuals due to their massive or otherwise notable cognitive, behavioral and experiential effects. Not surprisingly, they came to be identified by labels (e.g., mystical state, alternate states).

Another evolutionary line concerns control of the evocation of the organizational transformations. Initially there may have been no possibility for an intentional evocation. One had simply to wait until an organizational transformation of a familiar or unfamiliar nature happened. At some point a certain degree of control was gained by applying means, such as exposure to specific environmental conditions, hunger, sleeplessness, sensory deprivation, drugs, or particular behavioral or cognitive techniques, sometimes applied by particular people, such as Shamans and priests. The next developmental stage has been attained more recently with the possibility of evoking these states at will, for example, by purely cognitive means of the kind devised by Kreitler (2002, 2009; Kreitler, Kreitler and Wanounou, 1987-88).

Finally, the third evolutionary line concerns the number and variety of the organizational transformations in the cognitive system. Initially there must have been only one or two such identified states. In the course of time many more different states came to be evoked and recognized. They were mostly considered as cognitive states, but as noted, some came to be identified as states of consciousness. The wealth and variety of these states increased appreciably the modes of cognizing and experiencing external and internal realities.

Each state of consciousness promotes certain cognitive and experiential responses. These responses may promote and sometimes enhance success in specific cognitive and other tasks. For example, in a state of consciousness focused on the subjective-personal meaning individuals function better in tasks that require visual thinking but worse in tasks that require formal-logical thinking than in a state of consciousness focused on the interpersonally-shared meaning (Kreitler and Kreitler, 1999). Accordingly, an important advantage of the increase in the availability and identification of states of consciousness is that it provides the possibility of adapting the optimal state to each type of cognitive task.

In view of the availability of cognitive means for controlling the evocation of the states of consciousness, controlling the evocation of these states enhances the range of possibilities for controlling ourselves and our behaviors. The next step in this evolutionary development would be attained when different states of consciousness are invented and defined in view of definite cognitive, emotional and other goals. This stage is already under way due to developments in the sphere of virtual reality and the means provided by the system of meaning.

From Bicamerality Onward

A different and highly original approach to the psychological evolution of consciousness has been proposed by Jaynes (1976), based partly on his studies of classical Greek texts. Jaynes views consciousness primarily as an operation, rather than a thing, a repository, or a function. Further, he assumes that it operates by constructing on the basis of language an analog space in which one can move in a metaphorical sense. Thus, "conscious mind is a spatial analog of the world and mental acts are analogs of bodily acts" (Jaynes, 1976, p. 66). In other words, consciousness is a cognitive representation of the world and of our acts in it. It does not represent the thoughts or beliefs that we have but rather thoughts about thoughts, or beliefs about beliefs. Hence, consciousness is a kind of meta-awareness or meta-cognition. Most importantly, this kind of consciousness was made possible by language, which is itself based on representation in the form of the sign-signified relationship. On the basis of these assumptions, Jaynes concluded that our modern kind of consciousness emerged no earlier than 3000 years ago. Prior to that, human beings functioned in terms of what Jaynes called metaphorically a 'bicameral mind', reflecting a dissociation between the left and right hemispheres of the brain. Individuals with a bicameral mind acted in line with automatic, non-conscious habitual schemas, which were not reported, reflected upon or controlled by consciousness. Due to the absence of consciousness, these individuals did not have the sense of an ego or autobiographical memory, nor the capacity for introspecting, deliberating or reflecting about their options and decisions. Hence, they could not explain to themselves and others the reasons for their acts. When they acted on the basis of familiar habits, no explanations were required or expected. But when habit proved to be insufficient for handling novel situations and when new decisions had to be made, acts in the "dominant" (left) hemisphere were accompanied by auditory verbal hallucinations originating in the "silent" (right) hemisphere, which were heard as commanding voices of a figure in authority ordering immediate compliance. There was no separation between the command and the action, so that "hearing" the order automatically produced obeying. Jaynes claimed that these "voices" came from locations in the right hemisphere which corresponded to the speech areas in the left hemisphere (Wernicke's area and Broca's area). These areas are dormant in most modern humans, but show some activity in cases of auditory hallucinations (Jaynes, 1986a, 1986b, 1990).

The change from bicamerality to consciousness was assumed to have taken place over a period of many centuries, from about 1200 BC. It was spurred at least partly by disintegrating social organizations caused by massive environmental changes, such as the big Mediterranean earthquakes in the second millenium BC. The new pressures on survival necessitated creativity, reflection and flexibility. Hence, consciousness emerged as a kind of physical adaptation to social pressures in a changing world. The development of writing has aided in the process of enhancing consciousness. Still some remnants of the bicameral mind have survived in schizophrenia, shamanism, hypnosis, and divination practices.

Jaynes supported his theses by analyzing primarily Homer's *Iliad*, which he viewed as a kind of psychological document, providing an accurate description of how people at that time actually experienced themselves and the world (Dodds, 2004). In the Iliad the heroes do not engage in deliberations that culminate in decisions attributed to themselves. Rather, they are described as acting upon plans and decisions conceived without consciousness, which are 'announced' to them, often by an hallucinated god or some other authority figure, sometimes by a voice alone. Notably, in the context of poetry and music, artists used to call this "voice" "the muse". Similar phenomena can be identified in certain texts of the Old Testament and early Mesapotamian literature.

In this context Jaynes emphasized the mistaken translations of Greek words in the Iliad as denoting various aspects of consciousness. These are mainly the following terms: *psyche, thumos, phrenes, noos,* and *kradie* originally referring to blood, movement, breathing, seeing and heart, respectively, but translated erroneously as denoting the mind, spirit or soul. Jaynes noted that in the period from about 850 to 600 BC the frequency and the meaning of these words changed in a systematic manner. Thus, in the bicameral age they denoted originally simple external objects, then internal sensations, followed by mental processes, and finally by concepts representing self-consciousness, reflection and introspection, which are the earmarks of consciousness.

The general claims proffered by Jaynes have been evaluated positively by major authors. Thus, Damasio (1999) accepted the suggestion that the concept of consciousness differed greatly from ours even in the times of Plato and Aristotle. The modern concept of consciousness emerged only about three to four centuries ago, so that in the 20th century it could become a construct of major interest for the sciences. Similarly Dennett (1986) appreciated Jaynes' thesis that in order to understand the present state of consciousness we need to assume that a change has occurred in the organization of our information-processing system, despite the fact that the hardware of the human brain has evidently remained unchanged in the last thousands of years.

However, some of Jaynes' specific claims evoked criticism on the part of experts. For example, the assumption that consciousness did not exist prior to 3000 years ago was contested by evidence that in some ancient texts, such as the epic of Gilgamesh or the Old Testament there are references to introspection and mental deliberation of the kind Jaynes defined as criteria for consciousness. Further, the assumption that schizophrenics have a bicameral mind seems to be unfounded. Likewise, the claim that the two brain hemispheres were dissociated in the Greeks of the Iliad period seems to be wrong since it is unlikely that the hemispheres have become so-well connected as they are nowadays in the course of only three thousand years.

On the other hand, there is some evidence favoring other claims made by Jaynes. Thus, findings based on neuroimaging provide some support for Jaynes' neurological model about the involvement of the right hemisphere in auditory hallucinations (Olin, 1999; Sher, 2000). Further, cultural studies shed light on Jaynes' claim concerning the salience of auditory hallucinations in earlier historical periods (Smith, 2007). There is also evidence that when speech is impaired artistic skills and production may gain in creativity (Humphrey, 1998; Mell, Howard and Miller, 2003).

Some Conclusions

The different presented approaches show that the question about the psychological evolution of consciousness is independent of the question about the biological evolution of consciousness. While biological evolution focuses on the environmental, physical and physiological conditions that render evolution of consciousness possible and necessary for the species, psychological evolution focuses on the internal, socio-cultural and psychological processes that constitute the development of consciousness in and for the individual.

Notably, all four approaches presented in this chapter support several common conclusions. The first and major one is that consciousness is a developing construct. This means that in the past there may have been a time when there was no consciousness in the modern sense of the term or that it was very weak or different. But it also means that at present consciousness is on the track of development and possibly on the verge of a great expansion. The development of consciousness denotes for human beings greater control over external and internal reality, and a broadened range of possibilities for creativity, cognitive achievements and flexibility. Ornstein would consider these as manifestations of conscious evolution.

Another conclusion shared by the described approaches is that the expansion of consciousness involves two major processes. One is emergence from dependence on the earlier more limited and sometimes primitive kind of thinking. In terms of the Jungian approach this means becoming freed from the domination of the archetypes and in terms of Jaynes – breaking free from the bicameral mind. The other tendency consists in preserving earlier developmental layers despite the advancement forward. Both the Jungian and the Jaynesian approaches forsee some maintenance of the initial origins: the former by integrating the subconscious contents with the conscious ones, the latter in the form of survival of vestiges of bicamerality in specific contexts (e.g., art, divination).

The Kreitler meaning-based approach clarifies how the preservation of forms of consciousness that appeared in earlier periods takes place. These earlier forms are organizational transformations of the meaning system that correspond to states of consciousness. As such, they dominate the cognitive scene and shape also manifestations in other domains, mainly the self, emotions, personality, and the view of reality. The bicameral mind is one such state of consciousness. The disadvantage of bicamerality did not reside in its nature or effects but in the fact that in earlier times it was probably the only state of consciousness, or at best one of very few that were available to human beings. According to the Kreitler meaning-based approach, the development of consciousness entails the emergence of a greater number of organizational transformations of the meaning system, and moreover the acquisition of psychological means to evoke their appearance. These developments constitute already the next phase of the evolution of consciousness, in which human beings would be able to select, even invent if necessary, and evoke at will the state of consciousness they consider adequate for any envisaged, desired or necessary act.

REFERENCES

Bentall R.P. (1990). The illusion of reality: A review and integration of psychological research into psychotic hallucinations. *Psychological Bulletin, 107*, 82-95.

Damasio, A. (1999). *The feeling of what happens*. New York: Harcourt.

Dennett, D. (1992). *Consciousness explained*. New York: Little Brown and Co. (Back Bay Books).

Dodds, E. R. (2004). *The Greeks and the irrational*. Berkeley and Los Angeles, CA: University of California Press.

Humphrey, N. (1998). Cave art, autism, and the evolution of the human mind. *Cambridge Archaeological Journal*, *8*, 165–191.

Jaynes, J. (1976). *The origin of consciousness in the breakdown of the bicameral mind*. Boston, MA: Houghton Mifflin.

Jaynes, J. (1986a). Hearing voices and the bicameral mind. *Behavioral and Brain Sciences*, *9*, 526-527.

Jaynes, J. (1986b). Consciousness and the voices of the mind. *Canadian Psychology, 27*, 128-148.

Jaynes, J. (1990). Verbal hallucinations and preconscious mentality. In M. Spitzer and B. H. Maher (Eds.), *Philosophy and psychopathology*. New York: Springer Verlag, pp. 157-170.

Jung, C. G. (1981). *The archetypes and the collective unconscious*. Collected Works of C.G. Jung, Vol. 9, Part 1 (2nd ed.), Princeton, NJ: Bollingen Paperbacks.

Kreitler, S. (1999). Consciousness and meaning. In J. Singer and P. Salovey (Eds.), *At play in the fields of consciousness: Essays in honor of Jerome L. Singer*. Mahwah, NJ: Erlbaum, pp. 175-206.

Kreitler, S. (2001). Psychological perspective on virtual reality. In A. Riegel, M. F. Peschl, K. Edlinger, G. Fleck and W. Feigl (Eds.), *Virtual reality: Cognitive foundations, technological issues and philosophical implications*. Frankfurt, Germany: Peter Lang, pp. 33-44.

Kreitler, S. (2002). Consciousness and states of consciousness: An evolutionary perspective. *Evolution and Cognition, 8*, 27-42.

Kreitler, S. (2003). Dynamics of fear and anxiety. In P. L. Gower (Ed.), *Psychology of fear*. Hauppauge, NY: Nova Science Publishers, pp. 1-17.

Kreitler, S. (2009). Altered states of consciousness as structural variations of the cognitive system. In E. Franco (Ed., in collab. with D. Eigner), *Yogic perception, meditation and altered states of consciousness*. Vienna, Austria: Oestrreichische Akademie der Wissenschaften, pp. 407-434.

Kreitler, S. and Kreitler, H. (1990). *The cognitive foundations of personality traits*. New York: Plenum.

Kreitler, S. and Kreitler, H. (1994). Motivational and cognitive determinants of exploration. In H. Keller, K. Schneider and B. Henderson (Eds.), *Curiosity and exploration*. New York: Springer-Verlag, pp. 259-284.

Kreitler, S. and Kreitler, H. (1997). The paranoid person: Cognitive motivations and personality traits. *European Journal of Personality, 11*, 101-132.

Kreitler, S., Kreitler, H. and Wanounou, V. (1987-1988) Cognitive modification of test performance in schizophrenics and normals. *Imagination, Cognition, and Personality 7,* 227-249.

Mell, J.C., Howard, S.M. and Miller, B.L. (2003). Art and the brain: the influence of frontotemporal dementia on an accomplished artist. *Neurology, 60*, 1707-1710.

Neumann, E. (1954). *The origins and history of consciousness* (Bollingen Series, 42). (11th paperback printing). Princeton, NJ: Princeton University Press.

Olin, R. (1999). Auditory hallucinations and the bicameral mind. *Lancet, 354* (9173), 166.

Ornstein, P. (1992). *The evolution of consciousness*. New York: A Touchstone Book, Simon and Schuster.

Sher, L. (2000). Neuroimaging, auditory hallucinations, and the bicameral mind. *Journal of Psychiatry and Neuroscience, 25* (3).

Smith, D. (2007). *Muses, madmen, and prophets: Rethinking the history, science, and meaning of auditory hallucination*. New York: Penguin Press.

Wulff, D. M. (2000). Mystical experiences. In E. Cardeña, S. J. Lynn, and S. Krippner (Eds.) *Varieties of anomalous experience: Examining the scientific evidence*. Washington, DC: American Psychological Association, pp. 379-440.

THE BIOLOGICAL APPROACH

In: Consciousness: Its Nature and Functions
Editors: Shulamith Kreitler and Oded Maimon
ISBN 978-1-62081-096-5

Chapter 4

CONSCIOUSNESS: THE PERSPECTIVE OF EVOLUTIONARY BIOLOGY

Karl Edlinger*
Museum of Natural History, Vienna, Austria

ABSTRACT

Biology cannot explain the emergence of mental phenomena and of the phenomenal experience. But biological and evolutionary thinking can represent the origin of those conditions which enable the formation of neural processes of varying complexity. The theory of Organismic Constructions provides the theoretical assumptions for that. This theoretical approach considers organisms as autonomous entities, and living beings as energy converters, that are driven from within and actively develop their environment. Orderly and efficient responses to environmental stimuli and features require a precise combination and cooperation of motorium and sensorium. This interaction is made possible by nervous systems or analogous structures. The structure of nervous systems depends on the mechanical construction of organisms. Brains, i.e, large complexes of nerve cells, can occur only in areas protected from mechanical disturbance. A minimal model should provide an approach for the explanation of the origins of qualia, i.e., sensory modalities. Its basis is the close functional relationship between and motorium and sensorium. Specific forms of sensation result from the interaction of motorium and sensorium. According to their construction and design, every sentient being, as well as every species, construct their special worlds, which cannot be transferred to other species. This special world constitutes their consciousness.

INTRODUCTION TO THE PROBLEM

No problem has occupied the natural sciences and philosophy of the last centuries to the same extent as the question of consciousness, commonly understood as awareness, an "inner experience" and perception of phenomena and ultimately "qualia". Attempts have been made

* E-mail: karlfranz.edlinger@gmail.com.

in the framework of different theoretical approaches to find a pervasive and accessible explanation for consciousness. Yet so far, no approach was presented, which provides a consistent explanation for consciousness. After numerous failed attempts on the part of philosophers to provide an explanation of the phenomena of consciousness, it was obvious that it was the turn of the natural sciences, especially biology, to try to solve the problem. There are many biological approaches to the explanation of mental phenomena, ranging from those provided by Herbert Spencer, E. Haeckel, and K. Lorenz to D. Campbell, R. Riedl and G. Vollmer. They can be subsumed under the title evolutionary epistemology.

It is evident that conscious experience brings an evolutionary advantage. This advantage is certainly the ability to focus on information, to combine and evaluate it, to decide which behavior is beneficial to the organism. Yet, this advantage alone cannot serve as an explanation for the emergence of conscious experience.

However, all statements, which are in accordance with the adaptational theories of the classic ideas of Darwinism and Lamacksism, refer only to the evolutionary benefits of conscious experience and cryptically introduce into the discussion an adaptation process, without any scientific explanation. So they are metaphysical in the sense of Popper. At best they may give reasons for some ways of understanding "cognition processes", but not for awareness and conscious perception in the strict sense.

It is undoubtedly the task of biology to present the evolutionary benefits of conscious perception. But at the current state of knowledge it must confine itself to presenting the conditions necessary for conscious experience, specifically the conditions that are given by the structure and design of organisms. In this context, the task of biology is not to explain the nature and the emergence of mental phenomena, but rather to present the terms and conditions under which conscious experience can take place. An undertaking of this kind requires at best a minimal model that demonstrates the basic characteristics of living organisms that constitute the prerequisites for conscious experience in an evolutionary framework.

THE SITUATION OF THE ESTABLISHED BIOLOGY

The attainment of the above-stated goal is possible only if biology has a useful and consistent model of the organism. Precisely this has not been the objective of Darwinism in its various kinds. Darwinism replaced the reflection on the organism and its place in the world by reference to natural breeding and adaptation.

Hence, the biological disciplines present a curious situation, which places an important distinctive mark between biology, on one hand, and astronomy, physics, chemistry, mineralogy, and geology, on the other. The development of the non-biological sciences gave rise to a high level of mathematical sophistication and physical explanation. They were considered from the perspective of experimental testing and reliable prediction. In contrast, the biological disciplines manifest this trend only in a restricted way and only in those sections which overlap with other sciences, especially physics and chemistry. However, the very subject of biology, the living organism as a functioning whole, is not taken into consideration; it bars the access of usual scientific procedures.

Attempts to solve the problem of the organism were made by formulating new models and theories on the basis of thermodynamics (Prigogine, 1979), synergetics (v. Bertalanffy,

1968; Gutmann and Weingarten 1987; Haken, 1981; Haken and Wunderlin, 1986; Meinhardt, 1978, 1987), systems theory (Haken and Wunderlin 1986), the theory of chaos (Gleick, 1988) or, last but not least, a new variant of vitalism, called holism (Elsasser, Marsh and Rubin, 1998; Harrington, 2002; Looijen, 1999; Meyer-Abich, 1989; Smuts, 1938).

However, these models and theories are helpful for explaining processes of physics and technology and are of general importance for a multitude of natural phenomena, but in no way specific for organisms. Systems theory refers to interactions and interdependencies of complex entities in general. It does not provide acceptable definitions of organisms and of the special features of living beings. Holism is also not acceptable because it must postulate forces and mechanisms which cannot be studied by the methods of natural sciences.

Darwinism and the Synthetic Theory

A new situation arose when in the middle of the 19th century Darwinism was established as the predominant theory of evolution. Both Darwinism and the Synthetic Theory as a subsequent stage of the theoretical development, ignore the organisms as specifically constituted entities. Darwinian theories are restricted to an extremely reductionist view of life and its functions. The exclusion of the organisms from reductionist concepts is liable to ignore the difficulties and the problems generated by the observable indisputable organismic properties. Only the particular aspects accessible to the reductionist approach can be objects of investigation.

The sum of particular processes is easily confounded with the organismic whole. This characterization applies to the methodology and theorizing in genetics and molecular biology. The reductionist methods, which use physical and chemical procedures, are helpful for the elucidation of the molecular and physiological mechanisms in organisms. Many of the elucidated molecular and physiological mechanisms can be simulated and demonstrated in the laboratory.

As a consequence, Evolutionary Synthesis (Dobzhansky, 1937; Mayr, 1967, 1979, 1984) claims to be the authentic scientific interpretation of life, living organization, and evolution. Its proponents consider themselves to be the only users of correct and adequate scientific methods. Some of them pretend that reductionist and Darwinian views provide the basic tenets of a general philosophy of nature (Mayr, 1991).

The notion of information is used in order to bridge the gap between the simplicity of biochemical and molecular mechanisms and organismic properties. In this manner it is pretended that complexity of organization might be reducible to simplistic rules. The sequence of a few molecular building elements is considered to be responsible for the complexity of organization. The chain of amino acids in their dependence on the nucleotide sequences is conducive to the formulation of analogies between biochemical structures, on the one hand, and letters or words, on the other. So it is not surprising if reductionists believe that they have found the essential structures of organisms within the nucleus and the translating mechanisms of the genetic apparatus. Thus, the genotype is assumed to be totally representative of the organism itself.

In line with this view, genes are considered to be interacting elements which are organized as networks of hierarchic systems (Franklin and Lewontin, 1970). The structures beyond the genetic apparatus, which is supposed to contain the information of the organismic whole, are considered to be of relatively minor importance or of no real importance at all. The organisms lead a ghost-like life. However, even in the perspective of reductionism, the integration of the genetic mechanisms in the living entity cannot be ignored. It is too obvious that genes can only function over the structure of the organisms. The outer framework, for which specific principles and laws are not given, are called phenotype. The phenotype is thought to consist of the sum of gene-dependent structures which are built by genetic information.

Darwinism tries to convince a benevolent audience that the phenotypes are exposed to the selective influences of the environment (Darwin, 1899, 1906). Under selective pressure some variants will survive and others will succumb. In line with this view, the selective influence on the genetic level is exerted indirectly over the phenotype, because only the phenotype is in a direct contact with the environment. In the last resort, the genotype is selected from the environment. Again, the organism and its specific properties are ignored, while the reductionist view prevails. Only the genotype finds explicit recognition. The environment gains a determining influence on the molecular mechanisms and thereby on living organization. This concept refers specifically to describing an external selective influence on the genetic apparatus. The linkage function of environment and genetic apparatus results from a very reduced view of the phenotype and of the organism as well. The term "organismic" is only an empty game of words.

Thus, organisms seem to be well defined by the genotype. Biological fitness, as the most important idea and as the driving mechanism in Darwinian concepts of evolution, presents evolution as a process resulting from hereditary disposition, given by the genetic structures, in relation to external environmental mechanisms. This theory-based dissolution and liquidation of the organism, committed by the proponents of "Synthesis", continues when evolution is defined in the terms of population dynamics. Evolution, in this view, consists of changes of gene-frequencies over time. The pressure of selection results in so-called "well adapted" genotypes characterized by gene-frequencies within the populations.

Models of Origin and the Early Evolution of Life

The described ideas of molecular reductionism exerted a decisive influence on models about the origin of life (Eigen, 1971, 1987; Eigen, Gardiner, Schuster and Winkler-Oswatitsch, 1983; Eigen and Schuster, 1977-1978; Eigen and Winkler, 1973, 1976; Schuster, 1987; Küppers, 1979, 1980, 1986a,b). Most of these models focus on the emergence of the genetic apparatus and ignore the constitution of all other basic structures of organisms, including those responsible for the functioning of the chemical and physical mechanisms described on the molecular level of the genetic apparatus. Organic substances like amino acids, nucleotides or lipids can be synthesized by technical arrangements in experimental machineries, which seem to simulate the conditions of the early time of the planet earth. It is also possible to observe alterations of polymerized organic substances, which are self-reproducing with different speed and intensity. The rate of self-reproducing activity depends on the conditions constituted by the apparatus. The continuous input of energy and material and the removal of the waste material are indispensable for the continuation of all of these processes.

On this basis it seems possible to reconstruct the early processes of spontaneous generation, alteration and adaptation of organic substances as prime elements of living beings. In addition, "cooperation-like" interactions between organic molecules could be set going. These artificial procedures provide the experimental foundation of the theory of the so-called hyper-cycles presented by Eigen and his group. The apparatus figures as the early environment.

CRITICISM OF REDUCTIONISM

In previous years these reductionist views and methods were frequently criticized. It can be shown, that most of the models of population dynamics are hypothetical and do not concern natural and real entities. In regard to early chemical evolution, Vollmert (1985) argued that the physical and chemical conditions of the early period of earth, called "primeval soup", give no chance for the formation of long chains of polymerized RNA and DNA-molecules. Hydrolysis would immediately destroy them. Similarly, self-reproduction, cooperation and enzymatic reactions are dissipated and rendered functionally infeasible. The conditions which give freedom to this thermodynamic tendency must be counteracted by sophisticated technical arrangements or by compartmentalizing and enclosing structures of protobiotic entities. These can delimit the space of molecular mechanisms and exert sufficient control on the enclosed processes. All of these presuppositions cannot be found in a non-living nature.

As a result of critical studies, Eigen (1987) was forced to concede that the hypercyles could only evolve in enclosures and necessitate compartments for an undisturbed function of the chemical mechanisms. These conclusions can be generalized. All physiological and genetic functions appear to depend strongly on surrounding and supporting structures, which are provided either in the apparatus or in the structure and organismic set-up of living beings. There is no reasonable way to reconstruct prebiotic and the early stages of protobiotic evolution by concentrating on molecular mechanisms alone. It is necessary to deliberately consider the role of the living apparatus and its structure. Here the principles of hydraulics must be invoked.

Thus, each theoretical and practical approach to the question of organization and evolution must regard organisms as systems, consisting of interdependent different structural elements. Physiological and genetic mechanisms are conceivable only as integrated but not as the dominating components of organisms.

If Eigen and other authors want to reach a proper evaluation of their prebiotic models they will have to stop regarding the molecular mechanisms alone and would have to take account of the structural aspects as constitutive for evolutionary alteration. It is necessary to consider that evolution is a change of a complex structural whole, containing numerous highly ordered elements. Genes and enzymes belong to these. As a consequence, for Eigen, the apparatus, not the "evolving" substances it contains, must play the role of the organism.

The theoretical deficiencies of the reductionist approach did not go unnoticed. Some other concepts from non-biological sciences tried to figure as the theoretical basis for the definition of organisms and provide premises for the explanation of their specific properties, such as self-organization, spontaneity, autonomy and the unsuppressable tendency to evolve.

However, none of these approaches, which included systems theory, thermodynamics of open systems, the theory of dissipative structures, synergetics, and chaos-theories, were able to present a useful model or consistent concept. They are conducive to different incongruous experimental arrangements and models, demonstrating the inability of a special discipline of physics or chemistry to construe an adequate concept of living organization. As pointed out above, the relation of the hypotheses to living beings is undefined.

In this situation it would be a mistake to fall back on inappropriate and old-fashioned stages of biological research. Criticism of the kind formulated here is fully applicable to the understanding of organization as propagated by classical biological disciplines of comparative anatomy, morphology and systematics. It is deplorable but necessary to state that in these fields no reasonable concept of the organism and its activity could be developed, because the subject of investigation was and continues to be the dead organisms, the cadavers, depleted of their living functions. However, there can be no doubt that a reconstitution of organismic biology can emerge and prosper only on the basis of morphological knowledge. Organismic and constructional concepts will rejuvenate morphology.

TOWARDS NEW FOUNDATIONS OF ORGANISMIC THEORY

For a really consistent model of the organisms we need first to clarify the theoretical approach to the problem of organism and organization. The second step will consist in elucidating the solid foundation of our methodological procedures. In accordance with some constructivist positions, especially the "School of Erlangen", we must call for a scientific theory of action and behavior. We have to be aware that useful behavior and action are dependent on the precise sequence of operations treating the objects of investigation. But each operational step, which comprises a manipulation of an object, presupposes a consistent hypothesis about the nature and constitution of this object. It is possible to argue, that theoretical presuppositions, i.e., hypotheses and useful manipulations or experiments serve as necessary supplements.

In physics and Euclidian geometry consistent basic theorems, for example, for the functioning of forces or for the three-dimensional space, were offered on the basis of some technical, particularly grinding operations (Janich, 1989, 1992, 1993). In the case of biology and organismic theory we cannot figure out any kind of such operations, neither in the descriptions and comparisons of traditional morphology nor in the explanations of physiology or, as pretended by several authors, in the practice of breeding and cultivating. Morphology is focused on cadavers and artefacts. The same holds for physiology which uses preparations and artificially isolated partial models of organisms. Selection, as performed in breeding domesticated animals or plants, treats complete living organisms. Manipulation is confined to positively influencing the reproduction of appropriate individuals. Thus, the breeders take the basic properties of organisms as given; they may even be unaware of the presupposed indispensable properties of life. By ignoring but presupposing the autonomy of life they produce an "unnatural", artificial situation, in which some essential aspects of life under natural conditions are eliminated and compensated.

It follows from this that selection procedures and cultivating techniques are in no way sufficient to serve as premises for the constitution of living organisms. Darwinism is a basically misconceived approach to living organization and is in no way appropriate to justify the assumption of evolution.

MEDICINE

Only medicine in its basic handling of patients provides the basic procedural requirements on which organismic concepts might be founded (Edlinger, 1995). Rothschuh (1957), as a physician, developed a specific organismic theory which has no counterpart in the field of biology. It is based on the experience of a medical understanding of the organism.

Interventions of physicians have to maintain and to support living functions of organisms. Medical descriptions of the living beings and their functions are mostly formulated in other terms than those considered appropriate in biology. Technical models and technical terms are of utmost importance. Most organismic functions can be described in the jargon of machine building and according to concepts of engineering. In many instances machines and machine-like devices are used during life-saving medical interventions and treatment of diseases. This can be useful only if the machines function as real supplements of organisms and if the physician takes account of specific machine-like features of the organisms. Operational closeness of the body and of specific organs have to be observed: measurements of blood-pressure and other body fluids are essential indicators which can only be properly evaluated in the framework of a constructional understanding. The intactness of body cavity and the circulatory system have to be ensured. The turgescence and the hydraulic pressure of the tissues and cells, the character of the heart as a pump and the mechanical construction of the locomotion apparatus are all objects of diagnostic evaluation. In addition, it is necessary to evaluate the possibility of utilization of energy-conversion, the unimpaired and frictionless functioning of all deforming structures, the necessity of harmonic activity of all mechanical elements, mechanical coherence, the cybernetic functions of the nervous system and so on.

Only after the acceptance of the aforementioned aspects can physiological and biochemical mechanisms be taken into consideration. It is possible to monitor chemical and physiological processes of the living functions. Only on the basis of preconceived constructional knowledge it can be legitimate to argue about gene-expression and other molecular processes. All of these depend on a pre-established structural and organizational constitution of the living entities.

In the case of experiments, the structural and organizational conditions are given by the set-up of the apparatus; in the case of organisms, by the scaffoldings, channels and pumping systems of the cells, tissues, organs and constructional wholes. It is obvious that the models of technology and engineering are fully applicable for organisms.

Chemical mechanisms are indispensable for the converting of matter and energy in the metabolism. Enzymes are responsible for the metabolism. Their structure is dependent on the DNA. In a similar way, all proteins building up scaffoldings of the cells and special tissues depend on the genetic machinery of the nucleus. But although the heredity of protein structures is accepted and well established in the biological sciences, we must consider the role of the DNA as an integral part of the organism and not as the organism itself.

In line with this view, organisms in general figure as hydraulic systems functioning like machines. It is hard to imagine what would happen, if physicians tried to save the life of a severely wounded person on the basis of self-organization theories, systems-theories, synergetics and thermodynamics or other models outlined above. Only adequate presumptions make us equal to the situation in this case. Yet this demand must not be seen as a postulate for a naive realism. It leads towards consistent theoretical organismic positions, which have to be proved by practice and corroborated by experiments. On the basis of these positions we can set up a model of the organism, which conforms to that of the theory of organismic constructions as presented by the Frankfurt group.

THE THEORY OF ORGANISMIC CONSTRUCTIONS

This theory conceives organisms as hydraulic systems (Edlinger, 1989, 1991a,b; Gutmann, 1988a,b, 1991; Gutmann and Bonik, 1981). Hydraulic systems are under pressure of the fluid filling, which is surrounded by dense and flexible membranes. This causes a tendency to assume a globular shape. All deviations from this rule, i.e., non globular shapes of organisms or of their parts are enforced by tethering fibers, by surrounding packing rings, also consisting of fibers, or by hard skeletal elements, secreted by glands or gland-like structures. Organisms are converters of energy and matter, which are deformed permanently by shortening of fibers, and by energy-consuming gliding of actin-myosin-complexes. They can regain their length only in a passive way (see Figure 1A).

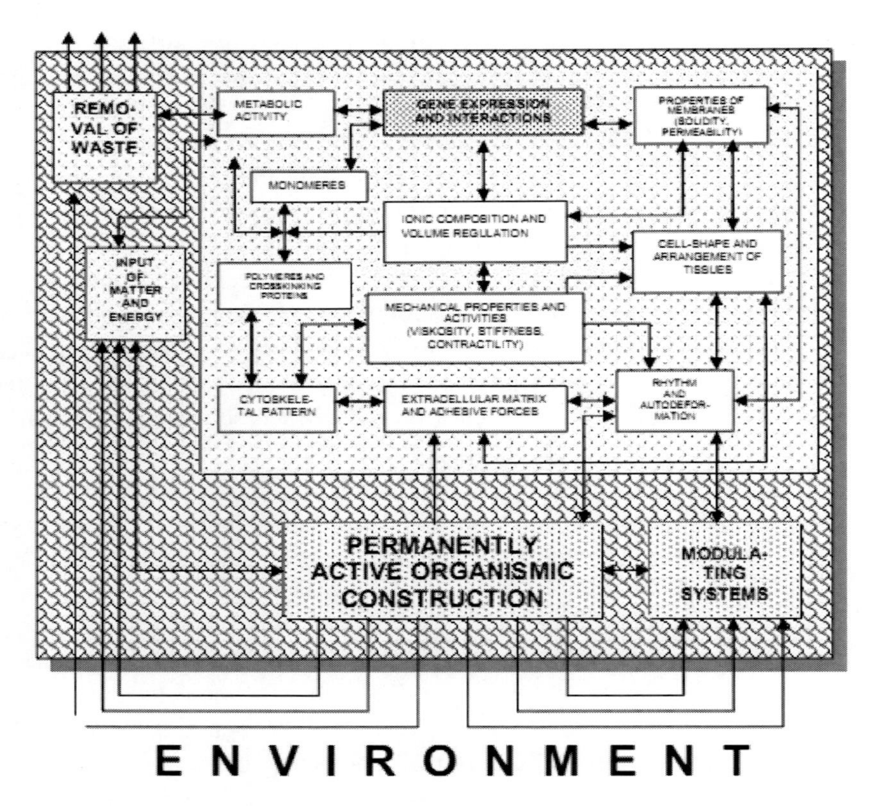

Figure 1A. Out of the theory of organismic constructions follows a new view of the organisms and of the various interdependencies between various components and functions of cells, organisms and the environment. (According to Bereiter-Hahn and Gutmann and Edlinger).

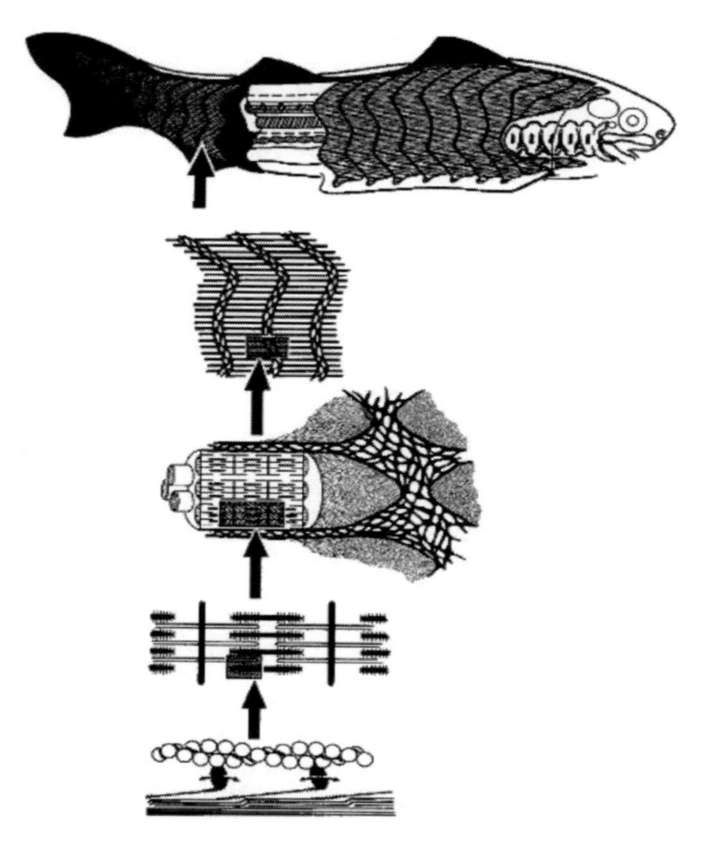

Figure 1B. Coherence in living structures. The energy transformation from the chemical to the mechanical stage is done on the macromolecular level and it consists of the displacement of the myosin heads when sliding on actin fibrils. A drive of the overall design can thus be effected only when the energy converter structures are suspended within a coherent mechanical structure, which is closed mechanically and operationally.

Fibers function in an antagonistic manner against other fibers or against the fluid filling and its surrounding membranes. Permanent deformations are generated by these mechanisms which cause the injection of energy into the environment. Contraction of fibers is elicited by stimulations, produced by pacemaker systems (Edlinger, 1991b). Contraction and restitutive expansion must be coordinated in a high degree. This is possible because fibers and also microtubules composed of proteins form scaffoldings, which have an additional effect in the enforcement of form (see Figure 1B).

The cellular and extracellular scaffoldings can be used also for anchoring of contractile fibers. Enzymes can bind to them and compartmentation, as a primary condition of an effective metabolism, is achieved. So we can conclude, that organisms are highly ordered arrangements of mechanical and chemical elements, which are strictly adapted to their function within the organisms (Edlinger, 1994). Mechanical structures function as frameworks for all other parts and components of the living machinery. This mode of internal "adaptation" is the only one which can be legitimately discerned in the kingdom of organisms. In the frame of the organismic machinery, genetic and physiological mechanisms depend on the mechanical framework, i.e., the organismic construction. They function as parts of its construction, supporting its activities and its permanent self-regeneration.

All reproduction activities must be seen as an aspect of energy conversion in the construction. The production of spermatozoans and eggs consists of energy-driven formation processes. It results in the formation of separate mechanical constructions, which are able to develop into complex organisms. Ontogenetic development is also an energy driven process in mechanical constructions that is strictly guided by mechanical constraints. The differentiation of cells and tissues and the constitution of body architecture are only conceived as enforced by mechanical stress.

New Results and Corroboration

Although the reductionist view of the Synthetic Theory is still dominating, new results are helpful in confirming the biomechanical theory of organismic constructions. A growing number of authors demonstrate the predominant role of the compartmentation (Bereiter-Hahn, 1987, 1991; Bereiter-Hahn and Strohmeier, 1987; Bissell and Aggeler, 1987; Clark and Masters, 1976) and the indispensable and constitutive influences of biomechanics on biochemistry and molecular mechanisms (Bissell and Aggeler, 1987). Contractile fibers are capable of opening or closing ion-channels. Ions may exert a considerable influence on biochemical and molecular mechanisms. As shown by Ingber and other authors, deforming activity caused by cellular motility can alter the stiffness of the cytoskeleton, which may result in the alteration of the physical and chemical properties of the protoplasm. There is also evidence that shearing forces are of great importance for the activation of some genes and for the generation of mechanical stress that results in the differentiation of muscle cells and other cells (Eimerman and Pitelka, 1977; Franke et al., 1990; Gutmann and Edlinger, 1994b; Ingber, 1993a,b; Ingber and Folkman, 1987; Ingber et al., 1994; Juliano and Hakill, 1993; Lansman, Hallam and Rink 1987; McClay and Ettensohn, 1987; Medina, Li, Obornn and Bissell, 1987; Opas, 1987, 1989; Resnick, Collons, Atkinson, Bonthron, Dewey and Gimbrone, 1993; Vanderburgh, 1988; Wang, Butler and Ingber, 1993).

Mechanical force exerted by the extracellular matrix of metazoans is transmitted as a signaling influence through the cell-membranes and gains effectivity in gene-expression. As shown, epithelial tissue can be seen as induced by the self-arrangement of fibers and other material adjacent to a membrane and of the resulting mechanical forces. The extracellular matrix and the cytoskeleton can be influenced by stretching forces, and can be mutually effective in restructuring subsequent structures. Also mechanical forces could be shown to deform the chromosomal structures, when fibers function as actuating structures.

Lastly, at the level of organic molecules, the distinctness of biomechanics and biochemistry disappears (Bissell and Aggeler, 1987). This can be demonstrated by the presence of "mechanoenzymes", which function simultaneously in molecular and mechanical mechanisms (Bereiter-Hahn, 1991). Ontogenetic development can be conceived as internal differentiation of mechanical constructions which enclose complicated chemical apparatuses. But the dominating structure in ontogeny, which becomes effective by constraining and enforcing the differentiation of cells, tissues, and organs, consists of the inner structural scaffolding. From this it follows that cells must not be seen as separate and isolated units, building up the organism by crystal-like arrangement; in reality they are integral components of constructional frames.

Some of these results form the basis of the tensegrity model of Ingber and his group (Ingber, 1993a,b; Wang, Butler and Ingber, 1993). The tensegrity model gives reasons for the alteration of some features of the cell-construction, such as the stiffness, by some deformation-activities. It is focused on the lattice-like inner scaffolding of the cells, the cytoskeleton. The architecture of the cytoskeleton is held responsible for a manifold of chemical processes.

This model may be seen as an advance towards a consistent theoretical basis for understanding living beings. On the other side, in its present form, it lacks the universality of application. The tensegrity model does not consider the role of hydraulics and the function of the different kinds of fibers for shape enforcement of cells, tissues and the organisms as wholes, which is given by the theory of organismic constructions. This theory shows us the interdependence of all levels and of all parts of living beings, functioning mechanically or chemically; it highlights the fact that organisms are ruled by mechanical principles, which influence all levels.

Thus, we can conclude that the distribution of various substances and genetic products, called "morphogens" (Nüsslein-Vollhardt, 1990; Nüsslein-Vollhardt, Frohnhofer and Lehmann, 1987; Nüsslein-Vollhardt and Wieschaus, 1980;Tautz, 1991) may result out of differentiation, but they are not the primary cause of these processes.

Autonomy of Organismic Constructions

The mechanical energy-driven framework of the organism has, in the case of animals, a great deferability, which becomes effective in motility, propulsion, capturing of food, expulsion of waste, and blood circulation. Locomotor deformations can display a variety of modes in dependence on the working construction. Motoric patterns are useful, if there is no conflict between the actions of different components. Similar to some machines, the propulsive apparatus can be used in different ways. The only demand is that deformations remain adequate to the constraints given by the organismic construction. The different deformations must be harmonized to a great extent. A great many options of locomotion become manifest in a high degree of organismic autonomy. Autonomy as resulting from energy-driven activity of constructions is not consistent with the basic tenets of molecular reductionism and is totally at variance with the Synthetic Theory of evolution.

Relations with the Environment

If we accept the hydraulic and mechanic nature of living beings, the interdependence of all organismic levels and the predominance of the organismic construction are apparent. The living being has the form of a well-defined apparatus which is moved by its own intrinsically generated deformations on the basis of energy consumption and under the influence of internal pacemakers. Actions of organisms have a predominant internal aspect. All actions are performed in relation to external and environmental factors (Gutmann and Edlinger, 1991a, b).

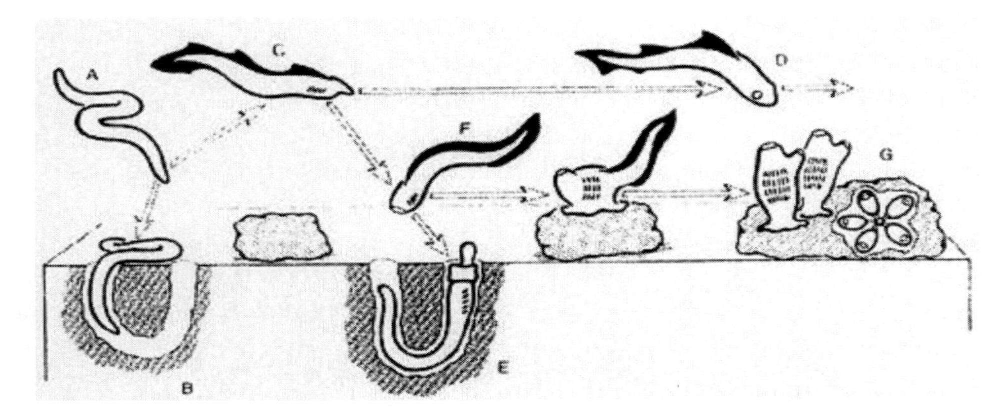

Figure 2. Habitats and environments are required by the active organisms. For that the particular design and the ability of living things are crucial. Not adaptation decides about the habitat, but the already existing capabilities and facilities. A. An ancestral freely moving worm construction. B. A floor inhabiting worm. C.Aa freely moving ancestral vertebrate construction. D. A freely moving vertebrate construction. E. A soil inhabiting vertebrate construction. F. A transitional form leading to E. and G. A sessile vertebrate constructions.

The constitution of the organismic construction and its mode of functioning determine the external "contacts" and the organismic interdependencies with the environment. The environment can only be actively conquered by the organisms. Environmental factors must never be understood as externally generated forces or constraints impacting the organismic constructions. There is no adaptation-generating influence of the environment (Edlinger, Gutmann and Weingarten, 1989, 1991). The constructional properties of the organismic units and their mode of functioning are responsible for survival, reproduction and even the death of organisms in their habitats. There can be no greater contradistinction to the Synthetic theory (see Figure 2).

Functional Interdependencies

The action of animal organisms is integral. It requires uniform pulse generation, coordination, and rigorous “control” of each sub-system through the entire construction. In small, simply organized animal constructions these controls may be done by enzymatic or mutual influence of mechanical stress fibers or tubules. Mechanical and chemical processes interact at this level.

In other words, biochemical reactions follow essentially from mechanical influences, which in turn are in interaction with chemistry, as already mentioned above. More complex animal structures must be controlled by highly specialized neural subsystems, i.e., nervous systems. The interconnections of nerve cells and the excitation produced by these interconnections must conform to the functional design of the mechanical construction. The patterns of neuronal excitation must be transferred successfully to the musculoskelatal system. This can only work with friction-less rhythmic activity.

The neural connections and the activation patterns produced by them are as they are if the biomechanical operation runs without friction and self-destruction is tied to the mechanical design. The randomness of the pattern formation and action are thus very tightly restricted.

This mutual relatedness, which is clearly dominated by the mechanical apparatus and its needs, is justified by the molecular biological reductionism of genetic determination.

The traditional justification for this mutual genetic relatedness solely by the molecular biological reductionism is not valid either. The reason is that because it requires precise matching of genetic activity, it produces unsolvable information-theoretic problems. For there is no plausible explanation for the sophisticated design of morphological and epigenetic development of complex systems of this kind, such as neural apparatus or the mechanical design.

MOTORIUM AND SENSORIUM

However, the mechanical apparatus with its structural constraints,as defined by Bereiter-Hahn (1987, 1991), Bissell and Aggeler (1987), Franke et. al. (1990), Lansman et al. (1987), McClay (1987), Opas (1987), and Van der Burgh (1988) is crucial for all functions of the organism, and for neural functions in particular.

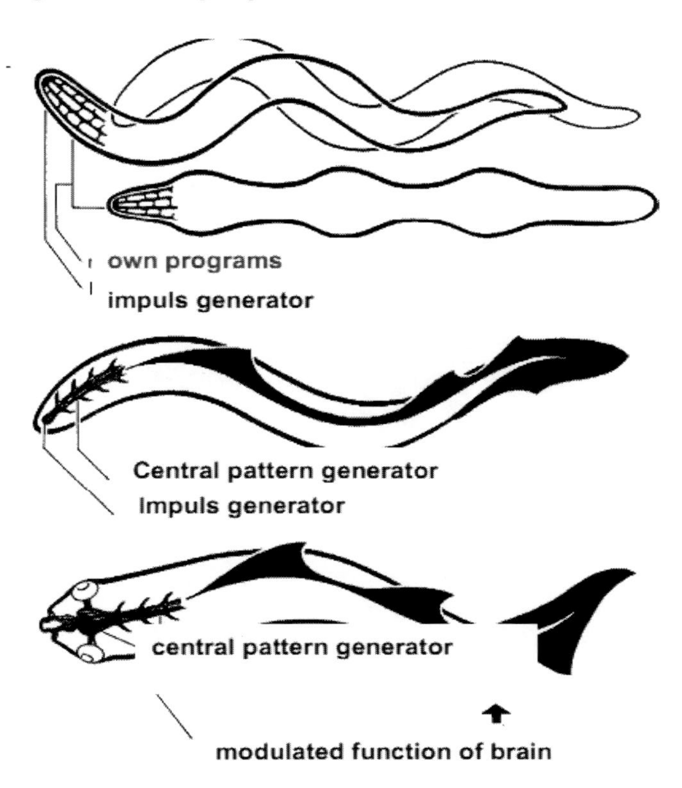

Figure 3. Organisms are autonomous movers and, as hydraulic systems, mechanically coherent. An organism can be active only in its entirety: head or the central nervous system and muscles all at the same time, cause the temporal pattern and the topographic distribution of the motor action. Due to this holistic nature and spontaneity the central nervous system is the central organ for all services. The primary reference of the nervous system consists of its own design and not directly or immediately of the outside world and its structure and resistance.

The neural apparatus is subjected primarily to those design constraints that are dictated by the mechanical apparatus and its function. They generate the rhythmic pattern of action, resulting in the rhythm of the motorium (see Figure 3).

MORPHOLOGY

The patterns of action require not only specific neuronal circuits, but also the rough morphology of nervous systems. The reason is that functional interdependence of motorium and the nervous system is impossible without a close spatial contact. Neural networks and ganglia complexes together make up nervous systems, but may be established only in locations where space is available among other hydraulic or contractile bloated subsystems. In principle, this can happen due to the nature of such spaces, in hydraulically and mechanically kneaded regions or mechanically closed, fluid-filled pressure-balanced and hydraulically-controlled sites.

NERVE NETS AND MEDULLAR STRAND SYSTEMS

The nervous system is bound primarily to accept a net-like or cord-like structure, because ganglion complexes, which have a complicated structure sensitive to mechanical influences, cannot exist under conditions of mechanical stress and kneading. These morphological restrictions determine the interconnections and thus the functional abilities of the nervous tissue.

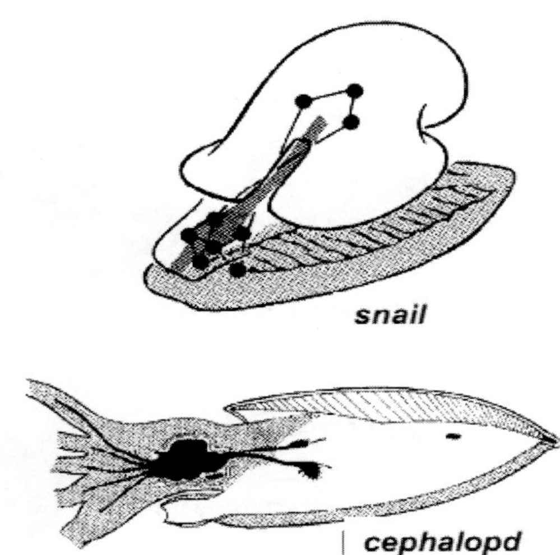

Figure 4. In most snails disused areas are distinguishable. Accordingly, the nervous system is structured differently. The kneaded foot portion developed a medulla-like nervous system..In coleoides cephalopods (cuttlefish), a cartilaginous head capsule is formed. Mechanically protected part in this location formed the highly compensated and organ-centralized nervous system in invertebrates.

The permanent body rhythms can be secured only by decentralized neuronal structures. On the other hand, the network-or cord-like organization prevents the formation of those interconnections that would be needed for higher integration services. More productive and flexible neuronal complexes require local concentrations.

In such designs, the functions of the nervous system are limited mainly to the maintenance of rhythmic motor activity, which leads to more uniform self-activity and can be modulated only by external mechanical resistances with which the animals come in contact out of their own initiative (see Figure 4).

Modulation is mechanical, but in these cases the rhythmic movements have to change by the interplay of body activity and external resistance. Both creep and tentacle movements are examples of this mode of change, but then, when the normal state is restored, once again the design-related continuous rhythm is restored.

The apparent primitiveness of these nervous systems, as we find them especially in coelenterates or Turbellaria, is by no means a sign of an original developmental status, but derive from the special construction of the respective motor apparatus.

Ganglion Complexes

Larger concentrations of neurons are possible only when no mechanical kneading is given. Moreover ganglia or, in the case of higher concentrations, brains, require a mechanical stabilization. Both will be achieved in one of two ways. One way is by mechanically stable axes around which a ganglion-like ring forms, with the surrounding fluid providing a hydraulic protection against kneading. Such an axis is mainly the foregut of various soft animal designs (see Figure 5).

The other way consists in the formation of closed capsules made of stiff materials, within which there is edgy material, that turn into complex units, according to available space and the potential number of neurons that receive synaptic connections.

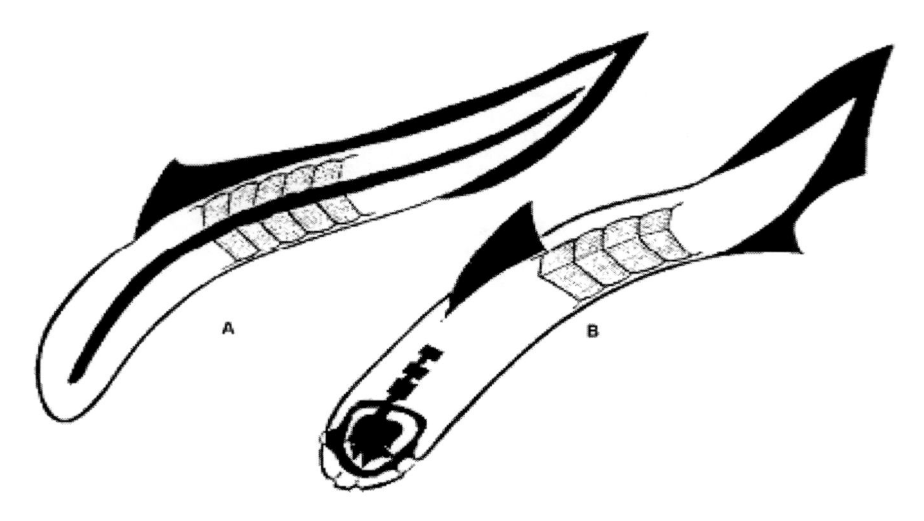

Figure 5. The situation of vertebrates. Cephalization allows the formation of a brain, surrounded by a bony capsule area in which to establish itself as a swelling of the spinal cord (B). Sensory organs that can only be bound together in with a continuous neural tube of uniform dimensions (A). Any relevant processing capacity in the central nervous system is connected, and therefore it would be uneconomical to be constructed in any other form.

In the first case we deal with a typical throat ring of numerous bilaterally symmetrical invertebrates; in the second, with the head capsules of cephalopods and vertebrates. Notably, these capsules do not serve only as a mechanical protection for the brain, but as insertion sites for muscles they form an integral part of the mechanical apparatus.

Increased Integration Units

Such ganglia complexes are now functionally tightly coupled with the rest of the nervous system and thus with the motorium. On the other hand, there is an opportunity for the generation of internal circuits. Thus, it becomes possible to intervene in these changing rhythms and modify them, however, only in so far as to allow the mechanically induced action potential of this motorium.

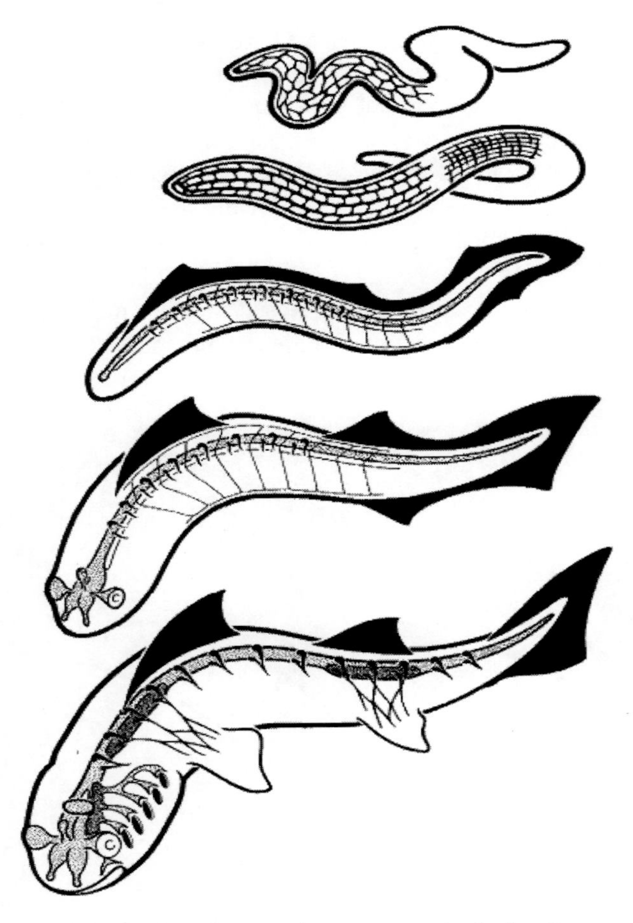

Figure 6. The nervous system was primarily spread across an entire hydraulic apparatus. By focusing on tortuosity in length, constancy could occur about the notochord, a first concentration of the nerve (with new wiring options). In vertebrates, this resulted in the formation of the head, a stiff and quiet location.. This lead to the formation of the brain. The conditions for the concentration of nerve structures as a precursor of brain development are always a biomechanical nature. Also, the preconditions of cognitive ability go back to biomechanical factors.

It also means that for ganglionic integration centers and brains a strict coupling with the musculoskeletal system is necessary. This coupling allows only certain degrees of freedom of organization, if the operation of the latter is secured.

Frequent modulations of rhythm and movement cause new patterns of behavior and, as a consequence, changes of the involved neuronal interconnections. On the one hand, new synaptic contacts are established; on the other hand, those which hinder the smooth movement of the excitement generated patterns are eliminated. Elimination can also affect all neurons and neurons complexes.

If successful and if repeated, these changes are not only fixed, but even strengthened. The more complex central nervous systems are organized, the more alternative modulation and movement patterns they have. It is in this manner that they can play a proactive role in developing responses to various environmental conditions (see Figure 6).

As a consequence, rough trial-error strategies in the motoric response to resistances of the environment become obsolete. The extent of these possibilities is a question of the evolutionary development and therefore of the design conditions.

ONTOGENESIS

In addition to the phylogenetic explanation that is offered in the organismic design theory, there is also the question of the ontogenetic development. It can be assumed that neuron-complexes receive their external morphology from their frame construction and their activity. Their internal organization, however, the growth of nerve cells and their interconnection, are problems that have been considered from a rather unilateral view in molecular biology. Nervous systems are treated separately and the outgrowth of neurons and synaptic connections can be seen from a perspective that allows almost only chemical influences as explanations. Obviously, it can be proved that chemical attractors influence the formation of nerve tissue or the growth direction of nerve fibers.

On the other hand, the differentiation of their mechanical construction and its chemical properties as well as the selective expression of specific gene products are not imaginable without mechanical stress. Only that which fits into the overall mechanical frictionless association, may ultimately endure also as a subsystem of the organism. This is in accordance with the ideas of Edelman (1987) about “neuronal Darwinism”.

Therefore, the ideas of Opas (1987), Lansman et al. (1987), Franke et al. (1990), Vanderburgh (1988) and Bereiter-Hahn (1991), result in a new model, which shows us the biochemical processes in the organismic context and thus provides a biomechanical reason for that context and all those epigenetic processes that lead to the formation and organization of nervous systems. Neurons will be considered as viable, mechanically active units that establish themselves under various mechanical and chemical influences in the space systems of animal designs and are constantly in close interaction with each other and with the muscles.

This interaction, which affects the role of neurons, especially in the generation of excitation patterns, leads to activity, and thus also to constant re-formation of the motorium and the mechanical frame construction. The lattter in turn reacts by morphological design constraints and mechanical stress on the nervous system and its individual components when there are dysfunctional disorders of the interconnections to perform mechanical operations. Thus, the biochemical influence is not denied but placed in a new context.

Receptors and Sense Organs

The active animal organisms do not consist only of the motorium and its associated nervous system, but also of external structures, which provide for energy flows from the outside world and, in addition, contribute to the modulation of motoric processes of the musculoskeletal system. They represent the external boundary areas of the nervous system and have an organism-specific design, so that they also form an integral part of the organism and follow the design principles.

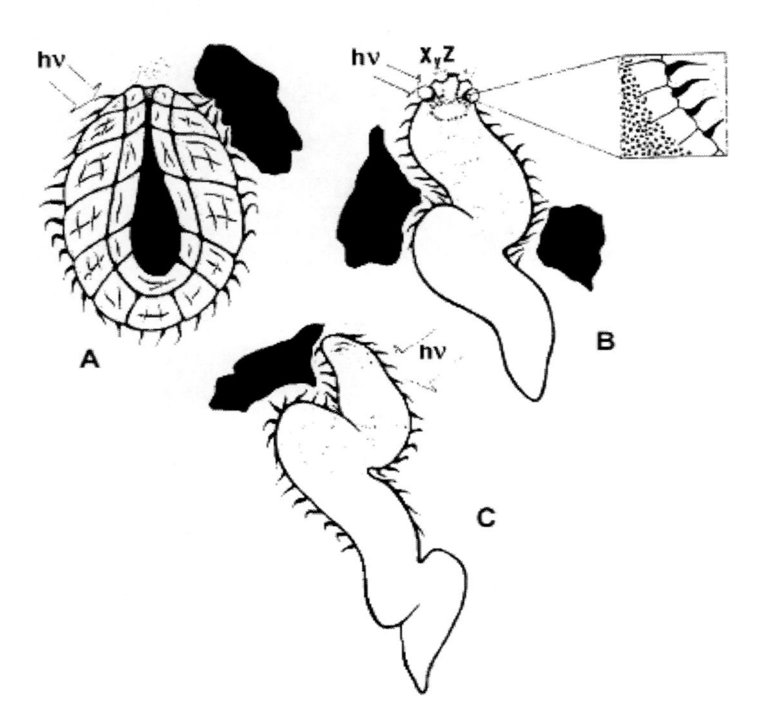

Figure 7. Formation of a complex sensory system through evolutionary change of motorium and sensorium.(A) Low specific reception, as shown by the example of an early Gallertoiden. These organisms are only capable of ciliary movement and slight deformation by contractile tension. The number of possible states and responses is limited, since the ontogenetic acquisition of distinctive opportunities between stimuli of different modalities and qualities is severely limited (B). The expansion of opportunities through constructive action necessarily forms deformations that allow for an increase in the associative connection between movements and thereby produce actively induced changes of state. Ciliary structures are working against mechanical and chemical resistances, which can be distinguished by the success of reactive action. This leads to the formation of ontogenetic-related patterns of discrimination. (C) The training of organismically preconstructed and related outdoor structures that allow only specific energy flows of the environment lead to increased possibilities of distinction that result in enhanced sensory-specificity.

Irritating "sense impulses" that run on them, have no specific quality, i.e., they represent the same depolarisation and polarisation of membranes, which is generally observed in neuronal activity. In the words of Roth (1987), the language of the nervous system is, in this respect, very monotonous. It consists only of the signals "click click".

The interconnection with other neural structures enables comparing the effects of responses and consequently the appearance of very specific modulations of motor performance. This leads to the establishment of different sensory modalities and qualities.

A rigorous selection of possible influences from the outside arises only by the special structures of these sense organs, also known as outer portions of the nervous system. They usually consist of near-surface hydraulic deformable bodies, such as living cells or cilia on the apical cell processes, that extend into the medium and interact with it.

By means of a simultaneous action of sensory structures (i.e., sense organs) and the motorium, mechanical resistances, as light, sound or chemical substances contained in the external medium, cause other modulations of rhythmic movement. This triggers neural signals.

Concurrent motor action will also lead to various influences which expose the organism to a manifold of diverse situations and mechanical stresses. The interconnection with other neural structures results in a specificity in the modulation of motor performance, which ultimately leads to the establishment of different sensory modalities and qualities.

The use of a stimulus depends on whether it can be utilized in an economical manner and whether the processing of livelihood-sustaining causes modulations of motor actions. This result shows up especially by means of reinforcing and fixative confirmation through repetition and associated wiring. Gradually there arises the specificity of different modalities and qualities.

The more monotonous the effects are while the receptive structures are formed, the faster these specification processes will happen; in extreme cases, they may even eliminate the specificity of sensory structures altogether (see Figure 7).

SPECIFICITY IN SENSORY STRUCTURES

As mentioned above, the sensory structures usually consist of near-surface hydraulic deformable bodies, such as cells (from cilia or stereocilia apical cell processes, microvilli) that extend into the medium and interactions can occur with it. This interaction in turn triggers neural signals in the nervous system and affects the motorium.

The possibilities for influence by external factors depend entirely on the receptor structure. So, what is a stimulant, and especially what is an adequate one, and what is not, is decided by the organism and its construction.

Whether the stimulus may eventually continue to act as such depends on whether it can be used in an economical manner that leads to processing it in terms of its livelihood-sustaining arousal modulation and motor actions.

The best example for this is provided by the optical spectra of different animal groups that have constructed different eyes.

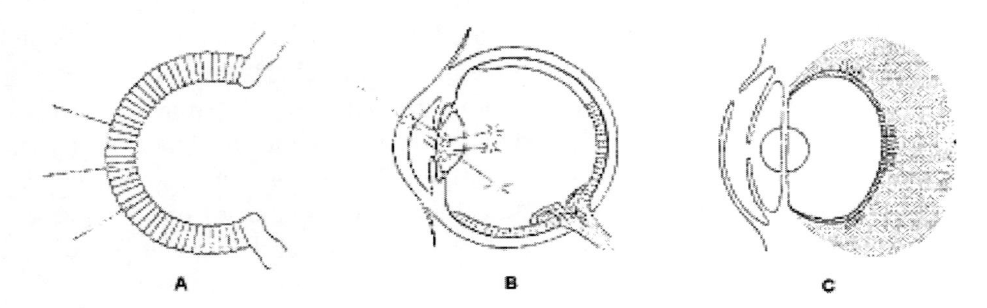

Figure 8. Organismic contingency of photoreception by the example of the compound eye (A), the vertebrate eye lens (II) and the cephalopodenauges (C). Ultraviolet light stimulates the receptors in facets that are arranged with their receptor poles in the form of a hollow ball outwardly.There is a lot of scattering at the short wavelength light. This is due to UV blur. For structural reasons, less energy light is used.

By considering the usual adaptation ideas, one would have difficulty to understand why vertebrate lens eyes perceive with no ultraviolet. This will become clear quickly when one considers that exorbitant increase with decreasing wavelength, so that the scattering effects of light passing through the cornea, lens and vitreous body impede the creation of sharp images with clear contours (see Figure 8).

The human visual window, but also that of other species, is understood not as an adaptation to certain environmental characteristics, as Vollmer (1975) suggested, but as due to the structure and functions of the organism. In the same way, the UV-perception can be explained by compound eyes that are not affected by their construction of scattering effects.

As a consequence, by considering the theory of organismic constructions the models of the traditional neuroscience are in need of revision. It does not follow that material presented by these models cannot be used. However, conceptual models that ignore the conditional nature of neural organismic performance, such as the "radical constructivism", are dispensable.

ORGANISMS AND ENVIRONMENT

Accumulations of neurons give the chance to evolve medullar cords, ganglia and brains (Edlinger, 1992, 1994b, c). These are the prerequisites for effective interactions between the organism and its environment. Responses of more complex sensoric patterns indicate advanced stages of evolution. Such patterns make coordination and inner representation in more complicated nervous systems possible.

Nevertheless all these neuronal structures, which seem to be superordinated and superimposed over the other parts of the nervous system and over the musculature functionally, are the result of an evolution of the locomotory apparatus. They function first as pacemakers for the locomotor apparatus, even when they fulfill other functions (Edlinger, 2009; Edlinger, Gutmann and Weingarten, 1989, 1991).

The environment and its energy flows must first be seen as drags or obstacles for the autonomous spontaneously self-deforming and moving organism. They can hinder its rhythmical locomotion and disturb the rhythmicity.

The reactions of organisms to such disturbances are attempts at compensation by changes of their own rhythmical activity. The compensatory activity continues until the former situation, that is, the organism's own internal generated rhythm is re-established.

Thus, by their deformation, propulsion and locomotion animals bring themselves into the energy-conversion of their environment, which are retroacting in regard to the organisms, whereby retroacting means irritations for the organisms.

Excitations and irritations always mean un-normal situations, which are to be compensated for. The organisms must compensate for various irritations provided by drags, by means of energy conversion in the environment or by means of chemical stimulation.

Thus, organisms stimulate themselves indirectly over the environment. This retroaction can be perceived as sensoric input. In principal, we can assume a sensory input, which is reducible at least to autonomous actions of the organisms themselves. Evidence is provided by the physiological sciences. The best example is the saccadic rhythm of vertebrate eyes.

The sensory input causes excitations by reaction-potentials of nerve cells. These potentials are the same in the whole nervous system. Thus, one can say that the language of the nervous system is common in all parts.

We have to ask, how the nervous system can produce this manifold of different modalities and qualities known by humans and, as we can suppose, also by other members of the animal kingdom. The only reason for these phenomena is a new view of the excitation, resp. irritation and of the compensatory reactions.

These new views of the sensory input and of the organisms put Holst's (1974) principles of permanent rhythmicity and in particular of reafference in a favorable light. His claims can be supported by the theory of organismic constructions.

Special structures of receptors decide which modalities are sensed. Many reactions are thinkable and possible, but in the special situation, given by some very specific modalities of excitations, only a few of these reactions are successful for a particular animal.

Success or failure of compensatory actions lead to more and more specific reactions and, as a consequence, to increasingly differentiated inner representations. As a consequence, at first a primitive differentiation of modalities and qualities takes place.

THE THEORY OF THE FUNCTIONAL CYCLE

This view of the differentiation of modalities and qualities leads us to J. v. Uexküll and his theory of the functional cycle. Von Uexküll was a non-Darwinian biologist in the first half of the 20th century, who focused on the autonomous activity of organisms, in particular of animals. In contrast to the dominating Darwinian view, Uexküll found out, that every contact between organisms and their environment follows primarily out of the organisms' activity.

Living beings require that sectors of the environment can be used in accordance to their special nature and needs. Their activity and impact on the environment cause effects on the organisms. On the other side, one can say in modern language, action and reaction constitute some kind of self-reference.

Uexküll wanted to draw far-reaching conclusions from the anatomical type to which an animal belongs, and he insisted on the validity of these conclusions. He claimed that if we know the characteristics and the activities of an animal species, we thereby look into its "inner life", for everything that undergoes inside is strictly dependent upon the way the animal is able to accept and act on external stimuli. An animal can receive only those impressions for which it is prepared by its structure and can react to stimuli only insofar as it possesses the appropriate organs.

This highly selective approach to the environment resp. to some special sectors of the environment underlies Uexküll's suggestions that every organism fits its special environment perfectly from the beginning, always in accordance with its internal needs. As a consequence, in Uexküll's view, Darwinian adaptation cannot happen. Animals are fitting to their environment, but they are not adapted.

In the words of Cassirer (1944, p. 22): "Every organism is, so to speak, a monadic being. It has a world of its own because it has an experience of its own. The phenomena that we find in the life of a certain biological species are non transferable to any other species.....In the world of a fly, says Uexküll, we find only "fly things"; in the world of a sea urchin we find only "sea urchin things."

In accordance with his non-adaptationist conception Uexküll claims that every kind of animal has an inner world of its own, which depends on the special structure or construction of the animal and to which there is no approach for others.

Because effectors and receptors are connected by an internal network of nervous structures on the one hand and by the environment on the other, Uexküll calls this circular arrangement of various activities and perceptions - the arrangement of the perception system (in German Merknetz or 'sense net') and the effector system (in German Wirknetz or 'effect net') - the functional cycle (Funktionskreis).

For Example: Three-Dimensional Space – Is It a Natural Entity or a Construction?

The bilateral symmetry with special chances for effective locomotion causes a new structure, namely, new dimensions of the animal's environment, in particular its space. Only elongated and bilateral symmetric organismic constructions show us a unidirectional locomotion. As a result of their bodily construction plan it is possible for these organisms to distinguish the front and the rear of their body, and also to distinguish between left and right and at least up and down. By special sensory organs, which are evolved at a high level of organization, the situation relative to the space and to gravitation can be discerned and compensated for in a second step. Globular shaped organisms have no possibility for a comparable differentiation.

Environmental Adaptation of Perception? - Refutation of Evolutionary Epistemology

We present the background of biological reasons for the organismically determined perception and for an internal three-dimensional construction of the space, in which bilateral-symmetrical animals live and move. But before that we have to criticize any suggestion that this organismic three-dimensionality is an inner representation of a real situation in space, namely, in the environment. As a consequence of the importance of the internal and constructional needs of the animal's and also of human's activity on the one side, and of spontaneity and autonomous action, on the other side, various attempts of evolutionary epistemologies, for example by Spencer, Haeckel, Lorenz, Campbell, Riedl and Vollmer must be rejected (Edlinger, Gutmann and Weingarten, 1989, 1991).

As a consequence of the theory of organismic constructions, it is evident that organisms are stimulating themselves indirectly when they are acting in their environment. All sensory inputs coming from the environment are consequences of the organisms' own activities.

Thus, we are led to suppose that the modes of perception in any case depend on the organisms themselves. We cannot have any knowledge of real things or processes. All we can perceive is designed and constructed by nervous systems, sense organs included. In the case of humans, the so called "reality" follows out of organismic constructions and of our minds which are hitched up to the material basis of the organisms.

Evolution of Cognition

On the basis of these presuppositions it is possible to reconstruct not only the evolution of organismic constructions but also that of mental and cognitive abilities too. These are, in contrast to Darwinian evolutionary epistemology, the outcome of special constructional properties of autonomous organisms. They cannot be considered without a very close connection with their organismic basis.

This is true also for the evolution of complicated nervous systems and their abilities. In some special groups of animals, the vertebrates in particular, by reason of the occurrence of some inner spaces which are undisturbed by movement and deformation, some concentration effects of nervous tissues take place and make many additional neuronal operations possible. With increasing concentration primitive anticipatory operations take place increasingly. They cause the origin of some kind of inner world, but a world, which depends, in accordance to Uexküll, on organismic properties.

The consequence of evolution of increasingly complicated nervous systems is a more and more complicated inner world of organisms, which becomes more and more autonomous. That means an increasing emancipation from the organismic construction too. This emancipation happened with the increase of a mechanically undisturbed brain in vertebrate head-capsules and, in particular with the evolution of special ape-constructions, whose emancipation enabled them to evolve a differentiated symbolic language.

At this point the beginning of cultural development is given. Although some indispensable biological and constructional needs limit every activity, no direct biological limitation is given. It follows from this, that cultural activities and development must be conceived as an other level of evolution for which no reasons can be given by biological theories.

Cultural Implications

As a consequence of the presuppositions of the theory of organismic constructions and of the autonomy of cultural development only a constructivist view of knowledge and cognition can be accepted. Constructivism can be seen as the outcome of various traditions of thinking, in European tradition - of Kantian thinking in particular.

On the other hand, the suggestion of organismic autonomy has very radical and far-reaching consequences not only for thinking about the relation between organisms and their environment but even for thinking about the approach to nature in general. Although we have to concede that every ontological description of nature must be invalid we can say that a mechanistic approach in the tradition of European sciences must be rejected. Autonomous organisms must act very independently in their environment and must be pushed to activity by their own inner mechanisms. Hence, they cannot be seen any more as some kind of cog-wheels of a gigantic clockwork, as is suggested by some models of traditional natural philosophy.

Consequently, the theory of organismic constructions accepts the philosophy of A. N. Whitehead (1934), which in fact disvalidates traditional biology. Whitehead conceived nature as an organic and organism-like structure which is permanently in motion engaged in change and development, but always spontaneously and pushed by impulses which come from the inside. Whitehead's world consists of organisms. Organisms in the sense of Whitehead are very manifold and occur on various levels of complexity. Because of the organismic nature of the world this philosophy does not need a god like the supernatural clockmaker, as European tradition does, but it refers to an all-embracing permanent process.

Some "physicalistic" theories, such as the chaos-theories, synergetics, the theory of dissipative structures and others which intend to provide a better understanding of the complexity of the world, show us very similar ways of thinking but they cannot present a consistent view of living organisms.

On the other hand, there are many more points of contact of the organismic approach with non-European views of world, in particular with the thinking of Chinese Taoists. Joseph Needham (2011) showed us that especially for Taoists the suggestion of a permanent change and development of world was obvious. Needham presents a view of nature in ancient China, which lacks a personal creator as the god of the Jewish-Occidental tradition. The outcome of this view was the possibility of conceiving the world as an organismic-like system, which is developing and evolving itself without an stimulus from the outside. Thus, about 300 years BC some Taoists had very consistent ideas not only about permanent change of the world as a whole but about an evolutionary change of organisms too. We must consider that these ideas are not identical with Darwinism, which represents itself as a mechanistic idea, but in many respects is very similar to the view presented in this chapter.

REFERENCES

Barash, D. P. (2003), Dennett and the Darwinizing of free will. *Human Nature Review*, *3*, 222–225.

Bereiter-Hahn; J. (1987). Mechanical Principles of Architecture of Eukaryotic Cells. In J. Bereiter-Hahn, O. R. Anderson and W.-E. Reif (Eds.). *Cytomechanica:. The mechanical basis of cell form and structure*. Berlin-Heidelberg-New York: Springer Verlag, pp. 5-28.

Bereiter-Hahn; J. (1991). Cytomechanics and Biochemistry. In N. Schmidt-Kittler, and K. Vogel (Eds.). *Constructional morphology and evolution*. Berlin- Heidelberg-New York: Springer Verlag, pp. 359-374.

Bereiter-Hahn, J. and Strohmeier, R. (1987). Hydrostatic pressure in metazoan cells in culture: Its involvement in locomotion and shape generation. In J. Bereiter-Hahn, O. R. Anderson and W.-E. Reif (Eds.). *Cytomechanica: The mechanical basis of cell form and structure*. Berlin-Heidelberg-New York: Springer Verlag, pp. 261-270.

Bertalanffy, L. v. (1968). *General system theory: Foundations, development, applications.* New York: Braziller.

Bissell, M. J. and Aggeker, J. (1987). Dynamic reciprocity: How do extracellular matrix and hormones direct gene expression? In M. C. Cabot and W. L. Mckeehan (Eds.). *Mechanisms of signal transduction by hormones and growth factors*. New York: Liss, pp. 251-262.

Blackmore, S. (2005). *Conversations on consciousness. Interviews with twenty minds.* Oxford, UK: Oxford University Press.

Campbell, D. (1974). Evolutionary epistomology. In A. Schilpp (Ed). *The philosophy of K. R. Popper*. Lasalle: Open Court, pp. 413-463.

Cassirer, E. (1945). *An essay on man*. New Haven: Yale University Press.

Cassirer, E. (1994a). *Versuch über den Menschen: Einführung in eine Philosophie der Kultur.* Frankfurt am Main: S. Fischer.

Cassirer, E. (1994b). *Philosophie der symbolischen Formen* (4 Vol.). – Darmstadt: Wiss. Buchges.

Churchland, P. (2002). *Brain-Wise: Studies in neurophilosopy*. Cambridge, MA: MIT Press.

Clark, F. M. and Masters, C. J. (1976): Interactions between muscle proteins and glycolytic encymes. *Int. J. Biochem. Bristol. 7*, 359-365.

Damasio, A. R. (1999). *The feeling of what happens: Body and emotion in the making of consciousness.* New York: Harcourt Brace and Comp.

Darwin, Ch. (1899). *Über die Entstehung der Arten durch natürliche Zuchtwahl.* Stuttgart, Germany: Schweizerbart`sche Verlagshandlg.

Darwin, Ch. (1906). *Das Variieren der Tiere und Pflanzen im Zustand der Domestikation.* Stuttgart, Germany: Schweizerbart`sche Verlagshandlg., 2 Vols.

Eigen, M. (1971): Self-organization of matter and the evolution of biological macromolecules. *Naturwiss., 58*, 456-522.

De Waal, F. (2010). Towards a bottom-up perspective on animal and human cognition . *Trends in Cognitive Sciences, 14*, 201-207.

Dennett, D. (1991). *Consciousness explained* . London: The Penguin Press

Dobzhansky, Th. (1937). *Genetics and the origin of species* (1st. Ed.). New York: Columbia University Press.

Edelman, G. and Tononi, G. *(2001). A universe of consciousness. How matter becomes imagination* . London: Penguin.

Edelman, G. (1987). *Neural Darwinism: The theory of neuronal group selection.* New York: Basic Books.

Edlinger K., Gutmann, W.F., and Weingarten, M. (1989). Biologische Aspekte der Evolution des Erkenntnisvermögens - Spontaneität und synthetische Aktion in ihrer organismisch-konstruktiven Grundlage. *Natur und Museum, 119*, 113-128.

Edlinger, K. (1991a): The mechanical constraints in mollusc constructions -the function of the shell, the musculature, and the connective tissue. In N. Schmidt-Kittler, and K. Vogel (Eds.), *Constructional morphology and evolution.* Berlin-Heidelberg-New York: Springer Verlag, pp. 359-374.

Edlinger, K. (1991b). Organismus und Kognition - Zur Frage der biologischen Begründung kognitiver Fähigkeiten. In M. F.Peschl (Ed.), *Formen des Konstruktivismus in Diskussion - Materialien zu den "Acht Vorlesungen über den Konstruktiven Realismus".* WUV: Universitätsverlag Wien, pp. 108-150.

Edlinger, K. (1994). Das Spiel der Moleküle - Reicht das Organismusverständnis des molekularbiologischen Reduktionismus? *Natur u. Museum, 124*, 199-206.

Edlinger, K. (2009). *Darwin auf den Kopf gestellt - was bleibt von einer Ikone?* Wien/Klosterneuburg: Ed. VaBene.

Edlinger, K. (Ed.) (1989). *Form und Funktion - Ihre stammesgeschichtlichen Grundlagen.* WUV: Universitätsverlag Wien.

Edlinger, K., Gutmann, W. F. and Weingarten, M. (1991). Evolution ohne Anpassung. In W. Ziegler (Ed.), *Aufsätze u. Reden Senckenb. Naturforsch. Ges.* 37.

Edlinger; K. (1995): Elemente einer konstruktivistischen Begründung der Organismuslehre. In W. F. Gutmann and M. Weingarten (Eds.), Die Konstruktion der Organismen II. Struktur und Funktion. *Aufs. u. Reden Senckenb. Naturf. Ges.*, 43, 88-103.

Eigen M., Gardiner, W., Schuster, P. and Winkler-Oswatitsch, R.(1983).Ursprung der genetischen Information. In E. Mayr (Ed.), *Evolution - Die Entwicklung von den ersten Lebensspuren bis zum Menschen.* 3rd Ed.. Heidelberg, Germany: Spektrum, pp. 61-81.

Eigen, M. (1987). *Stufen zum Leben - Die frühe Evolution im Visier der Molekularbiologie.* München/Zürich: Piper.

Eigen, M. and Schuster, P. (1977-78). The hypercycle. *Naturwiss.,64*, 541-565; *65*, 7-41; 341-369.

Eigen, M. and Winkler, R. (1973). Ludus vitalis. In H. v. Ditfurth (Ed.), Mannheimer Forum 1973/74, *Studienreihe Boehringer Mannheim*, pp. 53-140.

Eigen, M. and Winkler, R. (1976). *Das Spiel - Naturgesetze steuern den Zufall.* München/ Zürich: Piper. .

Elsasser, W. M., Marsh, B. D. and Rubin, H. (1998): *Reflections on a theory of organisms: Holism in biology.* Baltimore: The Johns Hopkins University Press.

Emerman, J. T. and Pitelka, D. R. (1977). Maintenance and induction of morphological differentiation in dissociated mammary epithelium on floating collagen. *In Vitro, 13*, 316-328.

Franke, R. P., Gräfe, M., Aauer, U. D., Schittler, H. and Mittermeyer, C. (1990). Stress fibers (SF) in human and endothelical cells (HEC) under shear stress. In H. Kiesewetter, and F. Jung (Eds.), *Blood fluidity and endothelial influences on microcirculation.* Berlin-Heidelberg-New York: Springer, pp. 989-992.

Franklin, J. and Lewontin, R. C. (1970). Is the gene the unit of selection?*Genetics, 65*, 707-734.

Gadenne, V. (2002). Warum tun Schmerzen weh? Zur Problematik der Qualia in der Philosophie. In K. Edlinger, G. Fleck and W. Feigl (Eds.), *Organismus - Evolution - Erkenntnis. Perspektiven mentaler Gestaltungsprozesse.* Frankfurt/M.: P. Lang - Europ. Verlag der Wissenschaften.

Gleick, J. (1988). *Chaos- die Ordnung des Lebens.* München, Germany: Droemer-Knaur.

Gutmann, W. F. and Weingarten, M. (1987). Autonomie der Organismischen Biologie und der versklavungsversuch der Biologie durch Synergetik und Thermodynamik von Ungleichgewichtsprozessen. *Dialektik, 13*, 227-234. Pahl Rugenstein, Köln.

Gutmann, W. F. (1988b). *Die Evolution hydraulischer Konstruktionen. Konstruktive Wandlung statt altdarwinistischer Anpassung..* Frankfurt am Main, Germany: W. Kramer.

Gutmann, W. F. (1991): Constructional principles and the quasi experimental approach to Organisms. In N. Schmidt-Kittler and K. Vogel (Eds.), *Constructional morphology and evolution.* Berlin-Heidelberg-New York: Springer Verlag, pp. 91-112.

Gutmann, W. F. and Bonik, K. (1981). *Kritische Evolutionstheorie.* Hildesheim: Gerstenberg Verlag..

Gutmann, W. F. (1988a). The hydraulic principle. *Amer. Zool., 28,* 257-266.

Gutmann, W. F. and Edlinger, K. (1991a). Die Biosphäre als Megamaschine - ökologische und paläo-ökologische Perspektiven des Konstruktionsverständnisses der Organismen l. *Natur u. Museum, 121 (10)*, 302-311.

Gutmann, W. F. and Edlinger, K. (1991b). Die Biosphäre als Megamaschine - ökologische und paläo-ökologische Perspektivendes Konstruktionsverständnisses der Organismen ll. - *Natur u.Museum, 121 (12)*, 401-410.

Gutmann, W. F. and Edlinger, K. (1994b). Molekulare Mechanismen in kohärenten Konstruktionen. In. W. Maier and Th. Zoglauer (Eds..), Technomorphe Organismuskonzepte. *Problemata fromann-holz-boog*, 174-198.

Haeckel, E. (1904). *Lebenswunder. Gemeinverständliche Studien über Biologie*. Stuttgart, Germany: Kröner.

Haken, H. (1981). *Erfolgsgeheimnisse der Natur*. Stuttgart, Germany: DVA.

Haken, H. (1990). Synergetik und die Einheit der Wissenschaft. In W. Saltzer, *Zur Einheit der Naturwissenschaften in Geschichte und Gegenwart*. Darmstadt: Wiss. Buchges, pp. 61-78.

Haken, H. and Wunderlin, A. (1986). Synergetik: Prozesse der Selbstorganisation in der belebten und unbelebten Natur. In A. Dress, H. Hendrichs and G. Küppers (Eds.), *Selbstorganisation - Die Entstehung von Ordnung in Natur und Gesellschaft*. München, Germany: Piper, pp. 35-60.

Harrington, A. (2002). *Die Suche nach Ganzheit. Die Geschichte biologisch-psychologischer Ganzheitslehren. Vom Kaiserreich bis zur New-Age-Bewegung*. Reinbek bei Hamburg: Rowohlt.

Horgan, J. (1999). *The undiscovered mind: How the human brain defies replication, medication, and explanation.* New York: Free Press.

Ingber, D. E. (1993a). The riddle of morphogenesis: A question of solution chemistry or molecular cell engineering? *Cell, 75*, 1249-1252.

Ingber, D. E. (1993b). Cellular tensegrity: Defining new rules of biological design that govern the cytoskeleton. *J. Cell. Sc.,104*, 613-627.

Ingber, D. E. and Folkman, J. (1987). Regulation of endothelial growth factor action: Solid state control byeytracellular matrix. In D. E. Ingber (Ed.), *Mechanism of signal transduction by hormones and growth factors.* New York: Alan. R. Liss, pp. 273-282.

Ingber, D. E., Dike, L., Hansen, L., Karp, S., Liley, H., Maniotis, A., McEe, H., Mooney,H., Plopper, W., Sims, H. and Wang, N. (1994). Cellular tensegrity: Exploring how mechanical changes in the cytoskeleton regulate growth, migration, and tissue pattern during morphogenesis. *Int. rec. Cytol., 150*,173-224.

Janich, P. (1989). *Euklids Erbe. Ist der Raum dreidimensional*? München, Germany: C.H. Beck..

Janich, P. (1992). *Grenzen der Naturwissenschaft*. München, Germany: C.H. Beck..

Janich, P. (1993). Biologischer versus physikalischer Naturbrgriff. In G. Bien, Th. Gil and J. Wilke (Eds.), "Natur" im Umbruch - Zur Diskussion des Naturbegriffs in Philosophie, Naturwissenschaft und Kunsttheorie. *Problemata frommann-holzboog, 127*, 165-175.

Jonas, H. (1987). *Macht oder Ohnmacht der Subjektivität?* Frankfurt/M.: Suhrkamp.

Juliano, R. L. and H. Haskill (1993). Signal transduction from the extracellular matrix. *Cell. Biol., 120/3*, 577-585.

Kant, I. (1839). *Kritik der Urteilskraft. Collected Works, Vol. VII*. Leipzig: Modes u. Baumann.

Küppers, B. O. (1979). Towards an experimental analysis of molecular self-orgnization and precellular Darwininan evolution. *Naturwissenschaften, 66*, 228.

Küppers, B. O. (1980). Evolution im Reagenzglas. In H. v. Ditfurth (Ed..), *Mannheimer Forum*. Studienreihe Boehringer Mannheim, 1980/81, pp. 47-114.

Küppers, B. O. (1986a). *Der Ursprung biologischer Information - Zur Naturphilosophie der Lebensentstehung*. München-Zürich: Piper.

Küppers, B. O. (1986b). Wissenschaftsphilosophische Aspekte der Lebensentstehung. In A. Dress, A., H. Hendrichs and G. Küppers (Eds.), *Selbstorganisation - Die Entstehung von Ordnung in Natur und Gesellschaft*. München Zürich: Piper, pp. 81-101.

Lansman, J. B., Hallam, T. J. and T. J. Rink (1987). Singlestretch activated ion channels in vascular endothelial cells and me-chano-transducers? *Nature, 325*, 811-813.

Looijen, R. C. (1999). *Holism and reductionism in biology and ecology: The mutual dependence of higher and lower level research programmes*. Dortrecht, The Netherlands: Springer.

Lorenz, K. (1941). Kants Lehre vom apriorischen im Lichte gegenwärtiger Philosophie. *Blätter dtsch. Philos., 15*, 94-125.

Lorenz, K. (1973). *Die Rückseite des Spiegels*. München, Germany: Dtv.

Matthews, B. D., Thodeti, C. K., Tytell, J.D., Mammoto, A., Overby, D.R. and Ingber, D. E. (2010). Ultra-rapid activation of TRPV4 ion channels by mechanical forces applied to cell surface beta1 integrins. *Integr Biol, 2*, 435-442.

Mayr, E. (1967). *Artbegriff und Evolution*. Hamburg: P. Parey.

Mayr, E. (1984). *Die Entwicklung der biologischen Gedankenwelt*. Berlin- Heidelberg-New York: Springer Verlag.

Mayr, E. (1991). *Eine neue Philosophie der Biologie*. München, Germany: Piper.

Mayr, E.(1979). *Evolution und die Vielfalt des Lebens*. Berlin-Heidelberg-New York: Springer.

McClay, D. R. and C. A. Ettensohn (1987): Cell adhesion in morphogenesis. *Annu. Rev. Cell. Biol., 3,* 319-345.

Medina, D., Li, M. L., Obornn, C. J. and Bissell, M. J. (1987). Casein gene expression in mouse mammary epithel cell lines: dependence upon extracellular matrix. *Exp. Cell Res. 172,* 192-203.

Meinhard, H. (1987). Bildung geordneter Strukturen bei der Entwicklung höherer Organismen. In B. O. Küppers (Ed.), *Ordnung aus dem Chaos - Prinzipien der Selbstorganisation und Evolution des Lebens*. München, Germany: Piper, pp. 215-242.

Meinhardt, H. (1978). Models for the ontogenetic development of higher organisms. *Rev. Physiol. Biochem. Pharmacol., 8,* 48-104.

Metzinger, Th. (2008). Empirical perspectives from the self-model theory of subjectivity: A brief summary with examples. In R. B. Bikas and K. Chakrabarti (Eds.), *Progress in brain research.* The Netherlands: Elsevier, 168, pp. 215, 246. .

Metzinger, Th. (2010). The no-self-alternative. In S. Gallagher (Ed.), *Oxford handbook of the self.* Oxford, UK: Oxford University Press, pp. 277-294. (Chapter 11).

Meyer-Abich, K. M. (1989). Der Holismus im 20, Jahrhundert. In G. Böhme (Ed.), *Klassiker der Naturphilosophie. Von den Vorsokratikern bis zur Kopenhagener Schule.* München, Germany: C. H. Beck, pp. 313-330.

Nagel, Th. (1974). What is it like to be a bat? *The Philosophical Review, 83/4*, 435-450.

Needham, J. (2011). *Science and civilization in China.* New York: Cambridge University Press.

Nüßlein-Vollhardt, Ch. (1990). Determination der embryonalen Achse bei *Drosophila. Verh. Dtsch. Zool. ges., 83*, 179-195.

Nüßlein-Vollhardt, Ch. and Wieschaus, E. (1980). Mutants affecting segment number and polyrity in *Drosophila. Nature, 287*, 795-801.

Nüßlein-Vollhardt, Ch., Frohnhofer, H. G. and Lehmann, R. (1987). Determination of anteroposterior polarity in *Drosophila. Science, 238*, 1675-1681.

Opas, M. (1987). The transmission of forces between cells and their environment. In J. Bereiter-Hahn, O. R. Anderson, and W.-R. Reif (Eds.), *Cytomechanics.* Berlin-Heidelberg-New York: Springer, pp. 273-285.

Opas, M. (1989). Expression of the differentiated phenotype by epithelial cells in vitro regulated by both biochemistry and mechanics of the substratum. *Dev. Biol., 131*, 281-293.

Pauen, M. (2007). *Was ist der Mensch? Die Entdeckung der Natur des Geistes.* München, Germany: Deutsche Verlags-Anstalt.

Penrose, R. (1997). *The large, the small and the human mind.* Cambridge, MA: Cambridge University Press. [Contributors: A. Shimony, N. Cartwright, S. Hawkins; M. Longair, Ed).

Popper, K. R. and Eccles, J. C. (1984). *The self and its brain. An argument for interaction.* London: Routledge.

Prigogine, I. (1979). *Vom Sein zum Werden.* München/Zürich: Piper.

Resnick, N., T., Collins, W., Atkinson, T., Bonthron, C. F., Dewey J., and Gimbrone, M. A. (1993). Platelet-derived growth factor B chain promotor contains a cis-acting fluid shear-stress-responsive element. *Proc. Natl. cad. Sci., 90*, 4591-4595.

Riedl, R. (2000). *Strukturen der Komplexität. Eine Morphologie des Erkennens und Erklärens.* Berlin-Heidelberg-New York: Springer.

Roth, G. (1987). Erkenntnis und Realität. In S. J. Schmidt (Ed.), *Der Diskurs des radikalen Konstruktivisms.* Frankfurt/M.: Suhrkamp, pp. 229-255.

Rothschuh, K. E. (1957). *Theorie des Organismus. - Bios Psyche Pathos.* München/Berlin: Urban u. Schwarzenberg.

Schuster, P. (1987). Molekulare Evolution und Ursprung des Lebens. In B. O. Küppers (Ed.), *Ordnung aus dem Chaos - Prinzipien der Selbstorganisation und Evolution des Lebens.* München-Zürich: Piper, pp. 49-84.

Searle, J. (1997). *The mystery of consciousness.* London: Granta Books.

Smuts, J. Ch. (1938). *Die holistische Welt.* Berlin: Alfred Metzler.

Spencer, H. (1901). *Grundsätze einer synthetischen Auffassung der Dinge*. Stuttgart, Germany: Schweizerbart'sche Buchhandlung.

Tautz, V. (1991). Genetic and molecular analysis of pattern formation process in *Drosophila*. In N. Schmidt-Kittler, and K.Vogel (Eds.), *Constructional morphology and evolution*. Berlin-Heidelberg-New York: Springer Verlag, pp. 273-282.

Uexküll, J. v. (1980). Der Funktionskreis. In J.v. Uexüll, *Kompositionslehre der Natur: Biologie als undogmat. Naturwiss.: Selected writings* (German Edition). Frankfurt: Ullstein, pp. 226-290.

Vanderburgh, H. (1988). A computerized mechanical cell stimulator for tissue culture: effects on skeletal muscle organogenesis. *In Vitro Cell Dev., 24*, 609-619.

Vollmer, G. (2002). *Evolutionäre Erkenntnistheorie*, 8. Ed. Stuttgart, Germany: Hirzel.

Vollmert, B. (1985). *Das Moldekül und das Leben. Vom makromolekularen Ursprung des Lebens und der Arten: Was Darwin nicht wissen konnte und Darwinisten nicht wissen wollen.* Hamburg, Germany: Rowohlt, Reinbek.

Wang, N., Butler, J. P. and Ingber, D. E. (1993). Mechanotransduction across the cell surface and through the cytoskeleton. *Science, 260*, 1124-1127.

Whitehead, A. N. (1934). *Nature and life*. Chicago, IL: University of Chicago Press.

THE ANTHROPOLOGICAL APPROACH

In: Consciousness: Its Nature and Functions
Editors: Shulamith Kreitler and Oded Maimon
ISBN 978-1-62081-096-5

Chapter 5

Consciousness and Indigenous Healing Systems: Between Indigenous Perceptions and Neuroscience

Diana Riboli
Panteion University of Social and Political Sciences, Greece

Abstract

For the last four decades or so, anthropologists have been concerned with the concept of consciousness in indigenous healing systems, in particular with altered states of consciousness (ASC). This text aims to define actual perceptions of consciousness or other aspects of it in certain indigenous cultures, using epic anthropological case studies as well as material from research conducted by myself in Nepal and Malaysia.

The greatest problem encountered is that in most cases indigenous concepts and perceptions of existence and the functions of consciousness have been translated –given the failure to find a more appropriate term- using the terms soul or vital breath.

This text attempts to demonstrate that many of these indigenous perceptions are closer to what is defined in neuroscience as consciousness than concepts more pertinent to western culture such as soul. In the light of this affirmation, therapeutic practices performed in most indigenous healing systems are presented as a supreme play between different types of consciousness (practitioner, patients and bystanders) in a symbolic process involving the knowledge, deconstruction and reconstruction of the self and the world around one.

Prologue

We had left the camp at dawn with *halak* Machang – the most respected shaman of the Batek hunter nomads in the Taman Negara (Peninsula Malaysia) to penetrate the jungle in search of medicinal plants. After some time, the combination of Machang's advanced age, my limited agility in the rainforest, and the suffocating, sticky heat led us to decide on a break and rest ourselves on the trunk of a tree which had fallen during a storm.

After a few minutes, *halak* Machang, an extremely gentle and intense man of few words, began to move his arms about slowly and rhythmically, muttering one of the therapeutic chants he had received in a dream. I noted that during the chant his gaze was fixed on the top of a tree covered in parasitic plants with beautiful flowers. After the chant had finished, the shaman fell silent, as if entranced. Without being able to explain why, I realised that Machang was no longer fully with me: his body and part of his consciousness continued to interact with me and his surroundings, even verbally, but there was something missing. *Halak* Machang appeared to be perfectly at ease during this altered state of consciousness, and in full possession of his senses.

A few dark rain clouds on the horizon and a slight fear of a storm breaking out prompted me to ask Machang if he wanted to turn back to the camp. With a smile and a few words, the shaman resolved any doubts I had on the change I thought I had seen in him:

> Yes, my daughter, it would be better to get back before the rain starts. Just give me a few moments to recall my *bayang* which is still in that lovely flower at the top of the tree. It is also a flower at this moment. (*Halak* Machang, personal communication, July 2005)

Bayang, translated by Endicott as the shadow-soul is believed by the Batek to be one of the essential components of mankind, together with the body and the *ñawa'* or life-soul, a sort of life breath that animates all mortals (Endicott, 1979). In contrast, the *bayang* appears to belong exclusively to human and supernatural beings. It is described as a sort of invisible entity which diffuses through the whole body which contains it. Though invisible, the *bayang* is represented by the shadow of its owner.

Shadow-souls are believed to leave the body of their owners when they sleep and anything they encounter, see and learn is seen by their sleeping owners in the course of dreams. Other shadow-souls can also be encountered during dreams and shamans can also see shadow-souls during altered states of consciousness.

While the *ñawa'* is a similar essence homogeneous in all living beings which can even be recycled and pass from one body to another, the shadow-soul, just like a shadow, is strictly personal and differs from one individual to the next. Dreams are just a different aspect of reality, whereas the *bayang* represents perhaps one of the possible indigenous interpretations of what we know as consciousness.

INTRODUCTION

After decades of discussions in different scientific fields, the concept of consciousness still appears to be somewhat confused. One of the main difficulties lies in understanding how the subject-object (Baruss, 1992) and individual-social dichotomies act, given that consciousness appears, at least on the surface, to have a strongly suggestive nature. In many cases consciousness becomes almost synonymous for awareness, or is confused with the concept of conscious, in contrast to what is termed unconscious.

These concerns become even more evident during the study of the definition, role and characteristics which consciousness assumes in indigenous healing systems throughout the world. Anthropological approaches to the study of consciousness are mainly dedicated to the analysis of altered states of consciousness (ASC), in particular, to phenomena such as shamanism and possession (Bourguignon, 1973; Bourguignon, 1968; Lewis, 1989). As highlighted by Winkelman:

> Shamanism's practices for altering consciousness produce one pole in the various dualisms of human consciousness that contrasts the experiences of ordinary awaking consciousness and its intellectual, verbal, rational, externally oriented functioning with a transformed consciousness that is characterized by an internal orientation, intuition, and holistic, tacit, nonverbal perceptions and meanings. These perceptions and meanings have been known as unconscious and intuitive, as special forms of consciousness that are contrasted with the ordinary egoic waking models of consciousness and are manifested in dreams, ASC, and transpersonal experiences (Winkelman, 2000: 3).

The role played by ASC in many indigenous cultures is vital, and altered or alternate states of consciousness appear to be the most common means used during therapeutic ceremonies. These alterations of ordinary consciousness may only involve traditional therapists, or, in certain cases, patients. The essence of the treatment and probably its efficacy, is determined in a way difficult to explain using scientific terms by the encounter, comparison and interaction between different forms of consciousness: that of the practitioner; that of the patient and that of those participating in the therapeutic séance. Winkelman holds shamans to be "technicians of consciousness" who can use this potential for healing and for personal and social transformation.

The purpose of this chapter is to attempt to throw light on how the same indigenous cultures interpret the concept of consciousness and the complex interaction of these different therapeutic forms of consciousnesses using case studies from authoritative and now epic anthropological texts as well as my research in Nepal and Malaysia.

Shadow-Souls, Dream Agents, True Souls. Indigenous Interpretations of Consciousness?

Despite the proliferation of anthropological texts published in the last four decades on consciousness and indigenous healing systems, most scholars do not appear to be particularly concerned with how a concept similar at least in part to the western definition of consciousness can manifest itself in indigenous cultures. As this is a difficult phenomenon to quantify and catalogue, perhaps due to a more or less conscious prejudice according to which peoples with inferior technological development do not have the faculty for such speculations, the debate on the definition of consciousness remains a mainly academic and western prerogative.

However, many ethnographies on the many different cultures of the world present interpretations and concepts –in certain cases even very refined- of what we call consciousness or certain fundamental aspects of the latter. The problem lies perhaps in the fact that in many cases the indigenous versions of consciousness or aspects related to it have been translated –for obvious and often unresolvable practical reasons- with terms such as soul, spirit, vital essence and so on, or with terminology firmly anchored with concepts which have a particular meaning only for western languages and cultures.

Another significant problem is encountered in that when analysing indigenous cultures and therapeutic practices where altered states of consciousness play an important role, and/or the oniric dimension held by many peoples to be more real than daily reality, different elements which in our cultures are held to be part of the unconscious move arrogantly out of this sphere and contribute to the formation, development and very existence of what we have defined as consciousness. As stated by Cohen and Rappaport in particular, the study of abnormal or altered states of consciousness demonstrates that different cultures have different ways "in which people can be conscious" (Cohen and Rappaport, 1995:13).

Some scholars suggest that in general the concept of consciousness should include the unconscious and this may be one of the best solutions to avoid, as often happens, that consciousness and conscious become almost synonymous. As Peacock says:

> Consciousness must also be considered, for our purposes, to embrace the unconscious. The repressed and suppressed memories, feelings, and motives that psychoanalysts have so powerfully shown to underlie the symbolism of their patients may usefully be regarded as unconscious in that the patient can not verbalize them, at least not before years on the couch. Nevertheless, these unconscious ruminations cannot be entirely outside the patient's awareness, for he expresses them in his dreams, wit and delusions. The very fact that the unconscious can be so symbolized proves that it is not entirely unconscious....
>
> ...In short I propose that we understand the notion 'consciousness' to include the 'unconscious', at least until we can find a better term to encompass both the lucidly conscious and the murkily unconscious thoughts, feelings, sensations, motives, beliefs, and memories which are expressed through symbolic forms (Peacock, 1975:6).

In the famous work by Evans-Pritchard dedicated to witchcraft, oracles and magic amongst the Azande of Sudan and Congo, the anthropologist notes, without being in a position to provide further explanation, that amongst these populations man is believed to have two souls: one for the body, and one for the spirit (Evans-Pritchard, 1976).

George Devereux describes the belief of the Mohave in four souls (*matkwí'cà·*) which resemble the body and in some way determine its actions from the fetal state during which these souls follow the acts of the mother and dream of how to be born. The *matkwí'cà·*, which act independently of each other, can be seen by shamans during altered states of consciousness and in certain cases even by their owners in the course of dreams.

The first soul (*hlăkù' ·ytcitc*) is the real shadow, the second self of a person and the core of his identity. This is the only soul to survive after death and it plays an important role during therapeutic shamanic séances, as it can be seen by shamans who can thus have a clear picture of what happened to the sick person. The second soul (*cúma'·tc măhò'·tvetc*) is the power soul, a sort of energy which determines either the fortune or misfortune of its owner. As pure energy, it can confer special powers, shamanic powers in particular. This is also the soul the shaman sends to other cosmic zones to recover a departed soul. Though it dies during the cremation, in some way its effects live on, in the sense that someone who was a shaman during his lifetime will remain a shaman in the land of the ancestors. The third soul (*cúna' ·kavokyé'ttcitc*) is the soul of worldly wealth, or a soul through which its owner acquires wealth. This soul never creates any problems and dies with its owner during cremation. The last soul –the fourth- (*matmakwí'·ca: cúma'·tc mítce'·mvetc*) only appears just shortly before death and in some way notifies or confirms their imminent decease to their owners. This soul also dies during cremation (Devereux, 1937).

It appears to be clear that the first and the second souls defined by Devereux as the real soul and the power soul are interpretations of what we define as consciousness.

Similarly amongst the Manchus, a Tungusic people studied by Shirokogoroff, perhaps one of the first scholars to relate the concept of consciousness to the perceptions of non western ethnic groups, it is believed there are three different souls (Shirokogoroff, 1935). The Manchus believe that human beings are made up of material elements –such as the body- and invisible elements which are the real entities responsible for their actions and thoughts. The invisible elements are called *fojeŋo*, translated by Shirokogoroff as soul, though readers are warned that the translation is fairly inappropriate. The first *fojeŋo* is the true soul (*wuneŋi fojeŋo*), followed, in order of importance, by the *čergi fojeŋo*, the soul which precedes, and the *olorgi fojeŋo,* the external soul. These three elements are not understood as completely independent from each other, and to describe the interrelation between the three, the Manchus compare them to a finger which is a finger because it has a nail, bone and flesh. According to the Russian scholar, the first or true soul is that which may be compared with "consciousness" and "self-cognition" though not directly with thought (*gonin*) which has to have the interrelation of all three souls in order to form. The second soul, after the death of its owner, is in some way recycled and given by supernatural beings to babies just about to be born, while the third soul, the external soul, can reincarnate again in either human or animal form. According to Shirikogoroff, in some way the second soul is responsible for all the physical functions and represents the biological force of the continuation of the species, characteristic of both humans and animals[1]. In contrast, the first soul is independent and according to most Manchus, only belongs to humans. The third soul, the external soul, is closer to the western concept of soul in that this is a migrating soul which goes to the land of the ancestors after death.

With this interpretation, the Manchus closely link the psychic and physical functions of humans. As noted earlier, thought is only formed by the concurrence of the three souls which also determine the state of health and wellbeing of the individual.

In fact it is believed that the three souls, which are located in a system of perennial movement, must be extremely balanced. The souls can be in different parts of the body and are in some way anchored to a plank with 7 holes which is located in a circle. The first soul remains stationary, while the other two move without touching each other and the first is always in front of the second. If this movement is regular, the owner of the three souls is in good health, in a state of well being and sleeps very well. Any type of irregular movement or unbalance brings problems of different types to the owner, which are initially manifested in sleep disturbance.

The Manchus demonstrate the existence of these souls and their complex organisation. For the purposes of our discussion, which attempts to create a link between indigenous concepts of consciousness and how these intervene in the different healing systems, the proof offered by the Manchus of the existence of the true soul appears to be particularly relevant. Examples of evident proof cited are: when loss of consciousness is not followed by death; travel during dreams or during ASC; telepathy and the intrusion of a soul into a human other than its owner. All elements which take on a particular value, especially during shamanic séances, which, according to this interpretation, are the true souls/consciousnesses of the

[1] The Manchus do not attribute this property to plants.

shaman, patients and spectators, play the role of the protagonist, usually in order to restore the equilibrium between all the psychic and physical elements which make up the human being.

In many indigenous perceptions of the invisible aspects and elements which make up an individual, importance is attributed to dreams and some element –once again generally translated by anthropologists as soul- which is held to be responsible for the interrelation between an individual and his/her oniric sphere. For many people in many parts of the world, dreams are held to be more real, in that humans can communicate with all beings and with the world of nature, learning elements from their invisible reality of their self or the external world. It is no mere coincidence that in many parts of the world, the knowledge required for the shamanic profession is imparted through the medium of dreams and many cultures associate the potentiality of dreams with those experienced during ASC. Amongst the Chepang of southern central Nepal, where I conducted in depth research, instructions for shamans (*pande*) are only imparted during dreams (Riboli, 2000). The most powerful *pande* state they can clearly recall the dreams they had when they were still in their mother's belly, before they were born.

It appears fatuous to note the importance attributed by psychoanalysis to the world of dreams, also an important component of the personality of an individual.

The Wana of Sulawesi (Indonesia) distinguish between *lemba*, *koro* and *tanuana* (Atkinson, 1989). The first term describes the external body and could also mean corpse. *Koro*, on the other hand, always defines a living state and refers both to a live body as well as to its sentiments, emotions and way of being. In some way, therefore, *koro* also takes on the meaning of soul, which is believed to reside in the back and lives on after the body is physically dead. According to the testimonies of Atkinson, the Wana believe there are seven *koro*, perhaps more of a symbolic number as no-one appears to be able to define the difference between them. When *koro* indicates the concept of soul rather than body, different expressions are used to define the qualities: *une koro*, which literally means the inside of the body; *koro tongo* or middle soul; and *koro nonong* or genuine soul. Apart from the *koro*, one of the essential elements of humans is the *tanuana* or dream agent. The *tanuana* is believed to be a perfect miniature reproduction of its owner, and resides in the head, in the fontanelle. The Wana say that the *tanuana* is different from all other invisible elements which make up the individual in that this is the element that most closely resembles its owner, both physically and in terms of behaviour and sentiments. In a way, the *tanuana* could be interpreted as a double of its owner, or, to take the analysis even further, as the essence of his/her self. It is also the only vital element which can leave the body of a person without necessarily provoking unbalance and illness. While its owner sleeps, the *tanuana* travels and encounters the *tanuana* of other humans or supernatural beings and all this manifests itself to the sleeping person in the form of dreams.

As dreams are a reality which transcends its daily form in which an invisible aspect of one's self acts (the *tanuana*), they are interpreted by the Wana as a means of accessing and communicating with the supernatural world and better understanding aspects of their own way of being. Recounting dreams in which they have seen themselves, the protagonist of the dream is never referred to in the first person, and the expression my *tanuana* is used instead. The *tanuana* can abandon the body when awake, after a shock (such as a big fright) or surprise, making its owner physically weak and could also bring on serious illness. In this case, the intervention of a shaman is required to bring the dream agent, often reluctant to return to a sick body, back to its owner.

The dream state is associated with altered states of consciousness in the same way as the experience of a performing shaman is considered to be similar to a dream state. Both dreams and ASC promote the development of a sort of second sight and both dreamers and shamans use particular agents in order to contact hidden realities. The difference lies in the fact that during séances and altered states of consciousness, shamans are clearly not asleep and are considered to be perfectly conscious. In this case, therefore, the Wana believe that the entities responsible for these states are not so much the *tanuana*, but rather the agency of the shaman's spirit familiars, who in some way transform his/her consciousness and perception of the world. In both states, it is believed that many things can be learnt which up to that moment are unknown.

In the introduction, we mentioned the Batek in Peninsular Malaysia, who believe that humans are made up of three elements, one of which –the *bayang* or shadow-soul- shares many features in common with the *tanuana* of the Wana. For the Batek too, the dream state and the altered state of consciousness are analogous, so much so that the latter is known as "walking in the dreams" (Riboli, 2010). In this case, dreams and ASC are powerful means for getting to know oneself and visible and invisible realities both for common mortals as well as for shamans (*halak*) and it is often difficult to distinguish between the two. The young *halak* B., who belongs to the nomadic Batek hunter gatherer group in the jungle in Taman Negara, often wanders off on his own into the rainforest and, when he finds a particularly beautiful spot, sits down and sleeps. During this state he transforms into a scorpion and goes to the river to catch crabs, which he is fond of. After he has finished eating, he lets the waters of the river transport him into the earth and climbs up into the stems of the plants, learning of their therapeutical properties on the way (Riboli, 2009).

The Batek still feel very much part of the world of nature and its flora and fauna. The different spheres – human, animal and vegetal – are very difficult to separate and are in a certain sense interchangeable. The most powerful *halak* are believed to be able to transform themselves into plants and animals. In this transformation the shadow-soul enters the body of other beings and acquires certain powers. All the knowledge relative to the shamanic profession is received during dreams, especially therapeutic chants, which are strictly personal and mostly secret. Dreams, like ASC, reconnect to the perfect world of the primordial, where all spheres –human, supernatural, animal and vegetable- co-existed in perfect harmony and could communicate amongst themselves (Riboli, 2010).

These states are therefore perceived as connected to an absolute supreme reality which is not daily life. The shadow-soul transcends the physical and permits communication between individuals or their *bayang*, their consciousnesses, even from a distance. Every time I leave a Batek camp, the most common greeting is that though my body will be in far off Europe for some time, our shadow-souls will meet and communicate often in dreams. As noted by Endicott:

> The life-soul animates the body, and the shadow-soul helps to define the individual person. The role of the shadow-soul in dreams and after death suggests that it is the agent of consciousness and perception. Together, the two kinds of souls transform the body into a human being. (Endicott, 1979: 96)

After the death of a human being, the shadow-soul reconnects with primordial harmony between the different spheres, and takes on supernatural characteristics. As soon as it reaches the world of the ancestors, it is given at least two bodies, one of which is similar to the one it inhabited during life, at its best moment, therefore in its youngest and most vigorous form, and the other is that of a tiger. The two bodies can be changed just like clothes, and the body not in use is always stored inside the other.

The Jívaro, a warrior people studied by Michael Harner (1972), like other Amazon peoples whose cultures and shamanic complexes use hallucinogenic plants, believe that the fundamental elements which determine the life and death of an individual can only be seen with the use of these substances (Harner, 1972). The daily world, when one is awake, is considered to be false, untrue, whilst the real world inside and outside the individual is the world which can only be perceived and experimented with after the consumption of *maikua* (*Datura arborea*) juice, a plant with strong psychotropic powers. During these states, the individual can acquire an *arutam wakanï*, which Harner translates as ancient specter soul, which produces the visions and only occasionally appears to its owners for very limited periods of time. Though invisible, once created, the *arutam* continues to live eternally, outliving its owner. Those who manage to acquire one of these souls become almost immune to any form of physical violence and sorcery attacks, despite the fact that there is no protection against contagious illnesses such as measles and smallpox. The attempt to obtain an *arutam* usually takes place during childhood, at around six years of age, through special rituals which also involve the consumption of *maikua*, but also often takes place at a more advanced age. In most cases, the *arutam* appears in the form of a jaguar, the most powerful animal, which is also the most closely linked to shamanic groups in the Amazon in the same way as the tiger is to the Batek of Malaysia.

What is important here and what suggests a strong link between consciousness and *arutam*, is that individuals who manage to achieve this, though this is never admitted publically, can generally be recognised easily as their personality changes quite radically. The *arutam* is believed to increase one's power in all senses. This particular power is termed *kakarma* and increases ones physical, psychological and moral strength. Those who acquire an *arutam* become more intelligent, have great faith in themselves, and all of a sudden feel it is impossible to commit any immoral acts or lie.

Persons who possess an *arutam* can form a second type of soul –known as the *muisak*, or avenging soul (Harner, 1972). The *muisak* is a soul which only appears if its owner is killed by natural or supernatural causes and the only reason it exists is to vindicate the death of the individual. Apart from these two elements, the Jivaro believe in the ordinary soul (*nekás wakanï*), born during birth and part of the blood. This soul does not appear to interest the Jivaro much, as it merely represents a sort of vital force and initial nucleus of knowledge and conscience. For a people for whom war, headhunting and continuous threats from witchcraft or other populations play an important part in their culture, the soul closest to the concept of consciousness, which confers power and knowledge on all non visible aspects, is the *arutam*. All shamans have to have an *arutam* in order to be able to practise their profession.

In the beliefs of many populations, dreams or ASC are the real means through which individuals can communicate with themselves or their surrounding visible or invisible reality. In many cases, it could be said that altered states of consciousness, independently of whether these have been induced by hallucinogenic plants or not, are first of all expressed in a revelation of the conscious and unconscious aspects which form consciousness.

The Kung, who reside in the Kalahari desert in Africa, clearly express the above sentiment. According to research carried out by Richard Katz, the Kung represent an exceptional case in the different indigenous healing systems, as their group does not have any specific traditional therapist figures: during ceremonies and therapeutic dances anyone can act as healer, as it is believed that everyone has a particular spiritual energy –*num*- which is activated in the course of therapeutic dances (Katz, 1982). The *num* resides in the pit of the stomach and the base of the spine. During the dance, the *num* gets hotter and hotter until it boils and becomes vapour. In this form it goes up the spine to the base of the skull, at which point its owner goes into an altered state of consciousness, which induces a very intense and often fairly painful state called *kia*. As soon as the individual arrives in this state, s/he communicates with the reality of their being, the supernatural world and the world that surrounds it, acquiring healing powers which permit them to heal any of their companions that should require it.

In a certain sense, it is the *num* itself, which is present at birth and grows as the individual gets older and more experienced, which provides the means to access even the most secret aspects of consciousness, as illustrated by the testimony of Toma Zho, perhaps one of the most powerful healers encountered by Katz:

> I want to have a dance soon so that I can really become myself again (p. 43).

This feeling that one really becomes oneself again during the *kia* is expressed using a special term in the Kung language: *hxabe*. It should be noted here that *hxabe* is an experience felt deeply by both healer and by the one being healed. In another interview, Toma Zho expressed this concept as follows:

> I want to *hxabe* so I can feel myself again. When I *hxabe*, I feel my body and my flesh properly; I unwind and unfold myself, I open myself up in dance. I feel lousy when there is no dancing. The singing and *num* lets you *hxabe* yourself. I want to pull myself so that I can *hxabe* myself (p. 194).

From South America to Siberia; and South East Asia to Africa as indicated in the abovementioned cases, elements often inappropriately translated as souls express indigenous perceptions of what is consciousness and these elements are also the principal actors which interact in the indigenous healing systems: at the same time causing illness as well as being those responsible for healing.

FROM DREAMS TO NEUROSCIENCE

After this presentation of the perceptions of consciousness in different indigenous cultures, we take a step back, attempting to return to what is defined as consciousness in an academic and western environment in order to attempt to understand the points where the two possible versions meet and interpret indigenous healing systems from this element.

As already indicated, there are many different definitions of consciousness, and –despite the many studies dedicated to it- it cannot be said that an exhaustive and universally applicable result has been reached. The definition of this concept obviously does not only require the deliberations of anthropologists. I believe that one of the most fascinating and illuminating works in this sector is the work compiled by the neuroscientist Antonio

Damasio. This work does not focus on ASC, but more on the role of the body and emotions as consciousness is formed and acts (Damasio, 1999).

The first problem is the study of how the human brain creates “the image of an object” where object can mean a person, place, music, physical pain or state of happiness, and image is the mental pattern present in the sensory modalities (eg. a sound image, a tactile image, the image of a state of well-being).

Above all consciousness is “the unified mental pattern that brings together the object and the self”(p. 11) without this meaning necessarily that consciousness should be associated with awakeness, given that even clinical investigations have clearly indicated that certain patients can be awake and attentive without having a normal consciousness. Another important element is that consciousness and emotions are inseparable and therefore, as emotions are directly linked to the body and the senses, it is clear that consciousness, body and emotions make up a single unit.

This indissoluble unit can clearly be seen in most of the therapeutic practices adopted in indigenous systems.

Moreover, according to neurological studies, consciousness “is not a monolith” and can be separated into simple or complex types. The simplest nucleus of consciousness which Damasio calls core consciousness (p. 15), gives the individual a sense of self limited to the “here” and “now”. Core consciousness, not an exclusively human prerogative, is a simple biological phenomenon and remains stable during the life of the organism, is not concerned with the future or the past and has no now, after or other place. The most intricate consciousness is the extended consciousness, a complex biological phenomenon which evolves during the life of the organism, takes place on different levels, gives the organism a highly elaborated sense of the self and therefore an identity, positions the individual in a historical time by making them mindful of the past, anticipating the future and interpreting the external world.

Extended consciousness confers on human beings the faculty to reach the peak of their mental abilities for:

- the creation of useful objects
- the ability to consider the mind of the other;
- the ability to sense the mind of the collective;
- the ability to suffer pain as opposed to just feel pain and react to it;
- the ability to sense the possibility of death in the self and in the other;
- the ability to value life;
- the ability to construct a sense of good and evil distinct from pleasure and pain;
- the ability to take into account the interests of the other and of the collective;
- the ability to sense the beauty as opposed to just feeling pleasure;
- the ability to sense a discord of feelings and later a discord of abstract ideas, which is the source of the sense of the truth (Damasio, 1999: 230).

Looking at the indigenous interpretations used as case studies in the light of the findings of the Portuguese neuroscientist, it appears to be clear that many define different levels of extended consciousness and even the idea of the existence of a core consciousness (in particular the *ñawa* or life-soul of the Batek; the *čergi fojeŋo*, or the soul which precedes of the Manchus or the *nekás wakanï* or ordinary soul of the Jívaro).

Many of the entities-soul-consciousness described by indigenous populations are descriptions of the extended consciousness, though there is a basic difference between western and non-western perceptions. As seen in the ethnographic cases, if we accept the idea that the souls described are actually expressions of consciousness, their main objective is communication with and knowledge of elements and kingdoms which we hold to belong to the sphere of the unconscious –such as the oniric dimension- whereas other cultures and populations see these as the supreme reality, the only dimension where it is possible to have a clear vision and perception of one's self and those around them. In this sense, the supreme level of potentiality of consciousness is expressed by the shadow-soul, or *bayang* of the Batek, *tanuana* of the Wana, *arutam* of the Jívaro, *num* energy of the Kung, true soul -*wuneŋi fojeŋo*- of the Manchus and *hlăkù' ·ytcitc* or real shadow of the Mohave.

One element which belongs to human consciousness and separates humans from other animals is the fact that people are the only living creatures who can perceive the fact that they have to die and that death is also inevitable for their companions. We know from many anthropological studies on indigenous healing systems, in particular, from those dedicated to shamanism, that death is a central theme both in the personal development of the practitioner, as well as during therapeutic ceremonies.

In order to be recognised, shamans are subjected to initiation tests which in many parts of the world correspond to processes of death and rebirth, where rebirth is accompanied by a new self, new potentialities and powers[2]. The fourth *matmakwí'·ca: cúma'·tc mítce'·mvetc* of the Mohave is an element whose only purpose is to notify the owner of their imminent demise, whereas the *arutam* of the Jívaro is a new element which confers a new understanding of the self and the external world, and provides clear ethical indications, which the individual acquires after particular rituals that involve symbolic processes of death and rebirth.

In the light of these initial suggestions, it seems right to attempt to examine how these elements act and interact in most indigenous healing systems.

Consciousnesses and Indigenous Healing Systems

In English and in particular in medical anthropology texts, there is a fundamental difference between the terms to cure and to heal. The concept of cure is applied to the treatment and removal of disease, where disease is understood to be a biological problem involving the body and its functions, generally treated by biomedicine. The concept of healing in some way counters the concept of cure. As stated by Winkelman:

> The concept of healing contrasts with cure in embodying a recognition of the need to recover one's well-being in other than just the health of the physical body. Healing involves processes of "whole-ing", putting one's psychological and emotional life back into balance (Winkelman, 2009: 141).

[2] There are many descriptions of the central themes of death-rebirth and of the relationship with death in general in almost all studies on shamanism and indigenous healing systems. One of the most extensive descriptions, in geographical terms, is the classical study by Mircea Eliade (1968).

In this sense it has been observed that the biggest problems with western clinical care are that the practitioner only cures diseases, whereas patients are really interested in healing the illness (Kleinman, 1980: 354).

The re establishment of a status of equilibrium in the psychological and emotional life of an individual in ethnomedical systems in many parts of the world also corresponds to a re establishment of equilibrium between the sphere of the individual/personal and the social, and between these spheres and the surrounding universe. According to Andrew Strathern the concept of consciousness in terms of the theory of healing and the use of ASC in particular during treatment sessions, is important as it transcends the dichotomy between psychogenic and sociogenic (Strathern, 1995). Again according to Sthrathern, one problem is that the term consciousness – as a mentalistic concept – could reinforce the idea of a dichotomy between body and mind which has no particular value when the concepts of health and sickness and the relative treatment practices in ethnomedical systems are analysed, unlike biomedicine. For this reason, the term embodiment is used in anthropology. As ASC are examples of embodied mentality, the concept of embodiment should always be borne in mind alongside the concept of consciousness. According to Strathern, these three elements –consciousness, embodiment and body- represent a sphere where the psychogenic and the sociogenic meet and overlap.

The union of these three elements reproposes the idea of the "mindful body", the union between three bodies- the individual, the social and the political- as defined by Scheper-Hughes and Lock (1987).

The greatest difficulty in understanding indigenous healing systems lies in the fact that these are not based on the rigid dichotomies which rule western thought such as body/mind, nature/culture, individual/social, ordinary states of consciousness/altered states of consciousness, real/unreal, natural/supernatural. The inconsistency of these contrasts is clear from the entities-soul-consciousness case studies cited earlier and peaks during therapeutic séances celebrated by traditional practitioners in the course of which, through processes of symbolic elaboration, the individual's self and the world in general is deconstructed and reconstructed.

During shamanic séances for example, even the rigid scheme according to which each of the five senses corresponds to a set category of sensations (tactile, visual, acoustic, olfactory or gustatory) has no absolute value. Studies on synesthesia, where sensorial stimulation leads to the perception of a category which usually belongs to another sense, as experienced during ASC by shamans or in certain cases by their patients, are still extremely limited, mainly to brief mentions (Bacigalupo, 1999). Despite this, we know that in many cases shamans can see a music or therapeutic chant, smell a colour, touch a smell or hear the taste of a sound.

According to the thought of Damasio, one of the first duties of consciousness is to create "the image of an object", using sensorial modes, and consciousness therefore represents "the unified mental pattern that brings together the object and the self" (Damasio 1999: 11). In the case of shamanic synesthesia, we can easily understand that consciousness extends to include the possibility of experiencing any object with sensorial fields not strictly pertinent to them, conferring greater power and awareness about one's self and the world around one.

A shamanic séance and other types of therapeutic ceremonies could be defined as a supreme play of consciousness which finds its expression, elaboration and possibility for enactment in a mythical time and place.

In his famous work, Dow located four fundamental modalities in the universal structure of symbolic healing.

- The experiences of healers and healed are generalized with culture-specific symbols in cultural myth.
- A suffering patient comes to a healer who persuades the patient that the problem can be defined in terms of the myth.
- The healer attaches the patient's emotions to transactional symbols particularized from the general myth.
- The healer manipulates the transactional symbols to help the patient transact his or her own emotions. (Dow, 1986: 56)

States of illness where the intervention of a traditional therapist is believed to be absolutely necessary are mainly attributed to disorders and disequilibriums created in the mythical world of the individual such as witchcraft attacks, detachment of the soul or one of the souls from the body of the patient, which could, as noted earlier, provoke disorders of the consciousness, disequilibrium and punishments from the world of the ancestors or supernatural beings. Symbolic therapeutic processes which mainly act through the soul-consciousness of the practitioner, patient or bystanders, revealing, analysing and recomposing the most secret and invisible aspects of the self, their emotional sphere and the visible and invisible worlds which surround them, acting extensively on the consciousness of the patient, his mind and consequently, also his body.

Communication with the invisible spheres of the different consciousnesses and natural or supernatural reality is mainly through altered states of consciousness. According to the different cultures studied, in some cases, these states are only experienced by practitioners such as in the Wana; in others, they also involve patients or bystanders (Manchus) or even the whole social group (Kung). In any case, the symbolic process, also activated by music, chants and dancing, is always collective, in that all those who participate in a therapeutic ceremony witness the representation and tangible manipulation of the mythical world common to all those who belong to the group.

We have seen how unity and harmony between the different soul-consciousnesses is held to be indispensible for one's health. The loss of one of these entities or disequilibrium between them provokes illness. We have already talked of the harmonic and perennial movement which should be maintained by the three *fojeŋo* of the Manchus (Shirokoroff, 1935). It is interesting to note that amongst these populations, one of the main causes of illness is attributed to shock, such as a great fright. In this case, the movement of the *fojeŋo* accelerates and the distance between them diminishes. As a result, the *olorgi fojeŋo*, or external soul could abandon the body of a person, and this person would start to feel tired, drowsy and dreamy for no reason. If the shaman does not intervene promptly and send his *fojeŋo* after the lost one, the individual will deteriorate and enter a state of confusion which could lead to complete unconsciousness or even death. The loss or flight of the first soul, the true soul (*wuneŋi fojeŋo*) which according to Shirokoroff corresponds to the consciousness of its owner, does not have the initial symptoms such as drowsiness or confusion, but immediately manifests itself in a loss of consciousness. In both cases, it is believed that shamans can bring missing souls back to their owners. The most serious case which usually ends in death, is when someone loses their *čergi fojeŋo*, or the soul which precedes, which we have defined as a possible interpretation of what Damasio defines as core consciousness, which for the Manchus also corresponds to the soul which presides over all the biological functions and therefore functions as a form of vital breath.

Similarly, amongst the Batek one of the most serious culture-bound syndromes[3] is related to the *ñawa'* or life-soul, fairly similar to the *čergi fojeŋo* of the Manchus. This condition is called *ke'òy* and corresponds to a state of depression accompanied by physical symptoms which could deteriorate and lead to death. Though this syndrome could be caused by a fright, provoking the loss of the *ñawa*, or vital breath of the individual and the first nucleus of his consciousness, in most cases this is caused by a feeling of being unjustly wrongly judged or criticised by another individual or by being frustrated at an unfulfilled wish (Endicott, 1979). As a result of one of these causes, the Batek believe that the wind of the *ke'òy* inflates the heart of the person affected, causing physical weakness, accompanied by fever, headaches and respiratory difficulties which symbolically express the weakening of the *ñawa'*, the vital breath. The most serious cases are depression and emotional turmoil caused by the feeling of being unjustly judged by a fellow comrade. We could say in this case that the disequilibrium is caused by the conflict between the two consciousnesses which are experiencing some disagreement.

Amongst the Batek, nomadic hunter gatherers, based on an acephalous political model, with no private ownership and absolute equality between all individuals, unity and harmony within the social group is essential (Riboli, 2010). The fact that two individuals could enter into conflict or that one may no longer feel appreciated or understood by another member of the groups is experienced as a situation of dangerous disequilibrium which could threaten the very existence, the being in the world of anyone affected by the *ke'òy* and therefore their *ñawa'*, at once core consciousness and life soul. The *ke'òy* is a social illness, which undermines the health of the whole group. Usually, for the purposes of healing, the person who more or less unconsciously and unjustly hurts the other person must change his way of feeling and express repentance through symbolic actions aimed at reinforcing the *ñawa'* of the offended person. Amongst the different procedures, the most interesting is that in which the individual who had provoked the unhappiness cuts his calf and collects the blood on a few green leaves which he will then use to massage the chest and back of the sick person in order to restore harmony and cast off the wind of the *ke'òy* to confer new vigour on his *ñawa'*.

Illnesses involving the *bayang*, the shadow-soul expressions of an extended consciousness, require the intervention of a *halak*, a shaman who can send his shadow-soul during an ASC in search of the cause of the state of disequilibrium. The pathologies involving the *bayang* and a possible flight from the body are not particularly common and mainly involve small children whose shadow-souls are still not completely formed, as they still have not acquired the force and stability of the adult form.

In any case, even adults are believed to be able to lose their *bayang,* which is particularly active when they are asleep, sending all information about the sphere not directly tangible for the self, or that of other individuals whose shadow-souls meet during dreams and of the world around them through dreams. As the Batek believe dreams are only the experiences and journeys of the *bayang* to a parallel reality as real as daily reality, to prevent sickness, no-one

[3] For ethnopsychiatry, culture-bound syndromes are illnesses which appear in certain geographical areas of the globe in certain cultural contexts, requiring the intervention of a traditional therapist (Simons and Hughes, 1985). Of the best known culture-bound syndromes is the susto (the flight of the soul in Central and Southern America), the amok (episodes of manic homicide in Malaysia and Indonesia) and the evil eye in different Mediterranean and Spanish speaking countries.

should ever be woken up with a start. In this case, his *bayang* may not have the time to return to the body of his owner. The Batek have told me that when this happens, the individual is 'no longer himself', has problems sleeping, pains throughout the body, fever, gastrointestinal disturbances, and confusion, which sometimes leads to total loss of consciousness. Only the intervention of a shaman during a collective ceremony where therapeutic chants, music and dance are important elements, during which a *halak* sends his *bayang* in search of that of the patient to convince them to come back, can resolve the situation.

Similar ceremonies can also be celebrated for sicknesses which have other causes, the first of which is the infraction of a taboo, when illness is brought on by a sense of guilt; or emotional disequilibrium brought on by the infraction (Riboli, 2010). These cases appear to be particularly interesting for our discussion. For the Batek, who do not believe in evil spirits, there are not many illnesses which can be put down to supernatural beings. Most of these appear to derive from completely human causes and concern the consciousnesses and emotional spheres of the individual.

The Wana appear to be fairly similar, the difference being that many pathologies deriving from the loss of one of the *koro* souls could have been caused by magic, whereas as seen earlier, the loss of the *tanuana* or dream agent usually happens after a shock, especially with children where, like in the case with the shadow-souls of the Batek, this element is still weak and needs training.

Another interesting case where the *tanuana* abandons the body in which it is contained, is when the individual is particularly sick or suffering, for example, after an accident, rendering the individual more fragile and weak. In fact, the *tanuana* represents a sort of double of the individual in his way of acting and feeling, and is subject to the same suffering as his owner. For this reason, it prefers to distance itself and the very vivid dreams that often accompany fevers are attributed to the fact that the *tanuana* is "living it up" in another place (Atkinson, 1989). Locating and convincing a reluctant *tanuana* to return to a sick body is one of the most difficult and tiring tasks for a shaman which takes place during ceremonies where the collective participates. As highlighted by Atkinson, the state of wellbeing of an individual is essentially due to "a fragile assemblage of hidden elements" and "when these elements are concentrated in their proper places, the person thrives; when they are dispersed grows weak and sick" (Atkinson, 1989: 118).

The most obvious case involving the interaction of the consciousness of the whole social group is probably that of the Kung, where all adults are both healers and healed. We have already mentioned the *num* energy, which, by boiling and becoming a sort of vapour which goes up the spine to the base of the skull, provides access to what the Kung call *kia*, an intense altered state of consciousness. The main reason for entering this state during collective dances is to heal one's companions. According to Katz, "healing in itself is a transcendent experience" (Katz, 1982: 101). In some way, the therapeutic potentialities are expressed when the Kung manage to go beyond their ordinary, daily self, exploring the most extreme possibilities of their consciousnesses. This provides collective access to what is believed to be the real world, things, people and surrounding realities can be seen and clearly understood. Complete knowledge of oneself and others is acquired. In the dances and therapeutic chants of the Kung, the social dimension becomes particularly important. According to Katz, "healing is a give-and-take process" (Katz, 1982: 102). The persons who participate in the ceremonies provided those dancing with the context required to activate their *num*. In exchange, the latter give everyone healing powers –both for therapeutical and preventative purposes- which activate as the *num* boils. Roles are interchangeable in the sense that during the ceremonies, everyone, according to their attitude and condition, participates at times as a healer, and at times as a patient, and at times can decide to be part of the chorus singing the therapeutic chants. Only the dancers go into a state of *kia*, but despite this, all participants share the common experience of finally feeling themselves and having a clear vision of their own self and others. In a way, the individual consciousnesses reinforce and are reinforced by a collective consciousness.

CONCLUSION

This chapter has attempted to revisit consciousness on the basis of indigenous interpretations which are fundamental when analysing the healing systems of different cultures in the world. In many of the anthropological texts compiled over the past few decades, the role of altered states of consciousness in these systems has been emphasised, without however making clear references to what in various cultures corresponds to what the western world calls consciousness. Most of the elements which express such concept or at least certain of its fundamental aspects have, due to the impossibility of literally translating into western languages, been translated with the term soul. However, from the ethnographic cases presented, it seems to be clear that traditional therapeutic practices are based on the encounter between what neuroscience defines as core consciousness and extended consciousness whose connotations and actions act on the deconstruction and reconstruction of the self, selves and the world required for therapeutic reasons in many of the indigenous healing systems in the world.

I would like to conclude quietly, with the words of a Nepali Chepang shaman, two Kung healers from the Kalahari and the neuroscientist Antonio Damasio:

> When I shake, it is not much different from dreaming. I see a sort of darkness, but as soon as I find what I am looking for, for example the cause of an illness, the darkness suddenly becomes light and my body feels very light. I fly into the Sky… (Narcing Prajā, Chepang shaman. Riboli, 2000: 120).

> When I pick up *num*, it explodes and throws me up in the air, and I enter heaven and then fall down.
>
> During the dance, when you look out beyond the fire, you see things. It's light, not dark, even though it is nighttime. You see camps; you see at a distance in the night. You see actual things, persons and objects.
>
> (Kung healers. Katz, 1982: 44 and 83).

> The scope of extended consciousness, as its zenith, may span the entire life of an individual, from the cradle to the future, and it can place the world beside it. On any given day, if only you let it fly, extended consciousness can make you a character in an epic novel, and, if only you use it well, it can widen the doors to creation. (Damasio, 1999: 195).

References

Atkinson, M. J. (1989). *The art and politics of Wana shamanship*. Berkeley, CA: University of California Press.

Bacigalupo, A. M. (1999). Studying Mapuche Shaman/Healers in Chile from an experiental perspective: ethical and methodological problems. *Anthropology of Consciousness*, 10(2), 35-40.

Baruss, I. (1992). Contemporary issues concerning the scientific study of consciousness. *Anthropology of Consciousness*, 3(3-4), 28-35.

Bourguignon, E. (1968). World distribution and patterns of possession states. In R. Prince (Ed.), *Trance and possession states*. Montreal: R.M. Bucke Memorial Society, pp. 39-60.

Bourguignon, E. (1973). *Religion, altered states of consciousness, and social change*. Columbus, OH: Ohio State University.

Cohen, A. P., Rappaport, N. (Eds.). (1995). *Questions of consciousness*. London: Routledge.

Damasio, A. (1999). *The feeling of what happens. Body and emotions in the making of consciousness*. New York, NY: Harcourt Brace and Company.

Devereux, G. (1937). Mohave soul concept. *American Anthropologist*, 39, 417-422.

Dow, J. (1986). Universal aspects of symbolic healing: A theoretical synthesis. *American Anthropologist*, 88(1), 56-69.

Eliade, M. (1968). *Le chamanisme et les techniques archaïques de l'extase*. Paris: Payot.

Endicott, K. (1979). *Batek Negrito religion*. Oxford: Clarendon Press.

Evans-Pritchard, E. E. (1976). *Witchcraft, oracles, and magic among the Azande*. Oxford: Oxford University Press.

Harner, M. J. (1972). *The Jívaro. People of the sacred waterfalls*. Berkeley, CA: University of California Press.

Katz, R. (1982). *Boiling energy. Community healing among the Kalahari Kung*. Cambridge, MA: Harvard University Press.

Kleinman, A. (1980). *Patients and healers in the context of culture. An exploration of the borderland between anthropology, medicine, and psychiatry*. Berkeley, CA: University of California Press.

Lewis, I.M. (1989). *Ecstatic religion*. London: Routledge. (Original work published 1971).

Peacock, J. L. (1975). *Consciousness and change. Symbolic anthropology in evolutionary perspective*. Oxford: Basil Backwell.

Riboli, D. (2000). *Tunsuriban. Shamanism in the Chepang of Central and Southern Nepal*. Kathmandu: Mandala Book Point.

Riboli, D. (2009). Shamans and transformation in Nepal and Peninsular Malaysia. In E. Franco, D. Eigner (Eds.), *Yogic perception, meditation and altered states of consciousness*. Vienna: Verlag der Österreichischen Akademie der Wissenschaften, pp. 347-367.

Riboli, D. (2010). Ghosts and paracetamol: Batek and Jahai shamanism in a changing world (Peninsular Malaysia). *SHAMAN*, 18 (1-2), 99-108.

Scheper-Hughes, N., Lock, M. M. (1987). The mindful body. A prolegomenon to future work in medical anthropology. *Medical Anthropology Quarterly*, 1(1), 6-41.

Shirokogoroff, S. M. (1935). *Psychomental complex of the Tungus*. London: Kegan Paul, Trench, Trubner and Co.

Simons, R. C., Hughes, C. (Eds.). (1985). *The culture-bound syndromes. Folk illnesses of psychiatric and anthropological interest*. Dordrecht: Kluwer Academic Publisher.

Strathern, A. (1995). Trance and the theory of healing: sociogenic and psychogenic components of consciousness. In A. Cohen, N. Rappaport (Eds.), *Questions of consciousness*. London: Routledge, pp. 117-133.

Winkelman, M. (2000). *Shamanism. The neural ecology of consciousness and healing*. Westport, CT: Bergin and Garvey.

Winkelman, M. (2009). *Culture and health. Applying medical anthropology*. San Francisco, CA: Jossey-Bass.

THE SOCIOLOGICAL APPROACH

In: Consciousness: Its Nature and Functions
Editors: Shulamith Kreitler and Oded Maimon
ISBN 978-1-62081-096-5

Chapter 6

CONSCIOUSNESS: SOCIOLOGICAL APPROACHES

Vasiliki Kantzara*

Department of Sociology, Panteion University of Social and Political Sciences, Greece

ABSTRACT

The chapter attempts to highlight common premises and views on consciousness in the discipline of sociology. It dwells in particular on issues involved analytically, theoretically and empirically in relation to the question, How is consciousness defined and studied? The term consciousness denotes a constitutive element of life and being of both individuals and society. It constitutes among other things a means by which humans understand themselves and the world around them that is mediated by culture, history and society. In sociology, the concept of consciousness has been employed mainly as denoting collective consciousness. The aim of sociological perspective has been to study and explain the relation of individuals to society, which has been the major question of the discipline since its establishment in the 19th century. Furthermore, the approaches that were developed in sociology relied upon philosophical approaches which inspired the founding theorists Durkheim and Marx.

Current theorizing and research conceptualize consciousness as referring to society as a whole, as a social and/or national collective, as well as in terms of the stratifying categories in line with class, gender and race consciousness. In sociology consciousness has been defined as degrees of awareness and knowledge that individuals have about the society they live in as well as about their own social position. Next, this kind of knowledge (or lack thereof) has been linked to attitudes, beliefs and social stance of individuals and social groups. The linking of consciousness to undertaking (or failing to do so) of socially transformative action is a question dealt with in the aforementioned approach. Consciousness is thus related to the structure of the organization of social relations, past and present, including the ways these are experienced, sustained or change following purposeful action. Currently, the question that remains open is how to define and study consciousness that is potentially both a constitutive process and a state of being, which is changing while creating the impression of being steady and unchanged.

* E-mail: vkantz@panteion.gr

The current chapter starts with presenting the foundation of consciousness in sociology, focusing first on the founding theorists and then on the interpretative paradigm. The following section provides a schematic overview of the various uses in study and theory after World War II in Western societies. A chronological order is implicit in the sections and sub-sections that follow, concluding with a few analytic notes and summative remarks.

1. Introduction

The term consciousness denotes a constitutive element of life and being, as well as of subjectivity, sense of self and identity. Consciousness forms a constitutive part of individuals and the organization of social life. Consciousness constitutes also a means by which humans understand themselves and the world around them. This understanding is mediated by culture, history and society and informs the ways the organization of social life and social relations are experienced. Consciousness is perceived through its impact on humans and their collective life. It has been reified to the extent that it has been conceived as an accomplished and finite entity, rather than an ongoing, constitutive process of knowing, feeling, perceiving and positioning oneself towards the world as well as being constituted by the world, in a powerful dialectic relationship.

As a concept consciousness has been employed in sciences since the 17th century. This was a significant century because it is then that philosophers have conceived that since human ideas express social issues and are influenced by the concerns of their era, they have to be understood in relation to the social context and not as some entity outside society. In addition, philosophers advanced the idea that humans differ from other living species because they possess consciousness and imagination. According to Marx, the difference between even the 'best of the bees and the worst of the architects' is that the latter conceives first with his imagination the edifice he wants to erect (Ritzer, 2000, p. 158).

In addition, the concept of consciousness was understood by philosophers as including the notion of conscience, a concept known since antiquity. Conscience originates from the ancient Greek word *syneidisis* (συνείδησις) and later from the Latin word *conscientia*, which means to understand together. Moreover, in the relevant bibliographies it is traditionally mentioned that the philosopher John Locke (1690/1975) was the first who defined consciousness as 'the perception of what passes in a man's own mind'. In this succinct and admirable formulation Locke links consciousness directly to individual perception, imagination and reflection and indirectly to society. It is by no means strange that philosophy first and cognitive sciences later have focused on the phenomenon called human consciousness in order to study, analyse and interpret what has been thought ever since as related phenomena of being, imagination, empathy, mind and perception.

In sociology, the concept of consciousness has been known to founding theorists of the discipline in the 19th century. Emile Durkheim (1915/1976, 1933) and Karl Marx (Marx and Engels, 1845-6/1998) drew upon philosophical conceptions and employed them in order to explain, on the one hand, the constitution of society and, on the other hand, the relation of individuals to this entity, which hitherto has been one of the main foci of the newly established scientific discipline. Durkheim's approach explicates collective consciousness, while Marx's raises questions about the formation of class consciousness and the role of alienation, a form of estrangement from one's true self or state of being.

Later, in the 20th century, the concept of consciousness gains a new momentum within phenomenology, a perspective and a method that focuses on studying the structure of human consciousness. Phenomenology spread as a movement in many disciplines and in sociology took hold, for example, in the sociology of knowledge, especially in the approaches of Schutz (Schutz, 1967; Schutz and Luckman, 1973), and of Berger and Luckmann (1966). In the meantime, another strand that was based on the writings of Marx, gave impetus to theorizing on class consciousness and the phenomenon of alienation. A related theme was picked up by the second wave of the feminist movement that used the term 'consciousness raising' as a way of gaining knowledge in regard to the origins of one's own oppression, which ideally results in emancipation and empowerment of subordinated groups, and in this case women. Today, the concept that emerges is 'critical consciousness' denoting a critical awareness and knowledge that inform the stance and attitudes humans assume towards the world around them.

The theme of consciousness has been taken up in the various strands, approaches, paradigms and sub-disciplines of sociology. For example, in the sociology of education the subject of consciousness has been used in an attempt to explain the socialization process of individuals, which is considered to be the foremost function of education. Some theorists consider socialization as having far-reaching effects on the maintenance of society whereas others consider it as suspicious in that it helps perpetuate current social arrangements, thus reproducing unequal organization of social relations. Different conceptions of the term consciousness point to its social aspect in attempting to uncover first, the missing link between individual and social positioning and second, the link between positioning and social action and praxis.

2. Founding Consciousness: Philosophical Underpinnings and Sociological Working out

Approaches to consciousness in sociology drew extensively on philosophy. The sociological turn has occurred, when founding theorists discerned and made explicit that a collectivity known as society is real and exists, and is moreover sustained by a number of structural characteristics. In sociological conceptions, consciousness is considered to be part and parcel of individuals, residing at the same time in culture and society and finding expression in human ideas and ideals as well as in sciences that express and represent the collectivity. Moreover, consciousness is not only the totality of these, but also a force that binds humans together obliging them by restraining them internally to act in ways socially approved and esteemed. It is therefore no coincidence that consciousness is often conflated with conscience, denoting moral codes that govern human behavior, feelings and thinking.

The modern roots of the scholarship on consciousness are to be found in Hegel's early work (1807). Hegel addresses the issue of the unity between 'subject' and 'object', a question that has been raised in his era. Hegel attempts to explicate the three phases consciousness undergoes in order to realize and succeed in accomplishing its universality and unity with the outside world. More particularly, consciousness 'moves', as it were, from 'sense certainty' where the object is perceived in a diffuse manner, to the phase of 'perception', where the object acquires certain qualities or properties. There follows the third phase, which Hegel calls 'understanding', through which the unity between subject and object, as it were, is realized. In this way self-consciousness arises, which amounts to a 'consciousness of consciousness', which according to Hegel further develops to accomplishing Reason and Spirit.

The absolute idea, the truth, science, totality, unity and universality are interlinked in Hegel, the ultimate goal being the unity of the subject and the object of inquiry. In doing so the unity takes place in revelation and in truth, which are proceeded by deep understanding. The complex way Hegel describes consciousness fits squarely with the subject matter. Notably, Hegel's work, and that of other philosophers on whom Hegel himself is based, has been a major point of departure for theorizing about consciousness, framing largely the work of subsequent scholars, some of whom are known as founding theorists of sociology. In fact, some of Hegel's ideas reach us today as though they are not altered for at least two centuries later. This however is a feature of consciousness to make things current, as the phenomenological approaches have pointed out (see below sub-section 2.3).

2.1. Collective Consciousness

The two sub-sections that follow deal with collective and class consciousness respectively, based on Durkheim's and Marx's approaches. Their ideas form the basis for much of subsequent sociological writings on this subject.

In sociology, Durkheim, a French sociologist, one of the three most important founding theorists, borrowed the notion of movement from Hegel in order to define an entity called society, which is not visible and can be perceived only through its impact on individuals. Regarding the constitution of society, Durkheim (1933) perceived two structural similarities that pertain to all known collectivities and act as a binding force: a division of labor among the members of the society, which is more or less extensive (viz. the famous distinction between 'mechanic' and 'organic solidarity') and a collective consciousness. The latter concept, better known as 'consciousness collective' is further perceived as consisting of two interrelated forces: on the one hand, common beliefs, customs and values and, on the other hand, a moral dimension, mostly known as conscience, which functions as a binding and restraining factor in regard to human beings. These two dimensions contribute to sustaining society and ought to be cultivated in humans.

Durkheim was particularly interested in the relation of education to society, as part of his teaching at Bordeaux University, his devastating experience of the First World War during which he lost his son, and the attempted secularization of education. The diminishing role of religious beliefs prompted him to unravel the bases upon which social cohesion is based. In his theory as to how education could contribute to sustaining society, he applied the concept of function and answered the question by attributing to education the function of 'methodical socialization'. By this concept he meant that education's purpose would be to cultivate in the individual student the social part of his personality and his consciousness, so that he could function as a member of his society (Durkheim, 1961). In further explicating his theory, Durkheim followed Kant, in considering morality as composed of three aspects: first, discipline, because through self-discipline a person learns to obey rules; second, attachment to groups, because it makes a person social by identifying with group characteristics in behavior and internalizing moral codes; and third, autonomy, a basic postulate of Enlightment, because it enables a person to function on his own, in an independent manner. The success of these dimensions of morality depends on the socialization process through which students internalize common beliefs and customs and embody them rather than simply learn them. According to Durkheim, the underlying reason is that society is characterized by patterned forms of behavior. The latter occur because individuals react to certain stimuli in similar manner, for example, if a person extends his/her hand to greet you, you respond 'automatically' by giving him/hers yours. Thus, what keeps society together is a consciousness collective which is not limited to common ideals and beliefs, but extends to common forms of behavior, action and praxis.

In theorizing about the 'consciousness collective', Durhkeim endorsed the double nature of the individual, namely, one private part that is idiosyncratic and particular to every individual and a second social part pertaining to society. This division resembles very much that of Mead's (1934) between 'I' and 'me', who in turn drew upon the work of the American philosopher Charles Sanders Pierce. Education then, according to Durkheim, socializes the social part of the person making society possible in a twofold sense: first, in terms of conscience and second, in terms of division of labor. The young student learns common norms and values and acquires a craft so that later s/he can exercise an occupation and occupy a position in the structure of the division of labor. Several questions arise here, such as 'what is the nature of the private part of the individual?', or 'is it not necessary for the private part to be socialized?' These questions however were not posed and I think that the leading question of Durkheim's era referred to the role of education as an institution in a newly secularized society.

Durkheim further related collective consciousness to 'collective representations', such as religious beliefs. He showed in his work how religious and cultural beliefs are the projection (as we would express it nowadays) of society about itself as a collectivity. This in turn moulds collective consciousness. Changing these beliefs and ideas then is not the result of the work of a single agent but of society, generations after generations, making use and altering given concepts, ideas, common representations, refining their meanings and level of abstraction (Durkheim, 1915/1976). Examples are provided among other things in folk music and other forms of fine arts.

Durkheim's approach to collective consciousness has influenced research and theorizing on relevant themes in education and other sub-disciplines of sociology. The notion of collective has been conceptualized as national or ethnic consciousness referring to group characteristics and its determining impact on forming individual identities, viewed in societal and collectivity terms.

2.2. Class Consciousness

Marx's approach to consciousness differs in important respects from that of Durkheim. Though the German scholar did not write extensively on this subject, his analysis sparked nonetheless a broad range of writings on subjects, such as class consciousness, alienation, false consciousness, and (dominant) ideology. Marx attempted to reverse the reasoning of Hegel and placed as it were concepts on their feet, that is grounded them in material circumstances. Thus, Marx referred to material circumstances and 'real', existing people and not abstract ideas. Moreover, he linked abstract notions and dominant ideas with ruling classes and in their interest to furthering their ruling.

Marx wrote in his characteristic style that 'it is not consciousness that determines the life of people but their condition of life that determines their consciousness' (Marx and Engels, 1845-6/1998, p. 42; see also Bottomore and Rubel, 1956, p. 24). In this postulate he gives priority and primacy to praxis as the generator and creator of consciousness, and attributes greater impact to material conditions of existence and living rather than to the Spirit, which according to him, Hegel was proposing. Furthermore in his work *The German Ideology* co-authored with Engels (first published in German in 1845-6), Marx postulates that 'consciousness is nothing else than the conscious of real people, who live in real circumstances' (1845-6/1998, p. 42).[1]

In his work Marx attempted to show the importance of social life and also of human beings themselves in creating institutions. Thus, he pointed to the primacy of praxis, the active aspect of individuals when creating institutions, which are based and built upon the ways people acquire the means for their subsistence in the material world (see also Antonopoulou, 2008). In so doing, humans not only create actively their relation to nature, but by appropriating nature's goods, they also change themselves in this process. Marx showed that society is man-made and not God or nature given, as was the dominant view in his era. He also showed that human beings are basically social beings, so that even in one individual one can see 'the total of social relations'. In addition, social relations are governed by social laws and the economic realm is a social realm. For example, laws in regard to property, which form the basis of exploitation, are dictated by society and not by nature or a divinity that stands above society.

Social classes are formed according to a certain structure which makes possible group positioning of individuals in the production process. Essentially two classes are formed that stand opposite each other: the capitalist class, consisting of those who own the means of production (materials and machines) and the workers, who own only their labor capacity and sell it to the first seemingly as free agents. In reality, workers sell their labor capacity in conditions of dependency and exploitation. Social institutions and the structure of social relations are based on this division in the production process, so that forms of culture, politics and societal organization 'correspond' to this form of capitalist production. These social relations are mirrored in individuals forming and informing their 'class consciousness', namely, their knowledge of their respective position in society. If individuals fail to see through these relations and do not understand the true basis of their exploitation then they are governed by a 'false consciousness'. The latter is due to alienation which human beings undergo in capitalism, the form of production characterizing our era. Furthermore, alienation, the estrangement of humans from their 'true' human nature, takes places in four important areas of social life. Humans are alienated in their relations towards: themselves, other human beings, nature and their own labor. One of the basis of alienation is fetishism, that is, that other people or labor are treated as a thing, a product, people think has a 'use value' by itself. The subject of alienation has been important ever since in philosophy and psychology, to name only a few disciplines (see also Lukács, 1968/1971; this point is discussed again below).

According to Marxist perspectives, man makes social institutions that 'correspond' to the modes of producing the means of subsistence. Individuals have acquired a class

[1] Most of the works of Marx and Engels are archived at www.marxists.org. In addition, there are different translations in printed books and here I chose the most common wording in English.

consciousness and ideology as part of their participating in the social world, namely, they have knowledge of the social positioning of people across class lines which is reflected in their inner world, in their consciousness, where knowledge of the true or contrary of the false knowledge is situated. In a way, it could be concluded that consciousness is not only a mental process but it informs a certain frame of mind, a social stance or positioning, on which the undertaking or failing to undertake of socially transformative action is based. In this social stance or attitude of the individual, social positioning, consciousness, ideology, knowledge and the undertaking of action are causally interlinked. Knowledge furthermore is linked with dominant ideology and the primacy, or more precisely the ruling of certain ideas that are interlinked with furthering class interests.

The concept of class consciousness refers not only to awareness of one's class position but to a deeper understanding of this position, which fuels the undertaking of transformative action in regard to society. In the opposite case, one speaks of false consciousness, when an individual does not recognize one's own true interests. Paradoxically enough, not only working class suffers from false consciousness, but also the middle class, the bourgeoisie. The bourgeoisie, according to Lukács, fails to 'see' and understand the real conditions of exploitation that members of this class impose on the whole of society. In his famous work *History and Class Consciousness* (1968/1971) Lukács distinguishes between a 'class in itself' and a 'class for itself', following Hegel's ideas, in order to define first, whether a class is formed independently of individual intentions and second, what class consciousness is. This situation of class consciousness is further obscured by alienation. Briefly presented, a person is alienated in her relations, when she is estranged from herself and her true nature. It becomes apparent when labor as well as nature or other people become the means to achieve ends that do not serve people's true interests, which are pursuing and achieving a society where there is justice and freedom, defined as lack of misery and exploitation. The dichotomy, or at least the notion of truth and real, returns in Marxist perspectives in an attempt to deal with the question of change of society. The agency of radical social change is presented in collectivity terms, for the protagonist of history is class not the individual. Therefore class consciousness is very important in realizing the bases of one's own oppression that ideally fuels the undertaking of transformative action. Though the analysis refers to social structures and collectivity as forces, somehow in the end it has to work out the relation of individuals to these forces.

In Marx's conception there is a direct link between the material world and individual consciousness, as the former determines the formation of the latter. Material is defined as the ways human beings have developed in producing the things that sustain them, and culture is included among these. In turn, these forms of producing and co-existing determine humans' consciousness and not the other way around. If, according to Hegelian inspired approaches, the Geist (Spirit) was the determinant of social life and its phenomena (which are manifestations of the Spirit), then for Marx it was the exact opposite: the material world determines the Geist, including the Time Spirit (Zeitgeist) of an era. Thus, consciousness is conceived as being generated by society, and the knowledge gained can be employed to change existing forms of exploitation, upon which the contemporary organization of social relations is based. However, Marx noted that the relation of material circumstances to consciousness is not unmediated. The question that arises is 'what about human free will?' 'Does it not exist?' Again, Marx's answer is characteristic of his way of thinking: 'people are born free, but in conditions or circumstances they find before them' (see also Marx, 1978, p.

9). In this phrase one can discern the relational aspect of phenomena, which to my view is the added value of Marx's theory.

In sum, according to Marx, consciousness amounts both to a process (of knowing, feeling and thinking) and to a state of being (positioning according to the prevalent class structure). It is informed by society and is a reflection of one's own position in it. This position is determined according to the relation of individuals to ownership (or lack thereof) of the means of production, broadly defined. These relationships are social relations and are mirrored in consciousness in a way that forms the basis of political awareness or a class consciousness that leads to undertaking collective action in order to change radically existing forms of exploitation. This linear description of cause and result is however obscured by the workings of the dominant ideology and the false consciousness, which means lack of awareness of ones' real interests.

Much of subsequent writing in sociology has been addressing the question of alienation, following the different forms explicated by Marx. Theorizing on class consciousness has led to numerous approaches that attempt to define unequivocally the number of classes by defining for instance the 'distance' between them. Marx's theory shows the power of relations in social life while subsequent approaches attempt to define substantively what is consciousness or class as things in themselves rather than as formative processes. In doing so, a certain degree of reification and fetishism seems to lurk, for characteristics and traits are attributed to class consciousness, as though it existed by itself and above the material conditions under which it is formed, informed or transformed.

In the Marxist writings consciousness is an embedded and embodied structure within the individual that mirrors his/her class position and her/his (alienated) relationship with the social world. This mirror could function like a 'camera obscura', to use a known metaphor, showing the object of inquiry upside down.

2.3. Structure of Consciousness

The phenomenological paradigm among the interpretative approaches is based not on the work of Hegel, as one would expect, but on that of Husserl. Phenomenology originated in 19th century Germany and has spread as a movement in various disciplines, informing various other paradigms.

The subject matter of phenomenology is the study of 'the structure of consciousness'. At the time it was known that time and space and more particularly the sense of time and space were structuring human consciousness. This however left open much space for research, for example, regarding the extent to which human experience is fundamental to and foundational of consciousness, especially the lived experience in the here and now which Heidegger (1996) formulated as 'Da-sein'(= to be present, to be there).

The notion that has influenced relevant writings was that of 'intentionality'. For Husserl (1900-1) consciousness amounts to awareness of something at the time, for instance, an object, or another person, that is wherever or whatever attention is directed to. In this sense, consciousness is always intentional. However, intentionality refers to being aware of something and thus it forms an aspect of consciousness among others, one of which is the 'stream of consciousness', as it was formulated later by the American philosopher James (1890/1983).

Schutz, a German scholar who fled the Nazi regime in Germany and eventually arrived in the United States, attempted to elaborate on the work of Husserl and Weber. Schutz (1967) argues that the notion of time means that an individual has the feeling of duration of time. This in turn includes the feeling of a point named zero, which is preceded by the past (that which is before this now present point) and followed by a future (that which is after this point experienced now). In a sense, consciousness amounts to having the feeling and experiencing it, which he calls deep consciousness. Some of these notions were employed later by Smith (1990), who used the terms 'bifurcated consciousness' or 'a line of fault' and 'social consciousness', when attempting to study relations of ruling (see below, section 3).

Notably, the reviewed approaches to consciousness end up again in Hegel, insofar as consciousness is the basis of the creation of the self and of human identity in current scientific terminology. Consciousness as reflection is suggested in the writings of Hegel, when he defines it in terms of movement and reason. In current terminology, reflection is a process by which selfhood and subjectivity are constituted, both of which are important in making science possible as well as in undertaking action. Briefly, reflection is important in acquiring knowledge both for oneself and scientifically. In a way, consciousness as a mental process is characterized by an ability to think upon itself, and to reflect, which amounts to a consciousness of consciousness.

Schutz, though working within the framework of the phenonemonological paradigm, was however not so much interested in studying structures of consciousness per se, but rather as an aspect of it, namely, as inter-subjectivity (Schutz, 1967; see also Ritzer, 2000). In his approach, Schutz builds on the theories of his teacher Husserl and others, especially the construct of stream of consciousness as used by James (1890/1983) and Bergson (1888). Schutz emphasized that knowing and knowledge depend on what is 'relevant' for the person in question, so that something becomes relevant because it is interpreted and is associated with something else familiar to the person in question. The 'scheme of relevance' is a well known concept that Schutz defined as the framework human beings have acquired by living in society and by which they associate things, frame and interpret them while learning new things or reflecting upon old ones. In addition to the principle of relevance, Schutz developed important concepts, such as 'stock of knowledge' that is available to members of society, and is preserved by tradition and culture; and 'social distribution of knowledge', indicating that persons have acquired a part of knowledge and not everybody knows everything about everything (Schutz and Luckmann, 1973). This in turn is related to sustaining or transforming existing relations of power.

Returning to the subject of inter-subjectivity, it is becoming possible because knowledge is being available in a form that resembles 'recipes' that inform people about what to do in certain circumstances. This kind of knowledge is furthermore objectified in myths and other knowledge carrier institutions through the medium of language. It could be concluded at this point that inter-subjectivity is an integral part of consciousness and that this explains the reason why conscience, which has a restraining effect on individuals, makes possible society, that is, the being and functioning of individuals as a collectivity.

In this context it is important to mention the earlier work of Mead, whose title *Mind, Self and Society* (1934) gives the impression that mind precedes self and society. In Mead's conception, mind, self and society are interlinked to such a degree that these three entities could not exist without each other. The self exists because of society and society because of the self. Furthermore, for Mead consciousness is not situated in the mind of the individual but in the interaction between them, at least as some sociologists interpret his writings (see Ritzer, 2000). To my mind, the thesis that consciousness is situated in interaction means that it is structured though at the same time is ongoing because interaction becomes possible and takes place due to the shared meanings and interpretations humans attach to their intentions and actions. Consciousness is then very closely related to society, being an integral part of a social (human) being.

Berger and Luckmann in their well-known work (1967) *The Social Construction of Reality* attempt a grand synthesis of the existing traditions in sociology in order to uncover the ways and methods whereby social reality is man-made. One of the major structuration aspects in this process is externalization. This notion, which is to be found already in Hegel and Marx, means that in or with praxis, such as labor, man externalizes himself and gives meaning to the world around him and to his own artefacts (and one of these is the 'use value' products have, according to Marx). The next step is objectivation and objectification, whereby meaning and value are attributed to things, although after many generations people do not remember any more that they have done it and believe instead that objects have these attributes by and in themselves. Berger and Luckmann further postulate that in its turn reality, that people themselves have constructed, 'imposes itself upon individual consciousness in a massive manner'. It is doing so, because it has been made an object, it has been objectivated. It is noteworthy that objectivation of social reality and social relations form partly the bases of fetishism as well. Fetishism, which is the belief that things have meanings in themselves that is independent of human attribution process, is a theme that recurs often in philosophy when analysing relations of human to others and to the world around them. The danger here is to consider consciousness in similar terms attempting to attribute to it characteristics and treat it as a thing, a substance existing outside individuals or society.

In sum, at the end of the 20th century, perception, mind and external reality come once again together, making Locke's definition of consciousness quite fresh, as though it had been formulated only yesterday. This movement of making things contemporary is also a feature not of time itself but of consciousness, which is characterized by the sense of duration and as though it is a 'stream'. However, the idea that consciousness is moulded into a certain social shape that relates the individual to society in terms of class consciousness has originated in the Marxist approaches.

In this section, consciousness has been conceptualized as part of social life informing interaction and governing social life.

3. Current Conceptions to Consciousness

In popularized sociological textbooks referring to basic concepts in sociology, one will hardly find consciousness as a heading among other 'rigorous' concepts, such as class, group, institution or structure. Such a cursory view gives the impression that sociology has not been concerned with the phenomenon of consciousness, while leafing through texts of known authors nowadays one come across chapters or section headings titled 'Sociology as a Form of Consciousness' (Berger, 1963) or 'Sociology as Constituent of a Consciousness' (Smith, 1987). From this use of the concept, one may get the idea that sociology is or could be as critical as consciousness can be; while at the same time one realizes that one of the most important concepts on which sociology as a discipline is based remains largely implicit as though hidden from open view, scrutiny and scientific dialogue.

During the 20th century, theorizing and research on consciousness draw on the founding theorists of sociology, but more in a diffuse than in a clear-cut manner. Relevant approaches could be distinguished between the quantitative and qualitative paradigm prevalent in social sciences; however even this distinction does not convey the feeling one gets that, on the one hand, there are numerous, actually thousand of articles written on consciousness, but, on the other hand, the theoretical insights gained from them are not apparent at first sight. In this part I attempt to give some examples of more recent work that refer to the various conceptualizations of consciousness in current scholarship.

A more recent attempt on relating consciousness to social relations is undertaken by Dorothy Smith, a feminist sociologist. Smith (1987) using the notion of 'bifurcated consciousness' or 'line of fault', refers to a particular point, so to speak, the juncture where an individual knows deeply the code, according to which two different if not contradictory worlds function and of which s/he is part. A little later, based clearly on Marx and Schutz's work, Smith (1990) studies 'relations of ruling' that are mediated by written texts. She concludes that social consciousness is a property of formal organizations, which by postulating beliefs and norms in a abstract, detached manner while remaining connected to other institutions and in a chronological sequence, impose certain forms of social consciousness, as for example stereotypical femininity, which, for Smith is part of social consciousness.

Thus, social consciousness is related to structuring and sustaining relations of ruling by, for instance, incorporating and imposing more or less stereotypical characteristics, so that a certain policy is legitimized and reinforced. Such theorizing attempts to connect society to individuals (and not vice versa), for it starts from the premise that forms of social consciousness preexist and influence social behavior. Relations of power are then related to social consciousness as these form a web of relations that mutually influence each other, sustaining the status quo.

Furthermore, another strand of research scholarship during the 20th century is the attempt to specifically talk about consciousness, categorized into forms or kinds. Thus, authors refer to having a 'double consciousness' (in case of black people) or a 'bifurcated consciousness' or a 'line of fault consciousness' (in case of women). Implicit here is the idea that a person comes to know the functioning code of two worlds and tends to function properly according to both. The double code functioning may in certain conditions create problems and perhaps give rise to the dilemma of belonging. Such conditions are presented when functioning codes are not congruent, for instance, one world functions according to the code of being a slave the other of being or aspiring to be free.

Forms of social or collective consciousness that have been discerned include racial, ethnic, gender, black, or national consciousness. Empirical research done in this area deals first, with issues of measurement, secondly it strives to uncover factors that influence the formation of consciousness, and third, it tries to relate consciousness with political action or praxis. For example, the issue of awareness was studied in the relationship between class positioning and voting behavior, or between gender consciousness and showing solidarity to women politicians. Thus, consciousness conceptually became more or less synonymous to awareness and knowing of ones' own position in society, whether of class, color, or gender, which are the major stratifying social categories. Currently a quick search in the databases of journals provides thousands of links to consciousness research, some of them using the term in titles and others attempting to explain with this concept a different array of social behavior. However, the explanatory power of this concept depends very much on its conceptualization, that is, if it is defined as a substance that can be straightforwardly measured or if it is defined as a process being formed in certain social conditions.

To my mind, the concept of consciousness is being employed in order to explain homogeneity in social behavior or to explain awareness and change of behavior. In either case that which has to be explained, namely, the conditions under which it happens and the ways it is mediated by individual history and culture or available social meanings remain outside the research and the employed theoretical framework. In other words, consciousness is being used to refer to a state of being without conceptualizing it as a process, which is another important dimension of this concept.

Thus, one of the tasks set out in sociology was a quest for conceptual clarity in regard to an array of different concepts that gradually became influential in developing a body of theory and research. One prime example of this quest is the replacement of the construct collective consciousness by social consciousness, as it has already been suggested, by Cooley (1907) when he emphasized that the individual is aware that s/he is a social being. Another example is the replacement of the concept of race - which was challenged about its accuracy, and its inherent negative stereotyping about black people in the USA - by the concept of 'black consciousness', while for white people, the employed concept has been 'consciousness of a kind' or 'class consciousness'. In relation to women, the concept 'feminist consciousness' has been used interchangeably with 'gender consciousness'. In sum, stratifying categories of class, race, and gender are viewed as being reflected in the consciousness of individuals as well as of the organization of social life.

Thus, society is considered as being divided by various sorts of important differences and as being kept together by structural characteristics pertaining to work, culture and the sense of belonging to the same nation. National identity goes hand in hand with national consciousness, which not only constitutes a unifying factor, but also fuels patriotism, which plays an important role in defending one's nation against external and perhaps internal enemies. Education has been the institution which contributed more than any other to forging a national identity, but also to fostering a critical stance. In the discipline of pedagogy, the concept currently used for critical stance is that of 'critical consciousnesses. Accordingly, a main concern of education research has become the cultivation of critical consciousness in teachers and students alike.

Institutionalizing education systems in the Western world has been considered as a means to socialize individuals to common norms, providing them with the same elementary competencies for participating in social life. In order to present here another example of the use of the concept of consciousness in the sociology of education, let us recall Durkheim's conception of consciousness as referring to common beliefs, norms and values as well as to morality (see above sub-section 2.1). In the morality aspect of consciousness, Durkheim emphasized that the individual's autonomy is the target of education in the sense that the student internalizes society and then s/he can act on his own accord in a twofold sense: first, s/he does not follow others uncritically and second, s/he does not need direct social control in order to function according to society's needs of social cohesion and sustenance. Group identity fosters consciousness and sociability to individuals. Thus, society enters the individual by restraining his action through morality. In this manner consciousness turns gradually into conscience, which is the restraining factor that binds individuals to society, enabling the existence of society.

While for Durkheim education is the institutional means par excellence for sustaining social ties, for Marxists it has been the means par excellence for keeping working class kids down the social hierarchy. The function of education that fulfilled such a mission was called socialization, that is, the internalization of values, norms and rules, resulting in one's own acceptance of the status quo and its righteousness. In a way, the institution of education has an important mission to accomplish which is to socialize students and their consciousness in order to fit in their preordained and pre-given class position in the realm of the social division of labor. This functioning of education has been called 'reproduction', and it is based among others on the writings of Bourdieu (Bourdieu, 1977; Bourdieu and Passeron, 1977/2000). Moreover, in explaining the power of education, Bourdieu and other authors proposed that education 'moulds' the consciousness of pupils, 'instilling' in them common beliefs, norms and values that have far-reaching effects since these 'legitimize' existing social order and the structure of existing social arrangement and one's position within this order. Furthermore, this structure is characterized by power arrangements, which in turn are based on exploitation since social relations are relations of power and exploitation, which form the bases of social inequality. This effect is not only consciously achieved but also unconsciously, for common ideas and beliefs as well as culture permeate text books, teaching methods and the pupils' personality due to the workings of the hidden pedagogy or hidden curriculum as well. Pupils possess a 'habitus' (a concept employed also by Bourdieu, 1977), namely, a pattern of behavioral habits relating to culture that reveal one's social origins. The 'habitus', in the form of cultural capital (such as knowledge, opinions and 'taste') then accompanies an individual pupil at school and often is considered to be the cause of his school success or failure (Bourdieu and Passeron, 1977).

From the above conceptions it follows that consciousness is viewed as linking the individual to society by instilling in the former structures of personality and social positioning aiming at accepting this position so that existing social arrangements are not disrupted. In this context, the concept of alienation has been used in order to explain how people fail to 'see through' that existing arrangements are at fault. A related aspect of consciousness was picked up by the feminist movement in the middle of the 20th century onwards. In its quest for women's emancipation, the movement relied heavily on 'consciousness raising' processes, which would foster reflection and awareness of women's oppression, both in the private as well as in the public spheres of social life. This aspect of consciousness is employed nowadays by social movements in order to foster awareness and promote the undertaking of transformative action, a major example of this being the environmental movement. Consciousness is related then to uncovering one's true conditions of living and acting upon society with the help of this knowledge. Apart from a repressive aspect, consciousness acquires a liberating and emancipatory dimension as well.

The dichotomy between true and false or between collective and individual sense of consciousness seems to wither away, though it is still prevalent in many writings. In sociology, the concept of consciousness has been employed mostly in order to uncover and explain the link of individuals to society and vice versa. Moreover, the concept of consciousness has been categorized in various forms distinguishing thus the collectivities to which it refers to, such as class or gender. The concept is further employed in order to explain subject positioning, social behavior, relations of power, oppression but also acts of resistance.

4. Analytic Notes and Conclusion

Consciousness seems to defy direct accessibility to scrutiny and research when it is approached as a concrete substance. It could be best described both as a constitutive process and as the product or the outcome of this process; as such it partakes in the constitution of subjectivity (in today's terms) in humans and the world around them. Subjectivity indicates that humans learn to relate to themselves independently (as a third person) and to others, societal institutions, or nature and the world around them and furthermore learn to handle these relationships. Consciousness amounts to existence and life, while lack of it, namely, death or non existence are also part of consciousness. In an interrelated sense, organic, living elements whether of a stony or watery form have a consciousness. This view of the world which conforms to an animistic conception, on which some religious beliefs are based, has been abandoned early in social sciences. Yet, nowadays it returns framed in the form of respect for the natural environment. This attitude reflects partly a long-delayed impact of the movement of human rights that has been based on claims to dignity and respect between humans irrespective of social stratifying categories, such as class origins, beliefs, gender, age or ethnicity. These claims have been expanded to other realms, such as treating with respect the natural environment and other species living on earth, which today are clearly associated with accomplishing a high degree of culture, the most valuable quest of the human condition, according to many philosophers.

Consciousness amounts to a higher, complex order of thinking and being. Awakening, awareness, and reflection are processes associated with the functioning of consciousness, and at the same time are viewed as products of these processes. It seems to be difficult for humans to grasp with language that processes and outcomes are ongoing and changing, while at the same time they are patterned or stable. Sometimes, humans use metaphors borrowed from the natural environment or geometry, such as the word 'stream' or the word 'structure' in order to capture not physically existent entities. Human beings usually perceive consciousness by its outcomes, creating an array of other concepts related to consciousness but differing from it, such as spirit, mind, or soul. In this conception, consciousness materializes as opinion, ideas, social stance and attitudes, which are further operationalized and measured in quantitative research. In doing so, the study of consciousness even in sociology is bound with the framework attributed to Hegel of perceiving phenomena related to the materialization or the positive expression of Spirit (Geist), which in turn is viewed as the expression of consciousness.

In sociological theorizing, humans' collective consciousness departs from Marx's class consciousness and alienation and is mostly used as a descriptive concept (as, for instance, national identity) and not as an analytical one. The questions that remain open are 'how to define consciousness, in order to capture its multifaceted nature'; 'what role consciousness plays both in constituting social life and the life world of individuals'; 'how it is formed, informed, and how or under what conditions it is transformed or it changes'.

In sociology, the concept of consciousness is significant and though it has a distinct conceptualization it is still diffuse insofar as its uses defy an unequivocal meaning. It is still important today, as there are unexplored possibilities, though it is difficult to avoid reification and consequently fetishism.

From the sociological writings, one could conclude that consciousness is part and parcel of human beings constituting their self, mirroring society, being moulded by habits and other behavioral and thinking patterns, informing a person's habitus, conscious and unconscious subjectivity and his relations with others and the world around him or her. Consciousness is externalized and objectified in culture, language, science, fine arts and other human artefacts constituting and being constituted in turn by them, and thus constituting social reality and society.

It could be argued that consciousness is embodied and embedded, much the same way as are gender, class and race, the major 'deceptive distinctions' in current social stratification system. Consciousness resides in the individual as part of a larger structure created by existing society as a form of collectivity; it is articulated in interaction and it is transmitted through culture and societal institutions, whereby education and mass media are of paramount importance in crafting and sustaining a collective consciousness and the view on collective consciousness. The dialectic relation of individuals to collectivity is formed through consciousness that through interaction both at conscious and unconscious levels at the same time informs subjectivity, the sense of self and identity.

Regarding its conceptualization and in an attempt to pin it down substantively and ontologically, the concept of consciousness has been viewed as having different forms or kinds and as containing layers or levels. This shows at the same time the limits of language that cannot analytically deal with something literally if it is not made a substance described in terms of metaphors borrowed, for instance from geometry denoting space or distance. Consciousness could be better understood as a process, becoming manifest as the end-product of this process, which consists both of coming into being and becoming extinct, in a continuous cycle.

Today, it seems that consciousness has lost its appeal in everyday use as it has been associated with conscience. It has acquired an aura of ethics and morals of 'musts' and 'ought to', obligations that are considered to be hindering and limiting the expression of a certain type of individual freedom known as free choice. In addition, consciousness points out to several influences on the individual, making him/her actually a product of society, lacking a personality, or idiosyncratic traits, that make up that undefined uniqueness of every individual.

In addition, currently we witness a widespread feeling of fragmentation, of losing focus in the meaning that human existence has and the point in organizing a social life. Perhaps 'seeing through' in a sociological framework could mean to solve this kind of puzzle and by putting together all the bits and pieces we understand suddenly the big picture, and everything gets a meaning and a place in our lives. This activity is likely to be also the outcome of the workings of consciousness as well.

Consciousness could be used as a working concept making possible 'thick descriptions' of phenomena and the linkages between them, including phenomena pertaining to mind, conception and perception, experiencing and analysing, explaining in what conditions and how individuals in certain contexts act and react, forming their own conceptions for themselves and a different one for the public to see, based on existing discourses and/or its enemies. It could be also employed in understanding identity, affiliations, and subject positioning, having in mind that 'self and society go together, as phases of a common whole' (Cooley, 1907, p. 678).

Consciousness is a highly abstract and complex concept pointing to several processes. Indeed, this is not a novel idea. Yet, such a realization could be used as a starting point for further research and theorizing.

ACKNOWLEDGEMENT

I would like to thank Professor Dr. Maria N. Antonopoulou, Panteion University, for her advice on philosophical aspects of the term consciousness.

REFERENCES

Antonopoulou, M. N. (2008). *Oi klassikoi tis koinoniologias* [Classic theorists of sociology]. (in Greek). Athens: Savvalas.

Berger, P. L. (1963/1991). *Invitation to sociology. A humanistic perspective*. Middlesex: Penguin.

Berger, P. L. and Luckmann, T. (1966). *The social construction of reality. A treatise in the sociology of knowledge*. Middlesex: Penguin.

Bergson, H. (1888). *Essai sur les données immédiates de la conscience* (Electronic edition by J.- M. Tremblay) at http://mis-au-net.net/ebooks/philo/bergson/bergson_conscience.pdf).

Bottero, W. (2007). Class consciousness. In Ritzer, G. (Ed.) *Blackwell encyclopedia of sociology*. (Accessed, 05 June 2010).

Bottomore, T. B. and Rubel, M. (Eds.) (1956). *Karl Marx: Selected writings in sociology and social philosophy*. (Transl. by T. B. Bottomore). London: Watts.

Bourdieu, P. (1977). *Outline of a theory of practice.* (Transl. by R. Nice). Cambridge, UK: Cambridge University Press.

Bourdieu, P. and Passeron, J.-C. (1977/2000). *Reproduction in education, society and culture.* 2nd ed. London: Sage.

Cooley, C. (1907). Social consciousness. *The American Journal of Sociology*, *12*, 675-694.

Durkheim, E. (1933). *The division of labour in society*. (Transl. & introduction by G. Simpson). New York: MacMillan.

Durkheim, E. (1961). *Moral education: A study in the theory and application of the sociology of education*. (Transl. by E. Wilson and H. Schnuren). Glencoe, IL: Free Press.

Durkheim, E. (1915/1976). *The elementary forms of religious life.* 2nd ed. London: Allen & Unwin.

Hegel, G. W. F. (1807/1977). *Phenomenology of spirit*. (Transl. by A. V. Miller). Oxford, New York: Oxford University Press.

Heidegger, M. (1996). *Being and time: A translation of Sein and Zeit* (Transl by J. Stambaugh). New York: State University of New York.

Husserl, E. (1900-1/1973). *Logical investigations*. (Transl. by J. N. Findlay). London: Routledge.

James, W. (1890/1983). *The principles of psychology.* (Ed. by G. A. Miller). Cambridge, MA: Harvard University Press.

Kantzara, V. (2006). Society. In Fitzpatrick, T. et al. (Eds.) *International encyclopedia of social policy*. London: Routledge, pp. 1319-1325.

Kantzara, V. (2008). *Ekpaideysi kai koinonia: Kritiki dierevnisi ton koinonikon leitourgion tis ekpaideysis* [Education and society: A critical exploration of social functions of education]. (in Greek). Athens: Polytropon.

Kantzara, V. (2009). Social functions of education. In Ritzer, G. (Ed.) *Blackwell encyclopedia of sociology*. (Accessed, March 2009).

Locke, J. (1690/1975). *An essay concerning human understanding*. (Ed. by P. Niddich). Oxford, UK: Oxford University Press.

Lukács, G. (1968/1971). *History and class consciousness. Studies in Marxist dialectics*. (Transl. by R. Livingstone), Cambridge, MA: The MIT Press.

Marx, K. (1932/1964). *The economic and philosophic manuscripts of 1844*. (Ed. by D. Struik). New York: International Publishers.

Marx, K. (1978). *The eighteenth brumaire of Louis Bonaparte*. Peking: Foreign Language Press.

Marx, K, and Engels, F. (1845-6/1998). *The German ideology*. New York: Prometheus Books.

Matza, D. and Wellman, D., (1980). The ordeal of consciousness. *Theory and Society*, *9*(1), 1-27.

Mead, G. H. (1934/1962). *Mind, self and society: From the standpoint of a social behaviorist*. Chicago, UK: Chicago University Press.

Ritzer, G. (2000). *Classical sociological theory*. 3rd ed. Boston, MA: McGraw Hill Higher Education.

Schutz, A. (1967). *The phenomenology of the social world*. Evanston, IL: Northern University Press.

Schutz, A. and Luckmann, T. (1973). *The structures of the life-world*. Vol. 1. Evanston, IL: Northwestern University Press.

Smith, D. (1987). *The everyday world as problematic: A feminist sociology*. Milton Keynes, UK: Open University Press.

Smith, D. (1990). *Texts, facts and femininity. Exploring the relations of ruling*. London, New York: Routledge.

Wortmann, S. (2007). Collective consciousness. In Ritzer, G. (Ed.) *Blackwell encyclopedia of sociology* (Accessed, 05 June 2010).

THE EXPERIMENTAL APPROACHES

In: Consciousness: Its Nature and Functions
Editors: Shulamith Kreitler and Oded Maimon ISBN 978-1-62081-096-5

Chapter 7

TURNING ON THE LIGHT TO SEE HOW THE DARKNESS LOOKS

Susan Blackmore*
University of Plymouth, UK

ABSTRACT

Given a curious property of introspection, some common assumptions made about the nature of consciousness may be false.

Inquiring into one's own conscious experience "now" produces different answers from inquiring into the immediate past. "Now" consciousness seems to be unified with one conscious self experiencing the contents of a stream of consciousness. This implies a mysterious or magic difference between the contents of the stream and the rest of the brain's unconscious processing.

By contrast, looking back into the immediate past reveals no such unity, no distinct contents of consciousness or coherent stream, but multiple backwards threads of different lengths, continuing without reference to each other or to a unified self. From this perspective there is no mystery and no magic difference.

I suggest that the assumed difference between conscious and unconscious events is an illusion created by introspection into the present moment. So is the persisting self who seems to be looking. Most people are not introspecting this way much of the time if ever. Yet whenever they do the mystery appears. Looking into those times when we are not deluded is like turning on the light to see how the darkness looks.

INTRODUCTION

Whenever I ask the question "Am I conscious now?" the answer seems to be "Yes". But what about the rest of the time?

Here, it seems to me, is a gigantic clue to help us with the mystery of consciousness. It has been staring us in the face all the time but, like so many other useful clues, it seemed

* Prof Susan Blackmore, www.susanblackmore.co.uk; www.memetics.com.

either too obvious or too unimportant to take seriously. Perhaps you'd like to try it now. Ask "Am I conscious now?" and watch what happens.

Most likely you will look, listen and feel what's going on around you, and conclude that of course you are conscious: How could you not be? If you ask "*What* am I conscious of now?" you will find plenty of things springing to mind as the answer –sensations, perceptions, thoughts, feelings or just the sense of being someone who is inquiring.

But what about the rest of the time? What about when you are not asking these questions? How is it then?

It seems easy and natural to jump to the conclusion that all the rest of the time is like this too – that all day long, whenever you are awake and responsive, there are some things that you are conscious of and some that you are not. It is natural to jump to this conclusion because every time you ask yourself about consciousness it seems to be this way. This does not, of course, prove that it is.

I am here suggesting something very curious about the nature of consciousness – that looking into consciousness reveals only what it is like when we are looking into it – and most of the time we are not. So introspection on our own minds, which are, after all, the subject of our inquiry, is thwarted by the very fact that we are introspecting.

William James described something similar when he tried to observe the "flights" as well as the "perchings" in his "stream of consciousness". He said "The attempt at introspective analysis in these cases is in fact like seizing a spinning top to catch its motion, or trying to turn up the gas quickly enough to see how the darkness looks." If we do catch the moment, he said, "it ceases forthwith to be itself" (James, 1890, p. 244).

I love James's idea of trying to use light to see into the darkness. I imagine what fun it would be to show him electric lights so that he could turn them on even faster to see how the darkness looks. Or he might like the modern equivalent of looking into the fridge to see whether the light's always on (O'Regan, 2011). This analogy is perfect for the problem I am trying to describe. Asking "Am I conscious now?" or "What am I conscious of now?" can feel like turning on a light, but is that light always on? And if not then what is the darkness like inside the fridge?

This is the question I set myself to tackle and my explorations have led me through an intellectual inquiry into the science and philosophy of consciousness as well as personal inquiry into the darkness of my own mind. The result of this inquiry is that I have come to question some of the most conventional and ubiquitous assumptions that are made in the science of consciousness. I believe that much of what people take for granted about their own consciousness is in fact untrue and that we shall make real progress in solving (or dissolving) the "hard problem" of consciousness only when we abandon those assumptions and take a different tack.

There are obvious dangers in my making any claims based on my own introspection, but I do this with caution and in light of the history of introspection which shows how easy it is to be misled by one's own prior beliefs and expectations. Before describing my adventures I shall lay out the way I see the problem of consciousness, and some of the current attempts to solve it.

The Problem

Consciousness is a curious illusion. When I say this some people seem to think I mean that consciousness does not exist. So to be as clear as possible – I do not mean that it does not exist – at least, I do not mean that there is no problem to be solved – rather, I mean that consciousness is not what it seems to be. In this I am simply using an ordinary dictionary definition of "illusion". For example Webster's dictionary defines illusion as "the state or fact of being intellectually deceived or misled." The *Oxford English Dictionary* describes a "state involving the attribution of reality to what is unreal; a deception, delusion, fancy – something that deceives or deludes by producing a false impression." In other words an illusion is something that is not what it seems to be. Consciousness, I suggest, is not what it seems to be.

Our starting point, then, is how it seems. Here I sit at my desk contemplating the colourful, flickering flames and the roaring and crackling sounds of my wood-burning stove, the room full of books and papers around me, and the stiff coldness of my legs under my desk.

Here is the problem as it seems to me. As I look at the dancing flames I seem to be over here and they seem to be over there. I seem to be looking at them from somewhere inside my head. I can stretch out my feet and look at my own toes. Those are my feet down there, and this is my body and even my head. I can think about my heart and my brain even though I have never seen either of them. All this implies that I am not equivalent to this body but am more like some kind of owner or inhabitant of the body who experiences the world through its senses and who controls its movements.

I can change my perspective in some ways. For example I can mentally expand myself out so that I fill my whole body and even reach beyond it. Even so, with my eyes open there is a distinct and hard to eradicate feeling that I am in here and the world is out there, and the me in here is experiencing a stream of conscious impressions of that world and of my own inner thoughts and feelings.

If I stare at the orange, flickering tongues as they curl around the dark logs I can get extremely bothered about that orange sensation. This orangey, orange is surely private to me. This is what is meant by the philosophical term "qualia", those private and ineffable sensations of red or wood smoke or crackling. No one else can experience these flames exactly as I do and I cannot adequately explain how they look to anyone else. Indeed, there seem to be two distinct kinds of thing in the world: my private, ineffable experiences of the orange flames – and the physical flames themselves; my thoughts about the fire – and the fire itself; my inner self – and its physical body. The harder I stare into that orangey orange the more divided the world seems to be.

This is what creates the fundamental mystery of consciousness. It is the temptation of lapsing into dualism that bedevils every attempt to understand it. Dualism, in its many forms has been endlessly debated and widely rejected. Substance dualism, the idea that there really are two kinds of stuff in the world, as Descartes thought, fails largely because either the two worlds can interact, in which case they are not entirely distinct, or else they cannot, in which case there can be no explanation of why mind and brain, or subjective and objective, seem to correspond. As Dennett puts it "dualism is forlorn" (1991, p. 33).

The modern incarnation of this problem is what David Chalmers calls the "hard problem" of explaining consciousness *itself.* This "is the question of how physical processes in the brain give rise to subjective experience" Chalmers (1995, p. 63). Note that he uses the phrase "gives rise to". This already implies a kind of duality, in that one thing (the objective processes in a physical brain) gives rise to another (the subjective experiences). Chalmers (2010) is indeed a kind of dualist, although not a substance dualist. Others reject this idea altogether, claiming that subjective and objective must really be the same thing even though they do not appear to be. Many thinkers reject the whole idea of the "hard problem" even deriding it as a "Hornswoggle problem" (Churchland, 1996). Even so, many see its solution as the Holy Grail of consciousness studies and are avidly trying to solve it.

Another problem concerns the apparent unity of our experiences. As I stare into the flames it seems to me that I am having a unified stream of experiences – indeed that there is also one "me" experiencing them. Yet the more we learn about the brain the less possible it seems that everything ever comes together to create the unified experience I seem to be having right now, or that there is any conscious self that could be the recipient of the brain's perceptual workings. The brain is a massively parallel system, with endless streams of activity flowing from place to place, carrying out multiple functions of perception and action all at once. This suggests that they are never all brought together to create what we naturally think of as the "vivid picture I see in front of my eyes" (Crick, 1994, p. 159) or the "movie-in-the-brain" (Damasio, 1999).

As James explained long ago, there is no 'pontifical' neuron to which *our* consciousness is attached; "no cell or group of cells in the brain of such anatomical or functional pre-eminence as to appear to be the keystone or centre of gravity of the whole system" (James, 1890, i, 179-180). In more modern terms "there is no terminal station in the cortex" (Zeki, 2001, p. 60-1), no final integrator station in the brain, and no need for microconsciousnesses (as Semir Zeki calls them) to be reported to a 'center' for consciousness.

We know all this, and yet our own experiences seem to lead us, time and time again, into imagining our minds as like a kind of mental theatre in which our personal experiences appear for our benefit on the brightly lit stage of consciousness. When I turn my attention to those orange, flickery flames they seem to come *into* my consciousness when before they were not.

A century after William James, Dennett described this tempting fantasy as the "Cartesian Theater", that mythical place into which perceptions, sensations, thoughts and feelings come to be experienced by the audience of one. Nearly everyone rejects Cartesian dualism, he said, yet "When you discard Cartesian dualism, you really must discard the show that would have gone on in the Cartesian Theater, and the audience as well, for neither the show nor the audience is to be found in the brain, and the brain is the only real place there is to look for them." (Dennett, 1991, p. 134). Those who claim to be materialists while still hanging onto the Cartesian Theatre with all its alluring imagery, he says are trapped in "Cartesian materialism", "the view that there is a crucial finish line or boundary somewhere in the brain , marking a place where the order of arrival equals the order of 'presentation' in experience" (Dennett, 1991, p. 107). This is "the view that nobody espouses but almost everybody tends to think in terms of …" (Dennett, 1991, p. 144).

They do indeed. For example the simple phrase "the contents of consciousness" is used frequently, without comment, in both scientific and popular writing. Yet this idea of "contents" tempts us to imagine consciousness as a space or container into which perceptions, thoughts and feelings come and go – or a stage in the theatre of the mind on which perceptions, thoughts and feelings are illuminated by the spotlight of attention. We may think it more scientific to replace such images with the idea of a special process, or assembly or network of cells, but this does not change the basic conception. Ssurely consciousness is not some kind of container, and if it is not then this common phrase is deeply misleading (Blackmore, 2002).

Some Current Theories and Research

Theatre imagery is implicit in many theories and explicit in some. Possibly the most popular theories of all are variants on Global Workspace Theory (GWT) first proposed by Bernard Baars (1988, 1997) and later extended in various versions including neuronal global workspace theory (Dehaene, 2002). These suggest that the architecture of the brain includes a global workspace, something like a working memory space, in which some information is processed and then broadcast to the rest of the (unconscious) system. By virtue of this global availability the contents of the workspace are conscious, while the rest of what is going on in the brain is not.

Baars' version of GWT uses explicit analogies with a theatre, and the idea of the stage being lit by the spotlight of attention, but Baars insists that his is not a Cartesian theatre and he is not a Cartesian materialist. To work out whether this is true it helps to realise that there are two fundamentally different ways of interpreting GWT's proposed relationship between consciousness and global availability.

The more tempting interpretation is that when information reaches the global workspace and is broadcast then something else – something special – happens. Then, and only then, does the information become the "contents of consciousness" or turn into subjectively experienced qualia or in some other way *become* conscious. This interpretation leaves all the familiar problems in place, yet is undoubtedly the more popular.

The alternative is that being globally available simply *is* what we mean by being conscious. That is, information having access to verbal report or to other forms of behaviour is all there is to consciousness. This interpretation entails no dualism but for most of us it is difficult to accept – we *feel* that consciousness is something *more*. Dennett explains that on this alternative view the hardest part to understand is that global availability does not cause some further effect "igniting the glow of conscious qualia, gaining entrance to the Cartesian Theater, or something like that." "Those who harbour this hunch are surrendering just when victory is at hand." (Dennett, 2005, p. 134). Consciousness is like "fame in the brain", he says, or "cerebral celebrity"; fame is not something *in addition* to being well known and nor is consciousness.

Based on his original version of GWT, Baars urges us to adopt the method of "contrastive analysis". That is, he urges neuroscientists to compare the same activities or perceptions when they happen consciously with the same events when they happen unconsciously. We should, as he puts it, "consider comparable conscious and unconscious events side by side" (Baars, 1988).

This, I suggest, is a huge mistake. The proposal rests entirely on the supposition that there really is a difference between conscious and unconscious events (or perceptions or actions or feelings or thoughts). It may seem peculiar, if not downright bonkers of me to deny this, but I do deny it.

The same applies when the distinction is made between events going on in the brain. Since Baars first proposed this approach in the 1980s, the field of consciousness studies has progressed extraordinarily fast and the most popular experimental approach has become the hunt for the "neural correlates of consciousness" (NCC). The basic idea is to take a comparable conscious and unconscious event and look for neural correlates using EEG, functional MRI or other forms of brain scanning, to see what the difference is. This has proved enormously productive in the sense that we now know, for example, that when someone looks at an ambiguous figure and their conscious perception flips from seeing it one way to the other, no changes are seen in the early parts of the visual system but there are changes in higher visual areas (Lumer, 2000). This tells us where the processing that leads to verbal report is going on but what does it really tell us about consciousness?

I suggest that this whole enterprise is based on fantasy – on a form of unworkable Cartesian materialism – because looking for brain correlates of consciousness assumes a difference between "conscious" and "unconscious" events; between those brain events that "give rise to" consciousness and those that do not; between those neurons firing that produce (or create or give rise to) qualia and those that do not (Blackmore, 2010/2011). It is as though the hard problem has been shifted so that it applies only to some brain events and not others. Yet all brain events entail the same kinds of processes – waves of depolarisation travelling along axons, chemical transmitters crossing synapses, summation of inputs at cell bodies and so on. What could it mean for some of these to be "giving rise to" or "creating" conscious experiences while all the rest do not? If the hard problem really is insoluble or meaningless then shifting it to apply only to some brain events does not help at all. This is why I refer to this distinction as the "magic difference".

It seems to me that both the most popular theories and the currently most popular research programs are based on this false distinction and ultimately must fail. My purpose here is to explain why, and to think about how we might proceed in a different way.

A First Person Science?

All the problems I have discussed above really come back to one problem: the temptation to think of consciousness as something other than the workings of a complex brain, body and world – to think we have to solve the problem of "consciousness *itself*". Indeed the very noun "consciousness" tempts us into thinking of consciousness as something independent.

This temptation and its consequences for understanding the mind underlies what is probably the greatest split between theorists working on consciousness. Daniel Dennett divides them into the "A team" and the "B team" with himself as captain of the A team (of course), and David Chalmers as captain of the B team. The distinction began with a disagreement over whether there can be a first-person science of consciousness. Chalmers argued that consciousness is a scientific problem quite unlike any other and requires a special kind of first-person science in which we collect first-person data. For him, and for others including John Searle (1997), first-person data (our own private subjective experiences) are irreducible to third-person data (our actions or the things we say about experiences). By contrast Dennett argues that there can be no first-person data. All we can ever do is observe what we and other people do and say about experiences. Science is intrinsically a public, shared activity and there can be no such thing as a special first-person science relying on first-person data.

Although the argument began this way, it nicely captures a fundamental split between theories of consciousness (Blackmore, 2010/2011). For the B team, including John Searle, Thomas Nagel, Joseph Levine and perhaps Jeffrey Gray, consciousness is something separate from the processes of perception, learning, memory, and cognition. There really is a "hard problem" that is distinct from the "easy problems"; if all the "easy problems" were solved and we really understood how learning, memory, perception and cognition worked, there would still be the problem of "consciousness *itself*".

For the A team, including Patricia and Paul Churchland, Andy Clark, Douglas Hofstadter, Kevin O'Regan, Alva Noë and many others, consciousness is not separate from all these processes. If ever we thoroughly understood perception, learning, memory, and cognition we would also know all we needed to know about consciousness, for it is not something separate from them. There is no such thing as "consciousness *itself*" and no "hard problem".

Those in the B team not only agree that we need a new kind of science to study consciousness, but think that studying third-person data leaves something out, that Mary the Colour Scientist learns something new when she steps out of her black and white room (Jackson, 1982; Ludlow et al., 2004), and that zombies are possible (Dennett, 1991; Chalmers, 1996). They agonise over the problem of the evolution of consciousness because they must find a function for consciousness *itself* apart from all the other processes and adaptations that did evolve (Blackmore, 2010/2011).

By contrast the A team tends to think that no special kind of science is needed, that nothing will be left out in a future third-person science of the mind, that Mary learns nothing new when she emerges from her room, and that falling for the "Zombic hunch" (Dennett, 2001) is understandable but wrong-headed, for zombies (though easy to imagine) are impossible. There is no special problem surrounding the evolution of consciousness because whatever consciousness is, it necessarily evolved along with all those functions and adaptations that did evolve for a reason.

I am firmly behind the A team. Given all I have said above it should be clear that I do not think that consciousness can be something separate from the workings of our brains and bodies in their complex environment. I do not think there is such a thing as consciousness *itself*, and I do not think the idea of a first person science of consciousness makes sense. Yet I am still deeply perplexed by consciousness. This appearance of a world when I open my eyes, this flood of thoughts and ideas that appear out of nowhere and seem to stream through my mind, this self who seems to be the subject of these experiences; what are they? I cannot (yet?) see how understanding perception, learning, memory, and cognition can explain all this.

My perplexity has only been increased by my years of Zen practice, which have taught me how to sit still and experience phenomena as they arise. I would like to suggest that the problem is so difficult that we might usefully go right back to the beginning and take a fresh look at subjective experience. This is the role that I believe disciplined methods of first-person exploration can play. The whole problem of consciousness concerns subjectivity, or "what it is like to be". So perhaps a serious attempt to look into "what it is like to be" might be useful.

Note that I am not here siding with the B team and suggesting a first-person science of consciousness. I do not think there can be any such thing because science is intrinsically a public activity. We all have to try things out, suggest theories, test them, carry out experiments, criticise those experiments and then agree, or not, on what we find. I am, however, suggesting a role for first-person methods in a science of consciousness. That role is a limited but potentially important one. It may allow us to gain a clearer picture of the phenomena we are trying to explain and from that to challenge some of the basic assumptions on which the current science of consciousness, with all its apparently insuperable problems, rests.

ASSUMPTIONS

The assumptions I have come to challenge can be roughly summed up by some of the most common phrases used in the literature of consciousness. These are:

The contents of consciousness
The stream of consciousness
The unity of consciousness
The neural correlates of consciousness

Elsewhere (Blackmore, 2009) I have spelled out further assumptions but I think these four, apparently innocuous, phrases reveal all the problems I have found in my personal explorations of consciousness. I will therefore describe how I have set about my investigations and then explain why I think all of these phrases are deeply misleading and need to be abandoned (or at the very least re-interpreted) if we are to make progress.

METHODS

If the aim is to explore consciousness through disciplined first-person methods, which method should we use? There have been many notable attempts in the past, including the notoriously failed introspectionism of the late nineteenth century, the methods of phenomenology based on the work of Edmund Husserl in the early twentieth century, and various currently more popular varieties of phenomenology (see e.g. Gallagher, 2007; Stevens, 2000; Thompson and Zahavi, 2007). The closest to my own approach is possibly the work of Francisco Varela who used meditation as part of his discipline of neurophenomenology (Varela and Shear,1999).The major problem facing these attempts is that each explorer can proclaim their own discoveries to be right and other people's to be wrong – their own minds to be typical and others' aberrant – a problem tackled in different ways by these various disciplines.

I make no such claims, for my purpose is different. I shall merely describe what I have done, what it seemed to me to reveal, and how what I found is relevant to some of the common assumptions I listed above. Others may decide whether or not they agree with me that this could or should have any impact on our current science of consciousness.

The method I used was meditation in the tradition of Chan Buddhism (the Chinese precursor of Japanese Zen). Since the term "Zen" is far better known, and the methods are very similar I shall refer to this as Zen throughout.

My training in Zen began in the mid 1970s when I attended meditation classes first in London, and then in Bristol with John Crook, a lecturer in psychology at the University of

Bristol and a Zen teacher. I went on my first Zen retreat at the Maenllwyd, John's farmhouse deep in the mountains of mid-Wales, during the exceptionally cold winter of 1982. Subsequently I have attended roughly one week-long retreat a year, although in some years I have done more, as well as several shorter weekend or day retreats. I have also worked with other Zen teachers including Reb Anderson and Stephen Batchelor. I began regular daily meditation practice in 1986. Although I have been training in Zen for more than thirty years, and to a lesser extent in Tibetan practices, I have never taken any vows or joined any Buddhist group. I do not consider myself to be a Buddhist (but see Crook, 2009).

In 1997 I did my first solitary retreat at the Maenllwyd, basing my routine on typical Zen retreats but with rather more sleep allowed! This means spending most of the day in half-hour sitting meditation sessions with ten minute breaks for walking meditation or exercises, as well as time for preparing food, doing jobs around the house or outdoors, and a walk in the hills each day. With no lectures, interviews, or ceremonies this amounts to rather more meditation than on a typical group retreat and obviously silence is easy when there is no other human being within miles.

In 2002 John initiated a new kind of "koan retreat". Participants are given a list of koans (short Zen stories or questions) and asked to choose one which they then work with for the whole week. The idea is to keep the koan constantly in mind, whether sitting in meditation, washing up, walking or doing anything else. Koans are not meant to be questions to be answered but are more of a stimulus to the mind's revealing itself. I attended two such retreats, working respectively on the koans "There is no time, what is memory?" and "When is this?" (Blackmore, 2009).

For more than ten years I taught a third-year course on consciousness first at Bristol University and then at the University of the West of England, Bristol. During these courses (the UWE course was 24 weeks long) I set the students homework each week which was to ask themselves a given question as many times as they could every day (Blackmore, 2010/11). I did the homework myself along with them and we discussed what happened at the start of the next week's lecture. I have also given many public lectures in which I have asked people these and similar questions.

When writing a book about Zen questioning (Blackmore, 2009) I did several short solitary retreats of a few days, both at Maenllwyd and at home, concentrating on questions including: Am I conscious now? What was I conscious of a moment ago? Who is asking the question? and How does thought arise? All these have contributed to the findings I wish to

Observations

"Am I conscious now?" appears to be too simple a question to provide much enlightenment but when I began giving it to students as their first week's question I quickly learned that it can have strange and interesting effects. Students told me that when I asked them "Are you conscious now?" they felt almost as though they were waking up, or becoming *more* conscious. They naturally began to wonder what was going on *before* they were asked the question, leading them on to ask "Was I conscious a moment ago?"

This second, apparently simple, question typically provokes two contrary reactions. One is "Yes, I must have been conscious because I am awake, alert, thinking and feeling, and I know I have been like that since I got up this morning." The other is "No, I can't have been conscious because when you asked me the first question it felt as though I was waking up, or becoming conscious in a way that I was not a moment before. Something changed."

How can we resolve these two apparently contradictory findings, bearing in mind that if the subject herself says she does not know the answer we have no independent way of checking? This is where looking into the darkness seems to be the appropriate analogy – the darkness is that of the immediate past moment.

Over many years I set both the students and myself two further questions, variants of the first two. They are "What am I conscious of now?" and "What was I conscious of a moment ago?". These have interestingly different effects from each other. Typically, if you ask about "now" what happens is that you latch onto some feeling, perception or thought and are sure that you are conscious of that. When your attention switches to something else you assume that the "now" has also moved on, so this readily gives rise to the sense of a "stream of consciousness" the idea being that at any moment there is something or other (or several things) that you are conscious of and these change as time flows on. So there is always something or other in the stream of consciousness.

Asking "What was I conscious of a moment ago?" has quite a different effect. When you first ask this question the answer does not seem to be too difficult. Something or other comes to mind as what you were thinking about or feeling a moment ago, and it is easy to imagine that this was the content of the flowing stream of consciousness back then. However, deeper inquiry throws this simplicity right out. A telling event can occur if someone else asks "Were you conscious of …. (that clock ticking, the hum of the air conditioning, the birds singing outside, the loud drill over the road)? What can happen then is that you listen for that sound only to get the distinct impression that someone or something had been listening to it for some time. You can remember the sound going along before the question was asked.

Once again the question provokes two contrary reactions. One is "No, I hadn't noticed that sound until you pointed it out to me, so obviously I *was not* conscious of it." The other is "Yes, now that you mention it I can remember how it sounded a little while ago, as though I, or someone, had been listening. So I *was* conscious of it (or someone or something was)."

Because of these peculiar effects I set myself the task of meditating systematically on this question. I did this in several solitary retreats at home in Bristol, where I spent approximately six hours a day in sitting meditation and some further time in walking meditation or mindfulness. I began each day with an hour or so calming the mind and then allowed the question to arise. With much practice the question would pop up from time to time and I would then watch what happened. I should add though, that it is difficult to write about this clearly because I find myself writing "I watched" when it might be more accurate to say that watching occurred. Doubt about who or what is watching arises naturally within this kind of practice.

The task of looking backwards in this way has very odd effects. At first it may seem as though there is a unified self looking back into immediate memory to answer a meaningful question about the past. After much practice this feeling dissipates and the effect is more like being several observers or several streams of experience at once. Indeed it seems that whenever I ask the question "What am I conscious of now?" there is only one answer – this. But when I ask the question "What was I conscious of a moment ago?" there are several answers.

Here is a typical example of an exercise practiced many, many times. I am sitting still with a calm mind, looking gently at the grass in front of me and aware of the walls of the hut I am sitting in, and the sounds of the birds around me. Now the question pops up "What was I conscious of a moment ago?". I look back, as it were, to a moment ago. The grass and birds remain but what else? It feels as though my mind is opening up to a myriad threads that reach back into "a moment ago". With surprise I note the drone of the traffic, which seems very obvious now I have noticed it. I can remember the sound stretching back into the past and even recall an especially loud lorry, now gone, as it laboured up the hill. It occurs to me that had I not asked "What was I conscious of a moment ago?" this brief sound would never have been remembered. Almost simultaneously with this I realise that I (or someone?) has been listening to the gentle sound of my cat purring at my side. Indeed now that my attention has turned to her, I feel her warmth against my leg, and that too seems to have been going on for some time. I can even sense a memory of her having slightly shifted position just a moment ago. And there is more: the feel of my bottom on the stool, that slight ache in my knee, the insect crawling across the grass (surely I have been watching that since it emerged into my field of view some time ago) and then, with some horror, I realise I am breathing. I can look back as though I have been watching the slow in and out of air through my nose as my chest rose and fell but I know that if I had not asked the question all this memory would be gone. Was I really conscious of any of these a moment ago?

All this takes a long time to write but happens in a fraction of a second. The sensation is as of realising that there were multiple backwards threads of sensation and perception. All seem to have been *conscious* in the sense that I can now recall them as though I had been aware of them at the time. All seem to have been *unconscious* in the sense that it took this question to provoke any memory of them, and when they were evoked they seemed to have been going on quite separately from the grass, hut and birds that I was aware of in the "now" before I asked the question.

There is a powerful tendency to grasp onto one of the backwards threads and call it "me", and to imagine that it is part of "my" stream of my consciousness, but this tendency weakens with time. Maintaining this peculiar state means continuously letting go of any temporary stream of experience and any temporary self or observer that arises with it. Indeed it is all about letting go.

And who is letting go? This natural question is, of course, just another aspect of the same old tendency to construct a self. A little trick I have used when this happens is to remember that this brain and body is a complex system doing multiple parallel things at once with no central controller or experiencer. Rather than there being anyone inside who is letting go, there is just a complex system learning how *not* to grasp onto one stream at a time or to construct the idea of a single experiencing self.

When practicing in this way, the sense of time and space sometimes changes so that streams of perceptions, thoughts or images seem to arise in their own time and space without reference to any others or to any underlying pre-existing time and space. I have not come to this very often, and I imagine that others may be more practiced at it than I am. From this extraordinary state the threads can then be gathered together again, reconstituting the sense of a single self experiencing particular things ordered in time. Normality is restored.

I have described here the consequences of asking just two main questions and their simplest variants. Elsewhere I have described the effects of several other questions (Blackmore, 2009), including more about self, space, time and action. Nevertheless I think these two provide sufficient basis for my intention here, which is to challenge some common assumptions and sketch out a new way of thinking about consciousness. I am very well aware of the limitations of claiming anything at all from one person's explorations. Yet I think it worth pointing out the possible implications of looking back into the darkness in this way.

IMPLICATIONS

These observations suggest a different way of looking at the problem of consciousness based on distinguishing between two different states of mind, only one of which creates the appearance of the hard problem. We might want to call these the "thinking about consciousness" state and the "ordinary state", but I think they are better described as the "self-reflexive state" and the "scattered state".

Self-reflexive Mind

Just occasionally something special happen. Although most of the time our minds are a scattered mass of barely interconnected ongoing processes, at these special moments some of the many threads are gathered together along with a model of a self who is experiencing them. When this happens it seems obvious that there is a self experiencing some things and not others. This change may take an intellectual form as when we ask ourselves "Am I conscious now?" or start wondering about the problem of consciousness or the nature of qualia or self. If we ask "What am I conscious of now?" one or more of the ongoing processes can be chosen to provide an answer. Whatever we do in this state, whichever way we direct our attention, we are sure that there is a self who is subjectively experiencing certain contents of consciousness. This is because a temporary self has indeed been constructed, and some of the streams are available to this constructed self while others are not.

In this state there seems to be a magic difference between conscious and unconscious processes; there seems to be a self who is separate from the conscious processes; and there seems to be a duality between the subjective world and an objective world. In other words, it is in this state that all the familiar problems of consciousness seem troublesome.

I suggest that this state is not a common state of mind for most people – or even for philosophers and consciousness researchers. Indeed it may happen rarely and last only a short time. Yet it causes all the trouble.

Scattered Mind

Most of the time our minds are not in this asking-about-consciousness or self-reflexive state. They are scattered. We go about our lives without worrying about the nature of consciousness, while our complex bodies and brains do lots of things at once; seeing, hearing, thinking, walking, talking, calculating, making decisions about what to do next and so on and on. If we had fabulously high resolution scanners we would be able to see the underlying neural activity of thoughts, perceptions and actions. We would see whole systems operating more or less independently of others (although nothing in the brain is completely independent of anything else).

For example, there might be streams of visual and auditory information leading to accurate walking over rough ground while other streams sustain a conversation with a friend. There might be streams of tactile information leading to controlled grasping of a cup of tea while other streams maintain body posture and yet others process the sights and sounds of a television programme. We might see the evidence of circles of repetitive thoughts, sudden ideas flickering and fading out, tiny rushes of activity in response to sounds too faint to be noticed or those blocked by attentional mechanisms.

There is nothing here to cause any concern about "consciousness itself" and no need for any worry about dualism, the magic difference between conscious and unconscious processes, or the hard problem. All these appear only when we flip into the self-reflexive state.

Yet those who are searching for the NCCs will ask which of these many streams is *really* "conscious" or which make up the "contents of consciousness". I suggest that this is a mistake, and there is no point in asking these questions because they have no answer. Dennett (1991) reached similar conclusions through philosophical argument, but I came to this conclusion by looking back into the darkness of the immediate past. This seemed to me to reveal lots of backwards threads of experience that appeared only because I looked for them. I could not say which ones I was conscious of and which I was not. If I could not say then who or what could?

From Scattered to Self-reflexive

Most people, most of the time, have scattered minds and do not think about self or consciousness, but there are many events that can provoke the switch to a self-reflective state in which both self and consciousness seem real. These can be quite ordinary events, not just difficult questions like "what am I conscious of now?". They include those that give rise to the familiar sensation that something has just come into my consciousness or that I have just become aware of something I was previously not aware of.

Here is a simple example. Suppose you are sitting comfortably by the fire, reading a book with your favourite music on in the background. As you become engrossed in the story the music suddenly changes pace and your attention is drawn attention to it. You have the distinct impression that the music has "come into consciousness".

What has really happened here, and what needs explaining? If someone could look inside your brain they would see all sorts of complicated streams of activity including two especially extensive and stable ones. I have used the word "streams" here but we might equally call them "coalitions" (Koch, 2004). Neither word quite does justice to them. Imagine a great spreading, branching, ever-changing octopus (or multipus, or mega-bush) of activation.

One corresponds to your reading, taking up parts of visual cortex, language areas, and other parts processing the meaning of what you are reading. The other is another large octopus evoked by the music, infiltrating not only auditory cortex but various emotional areas too, and even reaching out to the autonomic system, hormone levels and other bodily effects. These two may go on for some time, occasionally overlapping and influencing each other but mostly independently. This basic pattern might go on for five minutes, or ten minutes, or even longer, but eventually will break up. Here are two possible directions the change may take.

First, let us suppose that the story in the book reaches a particularly gripping point. The first octopus grows stronger and more extensive, invading the emotional parts of the brain, taking over influence of the autonomic system and spreading into new parts of the visual cortex as you visualise the dramatic events of the unfolding story. The other octopus grows weaker and its tentacles and their influences shrink. Then the music changes and you are surprised – sufficiently surprised to wonder what is going on and to ask yourself about consciousness and what you were aware of.

Now the two previously independent streams come together, along with new processes modelling a self as experiencer. From the perspective of the much larger book-reading stream, it seems as though you have just noticed the music. You might think "I was conscious of reading my book but suddenly the music came into my consciousness". That's how it feels. As soon as you construct a sense of self or start thinking about consciousness that is the story you will tell.

Alternatively, let us suppose that the music is coming to a particularly lovely and moving passage, while the book is getting a bit boring. The music stream grows in power and extent while the book-reading stream shrinks. Suddenly something draws your attention to the book. If you think about consciousness at that point you may say "Oh, I was enjoying the music so much that I forgot all about the story but then suddenly I became conscious of it again."

What has happened in the brain during what seem to be these changes in consciousness? Here is the critical point I want to make. I have described, in crude outline, all we need to know to answer that question. If we could see in great detail the two shifting coalitions we would not see one "becoming conscious" or one giving rise to consciousness or creating qualia. Such shifts happen all the time, often with multiple streams. Only occasionally do we also start thinking about consciousness or asking ourselves "what was I conscious of a moment ago?" On those rare occasions a new stream or octopus starts up, using verbal and self-modelling structures, constructing the idea of a self who was conscious of one stream and not another.

All these streams have neural correlates – or perhaps it would be truer to say that they *are* neural processes, for there is no duality here. And which of the streams was really conscious? This question makes no sense. Consciousness was only an attribution made after the fact by the self-reflexive processes. If we go back to the time before you wondered about whether you were conscious of the story or the music there is no answer to the question which was "in consciousness". There were just two streams using up more or less of the brain's resources and then a third that compared them. That's all.

Note that I have described this example as though when the critical question was asked the answer depended on the size or strength of the two streams. I think the decision may be made on other grounds, depending on what provoked the questioning. This may be close to what Dennett means by the effects of different probes. As he puts it "there are no fixed facts about the stream of consciousness independent of particular probes." (1991 p 138)

The important point here is that whenever you ask yourself "what was I conscious of a moment ago?" or have the experience of something "coming into consciousness", it is natural to think that something called "consciousness" has changed or that things that were previously unconscious have now become conscious. But consciousness is no more than an attribution made at that time. If you ask "but which was *really* conscious?" there is no answer.

I suggest that when we have methods for looking at brain activity with sufficient resolution in both space and time, we will be able to see all these processes occurring, including those associated with asking questions about consciousness. Then we will no longer ask such silly questions as "which process was *really* conscious?", "which was *in* consciousness?" or what were the NCCs? The duality between conscious and unconscious processes will have disappeared and with it the magic difference. In addition we will understand how the illusion of consciousness comes about. The curious stream, theatre and container-like qualities of consciousness will then make sense.

Conclusions

Consciousness is still a mystery, largely because dualism creeps into almost every attempt to explain subjective experience. Given the subject matter – subjectivity itself – it is understandable that people make assumptions based on looking into their own experience. Yet these assumptions may be false given a curious property of introspection on consciousness.

Inquiring into the nature of one's own consciousness in the present moment produces quite different answers from inquiring into the past. In the present moment consciousness seems to be unified. There is one conscious self experiencing the contents of consciousness. As the inquiry continues those contents change giving the appearance of a stream of consciousness. From this perspective the task ahead seems to be to explain the difference between the contents of the stream of consciousness and the rest of the brain's unconscious processing.

By contrast, looking back into the immediate past reveals no unity of either self or contents and no coherent stream. Rather there seem to have been multiple backwards threads, continuing without reference to each other and with no unified self experiencing them. From this perspective there is no mystery because there are no contents of consciousness and no difference between conscious and unconscious processes or events.

This suggests that the difference between conscious and unconscious events is an illusion created by introspection into the present moment. Since most people are not introspecting this way much of the time, or indeed ever, the science of consciousness is built on false premises.

References

Baars, B.J. (1988) *A cognitive theory of consciousness.* Cambridge, UK: Cambridge University Press.

Baars, B.J. (1997). In the theatre of consciousness: Global workspace theory, a rigorous scientific theory of consciousness. *JCS,* 4, 292-309.

Blackmore, S.J. (2002) There is no stream of consciousness. *Journal of Consciousness Studies,* 9, 17-28

Blackmore, S. (2009). *Ten Zen questions*. Oxford, OneWorld.

Blackmore, S. (2010). *Consciousness: An introduction*. (2nd Ed.). London: Hodder Education [2011, New York: Oxford University Press].

Chalmers, D.J. (1995). The puzzle of conscious experience. *Scientific American,* Dec., 62-68

Chalmers, D.J. (1996). *The conscious mind.* Oxford, UK: Oxford University Press.

Chalmers, D.J. (2010). *The character of consciousness.* Oxford, UK: Oxford University Press.

Churchland, P.S. (1996) The Hornswoggle problem. *Journal of Consciousness Studies,* 3 (5–6), 402–408.

Crick, F. (1994) *The astonishing hypothesis.* New York: Scribner's.

Crook, J.H. (2009). Response of a Zen master. In S. J. Blackmore, *Ten Zen questions.* Oxford, UK: OneWorld, pp. 166-174.

Damasio, A. (1999) *The feeling of what happens: Body, emotion and the making of consciousness.* London: Heinemann.

Dehaene, S. (Ed.), (2002) *The cognitive neuroscience of consciousness.* Cambridge, MA: MIT Press.

Dennett, D.C. (1991). *Consciousness explained.* London: Little, Brown & Co.

Dennett, D.C. (2001). The fantasy of first person science. Debate with D. Chalmers, Northwestern University, Evanston, IL, Feb 2001 http://ase.tufts.edu/cogstud/papers/chalmersdeb3dft.htm.

Dennett, D. (2005) *Sweet dreams.* Cambridge, MA: MIT Press.

Gallagher, S. (2007) Phenomenological approaches to consciousness. In M. Velmans and S. Schneider (Eds.), *The Blackwell companion to consciousness.* Oxford, UK: Blackwell, pp. 686–696.

Jackson, F. (1982) Epiphenomenal qualia, *Philosophical Quarterly*, *32*, 127–136.

James, W. (1890) *The principles of psychology*. 2 vols. London: MacMillan

Koch, C. (2004) *The quest for consciousness: A neurobiological approach.* Englewood, CA: Roberts and Co.

Ludlow, P., Nagasawa, Y. and Stoljar, D. (2004). *There's something About Mary: Essays on phenomenal consciousness and Frank Jackson's knowledge argument.*, Cambridge, MA: MIT Press.

Lumer, E.D. (2000) Binocular rivalry and human visual awareness. In T. Metzinger (Ed.), *Neural correlates of consciousness.* Cambridge, MA: MIT Press, pp. 231-240.

O'Regan, J.K. (2011). *Why red doesn't sound like a bell.* New York, Oxford University Press

Searle, J. (1997). *The mystery of consciousness.* New York: New York Review of Books

Stevens, R. (2000). Phenomenological approaches to the study of conscious awareness. In M. Velmans (Ed.) *Investigating phenomenal consciousness.* Amsterdam, The Netherlands: John Benjamins, pp. 99–120.

Thompson, E. and Zahavi, D. (2007). Phenomenology. In P. D. Zelazo, M. Moskovitch, and E. Thompson (Eds.) *The Cambridge handbook of consciousness.* Cambridge, UK: Cambridge University Press, pp. 67–87.

Varela, F.J. and Shear, J. (Eds.), (1999). *The view from within: First–person approaches to the study of consciousness.* A special issue of the *Journal of Consciousness Studies.* Also in book form, Thorverton, Devon; Imprint Academic.

Zeki, S. (2001) Localization and globalization in conscious vision. *Annual Review of Neuroscience*, *24*, 57–86.

In: Consciousness: Its Nature and Functions
Editors: Shulamith Kreitler and Oded Maimon
ISBN 978-1-62081-096-5

Chapter 8

BODILY AWARENESS, CONSCIOUSNESS AND MODIFICATIONS OF EXISTENTIAL VALUES IN OUT-OF-THE NORM EXPERIENCES: THE TRANSPLANTED PATIENTS' PERSPECTIVE*

***C. Piot-Ziegler*†**
CerPsa (Centre of research in health psychology), Institute of Psychology,
Faculty of Social and Political Sciences, University of Lausanne, Switzerland

ABSTRACT

Certain extreme situations, either positive or negative, associated with the course of transplantation, modify the experience of the body and/or of *bodily awareness*. In the context of extreme medical conditions (e.g., intensive care), *consciousness* of another nature is sometimes described as being positive and not only negative. For most patients, these experiences of illness are accompanied by a deep, long-lasting *existential reappraisal,* and modification of *existential values.* New knowledge and technologies enable to present some of these experiences in the form of objective images. These images prove their reality. However, they do not account for what the patients have experienced.

In this text, the discourse of the transplanted persons is presented as research data and as a basis for further theoretical and research developments.

Transplantation saves lives. Similarly to other difficult and sometimes invasive treatments, in the case of other severe illnesses (e.g. cancer), transplantation leads the patients and their significant others on difficult paths. Patients have much to teach us with regard to how they experience illness and treatments. In transplantation, *bodily awareness* and levels of *consciousness* are challenged and modified, whereby the consequences of the existential transformations are difficult to evaluate. The interviews performed in the framework of a longitudinal qualitative research project, contribute to a better understanding of intense *out-of-the-norm* physical and emotional experiences.

* The terms patient and person will be used without differentiating them.

† E-mail: Chantal.Piot-Ziegler@unil.ch.

1. Introduction

You must experience chaos to give birth to a star.
F. Nietzsche[1]

Certain extreme situations, either positive or negative, associated with the course of transplantation, modify the experience of the body and/or of *bodily awareness*[2]. In the context of extreme medical conditions (e.g., intensive care), *consciousness*[3] of another nature is sometimes described as being positive and not only negative. For most patients, these experiences of illness are accompanied by a deep, long-lasting *existential reappraisal,* and modification of *existential values.*

These experiences should be contextualized in terms of non-normative and non-judgmental perspectives (Piot-Ziegler et al., 2005; Piot-Ziegler, Fasseur and Ruffiner-Boner, 2007; Piot-Ziegler 2011).

2. Relation to the Body and Bodily Awareness in the Course of Transplantation

The crisis accompanying severe illness provokes a rupture in the familiar world of the ill person. This rupture has been described by many authors (Von Weizsaecker, 1939; Dutot and Lambrichs, 1988; Amiel-Lebigre and Gognalon-Nicolet, 1993; Deschamps, 1997) in the context of cancer, and in the context of severe illnesses (Santiago-Delefosse, 2002).

It is not new that medicine pays attention to bodily experience, and to the experience of the patient. It is one of the foundations of medicine centered on the patient, and was the medical approach of Hippocrates or much later of Balint (1957). The ill person is set in the center of the physician's preoccupations, where bodily experiences are important. However, in the reality of health care, and with time-limited consultations, physicians often have insufficient time to dedicate to the bodily experiences of their patients (Sicard, 2002).

The experienced body (in the phenomenological meaning, which is to be differentiated from symptoms listed or investigated in an usual medical anamnesis) is often distant from the medical description, and encompasses other significations, requiring for its expression a different language, not only physically, but also emotionally.

Starting from the perspective of patients throughout the course of transplantation, and taking examples from a longitudinal qualitative research[4], we shall develop the following aspects:

[1] Thus spoke Zarathoustra, 1883.

[2] We refer to De Vignemont's publication (2011) in using the term awareness.

[3] We refer to Kokozka's very detailed analysis (2007) in using the term consciousness, we however use it with a more general meaning. Both terms bodily awareness and consciousness will be defined later in this text, starting from the perspective of the patients.

[4] Project "Ajouts corporels" 2002, funded since 2003 by the project IRIS 8A, Health and Society, University of Lausanne, Switzerland. Transplantation: a qualitative longitudinal study following patients since their registration on the waiting-ling until 24 months after transplantation, direction of the project, C. Piot-Ziegler.

- The challenge of body integrity (2.1) and the modifications of bodily awareness during illness and treatments;
- The positive or negative modifications of consciousness or in bodily awareness (2.2) associated with extreme or temporary experiences (e.g., in intensive care or continuous care units);
- An opening toward another form of consciousness (2.3) with the transformation of time and space references, and out-of-the-norm experiences of unconditional love, spiritual, mystical or near death experiences[5];
- Finally, the long lasting questioning and modification of existential values (2.4).

2.1. The Challenge of Body Integrity and the Modification of Bodily Awareness

Even when the medical treatment of an illness follows its course without complications, the conditions and treatments focus the attention of the ill person on the diseased part of the body or the non-functional organ, strengthened by the confrontation with more complex technical instruments, and more specialized medical acts. When the patient follows particular treatments and is referred to different specialists, each of them deals with a specific body part or aspect of treatment.

Medical care, treatments and body modifications (arthroplasty, mastectomy-reconstruction or transplantation) also modify *bodily awareness*. The resulting transformations and their acceptance take place not only on physical, physiological or functional levels but also on a psychological level (Piot-Ziegler et al., 2010; Demierre et al., 2011).

This situation can lead the patient to experience the feeling of *being estranged* toward one's body. This feeling *of strangeness* toward the body is described in the psychosomatic literature (Freud, 1919; Raimbault, 1992). Freud uses the German word *unheimlich*: Being exterior to, strange, and different. In this perspective, the body has sometimes to be re-discovered. The continuity between *heimlich* and *unheimlich* was discussed by Freud, who emphasized the continuum between the two terms (Freud, 1919; Piot-Ziegler et al., 2009; Piot-Ziegler, 2011).

In the context of surgical incorporation of implants or transplants, the process of acceptance (integration) in the replacement of a body part, leads to challenges with regard to identity and body integrity. The following question arises: How are these different, *strange*[6] elements integrated physically and psychologically? (e.g., a prosthesis, a breast implant, or an organ coming from of another person; Zdanowicz et al., 1996; Piot-Ziegler, Sassi, Raffoul and Delaloye, 2010; Demierre, Castelao and Piot-Ziegler, 2011)?

Challenges to body integrity are described by sensations and feelings related to the loss of a body part or of a native organ, to the decrease in functionality of a diseased part of the body (articulation, tumor, organ) or even when experiencing pain (Schilder, 1950; Fisher, 1986; Demierre et al., 2011). Can some of these sensations be considered as phantom limb's pain,

[5] Near death experiences' scientific literature is not explored in this chapter. We shall not use terms such as wholeness, mindfulness, as they do not correspond, in our understanding, to the descriptions of the patients.

[6] An element not belonging to the native body.

which has drawn scientific interest for many years, and which is still a subject of theoretical and research developments (Weinstein, 1969; Fisher, 1986; Ramachandran and Blakeslee, 2002; MacLachlan, 2004)?

In organ transplantation, *bodily and organ awareness* does not appear only after transplantation; it is already present as illness progressively modifies the body functioning or when valves or pacemakers are implanted. A heart recipient describes the sensations of his heart at different stages of illness, before and after organ transplantation:

> ... the valve saved my life (...) Now, what is weird and what I have to get used to, I mean, I was living with this noise all time (...) You could hear it before (...) In a close space, in bed for example, or in the car. So I was living with this noise. Now I do not have this noise anymore and it is weird. So sometimes, I tell myself, is it going well? (...) And otherwise, when I feel tired, like in the evening, I feel it more. In a way, I... yes, I hear it but not in the same way (Colin[7], E2, six months after heart transplantation, p.11, l.20).

Cardiac and lung patients describe the relation between emotional states and the decrease of vital functions. The exhaustion and functional slowing down of the organ(s) generate symptoms, which are difficult to deal with because for the patients they mean life-threatening danger:

> I had still anxieties, yes. So, yes, it is normal, yes. It did not vanish. On the contrary, it became more and more serious I would say (...) To feel that you are fainting... and to tell yourself that the heart is worn out. And that you cannot breathe correctly anymore, hem (...) I had some sort of panic attacks, in fact, yes. You could say that (...) But in the end, I could, well, I could, sometimes I could handle that, because I knew it was at the end (of the episode). But one time or twice, or even twice or three times, I could not overcome them (...) fear for death (...) As if you, you feel really bad and that you are going to die. (Cyrile, E2, six months after heart transplantation, p.3, l.3)

For patients waiting for a lung transplant, mastering breathing is of great concern. Shrinking of pulmonary capacity is described along with the consequences of the progressive congestion of the lungs. Difficulties in breathing generate great anxiety:

> As soon as I make a physical movement, a little too violent, I loose energy, and I know I will pay for it later. I know it. So, if I climb two steps too many, after, I am folded in two and I look for my... I am searching for some air and I panic because, there are, two-three seconds, where it will become worse, and I know that it is, it is very difficult to bear in the sense that I know what to expect, if you will. And sometimes, when I feel bad, I cannot even look at, for example, scuba diving broadcasts or something like that, I mean (...) Or to be in small... spaces, I know I have to take my breath well in advance, before, before going into a car, because it will be a little, like a panic attack, or before going to bed, I know when I am lying down, like that, that there will be a moment which will be difficult. So, I know myself pretty well, I mean, I know the postures that are convenient for me, positions that I have to take in order to spare myself, in order to prevent the... the attack. (Pauline, E1, waiting for lung transplantation, p.13, l.30)

[7] All names are pseudonyms. The first letter or the name represents the organ C-coeur (heart), P-poumon (lung), F-foie (liver) and R-rein (kidney).

A person due to receive a heart transplant describes his questionings concerning the future transformations he will experience after transplantation, and his anxiety about the unconscious and uncontrollable consequences of the organs' exchange:

> I am listening to a heart, which is mine. But after, inevitably, I'll listen to a heart, which will not be mine... which will be... that I'll have to appropriate to myself in some way... And I don't know... I don't know how to make this reflection, I don't know... the question I ask to myself, is that everything which is on the conscious level, what you can talk about, you can discuss, it is something I can relatively master, but how can you master what is unconscious, how to master that... (Conrad, E1, waiting for a heart transplant, p.14, l.15).

Questionings about body integrity and the modifications of the native body through transplantation may arise:

> Personally, I am born with the body God gave me, organs God gave me and for me, it was whole. And suddenly, it is as if a limb has been amputated and it is replaced by a splint... made with another organ, which does not belong to me, at the beginning, I felt as though a limb had been amputated. And now, I try to learn again to live with a... like somebody who has been amputated. Simply, you don't see it, except a wonderful scar, but you don't see it from outside, you feel it. (Félicie, E2, six months after a second liver transplantation, p.14, l.19)

During his intensive care stay, one patient is very abruptly confronted with the presence of the graft in his body. The *vision* of the radiographies (X-rays) of his new lungs immediately after transplantation, provokes an overwhelming emotional reaction:

> ... the most critical moment, was especially when I was in the intensive care unit therapy, when I was confronted for the first time with my lungs, new... my new lungs. (...) I recall very well, when I saw them on the opposite side of my bed, on... in the hall, all my radiographies. And... I saw perfect lungs, magnificent, compared to the souvenir I had from mine... And I asked the nurse: "But this radiography, over there, was it my radiography? My first radiography?" And she answered: "Yes, it was your first radiography, with your new lungs". So I asked her if she could give me my glasses, which were there on the table beside me, in order to see better, because I had blurred vision. I could not see a distance of 5-6 meters away. So when she gave me my glasses and that I could see for the first time my new lungs, I had a nervous breakdown, quite violent. I could not breathe anymore. It was... it was quite hard... (Paul, E2, six months after lung transplantation, p.2, l.10).

Not only are *physical sensations* (bodily sensations, perceptions, visceral sensations) modified, but also the *noises* (sounds) and with them, the way of *feeling*, *experiencing* (pulsations, rhythms) changes after transplantation:

> I do not experience the anxiety I had before (transplantation), it has vanished, and if you want, the fact of not having a pacemaker anymore and the defibrillator is perhaps positive, because now I tell myself, I am a normal person again (...) and well, it is my heart, which is... when I experience an emotion and that my heart does not take the pulse, that it stays at a stable rhythm (it was raising up when doing efforts or when experiencing intense emotions)...

Now it has become perfect, I am confident. Concerning cardiac pains, I do not experience cardiac pains (anymore). (Colin, E2, six months after cardiac transplantation, p.12, l.18).

For another transplanted person, these sensory modifications of the *functioning* of the transplanted heart are still present a long time after surgery. The correspondence between sensations, thoughts and representations is difficult to achieve. These bodily modifications trigger thoughts about the relation between different levels of consciousness, as a new link between conscious and unconscious levels:

I don't think it is consciousness, because in the end, when I forget about myself, it is always like that. Precisely, it is not on the conscious level. It is even... I have the impression, I can be in the process of doing anything, and that this thing remains always here. In a small part of my brain, which I, so to say, am aware of. But it is a small thing somewhere, it is not dull, it is not oppressive, it is not... But it is just there... it stays there and... how can I express that... there are moments when it is not there anymore, these are moments when you feel the heart functioning. The heart, yes the heart and the body that comes along with it (...) the former heart becomes more and more like an object, becomes a very technical thing, very... well you have to prepare yourself to set apart from it, so well that's it. Whereas, the one that is grafted, is more, is rounder (she laughs) well, I don't know how to express that. It is less... it is less a thing, it is more the awareness of a thing. (Claire, E4, 24 months after a heart transplantation, p.10, l.37)

Feeling the new organ is reassuring as it acknowledges its presence. These sensations are a proof that transplantation has taken place. The new graft is a friend as long as no rejection or a medical problem occur:

Yes I can feel it, it moves (the kidney graft), I had bowel problems, they had to reposition, everything has to adapt, it creates a slight physical strain, no pains, but sensations, something new, which was not there before, but which is there now (...) you always think about it, without really thinking about it (...) but so-to-say, you live with it, I, I live with this third kidney (...) I am happy to feel it, if I would not feel it, it would not be there, so it is better to feel it... as it is going well, it is a friend. (Raoul, E3, twelve months after transplantation, p.5, l.21)

New sensations and pain experienced after transplantation sometimes take on another meaning, namely, that of a new organ, which the transplanted person must become acquainted with and accustomed to.

Sometimes, it's (the kidney) moving. If I stay seated too long... I don't know whether it is the kidney. I talked about that with my physician. He told me "it is not the kidney you feel, it is the scar" because he (the physician) presses on it, but I do feel it a little. (Robin, E2, six months after kidney transplantation, p.12, l.44)

When a kidney (graft) is added to the two native kidneys (left in the abdomen for medical precautions, because of polycystic disease), the new grafted organ creates a *protuberance* that cannot be ignored, neither *visually*, nor by the modification of the *internal sensations*. The fragility of the graft (just under the skin) induces protective feelings and behaviors, sometimes anxiety, in performing certain activities (professional, sport, leisure activities, etc.), through fear of hurting or harming the grafted kidney, and reinforces the feeling of vulnerability and of possible threat of rejection:

> I know it is there, I try to protect it as much as I can, if I have somebody in front of me who wants to punch me, I won't protect my face, I'll protect my kidney, but it is... I don't know, before, I would rather protect the upper part of my body, now I protect the lower part... well, I think, it makes... it is something new in me and I try to protect it perhaps more now, because it has less defences perhaps compared to the rest of the body, it is just at the surface under the skin. But otherwise, I don't think too much about it. (Robert-Paul, E3, twelve months after kidney transplantation, p.13, l.4)

This modified *presence* (bodily awareness) of the new organ is described and reported by several transplanted persons.

A *mutual habituation* takes place at a *conscious* level but it also includes all the consequences of what cannot be controlled or only incompletely mastered:

- the *physical, rhythmic, and noise* modifications, occurring after the change of organ or addition of a graft;
- the *position* and *place* occupied by the new organ, the modifications of the soft and hard tissues surrounding the graft, with specific *sensations*;
- the *physiological* and *functional* transformations;
- the *emotional experience* accompanying transplantation;
- the possible psychological (or multiple) consequences of the *immunological modifications* (Kradin and Surman, 2000).

When the different sensations and experiences are acknowledged by medical professionals, their explanations are based on, and refer to, the changes in function, to the healing process of the surrounding tissues (soft tissues, muscles, bones), to the new position of the organ, as well as to the technical aspects of transplantation and the connecting of the graft into the recipient's body:

> … the physician who made surgery, I asked him, what is this stuff, transplantation ? And he told me it is like plumbers, they change a pipe, and I told him, oh so it is like a mechanics, so you change the water pump. And he answered "yes it is exactly the same thing". (Renaud, E4, 24 months after kidney transplantation, p.10, l.1)

2.2. Modified States of Consciousness[8], and Modifications in Bodily Awareness: The Experience of the Transplanted Person in the Intensive Care Unit

The stay in intensive care generates experiences and anxiety, which leave long-lasting traces in three-quarters of patients (Green, 1996). Invasive medical care and constant physical

[8] Ludwig, 1966, talks about ASC (Altered States of Consciousness, cited by Kokoszka, 2007). We prefer to use the term *modified states of consciousness*, as some of these modifications refer rather to increased sensory perceptions or exacerbated sensations than to altered or diminished states of consciousness.

surveillance provoke feelings of danger, adding to the unpredictability of an aggressive and unknown environment (Jeannet, 1986; Daffurn et al., 1994; Hewitt, 2020; McKinney and Melby, 2002; Jackson et al., 2007; Davydow et al., 2008).

The evolution of the DSM to the DSM-IV (Diagnostic and Statistical Manual of Mental Disorders) shows that new classifications and new diagnostic categories are constantly needed (Caplan, 1995; Caplan and Cosgrove, 2004). The category *delirium* due to multiple etiologies and the term *disturbance* in level of *awareness*[9] are integrated into new classifications in the DSM-V, and describe short-term altered states of consciousness, troubled consciousness, or cognitive modifications (DSM-V). These modified states of consciousness are associated with extreme medical conditions and/or drug-related intoxications or side effects.

The words used in the nursing, medical, somatic or psychiatric literature are informative regarding these extreme experiences in the intensive care units, reported mainly by patients who had cardiac surgery and/or artificial ventilation. In these contexts, the terms *delirium* (Erikson et al., 2002), *hallucinations* (Hudsmith and Navapurkar, 2001), *intensive care unit syndrome*, *depressive anxiety* or *psychotic episodes* (Winship, 1008; Hewitt, 2002) are used. *Anxiety* is associated with invasive care (intubation, mechanical ventilation, pain), with the use of hallucinatory drugs or with feelings of vulnerability and life-threatening situations (Hewitt, 2002; McKinley et al., 2002).

In extreme situations, when experiencing intense pain, psychological withdrawal is described: Feelings of out-of-body experiences or of dissociation (Fisher, 1986). Discrepancies between evidence-based medicine general practice or data, and the individual experience of the patient, show the difficulty in understanding and responding to pain (Cahana, 2005).

Research in the context of intensive care environment is focused on the diagnostic evaluation of these reactions (Bergeron et al., 2001; Eriksson et al., 2002). Various methods are used to gather information (questionnaires, interviews) and different types of evaluations (Daffurn et al.1994; Hewitt, 2002; Jackson et al., 2007). The difficulties in comparing research in this particular domain are due to the various methodologies used, inherent in the unexpectedness of the environment, depending on the emergency priorities of a demanding environment in the intensive care and continuous care services, and on the vital priorities (Jackson et al., 2007; McKinney and Melby, 2002). Interviews appear to be the most appropriate method to elicit the expression of such experiences, since questionnaires report a high non-response rate (Andrykowski et al., 2005; Roberge, 2007).

In the context of a longitudinal research, exploring the experience of patients throughout the course of transplantation (Piot-Ziegler et al., 2005; Piot-Ziegler and coll., 2007; Piot-Ziegler 2011), qualitative interviews were proposed to patients at different points in time (after registration on the waiting-list, six, twelve and twenty-four months after transplantation), exploring their experience at each step of the transplantation process. One of the 37 patients, (from a sample which included 12 heart-transplanted, 14 lung-transplanted and 11 liver-transplanted patients), when interviewed six months after transplantation, described an out-of-the-norm experience in the intensive care unit, after the tape-recorder

[9] Consulted versions of the DSM-V from February to October 2010 and in October 2011. With time, the terms consciousness and awareness have changed. In our text, the term a*wareness* was chosen for describing bodily awareness defined as changes in bodily perceptions and sensations (De Vignemont, 2011) and *consciousness* for referring to different levels of consciousness or modified states of consciousness (Kokozka, 2007).

stopped. This report encouraged us to explore systematically the intensive care experiences in the following interviews six months after transplantation. The following question was used: "Have you had dreams or nightmares during your intensive care stay?"

A large proportion of the transplanted persons of our study (59% among the 37 heart, lung and liver transplanted patients) reported positive experiences (32%), and/or negative ones (49%) during this period of their hospitalization.

Our interviewees talked about their experiences in the intensive care unit and emphasized the difficulties encountered in communicating these destabilizing experiences. Recall of these experiences was deeply moving, and all patients expressed positive consequences of being able to talk freely about them in a non-evaluative and non-institutional setting. All expressed concerns when confronted with the (apparent) irrationality of these experiences[10], the reality of these long-lasting images, and their associated sensations. All interviewees described feelings of stigmatization when confronted with the misunderstandings (fears or uneasiness) of their significant others or of professionals, enhancing the difficulty of integrating these experiences in their illness course and in their lives.

a. The Fight for Life

The fight for life is reported by nearly half (49%) of transplanted patients: Descriptions of war scenes, battle-fields, of being taken hostage, of being removed or feelings of *imminent danger* (six lung transplanted, six heart transplanted and five liver transplanted patients). Feelings and experiences of constraints, restraints, confinement (two heart transplanted patients and one lung transplanted patient), and imprisonment, being put in a coffin or in a glass box (two lung transplanted patients) were described as well as body transgressions and aggressions.

> So well, hem, yes, I have dreamt that my bed was transforming. I was all black, I was all black, I had, I could see I was all black. Or when they were doing X-rays, I, I was always worried because they were doing them when, when I was really in a coffin. I was at the cemetery. In a coffin. I was put aside like that, there, and hem, I had this kind of, and I had the impression to be a little ice cold, like that, to be cold. Because of being warm, I never had the problem of being warm. But cold, yes. And I can tell you it is, when I was falling asleep, this is it, I saw, I saw my bed transforming, I saw things in a depth perspective, things that are not re-, which are not real, which are not, how can I say that? (Patrick, E2, six months after lung transplantation, p.5, l.26)

b. Physical and Psychological Suffering

Each patient tries to find his/her own solution as a means to keep death at a distance or reduce *physical* (n=14; heart n=5, lung n=4, liver n=5) *and/or psychological suffering* (as an overwhelming anxiety: Heart n=6, lung n=4 anxiety, liver n=2). It is important to emphasize that liver transplanted patients are reluctant to take pain relievers or morphine, and that they are ready to endure pain rather than to take the chance of reviving a possible dependence on medication (n=4).

One lung transplanted patient resists the extracorporeal circulation machine. Other two patients fight against their own automatic responses and try to motivate themselves to *"go*

[10] We could not confront these experiences with medical files and data, such as events, treatments, etc. However, this does not change or set in doubt the experience of the persons who participated to our study.

along with the machine". One person emphasizes the psychological suffering accompanying the necessity of giving up to let the machine control breathing.

> It is the fact of having this tube, and not being able to speak, and it was itching. Well, at the beginning, I was unconscious because they let me sleep, I think a few hours with this tube (...) The machine compensates a little, it is different than when you have nothing, well! You feel, that, well, as if you were suffocating. Whereas, in fact, they give you air through this tube. (Pablo, E2, six months after lung transplantation, p.2, l.19)

Surrendering to the command of one of the most vital and automatic functions of the body is experienced as extremely difficult and painful psychologically. High anxiety is experienced with ventilation and/or oxygen removal (lung transplanted patients, n=11).

A heart transplanted patient expresses the impossibility in the following words:

> ... I was completely paralyzed. I could not, I saw my hand, I could not do, I sent the command, I wanted just to raise my arm and I could not move only one finger... So over there, they immediately explained that it was a muscular paralysis, that I should not worry, that it was not neurological that... But nevertheless, I was anyway a little anxious... (Conrad, E2, six months after heart transplantation, p.5, l.35)

Another heart-transplanted patient describes how in a terrifying war scene, he makes alliance with an eagle he protects when feeling strong, and who protects him in moments of weakness:

> There were dead people all around me, I could not move... and there was a bird, some kind of eagle who was protecting me. And me at night, I covered him with a cow skin, because he was cold, but it was a bird that at the slightest alert was there. (Constantin, E2, six months after heart transplantation, p.3, l.10)

In this context the term "suffering" is used rather than "pain" to account for the physical and psychological components of pain experiences reported by the patients. Body image is created though various sensations, through pain and motor control (Fisher, 1986; De Vignemont, 2011). These reports are disturbing, confirming that the care provided must still be improved (Cahana, 2005; Schiemann, Hadziakos and Spies, 2011).

A sense of the limits of body boundaries and of bodily awareness (Fisher, 1986), through massage or soft touch can positively influence the experience of the transplanted patient in intensive care units, especially when consciousness levels are altered. This could help in integrating the newly transplanted organ and in reconstructing a sense of body integrity and body limits[11] (Anzieu, 1974; Montagu, 1979; Fisher, 1986; Heinrich and Marcangelo, 2009; Henricson, Segesten and Berlund, 2009; Gallace and Spence, 2011).

The elicitation of adverse reactions by certain analgesics and pain relievers, such as morphine or benzodiazepines, resulting in nightmares and post-traumatic syndromes is under evaluation (Ethier, Martinez-Motta, Tirgari, Djiang, MacDonald et al., 2010; Schiemann et

[11] Although care of the skin to prevent bedsores is already present in the intensive care units as part of the medical prescription and in the protocols, we talk here about another way of performing massages, in devoting to it more time and in privileging the quality of touch and wellbeing.

al., 2011), and new treatment protocols are tested to minimize the negative effects of these substances, (e.g., alternating periods of sleep and wakefulness).

c. The Awareness of the Modification of the Body Integrity

Some patients report having the impression that the transplanted organ was set next to their body or positioned in the wrong place (reported by two heart transplanted, one liver transplanted, one lung transplanted patient).

> Yes, but small silly things, like they had closed my body and that they had forgotten my liver, one time, I remember I dreamt about that, that they had forgotten... in fact, they had just closed me and they had turned their heads, and it was lying there on... so it is not... I woke up abruptly. (Frank, E2, six months after a liver transplantation, p.6, l.14)

These are complex sensations, with visual components and awareness (here taken in a broad meaning of having some sort of knowledge) of the exchange of the two organs:

> And I saw my transplant (...) I think these are the effects of the narcotics, hem, of this so-to-say therapy of medications they gave me, hem at certain times and, and hem, I have, for me I thought that, I had, hem, hem, three surgeries, three transplants. (Cyrile, E2, six months after heart transplantation, p.5, l.35)

One heart transplanted patient had the impression that his new heart had been grafted in the upside down position. Sometimes, the confrontation with the organs' exchange is very abruptly experienced, as in the case of this lung-transplanted person, when he was confronted with the images of his new lungs compared to the memory of his previous diseased lungs (see citation of Paul, E2, p.2, l.10).

d. Other Sensations and Bodily Perceptions: Functional, Visual, Olfactory, Auditory, Vestibular, Temperature

Sensations relating to the recovery of the *functionality* of the grafted organ and the possibility of being able to breathe again more freely, are mentioned by the lung transplanted patients, for whom autonomous breathing is an extraordinary experience, although accompanied by much *anxiety*. The automaticity of respiration has to be rediscovered, and at the same time it is the quality of the respiration, its depth that is surprising for most of the lung transplanted patients. It also involves regaining trust in organs that were not functional before transplantation.

A heart transplanted person reports the feeling of having to physically respond to a *physiological production stress*: The heart and its functioning, its rhythm, the associated noises of the intensive care environment, generate numerous questions.

> These memories related to a kind of delusions from the intensive care, and they were in fact delirium, somehow bizarre, because they were related at the same time I think to the mechanism of the extra-corporeal circulation, as it is represented in my imagination and perhaps also to my body experience, I don't know, this you cannot know for sure, and related to the context of the gift, to receive something, to give something, I mean I had always some sort of nightmare, which I think was related to the noise of the water that was present with the drains hem, I imagined myself like some sort of pump, which had to produce a fizzy beverage,

> I laugh because it is the first thing I asked when I woke up and that I did not want to do anymore (...) it was at the same time good and at the same time frightful because I always had to produce... (Claire, E2, six months after heart transplantation, p.2, l.41)

These reports remain stable over time and do not fade in the memories of the transplanted persons. The related representations are not disconnected from the experience of transplantation, from the symbolic relation to the gift and to the donor, and probably not from events in the intensive care units:

> Yes, yes, yes... I must say, it was weird, it was something... I had the feeling, well, when I was in these complete states of sleepiness with this nightmare, that I was a drinking pump, that I had to produce for other people, that I did not know where they were and that my body in fact was producing this beverage that was going to Holland, don't ask me why (laughs), and, and, that it was at the same time agreeable, because it was something useful, and frightening because there was this kind of constraints to produce that, and it was not so evident. And besides, from my point of view, it was a dream I had when I was under anesthesia. And my cardiologist has told me "no this (...) if you had such a long lasting surgery". But I had the impression that it was when (...) On the surgical table because in fact, it was, it was, the surgeon who was there, in fact, it was something like during surgery, like that. What was around, this corpse, which was a machine, like that, and that was functioning. So, I don't know if it was... And I was completely obsessed. Because when I had this dream, I was completely obsessed by the noise of thick things. And this, when they took off this, I felt at once better. But they did not take it off right away. (Claire, E4, 24 months after a heart transplant, p.8, l.36)

The functional modifications of the new organ, as well as the technology of health care, the body sensations or the associated representations confront the patient with disturbing experiences.

The loss of control of vital functions (cardiac and respiration), or of the ability to move (see citation of Conrad E2, p.5, l.35) provokes a high degree of anxiety.

Sensations of *aspiration*, of *spirals*, or of being drawn up into a *black hole* or a *light* are also described. They can be related to vestibular modifications and/or to medications. Other modified sensations of *temperature, noise, vision* and *odors* are also reported:

> I was very cold or I was hearing glug-glug. Oh yes I recall, it was something terrible. As if I was in the sea hem it was very loud. And I told them, don't you hear? And they were answering no. But I was telling myself that they were deaf, it is not possible. Or I was very cold, I was very, very cold. And they were telling me, but no, but no, it is not cold. And I remember one time there was a nurse, she came with a spray. And she told me, well you smell things so we shall spray the room, and after it was worse for me, because the odors were dreadful. Well it was difficult. Anyway (...) Your are overall completely disconnected, and you have the impression, you cannot find somebody who understands you, because you are completely out of phase, it is a fact that the people who perceive the reality as it is, well they cannot be on the same wavelength of what you live. And this, is absolutely terrible, so, you end up burying yourself in silence in taking dis(tance) hem (...) That's it absolutely, yes absolutely. In saying, yes, there is a huge problem. A terrible discomfort, a deep discomfort that makes you completely withdraw into your shell (Pauline, E2, six months after lung transplantation, p.8, l.41)

e. The Gift, the Debt: A Transaction, an Inheritance to Justify, to Pay or to Protect

Four persons (two heart transplanted, one lung transplanted, one liver transplanted) talked about dreams, in which monetary transactions with the donor took place, an inheritance to justify or to preserve, or the necessity to *pay for a purchase* (food) or even being *blackmailed.*

The sense of *being indebted* is already present, and appears very early in the mind of the patient. One heart transplanted person reports with humor a sentence that came to his mind during the troubled period of continuous care:

> My heart is a gift, but no inheritance, so I won't have any fiscal tax to pay. (Claude, E2, six months after heart transplantation, p.10, l.36)

f. The Presence of the Donor and Emotions Accompanying this Encounter

During the intensive care and in particular during continuous care stays, several patients describe experiencing the *presence* of the donor (three heart transplanted and one liver transplanted patients):

> ... there were a lot of dreams also, where hem, I was thanking ... well, I was talking to somebody, my donor in fact, I think, but without ever seeing him (...) During the intensive care. And then during this very period, impossible to talk about my donor without bursting out into tears (...) I knew I would be grateful to this person, but to this extent, I was afraid of myself... (Frank, E2, six months after liver transplantation, p.3, l.3)

Sometimes the anxiety towards the emotional atmosphere of the experienced *relationship* to the donor is reported:

> *Transplanted person:* Dreams, dreams that, that I dreamt... what did I dream?
> *Spouse:* I think a lot about the donor.
> *Transplanted person:* My new heart. Hem... It seemed, it was as if I knew this person, and after, sometimes, I dreamt it was somebody I did not like, you see. (Cesar, E2bis, seven months after heart transplantation, p.5, l.48)

Another heart transplanted person describes an *ambiguous encounter* with his donor, and also his presence with a smile and an *"image without image"*. Overall, the intensity and the strength of this encounter were reported:

> ... and I have a dream when I am in the intensive care, and then I have some kind of regression, I am lying down, in the intensive care and I have a whole lot of apparatus etc... and he is standing in front of me, at the foot of the bed and he is looking at me, smiling. But a smile that I... that it is difficult, it is, I have experienced it like an ironical smile, a little, not as an agreeable smile or gentle. Not sarcastic either or bad, or hem. But I don't know, there was some kind or irony in his smile (...) So, there was a young man at the foot of the bed, that I was seeing, and when thinking back about it, I am unable to tell you whether he had black hair, or whether he had blond hair, whether he had blue eyes. I cannot describe him. So it was an image like that (...) but at the same time, this image is empty. There is no (...) yes, there is no content. I cannot describe you the shape of his face, whether he was handsome, nothing (...) and however, he is very strong, he is very present. He is here and he has some kind of smile, a little, a little ironical, like that. So it is not really a nightmare, but it makes you uneasy. (Conrad, E2, six months after heart transplantation, p.12, l.33 and p.14, l.1)

Some transplanted persons keep thinking about the donor and his/her *family* along with acknowledgement of the *grief* accompanying the gift: An emotion difficult to deal with and the need to *create a bond*:

> And then, I started, in some way, a little to talk with hem the donor (...) And then, after, I have, I thought also to his family (...) I, I thanked them, so to say, to have left this organ for me, and then, I told them I would try to pray, to have a thought for them in my prayers, for them to accept better the grief of their child, hem, the fact that that they had the kindness of... And I must say, that it has gone well. I never had a nightmare, nothing at all (...) But I must say that I talk to him very often. (Colette, E2, six months after heart transplantation, p.8, l.21)

2.3. An Opening Towards Another Form of Consciousness: Unconditional Love (Transcendence or Person), Light (Mystical or Spiritual Experience), Near-death Experience (NDE)

Some patients (32%) talk at some point during their stay about positive experiences, of an indescribable and intense wellbeing. Some associate these experiences with a *turning point* at the occurrence of a *choice between life and death,* both frightful and extraordinary. These experiences are accompanied by a *light* of various colors (blue, orange, yellow, white, a rainbow, even paradoxically a black light), but always *dazzling* (three lung transplanted, four heart transplanted and one liver transplanted person):

> I saw a light... I have never seen something like that, it must remain secret (tears come out of his eyes), but I think, it was something like the beyond, it is unbelievable. It is not possible, because it is a light, I have never seen a light like that, I almost have tears coming to my eyes. Dazz(ling), you see, as some kind of door (...) So the door, has closed, I saw only the light, and I hear like voices (in the ICU).
>
> And I think sometimes about that, in spite of the sorrow (for the donor), I tell myself, you see again something, each time that I need something (...) always at the right time (...) I think sometimes about that. I am a believer. There is something. There is a support... It (the light) is always there at the right time (...) my wife has told me: don't tell that to anybody. (Pierre, E2, six months after lung transplantation, p.5, l.16)

Sensations of *aspiration* and of *spiral* or of *siphon* are reported. Loved or deceased persons, positive emotions, personal, symbolic, mystic or spiritual resources participate in contributing to the choice for life. What relates to life is a *thread*, a *bond,* sometimes a *hand* or a *loved one* (one heart transplanted person), a deceased person (one lung transplanted, one liver transplanted person), with whom it is possible to talk about philosophy or to communicate without talking, a grandmother or a deceased father, a loved dog at the lake bank.

One heart transplanted person finds himself in front of a wonderful woman. One liver transplanted person, when hospitalized for a hepatic coma before a second transplant, encounters an elderly woman with long white hair who closed the door and let her go. Situations where life and death coexist, and where deceased and loved persons come and reappear:

> I have been twice in the corridor, once unto the door, big white door, with a huge light. There was my mother, my grandmother, my friends, with a big white veil. It was wonderful, but everybody was telling no with their heads, so I went back at the bottom (...) it is difficult to find the words, but it was fairy, unimaginable, a whiteness... more dazzling and radiant than that... However, the faces, truly identical... (Chloé, E1, waiting for a heart transplantation, p.10, l.5)

One lung transplanted, one heart transplanted and two liver transplanted patients talk about an opening toward *another form of consciousness*, positive, extraordinary, but at the same time destabilizing and psychologically painful. One heart transplanted person describes how the experience he has lived through, has opened his mind and the upheaval experienced at the same time:

> ... I have more or less dreamt, that, well, I thought that it was at the beginning, that it was like a dream. And in fact, these were things that were really happening to me (...) so now, I am in the process of thinking about it all, and to analyze little by little, because it is still in a mess in my, hem, in my brain (...) I was all I had lived, it means, I was everything which is, the life cycle of, the, the imaginary experience, in an imaginary context, hem! It means, that I made the complete loop, and I saw my present. I saw my past. And I saw part of my future (...) I have experienced something quite important, good, good Lord, and I am conscious that I have mixed up a lot of, of, of, how do you call that, dreams and reality at the same time. Well, there were no borders any more (...) So my wife, at the beginning, hem, thought I was delirious. (Cyrile, E2, six months after a heart-transplantation, p.5, l.12)

For a lot of patients, these experiences were accompanied by a fear about sharing them, lest they would be considered mad. However, all discussed them freely in our interviews.

It is sometimes the fact of having *seen* death so close (or experienced it) that is the most destabilizing:

> The nights in general were a nightmare because I could not sleep. It was not (...) because I was all the time exhausted. I did not dare to close my eyes anymore, because it was frightening me. I could not visualize green spaces. I could not visualize even my children. I saw images which were not... they were like puzzles which were reproducing, I remember very well now. Doubled, tripled, quadrupled images. They were mineral images. Only mineral. So I was really, really scared form death at this time. I saw myself really in between the two of them (life and death). (Pauline, E2, six months after lung transplantation, p.3, l.47)

2.4. Forgetfulness or Persistence of these Experiences: The Relocation Syndrome, and Existential Questionings

After relocation in the continuous care unit, almost all patients expressed anxiety when falling asleep. This anxiety is described in the scientific literature as *relocation syndrome.* Research (Gardner and Sibthorpe, 2002) emphasizes that this anxiety is associated with *sleep disturbances* (frequent waking-up during the night), observed in three-quarters of patients. These disturbances last six months after hospital release (McKinley and Deeny, 2002; McFetridge and Yarandis, 1997; Moser and Dracups, 1996).

> I saw corpses of dragons with human heads I knew and that were falling on me, it was horrible (...) I don't stand morphine. Some (dreams) were very real, they remain very real, some of them have left deep traces on me. It is like my passage through the death tunnel as I say. Still now it remains very present. I cannot really describe everything I have seen, these are things that leave traces (Felicie talks about how she does not dare to take a morphine like substance - a pain reliever - at night). I don't go to bed right after, because I am afraid it could come back. I fight (not to fall asleep) but it is difficult. During the day, I can call for help. There is light (...) You cannot really describe, in fact, the fears you experience at that very moment. You can describe very precisely what you have seen, what has traumatized you, because it has traumatized you. There are no words to express the anxiety you experience; on the moment you are ready to scream. It is so deep. (Felicie, E4, twenty-four months after a second liver transplantation, p.25, l.9)

These experiences are often labeled *post-traumatic stress disorder (PTSD)*[12]. Many descriptions from transplanted patients do not correspond to these criteria (neither do they correspond to the CIM-10). To determine the presence of PTSD, follow-up is necessary. Despite the persistence and the importance of these experiences, and their association with intense emotions, they are not consistent with flashbacks[13]. For some, these experiences are paradoxically associated with positive emotional tone, although eliciting psychological suffering:

> It was, well, it was not frightful, well, they were there, they were eating fish (laughs). Finally it is funny. But there, I think it was induced by the anesthesia, this is for one, but no, when I was closing my eyes, I was, I blinked my eyes like that, and I was looking, hem, for example at television, and it made like, I saw like a flashlight, well, I don't know. And it happened again at the hospital, hem, when I had these 14 days there. I was closing my eyes like that. If I was looking at a light for example, not only television, but also a light, it was making tjou! A flashlight. So I don't know whether it was due to, I don't know, medications or what. (Frederic, E3, six months after liver transplantation, p.8, l.33)

It was not difficult for transplanted patients to discuss these experiences, but they feared misunderstanding. No patients reported a decrease in their reactivity to the surrounding world, but rather an increase in sensitivity or an *awakening of consciousness*, an *increase in emotional sensitivity* to nature, to authentic *existential values*, and *an opening* to *another form* or *spiritual level* of *consciousness.*

These data suggest the need for a comprehensive analysis to determine the relationship between these experiences and possible effects related to drugs' release (either regular or irregular) during treatments (Davydow et al., 2008; Kokozka, 2007). Some immunosuppressants may also generate neurological side effects (Lier et al., 2002; Zikowic et al., 2009; Penninga et al., 2010). Certain co-morbidities affecting brain function could have

---

[12] DSM-IV (F43.1[309.81]), criteria B, C, D, E and F.

[13] Two patients describe bright flashes, but their experiences do not correspond to the other characteristics of PTSD. They could also be induced by medication. More research explore the hypothesis of neurological, psychiatric impacts of some drugs used in the context of transplantation (Heinrich and Marcangelo, 2009; DiMartini et al., 2008).

an influence too. These symptoms must be clearly differentiated from other psychological symptoms and from PTSD (Ethier et al, 2010; Jackson et al., 2006).

The DSM-IV reports persistent avoidance or efforts to escape from thoughts, feelings or conversations related to the trauma or traumatic events (category D), but we found no evidence of this in our interviews[14]. The difficulty in studying such experiences must be underlined.

At this point, it is important to distinguish what has been described as:

- a fight for life, which seems to be associated with *extreme and negative experiences* (surgery, the awareness of having an open thorax, etc.) and experiences of *suffering*;
- from experiences described as *modifications of the bodily awareness*, or *modified states of consciousness,* and other forms or levels of consciousness. The latter are described in a positive tone although emotionally painful because of their intensity.

These *experiences of the extreme* question the psychological health of the person. The emotional and sensory reality is undeniable. What is certain is that these experiences lead the patients toward *existential and philosophical questionings.*

> He (the surgeon) told me: "there is something miraculous to see you like that because... We do master our technique... but there are certain things we cannot master"... It pleased me, he was referring to something which is beyond us, something... to a transcendence. And then you come to the point you talk about God, I mean, something that is beyond us, that we cannot understand and that's it... And after, on the psychological level, it has been hard, because long. (Conrad, E2, six months after heart transplantation, p.9, l.5)

The necessity of positive changes in life perspectives is also found in other illnesses or experiences with life threatening situations (Knaevelsrud, Liedl and Maercker, 2010).

Knowing that other people have gone through similar experiences makes this out-of-the norm reality become a reality that can be discussed and shareed. These experiences become acceptable for the person who has lived them. But also and most importantly, they become legitimate in front of their significant others, who have also experienced them through their own looking glass: As confused movements, hallucinatory episodes, words without significations, and also sometimes experiences beyond what is bearable to tolerate.

> *Transplanted person:* Yes I was wondering, but well, I have always told that is was morphine that caused everything, And apparently, I don't remember very well, but I had become very aggressive, I had a friend of my spouse who came to visit me and apparently, she just needed to tell me encouraging words to fight, for me to respond, apparently: "what do you know about what I am going through? Leave me alone, I don't want to see anybody". We come back a little to my personality, I mean, wanting to fight alone.
>
> *Researcher:* Do you have the impression that the persons around you have understood these behaviors?

[14] It is interesting to emphasize that historicity seems to influence the verbal reports and the content of the near-death experiences (but not for all), which seems to indicate that there might be an influence of the acceptability of such experiences due to cultural or societal norms (Athappilly, Greyson and Stevenson, 2006; Kokoszka, 2007).

> *Transplanted:* No, no! At this very moment, nobody has understood, even my spouse has not understood. Once or twice I had this violent reaction, and I am not a violent person, and once or twice, my brother has told me that she has been so deeply touched that he had found her on her knees in front of the door, crying. So, I have visibly hurt her in talking in this way, but sincerely, I do not remember that. (Paul, E4, twenty-four months after lung transplantation, p.4, l.44)

Finally, these experiences are legitimized in front of the health care professionals.

If these events that the patients have undergone are not integrated in their life-course so that they are given personal meaning, and contextualized as reactions to out-of-the-norm situations, they will very often retain a stigmatizing character for the transplanted person as well as for their significant others.

These experiences in intensive care units, and these descriptions must open the way to new research, in which the experience of the patients will systematically be aligned with medical events and with the medical treatments (medication, acts, care, etc.).

Could the medications given during this period disturb identity cohesion (Ramachandran and Blakeslee, 2002; Berthoz, 2003)? Do they open up a door towards another level of consciousness (Kokoszka, 2007; de Vignemont, 2011)?

PET-scan images have been compared to out-of-body experiences (DeRidder et al. 2007), grounding them in (objective) images. Nowadays experiences of de-corporation have found their physical explanations in voluntary or non voluntary modified states of consciousness, associated with the intake of substances or drugs, with neurological or medication-related modification, as well as with certain illnesses or pathologies (Blanke and Castillo, 2007; Kokoszka, 2007; DeRidder et al., 2007). These experiences are related to brain dysfunctions generating, in certain conditions, a delocalized body image (Blanke et al., 2004; Schwabe and Blanke, 2008).

For the persons who have lived through these experiences, it is illusory or unimportant to explain them rationally (InfoKara, 2004). Mostly, the transplanted persons wish their acknowledgement. These phenomena constitute the reality of the transplanted person in one phase of their course of illness, but they often generate deep, long-lasting existential questionings and transformations:

> True or not, this encounter is intimately part of my experience. In this perspective, it is indisputably 'true' for me. Its truth cannot be challenged (...) The somehow silly expression, "Today is the first day of what is left of my life", has received a deep meaning for me. It is not a grammatical construction anymore or abstract logic. The meaning of this sentence is deeply grounded in my experience. (Droz, 1996, p.43 and 46)

Distress is sometimes experienced in an extremely intense way, demonstrating the importance of an attentive follow-up (emotionally and psychologically) in the hospital environment during the post-transplantation period (intensive care and especially during the continuous care period for the transplanted patients) on the part of professionals. It is one of the most difficult times reported by patients:

> Fed up with everything, I did not want to see anybody... I wanted to jump out of the window (when he was in the continuous care unit). I would like to communicate the weight of what I have experienced. Not the physical modification, but on the psychological level... It is

> the acceptance, which is difficult. It is to have you put away, if on the physical level, things don't go well. It is as if I was bungee jumping... I am at the bottom... and how do I climb up again? I try to fill up my life. I must keep myself busy in a different way. They say they changed my heart... People cannot imagine... (Cesar, E2, six months after a heart transplantation, p.17, l.10)

Several transplanted persons have expressed a feeling of rupture or of crisis after their stay in the intensive care unit, one after a cardiac episode before transplantation occurred and two after transplantation (two heart-transplanted and one liver transplanted patients).

> I had a lot of weird dreams during this very period in the intensive care unit and then... well it stays in you mind, I don't know whether it is medication, surgery or all that together perhaps? it... Well, a lot of things have happened, at that moment, I changed a lot, and a lot of people did not like that... We'll say that they were surprised... Well, for one, the duration of illness, you see things differently, but after, yes there has been... like a big chasm (...) A big turmoil... complete... in the way of seeing things. To be with other people. Well... it changes life. (Frank, E2, six months after a liver transplantation, p.3, l.3)

Although experiences (positive or negative) do not always generate symptoms comparable to PTSD or lead to its development, their emotional intensity cannot be denied or ignored, and attentive follow-up needs to be available:

> I mean, the people who stay a long time in the intensive care unit have psychological reactions... out of anxiety... nightmares. They call it intensive care unit syndrome (...) So, depending on the nature of the person, if somebody is especially anxious, it can reach delirious proportions, even in a person who is not known to be particularly paranoid, well, it happens. It is, rather disturbing. It is a disturbing environment per se. And nobody comes out of it unharmed. (Conrad, E2, six months after a heart transplantation, p.29, l.13)

Experiences in intensive care units are partly due to medications, provoking hallucinations, and partly to invasive medical care. They are extensively described in the scientific literature along with the development of diagnostic tools and medical interventions (Bergeron et al., 2001; Eriksson et al., 2002; McKinney and Melby, 2002; Robergue, 2007; Ramachandran and Blakeslee, 2002; Berthoz, 2003; Blanke et al., 2004; Kokoszka, 2007).

PTSD is described as being accompanied sometimes with a personal positive evolution (*posttraumatic growth*), not occurring the same way in different situations or illnesses (Leugn et al., 2010; Andrykowski et al., 2005). When positive modifications are reported, they refer to a *positive and authentic manner* of considering the world and to a modified *relation to life and death:*

> (...) my (deceased) father... they were all smiling. They all seemed to be in peace, hem (...) Death. I was really afraid. Hem and now, I tell myself, no, it must be an enormous peace. Because the last, the last time I went away, I was going away in peace, because more I was approaching this light, more I felt in peace. So, hem, no, I would say, it does not frighten me anymore. For me suddenly, I tell myself, this is true, there must be life in the beyond hem, at least men who are in the beyond who are welcoming their significant others when they, they, they are ready to go across, in fact. Hem, he won't arrive there all alone in something, well, they are there, for, hem, there was my brother-in-law, who has committed suicide and who

had a face hem, like I have always known in fact, hem. And with a big smile in complete peace. Very, very calm. So, I have held on life, I must have had a strong will to live, and it was stronger than the call for peace. So, this is true that I, since the first transplantation, each time I stand up in the morning, I tell myself: "Hurray one more day! Well it is marvelous, hem, the sun is shining, well today it is raining, but it is beautiful anyway. I hear the birds, so it is marvelous." (Felicie, E3, twelve months after a second liver transplantation, p.17, l. 21 and 37)

Modifications in *existential values* and *priorities* are also mentioned. The confrontation with a severe illness, and here to transplantation, leads the transplanted person sometimes towards a higher level of *consciousness* or a lowered perceptive threshold, with increased sensitivity in regard to perceptions and sensations.

Yes all three days, one can admit! During which I had this (sigh), this wonderful sensation for me, it was a yes, I can tell you, a mind opening, hem, unimaginable. I have, I am still now, upset. And I felt I was drawn into this (...) enormous spiral. This is, my complete philosophy, which is in the process to go upside down. The meaning of life itself (...) And I must say that since this surgery, it has allowed me to make giant steps. Of giant. I have an understanding, an easiness. (Cyrile. E2, six months after a heart transplantation, p.6, l.42)

All these out-of-the-norm experiences place the person in the forefront of situations between rationality and irrationality, between life and death, between reason and madness, between joy beyond words and suffering. Existential theories provide a theoretical background for integrating the paradox in regard to its approach to human thought (Van Deurzen-Smith[15], 1997; Van Deurzen-Smith and Arnold-Baker, 2005; Piot-Ziegler et al. 2005). These theories consider that the human being lives with, and experiences simultaneously multiple truths, which may initially appear as contradictory.

Reker and Chamberlain (2000a) define the concept of meaning (action of giving a signification) as constituted from two distinct aspects:

- The *implicit* or *definitional meaning* referring to the personal significance to objects or life events;
- The *existential meaning* or *meaningfulness* referring to attempts to understand how events can be set in broader perspective (creating meaning, sense of coherence, sense of purpose; Reker and Chamberlain, 2000).

Reker (2000) integrates the process of transcendence and of transformation in this search for meaning, whereby *transcendence* consists of the process to rise above circumstances and reality, and *transformation* consists of enriching a given reality with new potentialities.

It remains an experience, which is deeply rooted inside of me (...) when I think about it, it is still very precise (...) It remains into my head more like a souvenir than like a (positive) dream (...) In fact, I associate that with the beginning of the transformation process (...) It is after these dreams that I told myself, there is something changing. (Frank, E4, 24 months after a liver transplantation, p.8, l.42)

[15] For a summary of different existential approaches.

3. Conclusion

It is important not to try to explain for the transplanted person an out-of-the-norm life-course, but to accompany him or her while it occurs, at the right time, and when needed, by emphasizing *continuity*, as well as the search for meaning; (Newmann, 2000; Santiago-Delefosse, 2002b; Piot-Ziegler et al., 2007; Leugn et al. 2010).

> It does me a lot of good to tell you this story. Because I'm rather a secret person. It is unavoidable, people question me, and I answer their question. I do not broach the subject too much. I avoid my colleagues... I don't like to tell about my life, to tell about my story. So the answers, are a little made up answers, stereotyped sentences. I talk to my wife, to my physician, he is the first person I talk to in an intimate manner, about what I experienced. With other people, I remain anecdotic... in order to protect myself... to keep some distance (...) But it does me a lot of good with you, to be able to do the contrary. Not to talk about ordinary anecdotes, but in fact about what I have experienced. (Conrad, E2, six months after heart transplantation, p.30, l.36)

These *out-of-the-norm* experiences and the patients' reactions confront the professionals with their own questions, their own fears, doubts, anxiety or suffering (Imperatori et al., 2001). Moreover, these experiences challenge professionals with regard to the care they provide (Schelling, 2008; Ethier et al., 2010).

> ... there was a doctor to whom I owe my life in some way, (...), who has not given up and who wanted to go on and who told: “no we cannot (...) we try one more, we try”, etc. And my wife was there, so the other ones did not dare in front of my wife (...) “we stop”. They have made a third one, but they did it reluctantly. But they told her afterwards, they could tell her, they have admitted in front of her, I mean, now we talk. She had very strong bonds with all this team who have taken her in their arms, I mean because at one point, there was a moment of extreme tension, they wanted to have her out (...) and she did not want to (...) she is a false gentle person, she is very, very, tough, she can, she can, afterwards she lets go, she cries like a little girl, etc. But on the moment, she can be of an extreme violence, I mean, she has resisted, she has confronted the physicians, I mean they wanted to have her out, but she stood steady in front of them completely equipped, also with a blouse and a cap and a mask, she stood steady in front of him (a physician) saying, he wanted her to step back, in saying 'you loose your time' (...) but she does not let people step on her feet. So of course, it created inside this team, great emotions. (Conrad, E2, six months after heart transplantation, p.14-15. l.18)

The body is where very intimate, complex feelings, sensations, sensuousness and suffering are experienced, grounded in the personal biography, in a form that is not always possible to communicate explicitly or to translate into words.

> It is hard, terribly hard. When you approach people, you have to talk to them very methodically, very specifically, the words you choose. The feelings you want to share. Because, if you do it literally, you damage it, you hurt yourself, and it is no use. Today, I have the opportunity to be able to share part of it. (Cyrille, E3, 12 months after a heart transplantation, p.52, l.5)

The transplanted persons have transmitted important events in their lives (regardless of the transplanted organ), without shame or reserve, sometimes being anxious to disclose for fear of telling banalities and of hurting themselves or of being misunderstood:

> ... for me, life, I am 57 years old, almost now. For me, life is not what we are building up now for people, it is not a goal anymore. I was bound to evolve, to follow-up technology, but this is not life, life... it is perhaps a banality what I say, but life is somewhere else... it is in the relationship with people, trust, the respect of donors, somebody... a different approach. (Roger, E4, twenty months after a kidney transplantation, p.11, l.14)

Intense emotions arise, while questionings about the value of life become more present. Thoughts about past experiences, when purpose of life was questioned, are placed in a different perspective, as transplantation has deeply modified the existential values of the transplanted person:

> I almost died because of something else... which did not have anything to do with hepatitis... However, I search for this death, it is unbelievable... How many times did I tell myself, if only I could disappear, die... suicidal attempts, I did so many in my life (...) I did not like myself, and when you don't like yourself, it is self-denial, your internal life, is a chaos, it is avoidance (...) Then you encounter a lot of different ways of being in the world, live, feel, see, and it is at this point that it becomes interesting, you don't feel alone (...) I need things which interest me (...) Now, I feel I am right, in my own truth. (Fiona, E1, waiting for a liver transplantation, p.12, l.12)

All transplanted persons know that life-duration of a graft is limited:

> Concerning my beliefs, they have not changed. Concerning my life, what was troubling me were the statistics, they (the physicians) were often reminding them to me and it was something disturbing. I told to the physician: "You'll see, I am part of the people who reach five years after transplantation!" And now, I come to that (...) it will be five years (since transplantation). However, it is true that regarding life, you know, I have always been a person who is a little sensitive, so it has reinforced the sensitivity I already had, when wanting to help other people, now I do it twice as much, but there is still this remaining thought, a little hazy in my head, from time to time fortunately: To tell yourself "well now I come to the five years, what will happen?" (Paul, E4, twenty-four months after lung transplantation, p.9, l.39)

One may wonder what it means for a person to be average, or to be in the (a) norm, whatever this norm may be? Does not the average, or the standardized human being (taken here in a statistical signification; Quetelet, 1835) miss the true or authentic human being? What is the meaning of statistics or norms when confronted with experiences generating such profound existential questionings, such intense emotions, and sometimes suffering, where the limits between life and death are so unforeseeable and narrow?

The images (descriptions, transpositions of the reality, metaphors[16]) and their interpretation belong to the patient. They represent real events and their symbolic expressions.

---

[16] Although they could be related to collective unconscious (Jung, 1953, 1964) or share common socio-cultural or religious backgrounds, they have individual meanings and are also context dependent and situated in historicity.

It has not been possible to compare these experiences to the care (medical acts) and to the treatments (medications) provided during intensive and continuous care stays, such as analgesics, pain relievers, immunosuppressant treatments and real events. These experiences are perhaps related also to the toxicity or side effects of certain treatments used during the hospitalization period (Heinrich and Marcangelo, 2009).

Reflexivity (in the sense of research reflection, thoughts, challenge of limitations) has led us to write these lines. We hope they will emphasize the importance of methodological and theoretical openness to this research field (Droz, 1984; Radley and Chamberlain, 2001; Anadòn, 2006), and to the value of what was feared by our interviewees to be banalities or isolated experiences[17]:

> Because the methodological reflection comes always behindhand compared to the effective practice of research, moreover, it does not solve the problems of research themselves or does not create tools, where impasse seems to be reached. (Droz, 1984, p. 28)

The progress of medicine goes faster than what is possible to imagine. Patients have much to teach us, regarding their physical illness, course of treatments and psychological reactions. Trying to better understand their experience will help define adapted support (Piot-Ziegler and Pascual, 2011). The context of our research has provided a protected space, where the persons we encountered had the desire to talk (Piot-Ziegler, Fasseur and Ruffiner-Boner, 2007; Piot-Ziegler 2011). They are the main observers and should be listened to.

> (...) at this moment you would feel a little frustrated, not being able to talk with somebody, because (in transplantation) questions arise: why, why? Because, it is just a new organ, so why does it change life like that (...) There are a lot of other things. I had a lot of pleasure to be able to share, and to be able to account for my experience, because for me, it is a testimony, because, well, it is extraordinary. (Roger, E4, twenty-four months after kidney transplantation, p.17, l.4)

In transplantation, *body awareness* and levels of *consciousness* are challenged and modified. The consequences of the following existential transformations are difficult to evaluate.

The most difficult for researchers is to pass on the experience of the transplanted persons, without betraying what has been given, without being counterproductive, with the purpose to better understand and make understandable intense *out-of-the-norm* physical and emotional experiences.

Acknowledgments

I dedicate this text to all the persons who have offered their experience as an invaluable present to my research, and who in the uncertain course of illness, have talked about the essence of life. I thank also:

[17] Car non seulement la réflexion méthodologique retarde-t-elle toujours d'un temps sur la pratique effective de la recherche, mais de plus, ce n'est pas elle qui pose et qui résout les problèmes de la recherche, et qui crée des outils, là où l'impasse semble être atteinte » (Droz, 1984, p. 28).

- The team IRIS 8A who has participated in the gathering of the citations presented in this text.
- Professor Manuel Pascual from the Center of Organ Transplantation (CTO) of the University Hospital of Lausanne (CHUV), Switzerland and the Coordination team.
- Professor Rémy Droz who encouraged me to tell difficult but true stories, and also to Professor Marie Santiago from the University of Lausanne (CerPsa, http://www.unil.ch/cerpsa), Switzerland, and Kerry Chamberlain, University of Massey, New Zealand.
- Professor Shulamith Kreitler, for the few words exchanged which helped me pose a wider perspective on my research data and for reviewing this manuscript, and who with Professor Oded Maimon from the University of Tel-Aviv Israël gave me the opportunity to publish this text.
- Elaine Sieff who helped with the translation. Language inaccuracies are certainly still present and are not attributable to the translator but to the author.

A summary of these data has been published in Piot-Ziegler (2011).

References

Amiel-Lebigre, F. and Gognalon-Nicolet, M. (1993). *Entre santé et maladie.* [Between health and illness]. Paris: PUF, Les cahiers de la santé.

Anadón, M. (2006). La recherche dite "qualitative": de la dynamique de son évolution aux acquis indéniables et aux questionnements présents. [The so-called "qualitative" research: From the dynamics of its evolution to the unavoidable present questions]. *Recherches qualitatives,* 26(1), 5-31.

Andrykowski, M.A., Bishop, M.M., Hahn, E.A., Cella, D.F., Beaumont, J.L., Brady, M.J., Horowitz, M.M., Sobocinski, K.A., Rizzo, J.D., and Wingard, J.R. (2005) Long-term health-related quality of life, growth, and spiritual well-being after hematotoietic stem-cell transplantation. *Journal of Clinical Oncology*, 22(3), 599-608.

Anzieu, D. (1974). Le Moi-peau. [The Moi-peau]. *Nouvelle Revue Psychanalytique,* 9, 195-208.

Athappilly, G.K., Greyson, B., and Stevenson, I. (2006). Do prevailing societal models influence reports of Near-Death Experiences? A comparison of accounts reported before and after 1975. *The Journal of Nervous and Mental Disease*, 194(3), 218-222.

Balint, M. (1957). *Le malade, son médecin et la maladie.* [The patient, his/her physician and illness]. Paris: Payot, rééd. 1996.

Bergeron, N., Dubois, M-J., Dumont, M., Dial, S., and Skobik, Y. (2001). Intensive care delirium screening checklist : Evaluation of a new screening tool. *Intensive Care Medicine*, 27(5), May, 859-864.

Berthoz A. (2003). *La décision.* [The decision]. Paris : Odile Jacob.

Blanke, O. and Castillo, V. (2007). Clinical neuroimaging in epileptic patients with autoscopic hallucinations and out-of-body experiences. *Epileptologie*, 24, 90-95.

Blanke, O, Landis, T., Spinelli, L., and Seeck, M. (2004). Out-of-body experience and autoscopy of neurological origin. *Brain,* 127, 243-258.

Cahana, A. (2005). Ethical and epistemological problems when applying evidence-based medicine to pain management. *Pain Practice*, 5(4), 298-302.

Caplan, P. J. (1995). *They Say You Are Crazy*. Da Capo Press.

Caplan P.J. and Cosgrove, L. (Eds.), (2004). *Bias in psychiatric diagnosis*. Oxford: Roman and Littlefield Publishing Group.

Daffurn, K., Bishop, G.F., Hillman, K.M., and Bauman, A. (1994). Problems following discharge, after intensive care. *Intensive and Critical Care Nursing*, 10(2), 244-251.

Davydow, D.S., Gifford, J.M., Desai, S.V., Needham, D.M., and Bienvenu, O.J. (2008). Posttraumatic stress disorder in general intensive care unit survivors: A systematic review. *General Hospital Psychiatry*, 30(5), 421-434.

Deja, M., Denke, C., Weber-Carstens, S., Schröder, J, Pille, C.E., Hokema, F., Falke, K.J., and Kaisers, U. (2006). *Social support during intensive care unit stay might improve mental impairment and consequently health-related quality of life in survivors of severe acute respiratory distress syndrome.* Critical Care, 1-12. http://creativecommons.org/licences/by/2.0.

Demierre, M., Castelao, E., and Piot-Ziegler C. (2011). The long and painful path towards arthroplasty: A Qualitative Study. *Journal of Health Psychology,* 16(4), May, 549-560.

De Ridder, D., Van Laere, K., Dupont, P., Menovsky, T. and Van De Heyning, P. (2007). Visualizing out-of-body experience in the brain. *The New England Journal of Medicine*, 357(18), 1829-1833.

Deschamps, D. (1997). *Psychanalyse et cancer: au fil des mots, un autre regard.* [Psychoanalysis and cancer: Following wordings, another perspective]. Paris: L'Harmattan.

DiMartini, A., Crone, C., Fireman, M. and Dew, M.A. (2008). Psychiatric aspects of organ transplantation in critical care. *Critical Care Clinics*, 24(4), 949-981.

Droz, R. (1996). *Mon cerveau farceur. Récits lacunaires d'une aventure banale.* [My mischievous brain. Lacunary narration of a banal adventure]. Vevey: Editions de l'Aire.

Droz, R. (1984). *Observations sur l'observation.* [Observations on observation]. In M.P. Michiels (Ed), L'observation. [Observation]. Neuchâtel : Delachaux et Niestlé, Textes de base en psychologie.

Dutot, F. and Lambrichs, L.L. (1998). *Les fractures de l'âme: du bon usage de la maladie.* [Soul fractures: Making good use of illness]. Paris: Robert Laffont.

Ethier C., Burry L., Martinez-Motta C., Tirgari S, Jiang D., McDonald E., Granton J., Cook D., and Mehta S., for the Canadian Critical Care Trials Group (2011). Recall of intensive care unit stay in patients managed with a sedation protocol or a sedation protocol with daily sedative interruption: A pilot study. *Journal of Critical Care*, 26(2), 127-132.

Eriksson, M., Samuelsson, E., Gustafson, Y., Aberg, T., and Engström, K.G. (2002). Delirium after coronary bypass surgery evaluated by the organic brain syndrome protocol. *Scandinavian Cardiovascular Journal*, 36, 250-254.

Fisher, S. (1986). *Development and structure of the body image*. London, Hillsdale, New Jersey: Lawrence Erlbaum Associates, Vol. 2.

Freud, S. (1919). *Das Unheimliche*. [The Uncanny]. In : Gesammelte Werke XII, 229-230, Trad. Française : L'inquiétante étrangeté et autres essais. Paris: Gallimard, 1985, pp. 213-214.

Gallace, A. and Spence, C. (2011). Touch and the body: The role of the somatosensory cortex in tactile awareness. *Psyche*, 16(1), 30-67.

Gardner, A. and Sibthorpe B. (2002). Will he get back to normal? Survival and functional status after intensive care therapy. *Intensive and Critical Care Nursing*, 18, 138-145.

Green, A. (1996). An exploratory study of patients' memory recall of their stay in an adult intensive care unit. *Intensive and Critical Care Nursing*, 13(5), 243-248.

Heinrich, T.W. and Marcangelo, M. (2009). Psychiatric issues in solid organ transplantation. *Harvard Review of Psychiatry*, 17(6), 398-406.

Henricson, M., Segesten, K., and Berglund, A.-L. (2009). Enjoying touch and gaining hope when being cared for in intensive care - A phenomenological hermeneutical study. *Intensive and Critical care Nursing*, 25, 323-331.

Hewitt, J. (2002). Psycho-affective disorder in intensive care units: A review. *Journal of Clinical Nursing*, 1, 575-584.

Hudsmith, G. and Navapurkar, V.U. (2001). Hallucination related to artificial ventilation in the prone position. *Anaesthesia*, 56(11), 1116.

Imperatori L., Gachet C., Eckert P., and Chioléro R. (2001). Dons d'organes et transplantation: qu'en pensent les soignants ? [Organ donation and transplantation: What do health care professionals think about it?] *Revue Médicale Suisse*, 628: publié le12/12/2001.http://revue.medhyg. ch/article.php3?sid=21788.

INFOKara (2004). *Des mots sur des maux: Témoignage de Monsieur M*. [Words on suffering]. 19(1), 24-29.

Jackson, J.C., Hart, R.P, Gordon, SM, Hopkins, R.O, Girard, T.D. and Ely, E.W. (2007). *Post-traumatic stress disorder and post-traumatic stress symptoms following critical illness in medical intensive care unit patients: Assessing the magnitude of the problem.* Critical Care, 11:R27. http://ccforum.com/content/11/1/R27

Jeannet, M. (1986). *Une expérience dans une unité de soins intensifs: De l'angoisse à l'agression.* [An experience in an intensive care unit; From anxiety to aggression] Connexions, 47, Intersubjectivité, 9-107.

Jung, C.G. (1953). *Métamorphoses de l'âme et ses symboles*. [The Metamorphoses of the soul and its symbols]. Geneva, Switzerland: Georg Editeur SA.

Jung, C.G. (1964). *Man and his symbols*. New York: Doubleday.

Knaevelsrud, C., Liedl, A., and Maercker, A. (2010). Posttraumatic growth, optimism and openness as outcomes of a cognitive-behavioral intervention for posttraumatic stress reactions. *Journal of health Psychology*, 15(7), 1030-1038.

Kokoszka, A. (2007). *States of consciousness. Models for psychology and psychotherapy.* In C. E. Izard and J. L. Singer (Ed.), Emotions, personality and psychotherapy. New York: Springer Verlag.

Kradin, R. and Surman, O. (2000). *Psychoneuroimmunology and organ transplantation* : Theory and practice. In P. Trzepacz and A. DiMartini (Eds.), The transplant patient. Biological, psychiatric and ethical issues in organ transplantation. Cambridge: Cambridge University Press, pp. 255-274.

Leugn, Y.W., Gravely-Witte, S., MacPherson, A., Irvine, J., Stewart, D.E., and Grace, S.L. (2010). Post-traumatic growth among cardiac outpatients: Degree comparison with other chronic illness samples and correlates. *Journal of Health Psychology*, 15(7), 1049-1063.

Lier, F., Piguet, V., Stoller, R., Desmeules, J., and Dayer, P. (2002). Neurotoxicité des inhibiteurs de la calcineurine et analogues. *Médecine et Hygiène*, 60(2382), 744-748.

McFetridge, J.A and Yarandis, H.N. (1997). Cardiovascular function during cognitive stress in men before and after coronary bypass grafts. *Critical Care Nursing*, 46, 188-194.

McKinley, S. and Deeny, P. (2002). Leaving the intensive care unit : A phenomenological study of the patients' experience. *Intensive and Critical Care Nursing*, 18, 320-331.

McKinley, S., Nagy, S., Stein-Parbury, J., Bramwel, M.,and Hudson, J. (2002). Vulnerability and security in seriously ill patients in intensive care. *Intensive and Critical Care Nursing*, 18, 27-36.

McKinney, A.A., and Melby, V. (2002). Relocation stress in critical care: A review of litterature. *Journal of Clinical Nursing*, 11, 149-157.

Montagu, A. (1974). *La peau et le toucher. Un premier langage*. Paris : Edition du Seuil.

Moser, D.K. and Dracup, K. (1996). Is anxiety early after myocardial infraction associated with subsequent ischemic and arrhythmic events. *Psychosomatic Medicine*, 58, 395-401.

Newman F. (2000). *Does a story need a theory? Understanding the methodology of narrative therapy*. In D. Fee (Ed.), Pathology and the postmodern: Mental illness as discourse and experience. Inquiries in social construction, ch. 12, pp. 248-263.

Nietzsche, F. (1883). *Also sprach Zarathurstra*. [Thus spoke Zarathustra]. Berlin: Ernst Schmutzer Verlag.

Penninga, L., Moeller, C.H., Gustaffson, F., Steinbrüchel, D.A., and Gluud, C. (2010). Tacrolimus versus cyclosporine as primary immunosuppression after heart transplantation: Systematic review with meta-analyses and trial sequential analyses of randomised trials. European *Journal of Clinical Pharmacology,* 66(12),1177-1187.

Piot-Ziegler, C. (2011). Le vécu des transplantés aux soins intensifs : regard normatif ou un autre regard sur les hallucinations ? [The experience of transplanted persons in the intensive care units: Normative perspective or another perspective on hallucinations]. *Annales Medico-Psychologiques*, 169, 361-366.

Piot-Ziegler, C. and Pascual, M. (2011). *Prise en charge psychologique dans la transplantation d'organes.* [Psychological support in onrgan transplantation]. Encyclopédie Médico-Chirurgicale, Elsevier Masson, SAS, Paris, Psychiatrie, 37-405-A-30. [Available on-line].

Piot-Ziegler, C., Sassi, M-L., Raffoul, W., and Delaloye, J-F. (2010). Mastectomy, body de-construction and impact on identity: A Qualitative Study. *British Journal of Health Psychology*, 15(3), 479-510.

Piot-Ziegler, C. and collaborateurs (2009). *Préoccupations et questionnements existentiels des personnes en attente d'une greffe d'organe: une approche qualitative.* [Concerns and existential questionings while waiting for organ transplantation]. Ouvrage collectif sous la direction de Marie-Jo Thiel (Ed.), Donner, recevoir un organe. Temps fort avec les deuxièmes journées internationales d'éthique de Strasbourg en 2007. [Donating, receiving an organ]. Presses Universitaires de Strasbourg, pp. 185-203.

Piot-Ziegler, C., Fasseur, F., and Ruffiner-Boner, N. (2007). *Quel(s) espace(s) de parole(s) dans la maladie grave ? En attendant la transplantation*. [What space(s) for disclosure in severe illness?]. Cahiers de Psychologie Clinique, 28, Numéro spécial, Soignants et Soignés, 133-165.

Piot-Ziegler, C., Ruffiner-Boner, N., Fasseur, F., Demierre, M., and Castelao, E. (2005). En attendant la transplantation : choix existentiels et paradoxes. [While waiting for transplantation: Existential choices and paradoxes]. In ouvrage collectif, *Eloge de l'altérité. Défis de société: 12 regards sur la santé, la famille et le travail.* [Praising otherness. Societal stakes: 12 perspectives on health, family and work]. Grolley : éditions de l'Hèbe, pp. 83-101.

Quetelet, A. (1835). *Sur l'homme et le développement de ses facultés, ou Essai de physique sociale*. Paris : Bachelier.

Radley, A. and Chamberlain, K. (2001). Health psychology and the study of the case: From methodology to analytic concerns. *Social Science and Medicine*, 53, 321-332.

Raimbault, G. (1992). *Morceaux de corps en transit*. Terrains, 18, mars, 15-26.

Ramachandran, V.S. and Blakeslee, S. (2002). *Le fantôme intérieur*. Paris : Odile Jacob, coll. Sciences, trad. Française (1e ed. in English, 1998).

Reker, G.T. (2000). *Theoretical perspective, dimensions, and measurement of existential meaning*. In G. T. Reker and K. Chamberlain (Eds.), Exploring existential meaning. Optimizing human development across the life span. Thousand Oaks: Sage Publications, ch. 3, pp. 9-55.

Reker, G. T. and Chamberlain, K. (Eds.), (2000a). *Exploring existential* meaning. *Optimizing human development across the life span*. Thousand Oaks: Sage Publications.

Reker, G. T. and Chamberlain, K. (2000b). *Existential meaning: Reflections and directions*. In G. T. Reker, and K. Chamberlain, (Eds.),Exploring existential meaning. Optimizing human development across the life span. Thousand Oaks: Sage Publications, pp. 199-209.

Robergue, M.-A. (2007). *Etat de stress aigu et état de stress post-traumatique après un infarctus du myocarde: prévalence et facteurs associés*. [Acute stress and post traumatic stress after myocardial heart attack: Prevalence and associated factors]. Thèse de doctorat en psychologie. Université du Québec à Montréal, pp. 1-213.

Santiago-Delefosse, M. (2002a). *Psychologie de la santé : Perspectives qualitatives et cliniques*. [Health Psychology: Qualitative and clinical perspectives]. Sprimont : Mardaga.

Santiago-Delefosse, M. (2002b). Analyse de l'activité du psychologue en milieu médical : Un nouveau Pharmakon? [Analysis of the psychologist's practice in a medical environment: A new Pharmakon?]. *Pratiques Psychologiques*, 3, 3-16.

Schelling, G. (2008). *Post-traumatic stress disorder in somatic disease*: *Lessons from critically ill patients*. In E.R. Kloet, M.S. Oitzl and E. Vermetten (Eds.), Progress In brain research, Vol. 167, ch. 16, pp. 229-237.

Schiemann, A., Hadzidiakos, D. and Spies, C. (2011). Managing ICU delirium. *Current Opinion in Critical Care*, 17(2), 131-140.

Schilder, P. (1950). *The image and appearance of the human body*. New York: International Universities Press, Inc.

Schwabe, L. and Blanke, O. (2008). The vestibular component in out-of-body experiences: A computational approach. *Frontiers in Human Research*, 2(17), 1-10.

Sicard, D. (2002). *La médecine sans le corps*. Une nouvelle réflexion éthique. [Medicine without body]. Paris: Plon.

Van Deurzen-Smith, E. and Arnold-Baker, C. (2005). Existential perspectives on human issues : A handbook for therapeutic practice. Palgrave: MacMillan.

Van Deutzen-Smith (1997). *Everyday mysteries*. Existential dimensions of psychotherapy. Hove and New York: Brunner-Routledge.

De Vignemont, F. (2011). *Bodily awareness*. Stanford Encyclopedia of Philosophy. http://plato.stanford.edu/entries/bodily-awareness/ Retreived October 18th 2011.

Von Weiszaecker, V. (1939). *Le cycle de la structure*. 2e éd. 1943, trad. franç., 1958.

Weinstein, S. (1969). *Neuropsychological studies of the phantom*. In A.L. Benton (Ed.), Contribution to clinical neuropsychology. Chicago, Illinois: Aldine, pp. 73-106.

Winship, G. (1998). Intensive care pychiatric nursing-psychoanalytic perspectives. *Journal of Psychiatric and Mental Health Nursing,* 5, 361-365.

Zdanowicz, N., Reynaert, C., Janne, P., Installé, E., Everard, P. & Delaunois, L. (1998). *Transplant and psychiatry, Psychosomatics*, 39(4), 390-391.

Zivkovic, S.A, Jumma, M., Barisic, N. & McCurry, K. (2009). Neurological complications following lung transplantation. *Journal of the Neurological Sciences*, 280, 90-93.

THE SPIRITUAL APPROACHES

In: Consciousness: Its Nature and Functions
Editors: Shulamith Kreitler and Oded Maimon
ISBN 978-1-62081-096-5

Chapter 9

Introduction to "The Spiritual Approaches"

Interview with Professor Charles Tart, August 3, 2011

Inteviewers: Professor Oded Maimon, Professor Shulamith Kreitler (in Tel-Aviv)
Interviewee: Professor C. Tart (in California)

Q. [Oded]: Your most recent book, based on a research career spanning half a century, was title "*The End of Materialism: How Evidence of the Paranormal is Bringing Science and Spirit Together*." When can you say that something has happened in a non-material manner?

A.: In order to claim that something has happened in a non-material manner one needs first to check carefully all the materially-based means available in the situation and whether they could be responsible for the occurrence of the event. If they can be responsible, then the ordinary, conservative explanation is that it is probably a materially-based event. But if you cannot identify anything material in the situation that could account for the event, then you can conclude that it has happened in a non-material manner. For example, in order to account for us speaking now across half the planet, we do not need to conclude non-materialism because the connection is through the telephone. But if we do not use the telephone or any other suitable material instrumentality, and still transmit information, we can conclude that it happened through telepathy or clairvoyance.

Q. [Oded]: Yet, telepathy could be accounted for by brain waves that occur during the transmission.

A.: No, because the electrical intensity of the brain waves is not sufficient for carrying the communication. The role of brain waves for telepathy has not only not been proven, it is materially impossible, given the minute power of the electromagnetic energies involved. Any electromagnetic signal emitted by the brain is lost in the noise within inches of distance, much less at thousands of miles.

Q. [Shulamith]: Why is it important to claim that there is non-materialism?

A.: It is vitally important. Because it is relevant to the spiritual drives that many people have. We need values to render our life meaningful. Just deciding about accepting values or endorsing them voluntarily does not suffice. In the modern world we need some evidence, some real basis for our spiritual values. And that is what non-materialism provides.

Q. [Oded]: How would you define consciousness?

A.: It cannot be defined in the way people try to define it. I cannot define it. I can only cite different aspects of consciousness pertinent to particular interests. And why should we expect the part to define the whole? The act of definition is only one aspect of many things that consciousness can do. Consciousness itself is very mysterious.

Q. [Oded]: Is it not dependent on the brain?

A.: It must be affected by the brain, but it is not clear that or how the brain *produces* consciousness. Consciousness, to my best understanding, is more than the brain. Consciousness is dependent on, or a manifestation of, the mind. The mind can do things that the brain cannot do. Take, for example, the case of the computer. You can explain everything about the hardware of the computer and how it works. But if you do not consider the human being who operates it for a specific purpose, you cannot understand much about what goes on. The analogy is that the computer is like the brain, and the person who uses the computer is the analog of mind. It is easy to define the brain, and we can increasingly understand what goes on with the brain in terms of electrical and chemical processes, but this has not brought us any closer to understanding the mind. We cannot point to the mind, it is not anything concrete. But the mind we ordinarily experience depends on the brain and uses it. The result is mind-brain which is a non-concrete entity but may appear as concrete. In my way of thinking about it, consciousness is a systems emergent of mind and brain interacting.

Q. [Shulamith]: What are the constituents of mind?

A.: By introspection, I can spot, right now, curiosity as a constituent, wondering how I am going to answer this question. Other constituents - far, far from a complete list - would be how I understand the world, motivation, and the whole range of emotions, like love. All these are aspects of mind. However, it is evident that you cannot look at the mind if it is not in interaction with the brain. The states of the brain change, for example in an emotion like anger, and mind changes then also, although the brain is not identical with the mind or for that matter with consciousness. What I'm saying, of course, is dependent on how one temporarily defines (or tries to define) "mind," "brain," and "consciousness" given the context of this discussion... . Since we are covering a lot of ground very rapidly in this interview, of course, we are not being very precise in our definitions.

Q. [Oded]: Is it always true that the brain and the mind are interconnected?

A.: Not necessarily. There are many accounts of two kinds of experiences which indicate that mind and brain can apparently be separated under specific conditions. These are the Near-Death Experiences (NDEs) and the Out-of-the-body experiences (OBEs). In these cases mind and body, including the brain, do not seem to function together in their usual way.

Q. [Shulamith]: What is the function of the mind?

A.: Going back to the analogy of the computer I mentioned earlier. The computer represents the brain, but the mind is that which activates it and puts it to use. If there is a person who types something into a computer, in order to understand what goes on you must consider the person and his or her goals, not just the computer. Another similar analogy is that of a car with a driver. You can look at the car which moves back and forth, along the road, but if you consider only the hardware of the car you are missing the essence. In order to understand what goes on, you cannot overlook the driver inside that car, which drives it where he or she wants to go. There is no way to understand the functioning of a car in motion without considering the driver.

Q. [Oded]: So what does that imply in regard to consciousness?

A.: It implies that consciousness is a non-material entity but that it co-exists together with the material, it requires that interaction with the material substrate to manifest. This kind of interaction is manifested also in clairvoyance, psychokinesis and in mind-body interactions in general.

Q. [Shulamith]: Could you explain somewhat your innovative idea about state-dependent science?

A.: State dependent sciences are a possibility, which, however, has not been developed. Consciousness is subject to individual differences. There may be large differences in style of functioning between different consciousnesses. In addition, consciousness may be in different states, such as hypnosis, and then it functions differently. Psychedelic experiences and other drug-induced experiences have shown us how the mind functions differently in these states. In proposing the creation of state-specific sciences, I am not talking about cases of delirium or a psychotic collapse; it is simply that the manner of functioning of the mind changes. These are alternate states of consciousness because there is internal validity to the manner of functioning in these states, they represent different forms of perceiving and experiencing.

Q. [Shulamith]: Are drugs the only way to induce an alternate state of consciousness?

A.: Not at all, there are many ways. I think that synesthesia, for example, is a case of alternate consciousness. It is manifested in having sensations that are not usually invoked by the given stimulus. It is said, for example, that some wine tasters have synesthesia, so that they perceive a wine with a particular taste as having the visual sensation of a jagged form, which let's them discriminate that wine from one that otherwise tastes the same to us ordinary non-synesthetes.

I would like to emphasize however that the perceptions in alternate states are not to be considered as *distortions* of ordinary consciousness. Good science starts with clear observations, and a word like distortion tends to carry implicit value judgments which confuse observation. On the other hand, altered state perceptions are also not to be considered as The Truth. They are not "true" or "untrue." They are simply *different* perceptions, different experiences. Some of them may prove to be useful, some not. What needs to be done is to check them out, to study them in line with basic science.

Q. [Oded]: What is basic science, in your view?

A.: There are four steps to essential science: The first step consists simply in accurately observing, and in improving your observations if you are smart, because any given individual may be insensitive or biased. The second step consists in trying to understand what your observations mean, following the intellectual curiosity of wondering about meaning. The attempt to understand may lead to the emergence of a formal theory that accounts for the observations. The third step consists in testing the theory's generality by making predictions on the basis of the theory and checking out if the predictions are confirmed. If they are, then this proves that the theory functions, it works. If the predictions fail to be confirmed, the theory must be revised or is simply discarded. The fourth step consists in sharing your findings and ideas with others. It is a social step, and its importance consists in that it can free your insights of individual biases and enhances the veridicality, objectivity and reliability of the observations and theories.

Q. [Shulamith]: Where is state-dependent science in all this?

A.: When I first proposed state-dependent sciences, state-specific sciences as I called them, the paper about it was published in *Science* (in 1972), there were over 100 letters-to-the-editor from readers, which is very unusual. Hardly any scientific articles draw any letters-to-the-editor. About 50% of the letters rejected the proposal as preposterous, and the other 50% judged it positively. Those who rejected the idea as “irrational”, "nonsensical," or "ridiculous" were, judging by names I recognized or high-level academic titles, well-established scientists, whereas those who were more tolerant seemed to be young and not yet established in their status or careers.

The most interesting letter came from a middle-aged psychiatrist, and it consisted actually of two letters: one in which he rejected the proposal as ridiculous and another, which his scientific integrity forced him to write in spite of contradicting his first letter. He had experienced some altered state of consciousness the previous evening and when he thought about my state-specific science proposal in that state in made perfect sense.

In this context it may be relevant to mention a recently published book *How the hippies saved physics*, by a historian of science, David Kaiser. It is an important book in the history of science because it shows how in the 1970’s, when physics was trapped in conservative ideas, a group of young physicists in the Berkeley area promoted completely different ideas about reality and physics (e.g., quantum physics) which could have had their origins in the psychedelic experiences to which they, as so many Californians at that time, had been exposed. Of course, insights one gets in altered states of consciousness need to be explored and checked before they have a chance to be integrated into mainstream science. Some of the insights may prove to be useful, some not.

INTERVIEWERS: ODED MAIMON AND SHULAMITH KREITLER

In: Consciousness: Its Nature and Functions
Editors: Shulamith Kreitler and Oded Maimon
ISBN 978-1-62081-096-5

Chapter 10

SCIENTIFIC AND SPIRITUAL PERSPECTIVE OF CONSCIOUSNESS: AN ANALYTICAL INTERPRETATION

Sanjay Srivastava *
Feroze Gandhi College, Bharata (India)

ABSTRACT

Consciousness, an integral component of the human existence, has become a central theme of study among scholars from diverse disciplines and different religions. Every soul, whether animate or inanimate, diverges from the same source of Infinite. Under a designated divine-plan, a soul comes to this world to traverse its journey—ultimately to meet its Source again. Consciousness is an important concept to follow this journey. According to the Hindu philosophy or Buddhist schools, consciousness, or more precisely, the 'expansion of consciousness' plays a fundamental and paramount role in the spiritual and scientific world. Scientists are in search of a 'unified theory of everything' or the 'most fundamental particle'—particle of God; while the spiritual quest is to find the ultimate-Truth or Supreme-Union—union with one's Higher-Self. Eventually, the findings of these two fields complement each other, in achieving their respective goals, with consciousness as a central theme of the entire process. In this endeavor, the fundamentals of classical and quantum physics; and the knowledge of scriptures of Hindu and Christian religions are explored on the basis of consciousness. Limitations of our physical approaches are identified with their lack of infallibility. Energy, as in science, is visualized in the background of Hindu scriptures such as Bhagavad Gita. The ideas and opinions of leading scientists and philosophers, on the esoteric nature of Nature, are analyzed in the light of science and spirituality—in search of ultimate-Reality.

Keywords: consciousness, spirituality, yoga, self-realization, ultimate-reality

---

* Author Note: Sanjay Srivastava, Department of Physics, Feroze Gandhi College, Rae Bareli 229 001, Bharata (India). E-mail: sanjaysphy@gmail.com.

INTRODUCTION

Consciousness, as a word, has become an integral part of our daily life. Over the years, it has become an explicit part of our spoken and written vocabulary. In the present connotation, it has gained momentum and reverence in the context of science and spirituality. At various platforms like meetings, gatherings, colloquiums, symposia, and conferences of science and spirituality—a *conscious* attempt is made to understand consciousness through deliberations, debates, and discussions.

Aspects like, what constitutes consciousness? How to define consciousness? What is the nature of consciousness? What is the origin of consciousness? How does consciousness function? And so forth, make scholars deliberate deeply on the depth of this soul driven word. Consciousness is primarily considered as a missing-*link* between God and man. Predominantly, it is believed that God is consciousness and vice versa.

Why all of a sudden consciousness has gained so much of importance? Why is everyone now *consciously* talking about consciousness? What we will gain by understanding consciousness? These thought provoking aspects intrigue us to probe consciously vivid aspects surrounding consciousness a little deeper, and comprehend its meaning; various dimensions, objective; and philosophy in a broader perspective.

Let us explore the subject from the view of both, modern- and spiritual-sciences, with consciousness as the central theme of study.

THE WORD

The word *consciousness* may have varied connotation, synchronizing with its usage. There could be a scientific inference of the word, or it may imbibe a 'deep' spiritual connotation. We will study both the aspects in detail; in finding the similarities between consciousness in the scientific and spiritual world.

The Scientific Connotation

Consciousness, in scientific parlance, is associated with the observation of phenomena. It may be motion of stars, or planets in the nature; or interaction of subatomic particles in the nuclear process, and so forth. Sagan (1980) writes:

> Today we have discovered a powerful and elegant way to understand nature, a method called science*: Observation and analysis* [italics added]; it has revealed to us a universe so ancient and so vast that human affairs seem at first sight to be of little consequence... . Science has found... that we are, in a very real and profound sense, a part of that Cosmos, born from it, our fate deeply connected with it. (p. xvi)
>
> Science is an ongoing process. It never ends. There is no single ultimate truth to be achieved, after which all the scientists can retire. (p. xix)

The Spiritual Connotation

In the spiritual context, the word consciousness carries 'deep esoteric' meaning. It is associated with awareness of God—such as one has consciousness of God or not. Hence, it has more subtle meaning in the spiritual language in comparison to scientific one. Essentially, spiritual-science is a science that discovers the deep esoteric human-nature by way of Self-enquiry—Who *am* I? Therefore, as contrary to scientific orientation, in spirituality one can realize *expansion of consciousness*. The only corresponding words we may find in science are higher 'brain power', or 'power of mind'.

FUNCTIONING OF CONSCIOUSNESS

The Wakeful State: A Gross Level

On close observation, we find that consciousness of 'touch' (skin), 'smell' (nose), 'sight' (eyes), 'sound' (ears), and 'taste' (palate)—these five-distinct sense qualities are governed by the five-different sense organs. These give us the feeling of touch, smell, and so forth. If any of the sense quality is not functional, we will not be conscious of that quality.

Let us consider an example of a hearing impaired person. Due to damaged auditory sense organs, the person is not able to perceive sound. Sound energy is actually there, *but* the person is not *conscious* of sound.

In another example: let us take into account the effect of anesthesia on pain. As anesthesia lessens pain; hence, it is considered as a pain relieving therapy. But in 'reality', anesthesia only blocks energy that carries pain sensation to the brain; and in turn, we feel relieved of pain.

Functional Energy

In above examples, energy that has to be carried for the processing of information in the brain is missing. Thus, a person with hearing impairment is not able to hear, and a person under anesthesia does not feel pain. Hence, consciousness may be *linked* to energy through which one works, or *awaken* in the physical-world.

Consciousness as a form of energy works through the medium of energy; and then, *in turn*, through the senses. When energy, or to say the *life-forces* are missing from our sense organs; these cease to function. So, when energy from the senses is withdrawn, one is oblivious of conscious sense of touch, sound, and so on. As a matter of fact, we perceive the world *only* through the functioning of five-senses. In the 'inert state' or non-functioning of the senses, the world is *dead* to us. Partly, we experience the inert state of senses in sleep.

> If thy hand or thy foot offend thee, cut them off... . And if thine eye offend thee, pluck it out, and cast *it* from thee. Matthew 18:8; 18:9, King James Version)

This transcends to the fact that how to withdraw consciousness—by withdrawing the life forces from senses. Our consciousness lies—where energy is.

The Sleep State

In the 'sleep state', life forces from the sense organs that carry information to brain are temporally withdrawn, and energy retires into the medulla oblongata. It makes sensing organs temporarily non functioning by ceasing the sensing operation. Thus, in sleep, ears are there *but* we are not *conscious* of sound. Eyes are there, *but* we are not *conscious* of sight. In sleep, we are not conscious of ourselves of being alive or not, having the male or female body, even the world exists or not.

The Dream State

In the 'dream state', one is awake in the dreamland. Dreams are associated and are thought to be the product of subconscious mind. As long as one is in dream state, he is in the delusion of *dream consciousness*. For an example, one might be playing a part of a king in the dream state, and may cry of losing his kingdom in the 'awaken state'.

Delusion: Maya

A spiritual person—a realized one considers this world as a *dream of Brahman* in his consciousness. Interestingly, as long as we are in the physical-world, we are considered to be in 'delusion'—under the spell of *Maya*—and in that sense we are in the *dream state*. When, by practice of meditation, we are able to raise our consciousness to say the higher-levels—above the consciousness of physical-world, we are considered to be in the *awaken state*. Thus, Self-realization in a 'real sense' is *not* a 'state of dream'.

Dream of Brahman

In scriptures, the 'whole of cosmos' is considered to be as a dream of Brahman in his consciousness. A cosmic view:

> The Hindu religion is the only one of the world's great faiths dedicated to an idea that the Cosmos itself undergoes an immense, indeed an infinite, number of deaths and rebirths. It is only religion in which the time scales correspond, no doubt by accident, to those of modern scientific cosmology. *Ironically, nothings happen in this world by accident or without the knowledge of Lord, "Are not two sparrows sold for farthing? and one of them shall not fall on the ground without your father"* (Matthew 10:29). Its cycles run from our ordinary day and night to a day and night of Brahma, 8.64 billion years long, longer than the age of the Earth or the Sun and about half the time since Big Bang.
>
> There is a deep and appealing notion that the universe is but the dream of god who, after a hundred Brahma years, dissolves himself into a dreamless sleep. The universe dissolves with him—until, after another Brahma centaury, he stirs, recomposes himself and begins again to dream the great cosmic dream. Meanwhile, elsewhere, there are an infinite number of other universes, each with its own god dreaming the cosmic dream. (Sagan, 1980, p. 213)

The Meditative State: A Subtle Level

Meditation is communion between 'man and God'—between 'finite and Infinite'—between 'soul and Spirit'. Meditation in a *subtle way*, directly and deeply, affects the consciousness of meditator (Truth seeker—who practices meditation and austerities). With its powerful 'technique of concentration', meditation causes the 'expansion of consciousness'. The expansion of consciousness accelerates the 'human evolution' on the 'ladder' of Self-realization.

The 'journey' of Self-realization in its own *natural*-way, may take thousands of years to come in terms of God-realization. Meditation—by its virtue of expansion of consciousness—only quickens the process. The scriptures say meditate now, and realize Him—or you may have to wait till thousands of years to come.

Limitations of Senses

We know that the mind can feel a *sense of touch*, if it is within the surface of body. Outside of it, mind is not capable of detecting a sense of touch. Similarly, our vision is restricted to the visible-spectrum of the wide range of electromagnetic spectrum, and our audio-faculty is restricted from 20—20,000 Hz of sound wave. These restrictions put limitation on consciousness through the senses.

Classical Example: A Test

An important point to observe and understand that even with great power of mind, information through the senses *may not always be true*. It is possible to draw wrong inferences from the inputs. A classical example from the elementary physics—and the experiment runs as follows:

Take three cups of water—one with cold water—one with normal water—and the last one with hot water. Now, if we put our two fingers simultaneously—one in cold water—and other in hot water for some time—then again, simultaneously put these two fingers into normal water—we might, invariably and unmistakably, have an experience—as the finger which was in cold water; feels the normal water warm—and the finger which was in hot water; feels the normal water cold. How could it be possible that the two fingers of the same person—register different results?

Even with the best of mind inputs from the senses may not be very reliable, and the mind may *fail to judge* the 'reality'—*to discriminate between the right and wrong answers*. On extrapolation, we may say that one can add any number scientific instruments in add to the senses, *but* still find at the 'end' there might be inadequacy of information.

Quantum Example: The Uncertainty Principle

In the world of atomic physics, we come face-to-face with the subatomic particles, and their interactions (Ford, 1965). These subatomic particles are unique and are not directly perceptible to the senses. On the measurement basis an interesting and amazing phenomenon came to light, and it took scientific community a long-time to accept the 'paradox' as fundamental to Nature.

"It seems that the uncertainty principle is a fundamental feature of the universe we live". (Hawking, 2006, p. 111)

That is, we cannot measure the position and momentum of a subatomic particle simultaneously, and accurately. It has been observed that if measurement in the position of an atomic particle is certain, then the measurement in momentum has become uncertain and vice versa. The quantum mechanical paradox is known as the uncertainty principle (Schiff, 1949).[1] The uncertainty in the measurement carries so small numerical value that it escapes its detection for the larger bodies.

A point of argument is that even with the best of available instruments; we, *but* come across with the 'limitation' imposed by Nature that on the analytical basis carries insignificant value. Generally speaking, our consciousness working through the senses is 'restricted' by Nature. We have no other choice, *but* to accept and believe the might of almighty Nature, and her *Lila–A divine-play of God.*

CONSCIOUSNESS IN SCIENCE

Observer, Observation and Analysis

In this context, I would like to mention the example of Newton's law of gravitation. It was not that before Newton, apples used to go up; *but* he was *conscious and observant* to think about this phenomenon; and thereafter, discovered the effect of pull of gravitational forces on the falling bodies. The outcome is laws of gravitation, subsequently, based on his observation and mathematical analysis.

Sunrises (daytime) and sunsets (night time) routinely govern their normal courses everyday in our lives. Most of the time and for many of us, these events never stir up our consciousness. *But*, Johannes Kepler by his painstaking conscious observation and analysis, over the years together, had opened an entire new field of science: The Astronomy.

A question, we may be asked to ourselves that if everyone is working through the senses, then why we differ in the analysis of inputs. The 'basic functions' of senses are observation of the phenomena. So, why one is more *conscious* of the same observation than others? What causes these differences, and who is responsible for it?

The Power of Mind

The difference may be explained and understood on the basis of 'power of mind', which is responsible for the analysis of inputs. How much we are *conscious* about our observation, also depends upon our power of mind to analysis, and draw inferences from the experimental phenomena. The power of mind is directly proportional to the expansion of consciousness—*higher the level of consciousness, greater the power of mind.*

---

[1] $\Delta x \times \Delta p \geq h$, where Δx is the uncertainty in position of a particle, Δp is the uncertainty in momentum of a particle, and h is Planck's constant (h = 6.626176 E - 34 Joule—sec).

Scientific Development: Expansion of Consciousness

Advancement of sciences has its own series of commendable achievements: Inventions—Electron microscope, computers; Scientific data—Speed of light, Planck's constant; Landmarks—Landing on moon—are some of the hallmark of man's scientific achievements. These achievements could be realized due to man's ability to acquire and realize content through the evolution of mind, or consciousness. And there are various forthcoming inventions and discoveries waiting to be unraveled by the man with the expansion of consciousness. Lord Krishna says, "There are infinite attributes of Me [Nature]" (Bhagavad-Gita X:19,40)—and even millions of incarnations are not sufficient to know all these attributes.

Evolution of Mind

The expansion of consciousness is fundamental and paramount; and the very basis of 'scientific advancement', 'evolution of mind', or 'progression of mankind'. The expansion of consciousness, scientific advancement, evolution of mind, or progression of mankind are all interlinked and complementary to each other. It is an ongoing continuous process—years after year—births after birth—and incarnations after incarnation—till, consciousness finds its ultimate-Destination—the goal of realization of Supreme-Power—the God. And, this is realized only after 'long' and 'arduous' 'journey' on the 'path' of Self-realization. Paramahansa Yogananda (2008) writes, in Autobiography of a Yogi:

> Through proper food, sunlight, and harmonious thoughts, men who are led only by Nature and her divine plan will achieve Self-realization in a million years. Twelve years of normal healthful living are required to effect even slight refinement in brain structure; a million solar returns expected to purify the cerebral tenement sufficiently for manifestation of cosmic consciousness. (p. 241)
>
> *Kriya Yoga* is an instrument through which human evolution can be quickened... . The ancient yogis discovered that the secret of cosmic consciousness is intimately linked with the breath mastery... . One thousand *Kriyas* practiced in $8^1/_2$ hours gives the yogi, in one day, the equivalent of one thousand years of natural evolutions: 365,000 years of evolution in one year. In three years, a *Kriya Yogi* can thus accomplish by intelligent self-effort the same result that Nature brings to pass in a million years. (pp. 238-239)

A Long Path

Everyone—'animate or inanimate', 'king or beggar', 'holy-man or beast'—have to trudge a 'long path'—whether 'good or bad', 'easy or tough'—to reach a state of higher-consciousness—that leads to God-realization; the same way, a river has to trudge a long path of plains—passing through 'smooth or rough' terrains to reach its 'source'—the ocean, and finally to get submerged into it.

The same holds true for scientific research and development. The present scientific developments are based on the sum of all past scientific researches—gross or subtle, matter or non-matter, success or failure.

CONSCIOUSNESS AND THE MODES OF PRAKRITI

We now discuss the meaning and philosophy of consciousness from the view of two-modes of Nature—*Apara* (outside, external) and *Para* (inside, internal) *Prakriti* (Primordial Nature).

Apara-Prakriti: The Physical World

The physical-world, in spiritual language, is known as *Apara-Prakriti.* It is manifested to us through the five-senses.

A Dual Nature

'Duality' is a hallmark of *Apara-Prakriti*. On close observation, we find that the 'entire creation' is based on the 'nature of duality'. Everything in the nature seems to have been created in the form of 'pairs'. The existence of pairs and their 'interaction' has become the very basis of further creation. Even on the atomic level, we find the creation of subatomic particles in pairs. Scriptures say that the Spirit—*an undifferentiated continuum of consciousness*—has descended in the forms of 'Half-male' and 'Half-female'. The Spirit has further descended in the 'male' and 'female' forms. It is a 'dual nature' creation of Spirit, in the consciousness of Brahman (Figure 1).

As formulated by Descartes, the Cartesian view of duality of nature is generally considered as the basis of modern science. Capra (1974) writes,

> The reduction of nature to fundamental is basically a Greek attitude … together with the dualism between Spirit and matter… . They are thought to be moved by some external forces which assumed to be of spiritual origin and was often identified with God. This marked the split between spirit and matter, between God and the world. (p. 16)

A Dynamic Interplay

We have learned that energy is 'indivisible', and total-sum-of-energy of entire cosmos is constant. The 'whole of cosmos' is *but* a 'one indivisible energy' of Cosmic-Consciousness; there is only transformation of energy taking place—from one form of energy—to another form of energy (e. g., Gita II:20-24). The similar transformations we find in 'creation and annihilation' of subatomic particles in the nuclear reactions. Or, to say, with any matter in the cosmos; and it is an ongoing ceaseless process (e. g., Gita III:23,24). *It is a dynamic interplay between Energy and Matter; between the Spirit and Nature.* A la Cosmic-Dance affair of Lord Shiva—The Nataraja (Figure 2)!

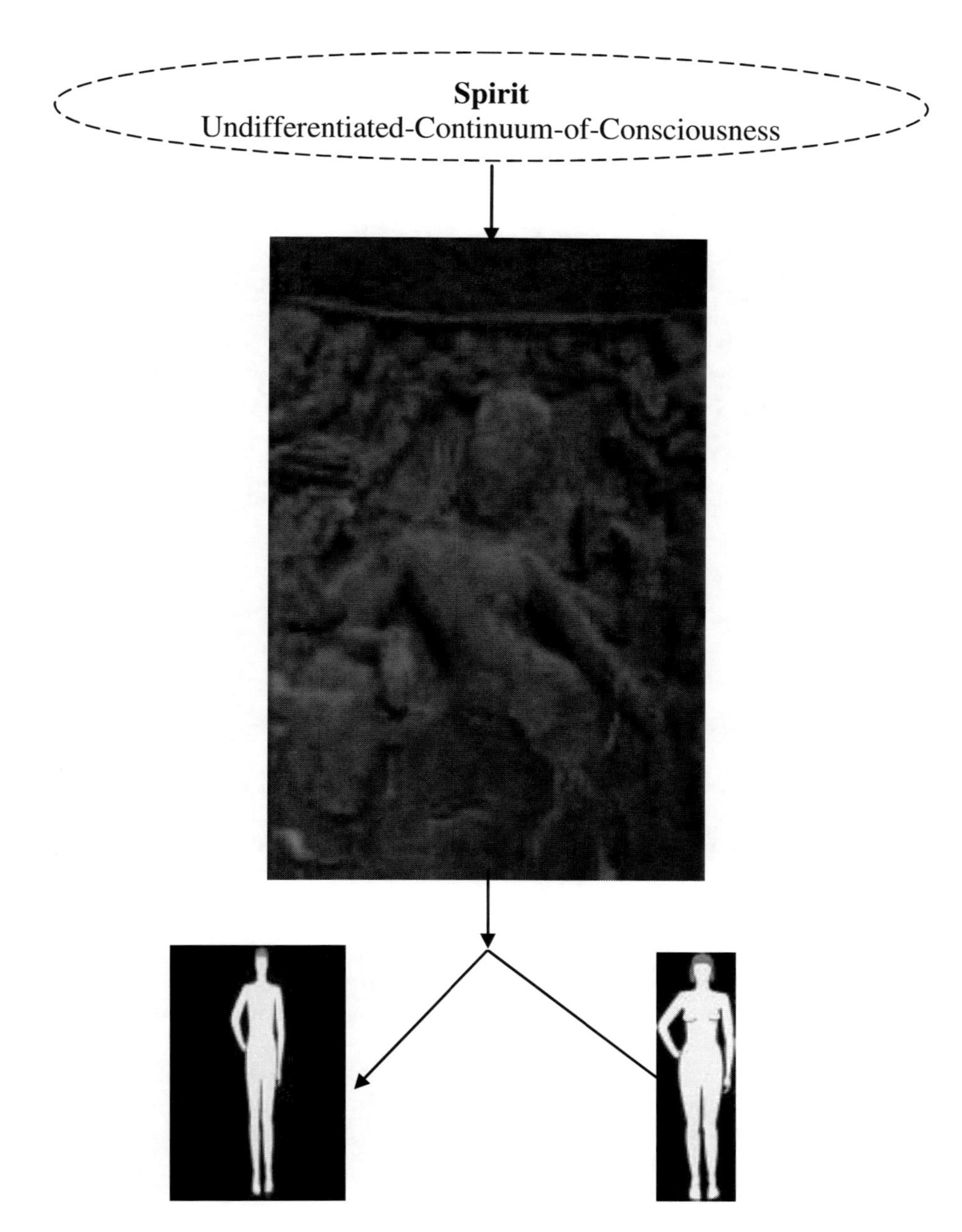

Figure 1. Ardhanarishvara ('Half-male'—'Half-female' God) at Elephanta-caves. The sculpture's left is 'female' and the right is 'male'—depicting Shiva and his consort 'Shakti/Pārbati' (Retrieved January 23, 2010, from http: //en.Wikipedia.org/wiki/Elephanta_Caves). The Spirit—an undifferentiated continuum of consciousness has descended in the 'Half-male' and 'Half-female' forms. The Spirit has further descended in the 'male' and 'female' forms. It is a creation of 'duality of consciousnesses' in His Consciousness.

Figure 2. Image of Lord Shiva as Nataraja—Lord of Dance—overlaid on particle-tracks in the bubble-chamber (Retrieved February 3, 2010, from http://mg-critic.blogspot.com/2009/09/anachronism-of-hindu-gods.html). The scriptures consider the 'whole universe' as 'playfield' of God—*A Dynamic Universe*. Capra (1991) photomontage the dancing picture of Lord Shiva superimposed with tracks of interacting subatomic particles.

> In India there are many gods, and each god has many manifestations. The Chola bronzes, cast in the eleventh centaury, include several different incarnations of the god Shiva. The most elegant and sublime of these is a representation of the creation of the universe at the beginning of each cosmic cycle, a motif known as the cosmic dance of Shiva. The god, called in this manifestation Nataraja, the Dance King, has four hands. In the upper right hand is a drum whose sound is the sound of creation. In the upper left hand is a tongue of flame, a reminder that the universe, now newly created, will billions of years from now be utterly destroyed. (Sagan, 1980, p. 214)

Scientific Approach: Observation and Analysis

In order to gain a constructive output from the scientific experiments, one needs to carry the entire experimentation process very carefully and methodologically. The procedure should be foolproof, and has to be performed in a systematic manner. For an example: in order to experiment the flow of electric current through a wire to light a bulb—the wire must be clean, free from rust, properly soldered to the terminal points. Only then, current will flow through the wire; and bulb will glow to its full wattage.

Para-Prakriti: The Spiritual World

Eminent sages of Bharata (India) had long discovered that each living being constitutes an equally glorious *Inner-world.* In spiritual language, it is called *Para-Prakriti*—the kingdom of God. Ironically, it is 'not manifested' and 'cannot be sensed' through the five-senses. It constitutes the world (e. g., Gita XV:6):

> Higher than the Moon—brighter than the brightest Sun that may ever be thought of, or discovered in the entire cosmos—higher than the highest mountain that one may ever be imagined, or measured—and deeper than the deepest known ocean that may ever be discovered, or measured.

A Non-dual Nature

Para-Prakriti in contrast to *Apara-Prakriti* is 'beyond' any kind of 'duality'. The spiritual-world is beyond 'space-time limitation', 'sorrow or happiness', 'life or death', 'light or darkness'; and beyond any duality of the physical-world. Inner-world—the kingdom of God is beyond comprehension of 'mind' or 'intellect'. Neither with mind or intellect, nor with reasoning one can know and reach Inner-world, and attain Self-realization. In this world, neither Vedas nor any other scriptures exist; as, this world is beyond *reach* of all these scriptures. The scriptures, at most, may only point to it—to show the 'pathway' (e. g., Gita XI:52-54). The Upanishad says, "There the eyes goes not, speech goes not, nor the mind. We know not, we understand not, how one would teach it" (Keno I:3)? A Blissful world:

> In this pure world, only pure love exists. A world full of Bliss exists. Pure, purer, and only purest form of love exist. Love that does not know any kind of selfishness; Love that does not know any kind of deceit; Love that does not know any kind of immorality; Love that does not know any kind of treachery; Love that only knows to love each other; Love that loves others in silence. Love knows that by loving each other truly—we only love God.

Ignorance

Why we are ignorant about *Para-Prakriti*—the Inner-world? Why the kingdom of God—the spiritual-world is hidden from our consciousness, and not manifested to us? How expansion of consciousness dispels *ignorance*, and raises consciousness—from the 'delusion of duality' of *Maya*-world—to the 'wisdom of non-duality' of spiritual-world?

Scientific Approach: Patanjali Yoga-Sutra

Like in the physical-world, there are set scientific procedures for the spiritual-world as well. Saint Patanjali (discoverer of Yoga-aphorism) in his famed treatise—*Patanjali Yoga-Sutra* has described the 'Eightfold-Path' to God-realization. It also constitutes the main theme of sacred yoga book: The Bhagavad-Gita (Yogananda, 2002). In Gita, Lord Krishna (Higher-Self) himself guides his beloved disciple, Sri Arjuna (Truth-seeker devotee), on the divine-path of God-realization.

The Eightfold-Path to God-realization, described in *Patanjali Yoga-Sutra*, are as follows: (1) *Yama* (avoidance of wrong-actions); (2) *Niyama* (right-actions); (3) *Asana* (right-posture for body and mind control); (4) *Pranayama* (control of *prana*, breath or life-forces); (5)

Pratyahara (interiorization of mind); (6) *Dharana* (concentration); (7) *Dhyana* (meditation); and (8) *Samadhi* (divine-union with God).

Expansion of Consciousness

Dhyana—meditation, described in *Patanjali Yoga-Sutra* before the stages of 'Samadhi', increases the magnitude of 'field of consciousness'. Meditation practice opens up (by activating) the dormant brain cells, and mystic 'psychic energy chakras' (plexuses) situated along the spinal column. During the ongoing spiritual practice, 'manifestation of negativism'—such as 'anger'; 'jealousy'; and so on, gradually wanes away of their own from the subconscious mind—as dry leaves get detached from the tree. The 'fire of meditation' burns up seeds of past, present, and future *karmas* (accumulated actions); and consumes all of the 'earthly binding' *karmic seeds*: whether, good or bad (e. g., Gita IX:28). There will be no regeneration of seeds that binds us to the cycles of 'birth and death'. Thus, in mediation, one advances many lives by eliminating the needs of future births, which is required to work out the karmic debts.

Cosmic Consciousness

Meditation, by virtue of expansion of consciousness, breaks down the domain of the 'narrow-self'—the 'ego'—to identify oneself with the body-consciousness. When meditator's consciousness begins to expand by breaking down the boundaries of his ego; gradually, a state appears when his consciousness begins to embrace the entire cosmos.

In the state of *Nirvikalp*-Samadhi, consciousness of a meditator is completely merged into Cosmic-Consciousness—*A state of undifferentiated continuum of consciousness.* A purest of pure divine-state—a pinnacle of divinity—when meditator becomes one with God—becomes one with his Higher-Self. The meditator may proclaim—as Lord Jesus pronounced, "I and *my* Father are one" (John 10:30).

Thereafter, for the mediator, deep-down-spiritually, ... there is no 'I or you'... 'electron or proton'... 'sun or moon'... *but...* 'pure consciousness'—pervading the entire cosmos of 'one whole' of 'indivisible energy'—sans any boundary or division—with no individual identity of his own. He perceives that his 'own self' is, but *an infinitesimal frozen part* of the 'Whole Self'. The entire cosmos is made in His consciousness, and exists only in His thought (e. g., Gita IV:35). The truth is beautifully portrayed in the following poem:

> I am the sky, Mother, I am the sky,
> I am the vast blue ocean of sky.
> I am a little drop of sky,
> Frozen sky. (Yogananda, 2003, p. 38)

He understands that (e. g., Gita VII:8,9; IX:16-19; 26; X:21-39; XI:5-7):

> Everything is Brahman, and Brahman has becomes everything. Everything belongs to Brahman, and Brahman belongs to everything. Brahman is One, and Brahman is Many. Brahman himself manifested in the form of many individual beings, both living and non-living, for his beloved children.

Capra (1974) writes,

> The highest aim for the Hindu or Buddhist is to become aware of the unity and mutual interrelation of all things, to transcend the notion of an isolated self, and to identify himself with the ultimate Reality. The acquisition of this knowledge, which is known as "enlightenment" is not a mere intellectual act but becomes a religious experience involving the whole person. (p. 16)

The Result

Meditation increases the 'field of consciousness'. As a result, instead of meditator under the possession of the senses, the senses are now under his command. The mind, with control over the senses, 'evolves'—and ultimately *itself* becomes a 'Guru' to guide the 'self'—on the royal-path of God-realization. The mind plays the role of 'charioteer' of 'bodily chariot', and driving it towards the divine-goal of Self-realization (e. g., Gita IV:39).

A spiritually advanced person is 'humble person'—a living symbol and embodiment of compassion, fellowship and humility. One cannot find any trace of manliness in him, *but* God-Alone. His existence is a manifestation of Godly deeds, as he feels hunger and pain of others. He suffers himself to alleviate the sufferings of others. He understands that by injuring others, a person is *in fact* only injuring himself.

Chakras : The Psychic Energy Centers

The sages of ancient India had long discovered that our 'body' is *but* a 'mini universe' in itself, which draws 'energy' in the form of 'breath' from Nature. Breath severs as a vital-*link* between the body-consciousness and Cosmic-Consciousness. Strangely, this very 'breath' *binds* us into the body-consciousness. Teachings of 'yoga', 'pranayama' and 'meditation techniques' are 'sciences' of mastery (stillness) over breath. Thus, Holy Bible declares, "Be still, and know that I *am* God" (Psalm 46:10).

The centers (plexuses) are situated along the length of spinal column of a human body. These, starting from the base, are as follows: Coccygeal-centre (Sanskrit: *Muladhara* chakra)—situated at bottom end of the spinal-column; Sacral (*Swadhishthan*)—situated small space above of Coccygeal; Lumber (*Manipura*)—situated opposite to the belly-button; Dorsal (*Anahata*)—situated opposite to the hearth-centre; Cervical (*Vishuddha*)—situated opposite to the joint of neck and shoulder; Medulla (*Ajna*)—situated opposite to the centre of forehead; and Cerebral (*Sahasrara*)—situated at the top of head.

Consciousness: As a Function of Astral Sound

Our consciousness when centered at Coccygeal—one is conscious about the buzzing sound of honey bee. With the Sacral—one is conscious about sound similar to the flute; while, with the Lumber—one hears sound like a harp or *Vina*. Consciousness when rises to Dorsal—one becomes conscious of Aum—sound of God's consciousness and manifestation—like a sound of a long-drawn astral-bell. With consciousness reaching up to the Cervical—one hears the sound of Aum—as of a roaring ocean. When consciousness gets merged with Medulla—one hears the symphony of sound of five lower-centers—and becomes one with Aum vibration. The meditator is now fully conscious of God. With consciousness synchronizing with the Cerebral—one attains the State of Cosmic-Consciousness—*A State of Nirvikalpa-Samadhi*.

Level of Consciousness and its Effect

Generally speaking, our consciousness is importantly governed by the consciousness of various centers. If consciousness is dominated by the lower-centers, we are ruled by the lower-nature with little discriminative faculty. Absence of consciousness at upper-centers predominantly leads to instinctive behavior. Fortunately, this sacred human-birth is far above on the 'scale' of 'evolution of consciousness'.

Human Birth: A Higher Level of Consciousness

Sri Shankaracharya edifies, "It is hard for any living creature to achieve birth in a human-form… . Only through God's grace we may obtain those rarest advantages—human-birth, the longing for liberation, and discipleship to an illumined Master" (Vivek Chudamani 2,3). A cosmic evolution:

> Single-celled plants evolved, and life began to generate… . One-celled organism evolves into multicellular colonies… . Plants and animals discovered that the land could support the life… . The ash of stellar alchemy was now emerging into consciousness… given fifteen billions years of cosmic evolutions. (Sagan, 1980, p. 282)

Only after millions of incarnations when God bestows his grace, we, His-children are born as human entities. A human-birth is rare, precious, and sacred; and at the same time an 'evolutionary consciousness' in terms of 'birth cycles'—and 'highly placed' on the 'path' of Self-realization (e. g., Gita VI:42). As only in the human-form it is possible to meditate and realize Him—and His divine qualities.

It is a spiritual-process of 'rising' of an individual's consciousness from the earthly-binding consciousness of 'self' to soul-liberating Cosmic-Consciousness of 'Divinity'—'A Journey'—from 'Coccygeal (*Muladhara*)' to 'Cerebral (*Sahasrara*)'—from the 'world of mortality' to the 'world of Immortality'—from 'confinement' to 'Freedom'—from 'bondage' to 'Liberty'—from 'ego of self' to 'selflessness of Higher-Self'.

Thus, we pray to God that He may shower and bestow his grace on his children—so by raising their consciousness His-children will achieve Him:

> 'Oh! Almighty and All Merciful God!
> Raise our consciousness and thus,
> Lead us from darkness to light of Thee;
> Lead us from ignorance to wisdom of Thee;
> Lead us from selfishness to compassion for All;
> Lead us from 'ego of self' to 'selflessness' of -Self'

SCIENTIFIC APPROACH TO CONSCIOUSNESS

There is renewed interest to study consciousness associated with the scientific and spiritual world. The study revolves around intrigues, like what kind of experiments may be set up to study the experiences that affect body, mind and consciousness during meditation? How a scientific approach may help to achieve higher-levels of consciousness? Mediation, in one-way, works as a cleansing-agent for the consciousness—how science can demonstrate it (e. g., Gita VI: 12).

How science may demonstrate that consciousness of the 'two worlds' synchronizes at 'deep esoteric' level? Consciousness as of quantum physics beautifully harmonizes with the spiritual-world; how science may pursue the subject? We, here, would draw our endeavors to analyze the aspect of consciousness from the view of modern and quantum physics.

Modern Physics

An important point to observe about the subatomic particles is that these infinitesimally small particles not only carry energy, *but* intelligence as well. These particles are the bundle of *consciousness of energy*, and carrier of *intelligent energy* as well as *consciousness*. There is a delicate *symbiosis* between consciousness and energy.

Exclusion Principle

Pauli Exclusion Principle says that no two electrons with identical set of quantum numbers can occupy the same subshell. Hund's rule says that only after filling up first the unoccupied subshells—pairing of electrons in the subshells would take place (Halliday, Resnic, and Walker, 2004).

How an electron, which is invisible to any known scientific instruments, is *conscious* to follow these rules? How an electron is conscious beforehand that another electron with particular spin (energy) is already there in a subshell, and other kind of spin (energy) is required to pair it up? Who gives electrons the *conscious intelligence*? Who provides these particles with *consciousness of energy*?

Consciousness of Electron

An electron possesses consciousness but has no *discriminative* consciousness of its 'own'. Consciousness is embedded there as 'consciousness of electron' by Nature. Electron has no other choice, *but* to follow the rules as ordained by Nature. We may call consciousness of electron as an *intrinsic consciousness*.

We, as individuals, *too* carry intrinsic consciousness—playing our part as ordained by God. The difference lays in power of *free will—the discriminative faculty*—to choose and follow the paths. Human beings have freedom to 'seek and follow' the 'path' of Self-realization that *ultimately* leads to 'liberation'. Or, alternatively, follow the path that keeps one engaged in the world of delusion (e. g., Gita VIII: 26,27).

Electron versus Human Consciousness: Observer or Object?

An electron may act as a 'subject' (or observer) to another electron which may act as an 'object' or vice versa. But in an individual capacity, electrons cannot be observer to themselves. *An observer is not observer to itself.* In world of physics, the 'subject' and 'object' are 'different' from each other. We only observe *action at a distance* due to the field created by an electron that acts on another electron.

In the spiritual-world subject and object are 'identical'. Thus, the scriptures declare that meditator progresses from I *am*—to—*Thou* are *That*—to—*I* and *Brahman* are *One*. Schrödinger (1984) observes,

> The same element composes my mind and the world. This situation is the same for every mind and its world… . Subject and object are only one. The barrier between them cannot be said to have broken down as a result of recent experiences in the physical sciences, for this barrier does not exists. (p. 79).

Atomic Orbits

We know that electrons revolve in the fixed orbits, and there are inter orbit transitions of electrons. An electron, when gains energy jumps into the higher orbits; and if loses energy falls into the lower orbits. At the same time if an electron gets sufficient energy, energy equivalent to the ionization potential of an atom, it will leave the orbit of atom forever.

Electrons that constitute the inner most cores of atomic orbits are known as the core or tightly bound electrons. Under the normal laboratory conditions, it may not be possible to dislodge these electrons. It may require an enormous amount of energy to dislodge these electrons from their orbits. However, electrons that constitute the valence band electrons of the outer most subshell require very less energy to get released from their orbit. These valence electrons are free to move within volume of the element.

Relation to Psychic Energy Centers

Our consciousness manifests itself as a function to these 'centers'. If consciousness is predominantly bounded to lower- (Coccygeal) centre, the person must be entangled badly in the worldly-affair, with little probability of getting out of it. At this stage, consciousness is tightly-bound to *Maya* of the world. The state of person's consciousness is like that of core electrons, which are tightly bound to the nucleus; with little chance of getting liberated from the entanglements of the delusive-world.

On the extremely opposite end, if a person's consciousness is established to the consciousness of Cerebral-centre, the person is an *avatar* (Divine incarnation—an enlightened person since birth); and is ever free. He takes 'birth' *only* to do tasks bestowed upon him by God. He has to work for mankind, as ordained by Supreme-Power, with no other desire. By doing His-work, he himself gains nothing (e. g., Gita III:18, 22). *Maya* is powerless before his will-power to attach him into the delusive-world. The person's state of consciousness is like that of valence electrons, which require little will-power to get freedom from entanglement of the delusive-world.

For other human beings consciousness, at any given time, clearly lies between these two-extreme states of consciousness.

Electric and Magnetic Field

We know that the charged particles exert electrostatic forces on each other—if these are in the vicinity of their electric fields. We may consider 'one' charged particle as an 'observer' (or subject), and 'another' charged particle as an 'object'. From the view of 'consciousness' the phenomenon may be understood and explained as follows:

The charged particles have their own respective electric field, *but* are not conscious of their 'own' *individual* electric fields. When the charged particles are brought together in the close proximity of their electric fields—these immediately become conscious of each other's presence—by the experience of electrostatic forces—that these particles exert on each other. Greater the magnitude of 'charge' on a particle—larger will be its 'field of action'—in terms of distances. *But its power of attraction or repulsion cannot act upon itself.*

We may observe a similar phenomenon in the case of magnetic field.

The Field of Consciousness

If a charged-particle comes into contact of another charged-particle's electric field—however, small magnitude of 'strength of field' these two charges may be carrying—these two become conscious of each other. Hence, if the 'strength of field' of first-charge is increased, in comparison with that of second-charge—the first-charge will always be *conscious* of the second-charge, even, at large distances and vice versa. On exploration—if 'strength of field' of first-charged-particle be made 'infinite'—the first-charged-particle as well as the second-charged-particle will always be *conscious* of each other—no matter, how far these 'two charged particles' are placed from each other.

Therefore, this leads to a tremendous understanding that—in principle and in practice—it is possible to expand the 'field of consciousness' to 'infinity'—that may cover the entire cosmos—where, the 'observer' and 'object' are always conscious of each other—no matter, how far they are placed from each other.

Quantum Physics

It would be interesting to compare the 'state of consciousness' to the probability of finding a particle in space, tunneling phenomenon, equation of continuity, uncertainty principle, and wave-particle duality (Schiff, 1949).

Normalization Condition

The normalization-condition[2] of a wave function $\psi(\bar{r},t)$ says that the probability of finding a particle in space is always one.

Quantum Tunneling Effect

Let a particle with energy E_o, approaches a potential-well of energy V, with $E_o \leq V$. It says that there will always be a non zero probability that the particle will penetrate the wall of potential well to reach the other side of well. From the classical physics approach, a particle cannot penetrate the potential-well.

Equation of Continuity

Equation of continuity[3] is associated with the position probability density $\overline{\mathrm{P}}(\bar{r},t)$[4] and probability current density $\bar{J}(\bar{r},t)$[5], in which there is no source or sink.

---

[2] $\int_s |\psi(\bar{r}.t)|^2 d^3\bar{r} = 1$, where, $\psi(\bar{r},t)$ represents the wave function of a particle.

[3] $\frac{\partial \overline{\mathrm{P}}(\bar{r},t)}{\partial t} + \overline{\nabla} * \bar{J}(\bar{r},t) = 0$, equation of continuity associated with the flow of fluid.

[4] $\overline{\mathrm{P}}(\bar{r},t) = \psi^*(\bar{r},t)\psi(\bar{r},t) = |\psi(\bar{r},t)|^2$, Position probability density.

The Inferences

Our consciousness has always a non-zero probability of its being conscious of divine-presence in each of us. We may consider 'consciousness of whole universe' equal to 'One' and to 'Consciousness of God'—the Brahman. A mathematical expression represents the truth beautifully as follows;

$$\int_s Sum-of-whole-conciousness \equiv \int_s Conciousness-of-God \equiv Brahman \equiv One$$

The 'whole purpose' and 'aim' of life is to become 'one' with God—to be 'one' with Consciousness of God— as, there is one and only one Reality—God—and God-Alone.

Our consciousness is scattered among the various centers as of a probability distribution. However, 'deep' our consciousness may be rooted to any of the lower-centers—consciousness may always *tunnel* through an impenetrable-fortress (≈ *infinite potential-well*) that may seem impossible to penetrate—from the view of human-consciousness—consequently, to get a rise to the higher-centers—to become conscious of divine-events.

A continuity view (e. g Gita VII: 6; X: 32):

> His Consciousness has—no-source, hence, no-sink; no-beginning, hence, no-end; no-birth, hence, no-death—but is ever flowing in continuum; with no past, present or future. He himself becomes the Source, and the Sink; He himself becomes the Beginning, and the End; He himself becomes the Birth, and the Death. He himself becomes the ever flowing in continuum— with himself becoming the Past, the Present, and the Future. He himself becomes pure-Bliss of Love, pure-Consciousness of Love—ever-flowing, ever-drenching, and ever-saturating his children of Love; with his nectar of Love.

The Uncertainty Principle

We will see how uncertainty principle, in a subtle way, plays its role with consciousness of the two-worlds, "Ye cannot serve God and Mammon [together]" (Matthew 6:24). An aphorism of Einstein, "As far as the laws of mathematics refer to reality, they are not certain; and as far as they are certain they do not refer to reality" (quoted in Capra, 1991, p. 49).

Relation with Consciousness

It is not possible to have consciousness of two-extreme 'centers' simultaneously. The person whose consciousness is bounded to the Coccygeal-centre; it would be hard to expect a saintly behavior from him. He is *only* conscious about the world, and is oblivious to the God or idea of Self-realization. He is far away from the concept of divinity. For him this world is the beginning and the end, with his consciousness heavily deluded by *Maya*.

A saintly behavior is a characteristic of a person, whose consciousness is established in the consciousness of Cerebral-centre. He is born with the predominant consciousness of divine-world. He is a Self-realized person; a lover of God, and mankind as a whole. It would be hard to expect an inhuman and instinctive behavior from him. It would not be possible for him to deviate from his saintly behavior, even, in worst of the situations. He is not oblivious

---

[5] $J(\bar{r},t)=\frac{\hbar}{2im}\left[\Psi^*\nabla\psi-\left(\nabla\Psi^*\right)\Psi\right]$, Probability current density.

of the world; *but* immune to *Maya* of the delusive-world, and unattached to it. He lives in the world like a lotus; with his consciousness surrendered and completely submerged into God. He consciously lives for God, and God-Alone—to serve and guide His-children. Ramakrishna Paramahansa—the saint of Dakshineswar was the living embodiment of divinity (Rolland, 2003).

In quantum physics parlance, it is possible to depict the expression as—If a person's consciousness is certain by the Coccygeal-centre, then his consciousness becomes uncertain to the consciousness of Cerebral-centre or vice versa.

Wave-particle Duality

Light exhibits the dual nature. It shows the particle and wavelike characteristic under different experimental conditions. The phenomenon is known as the wave-particle duality, and is fundamental to the nature (Beiser, 1969). The interference and diffraction are the phenomena, where light exhibits the wavelike behavior. Black body radiation and photoelectric are the experiments, where light shows the particle like behavior.

Dual Nature of Electron

We know that the electrons behave like a particle, and their motion produces electric current. *But* scientists have been able to exploit, and let electrons behave like a wave. The result is Electron Microscope; in which, the very notion of electrons as a wave is used to magnify an object. Similarly, there are other advanced techniques, where wave properties of the subatomic particles are used to probe deeper into the matter.

An electron, as a particle, is *localized*; subject to space-time limitation; and governed by the laws of motions. While an electron, as a wave, is *not localized*; and *its entity as a wave is spread out in the whole-space sans any boundary, or division.* As a wave, electron is neither bounded by the space-time limitation nor governed by the laws of motions. Electron, as a wave, seems to be altogether in a very different realm of its own. Scientists have to develop not a definite, but a *probabilistic* theory of quantum physics. Perhaps, a *holistic* approach to deal with this 'little invisible entity'!

Consciousness: As Duality of the World

We are born with inheritance of consciousness imbibing 'duality' as a gift of Nature. Duality is a hallmark of our consciousness by virtue of our existence as a human entity. The very 'birth' signifies its duality by fixing its opposite pole—'death'. The most meaningful and significant aspect is creation of an individual soul by separating it from the Spirit—Spirit and soul—two different concepts, but one in implication.

Maya—a powerful delusive force and the root cause of duality is fundamental to Nature. The dualities such as 'birth and death', 'heat and cold', 'evil and good', 'happiness and sorrow', or male and female' exist only in our consciousness; and cause of all miseries in the world. Lord Krishna counsels, "Arjuna raise your consciousness from duality of the delusive world. Be thou indifferent to these dualities, and raise yourself from this [oppressive] pairs of opposite—'cold and heat', 'joy and sorrow', 'honor and ignominy' and establish yourself in Me" (Gita II:45; VI:7-10).

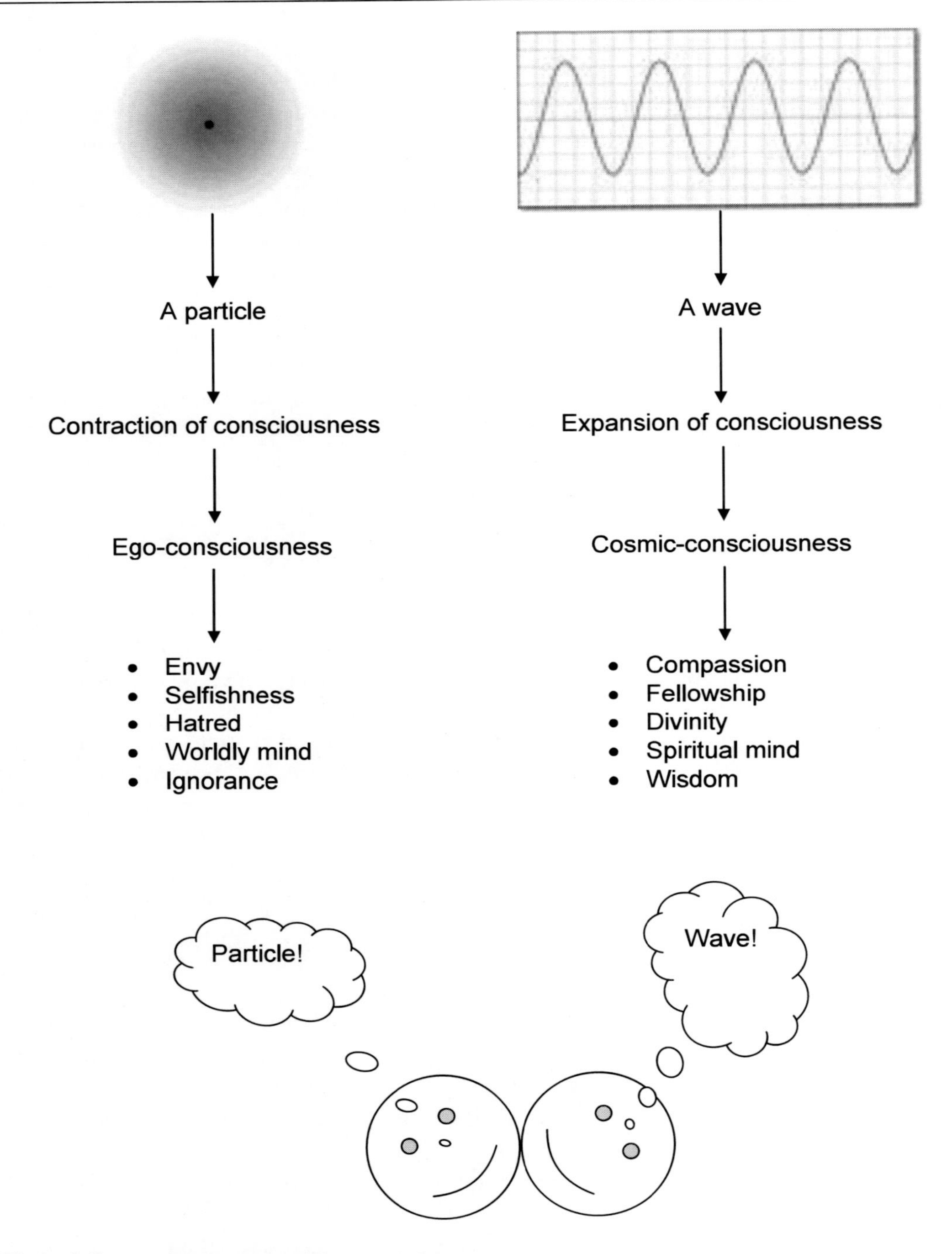

Figure 3. Is consciousness a particle or wave?

> Oh! Arjuna, conquer your lower-self that behaves as one's enemy by your Higher-Self that behaves as one's friend, which (Higher-Self) is established in Brahman—by raising your consciousness from the lower-self to that of Higher-Self. As, verily your lower-self is your enemy; and your Higher-Self is indeed your friend. (Gita VI:5,6)

Consciousness: A Particle or Wave?

Consciousness, when confined to an individual-soul, behaves like a 'particle'; and is subject to 'finite laws' (tribulations and turmoil) of the world (e. g., Gita IX: 8). When consciousness embraces Cosmic-Consciousness, it becomes free; and behaves like a 'wave' pervading the entire cosmos. It is free from finite-laws of the world, free from description of the language, and free from duality of Nature (Figure 3). We may express the *reality* as follows,

$$Partcile \equiv Confinment-of-Consciousness-to-an-Individual-Soul \equiv Ego-Conciousness$$

$$Wave \equiv Omnipresence-of-Consciousness \equiv Cosmic-Consciousness$$

CONCLUSION

The *conscious* searches, through physical- or spiritual-sciences, come to the conclusion that the purpose of life is to realize Supreme-wisdom or to attain Self-realization.

With physical-science search begins with 'observation and analysis'. And spiritual-science begins its search with the 'enquiry of Self'. Both the approaches are in search of finding the ultimate-Reality, the ultimate-Truth, and the ultimate-Goal. Or, the theory of everything: Hawking (2006) writes,

> We are trying to find a unified theory that will include quantum mechanics, gravity, and all other interactions of physics. If we achieve this we shall really understand the universe and our position in it. (p. viii)
>
> A complete theory, it should in time be understandable in broad principle by everyone, not just a few scientists... . If we find answer to that, it would be the ultimate triumph of human reason. For then we would know the mind of God. (p. 124)

In the recent years, scientist are in search of the elusive and most fundamental particle—'the particle of God'. *But* the particle, be it fundamental or not, has been derived from the same Source of energy—the Spirit.[6] A scenario, at the International Linear Collider Workshop:

> In the subatomic world, forces can also transform one kind of elementary particle into another... . There was only one kind of matter particle at the time of the Big Bang, which then took on many seemingly different forms as the universe cooled down... . Mean that the 45 different kinds of matter particles that are known today are really the same particle in different guises. (Discovering the Quantum Universe, 2005, p. 29)

Ironically, no matter how much we may have advanced scientifically... still, after years of 'evolutions' and 'advancements'—there will be innumerable higher-levels of realms would remain uncovered to be probed by the human being, in order to enhance his capabilities, and engage in the work of creation—through discoveries and further more discoveries. It is an ongoing and unceasing process. Science is a beautiful tool, and in its unique way attempts to uncover the secrets of the wonderful and beautiful creation. Yet!—The Creator is more

---

[6] $E = mc^2$, shows the equivalence of energy and mass, where c is the velocity of light.

beautiful—"But seek ye first the kingdom of God, and his righteousness; and all these things shall be added unto you" (Matthew 6:33).

Even, at the pinnacle of scientific achievements, the individual has 'unquenched thirst' for Self-realization. As if *something is missing* from his consciousness to make it 'whole', and to let it saturate with 'all pervading knowledge'. He realizes that virtually he has discovered nothing; rather, he is just working as an instrument; a fortunate chosen one *babe* of the Creator, and to let it (discoveries) happen through him. He is just a babe—a playing one, before the awesome mighty Nature. Nature herself uncovering her *secrets* before her babes—step-by-step—through the medium of science—to let her babes keep on playing! "Thou hast hid these things from the wise and prudent, and hast revealed them unto babes" (Matthew 11:25). Great Newton just before his death wrote:

> I do not know what I may appear to the world; but to myself I seem to have been only like a boy, playing on the seashore, and diverting myself, in now and then finding a smoother pebble or a prettier shell than ordinary, while the great ocean of truth lay all uncovered before me. (quoted in Sagan, 1980, p. 58)

Science, though a beautiful and elegant way to study nature—*just stops short of describing the ultimate-Reality*. In description of the physical-reality our consciousness *struggles* with 'duality' of the physical-world, till it gets freedom through 'spiritual approach' into the spiritual-world. 'The Copenhagen Interpretation… acknowledge [s] that a complete understanding of reality lies beyond the capabilities of rational thought… . The new physics was not based upon "absolute truths", but upon *us*' (Zukav, 2001, p. 63).

> Sooner or later we have to accept something as given, whether it is God or some other foundation for existence. Thus "ultimate" questions will always lie beyond the scope of empirical science as it is usually defined. It is probably impossible for poor old Homo sapiens to "get to the bottom of it all." Probably there must always be some "mystery at the end of the universe". (Davies, 1992, p. 15)

We then come face-to-face with the reality of world, and take a step towards Self-enquiry for the 'deeper scientific search', and seek refuge (solace) in the spiritual-science. Austrian physicist and Noble laureate, Ervin Schrödinger used to keep and draw inspiration from the ancient Indian scriptures—The Upanishads (Schrödinger, 1948). The renowned and learned German oriental philosopher, Max Muller used to say, "I take refuge in Lord Buddha" (quoted in Arnold, 1999, p. 3).

Who *am* I? What is the *purpose* of my life? Struggling through all these questions one 'follow and lead' to the 'path' of Self-realization. *But*, in order to to achieve Self-realization—a tremendous 'struggle' with *self*—immense 'will-power and determination'—unswerving 'faith' in Self and God, are essential and requisite. An infinitely times greater struggle is involved, in comparison, to get material success. *But* above all—*the Grace of Almighty*!

At the peak of intellectualization, sooner or later, our intellect and consciousness of its own make transition into the spiritual-world. How and when: it is—'His desire', 'His decision', and 'His choice'. Only then a 'real journey', a 'real research' and 'real science' of Self-realization commences. Science as advances, 'unknowingly and involuntarily' searches the deep esoteric human-nature. Scriptures say a human-body is in *itself* a mini-universe combining all the 'manifested and hidden' laws of cosmos. All the principles of physics, if we are able to look into their deeper esoteric meanings, points towards what is happening in our own physical body; and it, in turn, is governed by the spiritual-laws. These could only be understood if we have the consciousness of Higher-Self.

Sciences, or for that matter any study, is an exercise in intellectualization; and search for the Higher-Self starts after that. Higher-Self is beyond intellectual approach, and may be realized intuitively through the practice of spiritual-science. Gita says, "Senses are superior to the body, mind is superior to the senses, intellect is superior to the mind, but Me—the Self—is superior to all" (Gita III: 42). He is the Supreme-Reality: beyond—gross or subtle, matter or non-matter. 'Finding Him' is the highest and noblest kind of *search*. He is above the defilement of language; even, the scriptures can only points towards Him—in order to show the pathway.

The two-worlds are not *disjoint or mutually exclusive*, but are 'deeply and esoterically' 'interlinked' to each other. What is happening, in the one-world, is correspondingly influencing and affecting the other-world as well. Quantum physics may severe as a beautiful 'bridge' between these two worlds, and is in correspondence with the spiritual-world. The passage, from the physical-world to the spiritual-world, is through deep-tunnel penetrating through the spiritual-eye, "the entryway into the ultimate states of divine consciousness" (Yogananda, 2001, p. 440). An advanced yogi has a consciousness of the physical- or spiritual-world. He has a choice, using his will-power, to live in either of the worlds. A celestial way:

> The possibility of wormholes to get from one place in the universe to another without covering the intervening distances—through a black hole. We can imagine these wormholes as tube running through a fourth physical dimension… . But if they do [exist] , must they hook up with another place in our universe? Or is it just possible that wormholes connect with other universes, places that would otherwise be forever inaccessible to us? For all we know, there may be many other universes. Perhaps they are, in some sense, nested within one another. (Sagan, 1980, p. 221)

Gita echoes, "All manifestations of *Prakriti* are interlinked with each other like a pearl in the necklace in Me" (Gita VII:7).

A scientific mind helps to understand the nature rationally and logically. This, in turn, helps to understand and follow the spiritual-path in a systematic manner. A spiritual-path is a 'path'—*to seek and follow*—and to comprehend the deep esoteric human-nature of Higher-Self. This is a similar to the process in science to seek the truth by searching, probing and testing the physical-reality.

On extrapolation, these two paths come to the same point of realization of a 'Cosmic-Intelligence': An *Intelligence* which is hard to describe in language—or, to say, it is beyond the perimeter of language. This Cosmic-Intelligence is named *Sat-Chit-Anand*—Existence-Consciousness-Bliss—the God.

A Beginning

The *conscious* search—be it physical or spiritual will continue to be there—till we, nay everyone—reach the Supreme-Goal—the goal of Self-realization—the God, or by whatsoever name we may call it. He is omnipresent. He is keen on his beloved children to come to their Home—'Abode of God'. Lord Sri Krishna extols, "By offering the highest love and devotion to Me, My devotees shall come to Me" (Gita XVIII:65-68).

EPILOGUE

> We do not ask for what useful purpose the birds do sing, for song is their pleasure since they were created for their singing. Similarly, we ought not to ask why the human mind troubles to fathom the secrets of the heavens… . The diversity of the phenomena of Nature is so great, and treasures hidden in the heavens so rich, precisely in order that the human mind shall never be lacking in fresh nourishment—Johannes Kepler, *Mysterium Cosmographicum.* (quoted in Sagan, 1980, p. 32)

When one desires to enjoy Bliss—Joy of God, one has to follow the practice of meditation. This is the one Joy that never fades away, never saturates, and never deceives. And this Joy joyously increases day-by-day. The scriptures call this Joy—Bliss—*Ever-New-Joy*. The Upanishad echoes, "From Joy we have come, in Joy we live, move and have our being, and in that scared Joy we shall melt again" (Taittiriya III:6). hakur Rabindranath Tagore eulogizes the sweet love of God:

> Thou art my life, Thou art my love,
> Thou art the sweetness which do I seek.
> In the thought by my love brought,
> I taste Thy name, so sweet, so sweet.
> Devotee knows how sweet You are,
> He knows, whom You let know. (in Yogananda, 2003, p. 18)

ACKNOWLEDGEMENT

I am grateful to Yogoda Satsanga of Society of India and Self-Realization Fellowship, USA for its Teachings. My sincere thanks are to all those with whom I have thoughtful discussion on the manuscript.

I dedicate this work to the great Masters and my beloved Mother.

REFERENCES

Arnold, E. (1999). *The Light of Asia*. New Delhi: Srishti Publishers.

Beiser, A. (1969). *Perspective of Modern Physics*. New York: McGraw-Hill.

Capra, F. (1974). Bootstrap and Buddhism. *American Journal of Physics*, 42, 15-19.

Capra, F. (1991). *The Tao of Physics*. London: Harper Collins.

Davies, P. (1993). *The Mind of God*. London: Penguin Books.

Discovering the Quantum Universe: The Role of Particle Collider. 2005 International Linear Collider Workshop, Stanford, California, 18-22 March. (2005). Retrieved May 10, 2010, from http://www.interactions.org/quantumuniverse/dqu.pdf.

Ford, K. W. (1965). *The World of Elementary Particles*. New York: Blaidell.

Halliday, D., Resnic, R., and Walker, J. (2004). *Fundamentals of Physics*. New York: John Wiley and Sons.

Hawking, S. W. (2006). *The Theory of Everything*. Mumbai: Jaico Publishing House.

Rolland, R. (2003). *The Life of Ramakrishna*. Kolkata: Advaita Ashram.

Sagan, C. (1980). *Cosmos*. New York: A Ballantine Books.

Schiff, L. I. (1949). *Quantum Mechanics*. New York: McGraw-Hill.

Schrödinger, E. (1948). *What is Life. London,* Cambridge University Press.

Schrödinger, E. (1984). *Why Not Talk Physics*. In K. Wilber (Ed.), Quantum Questions. London: Shambhala.

Yogananda, P. (2001). *Journey to Self-Realization*. Kolkata: Yogoda Satsanga Society of India.

Yogananda, P. (2002). *God Talks with Arjuna-The Bhagavad Gita: Royal Science of God Realization (Vols. 1 and 2)*. Kolkata: Yogoda Satsanga Society of India.

Yogananda, P. (2003). *Words of Cosmic Chants*. Kolkata: Yogoda Satsanga Society of India.

Yogananda, P. (2008). *Autobiography of a Yogi*. Kolkata: Yogoda Satsanga Society of India.

Zukav, G. (2001). *The Dancing Wu Li Masters*. London: Random House.

In: Consciousness: Its Nature and Functions
Editors: Shulamith Kreitler and Oded Maimon
ISBN 978-1-62081-096-5

Chapter 11

TOWARD A MORE COMPREHENSIVE UNDERSTANDING OF MIND: A MUTUALLY INTERACTING NON-LOCAL DUALISTIC SYSTEMS (MINDS) APPROACH*

***Charles T. Tart*†**
Institute of Transpersonal Psychology, Palo Alto, California, US
University of California Davis, Davis, California, US

ABSTRACT

A useful scientific theory begins by finding or creating a conceptual model or framework which sensibly and efficiently organizes existing, relevant data and then goes on to propose a deeper understanding of phenomena that leads to prediction and, where possible, control. Major contemporary theories of consciousness are almost always set within an unquestioned belief that an exclusively materialistic (physical objects and forces, typically Newtonian) model of the world will eventually explain everything about consciousness as an aspect of brain functioning. While this kind of modeling is very useful for understanding many aspects of consciousness, particularly brain functioning, it cannot become an adequate and comprehensive theory of consciousness as it completely and irrationally ignores a variety of well-established psi phenomena (telepathy, clairvoyance, precognition, psychokinesis, and psychic healing, e.g.) which do not fit within a Newtonian physical model (although there is some speculation that they may fit within a quantum physics model). Thus a "dualistic" approach is here proposed, calling

---

* This chapter is a greatly updated version of a theory originally published in Brain/Mind and Parapsychology: Proceedings of an International Conference, held in Montreal, Canada, August 24-25, 1978, pp 177-200 under the auspices of the Parapsychology Foundation, New York, NY Tart, 1979). In this early version I called this approach Emergent Interactionism. Material from that earlier chapter is used here by permission of the Parapsychology Foundation.

† As this chapter was in press, I came across a major step forward in applying systems approaches to altered states and to NDEs, viz. David Rousseau's Near-Death experiences and the mind-body relationship: A systems-theoretical perspective, in the *Journal of Near-Death Studies*, Volume *29*, No. 3, Spring 2011, 399-435. Rousseau expands and depends the perspective I have taken in this chapter and applies it profitably to understanding Near-Death Experiences.

for recognition of these phenomena as inherent in consciousness and for research detailing their properties. This approach, *M*utually *I*nteracting *N*on-local *D*ualistic *S*ystems, MINDS, is intended to be scientifically useful and has empirical, testable consequences. It is pragmatic in emphasizing observable consequences of theory, posits mutual interaction between physical and mental aspects of consciousness, the latter of which have non-local, dualistic qualities, and the interaction of which, occurring in terms of modern system theory, results in consciousness as we ordinarily experience it, as well as altered states of consciousness (ASCs).

The MINDS approach calls for understanding consciousness in terms of two qualitatively different aspects of reality, what I have termed the *B system,* the brain, body, and nervous system and the physical laws which govern it, and the *M/L system*, the mental and life aspects of reality. Ordinary consciousness, as experienced and manifested, is then seen as a systems property, an emergent from the auto-psi interaction of the B and M/L systems. Further understanding of consciousness, then, while it requires further and extensive development of conventional approaches in the study of brain functioning and physical laws, also requires extensive development of our scientific knowledge of psi phenomena, as well as development of general systems theory, so principles of emergence in complex systems can be better understood. While this view is complex, it is more adequate to the reality of consciousness and psi than an overextended physicalistic monism, a promissory materialism, and exciting discoveries await us!

INTRODUCTION

Throughout my career I have been interested in a range of phenomena I usually group under the heading of altered states of consciousness, phenomena dealing with fascinating, large-scale, multitudinous changes that can take place in people's mode of experience. While changes in the manifestations of consciousness can be studied without asking any fundamental questions about the nature of consciousness, so avoiding dilemmas that have puzzled philosophers through the ages, I have, nevertheless, always been curious as to the ultimate nature of consciousness, and I have found certain parapsychological phenomena, *psi* phenomena, to be of great value in pointing to the direction of an answer to that question. What I will write about here is the beginnings of a scientific approach to understanding the basic nature of consciousness and altered states of consciousness, an approach that I call *M*utually *I*nteracting *N*on-local *D*ualistic *S*ystems, MINDS. This is an approach that would be classified philosophically as dualistic, and yet is grounded in empirical observables and has observable consequences, and so can be classified as scientific. I do not lay any great claim to originality in this approach, as I have drawn on multitudinous sources for the basic ideas, but I hope the particular way I have put these ideas together will be useful in understanding human consciousness. I shall also apologize in advance for the crudeness and gaps in these formulations, for, while we know enough to realize how much conventional, materialistic approaches to understanding consciousness need to be expanded, we still know very little about the actual properties of the mental aspects of explanation called for in the MINDS approach.

My observational base for trying to understand consciousness begins with my own experience, which is then expanded by my experiences of the world and others about me and by others' understandings. Perhaps the most striking thing about my own experience is the obviously different nature of my consciousness from the physical world about me. Despite difficulties in knowing precisely how to think about it or express it, it is simply a given that there is something fundamentally different about the experiences I call my mental processes from what I call the external, material world. This basic distinction has been drawn by multitudes of others, and in formal philosophy has been called a dualistic position, a formal postulation of some fundamentally different qualities of mind and matter, such that the nature of one cannot be adequately explained by or reduced to the other.

I have had no formal training in philosophy, but at times in the past I have attempted to study philosophical literature on the nature of consciousness, and, I must admit, I have always come away baffled and disappointed. Once the basic distinction between mind and matter is postulated, I get the feeling that most philosophers lose themselves playing word games, dealing with purely abstract and semantic distinctions, and end up with a dualistic position that might or might not be true; but the truth or falsity of that dualism does not seem to have any useful experimental or experiential consequences that I can discern. I am not comfortable with making distinctions that have no observable consequences, and I am strongly committed to the kind of scientific pragmatism that says observable consequences (whether physical or experiential) have ultimate priority over intellectual formulations. My understanding of the fundamental nature of scientific inquiry have been described elsewhere ((C. Tart, 1972, 1998a)).

Monistic Views

In terms of contemporary acceptance by the intellectual and scientific community, monistic materialistic philosophies, which postulate that mind and matter are basically manifestations of the same thing, that they are totally reducible to one another, are the accepted philosophies. This monistic materialism is particularly dominant in orthodox science. Figure 1, for example, an updated version of a figure from my *States of Consciousness* book (C. Tart, 1975a), diagrams this widely accepted scientific view of consciousness, what I have there labeled the "orthodox" or scientifically conservative view of the mind. The basic reality that is being dealt with in this diagram is physical reality, fixed physical reality, governed by immutable laws. As a result of these laws, a particular physical system comes into existence, the brain and its associated nervous system and body (which I shall just refer to as brain, for short, for the rest of this chapter). Many aspects of this brain are fixed in their functioning: instructions for your kidneys to work, for example, are encoded in the physical structure of the brain (and kidneys) and ordinarily never changed. We would compare this to fixed properties of the BIOSs of our computers nowadays. This physical structure also has many programmable capacities, so our culture, our language, the various events of our personal history, and our interactions with physical reality teach us a language, a way of thinking, values, and mores, etc. This large, computer-like physical structure then functions in a wide variety of complex ways. At any given moment we are aware of only a small fraction of the total functioning of it, and this tiny fraction of physical functioning that we are aware of is consciousness as we experience it. The exact way this electrochemical brain activity becomes consciousness is a mystery, now termed the *hard problem* ((Chalmers,

1996)), but monistic materialist approaches have faith that someday the hard problem will be solved in physical terms[1].

In presenting this model I have added something called "pure awareness" in the upper part of it, which can refer in a general way for our context here to those feelings of mental activity which do not seem tied to obvious bodily or sensory functioning, such, as certain meditative experiences or various altered states experiences. More formally. I have used the term “pure awareness” to mean that raw proto-experience of knowing you are conscious without any particular content (Forman, 1999)), knowing that something, your basic awareness, is happening before that experience gets absorbed into particular content and highly elaborated and articulated into semantic categories, where it has obviously been influenced by brain structure. In the figure, I show pure awareness as "emerging" from the physical structure of the brain. That is, Figure 1 is a representation of the monistic, *psychoneural identity hypothesis*, which says that while we might find it convenient to distinguish certain types of “mental” activity for semantic purposes, all experience is, in principle, completely reducible to physical activity within the brain. Whatever your mental experience, some day it will be totally reducible to something like “Activity of A quality in brain structures B, C and D),” and producing that physical activity by direct physical means (electrical or chemical stimulation, e.g.), will produce that mental experience. In terms of computer analogies, the psychoneural identity hypothesis predicts that any mental experience can, in principle, be produced by stimulating the appropriate areas of the brain in a certain way. In practice we are a long way from being able to carry out this reduction or control due to the sheer complexity of the brain, but in principle the orthodox, scientific view believes this is possible.

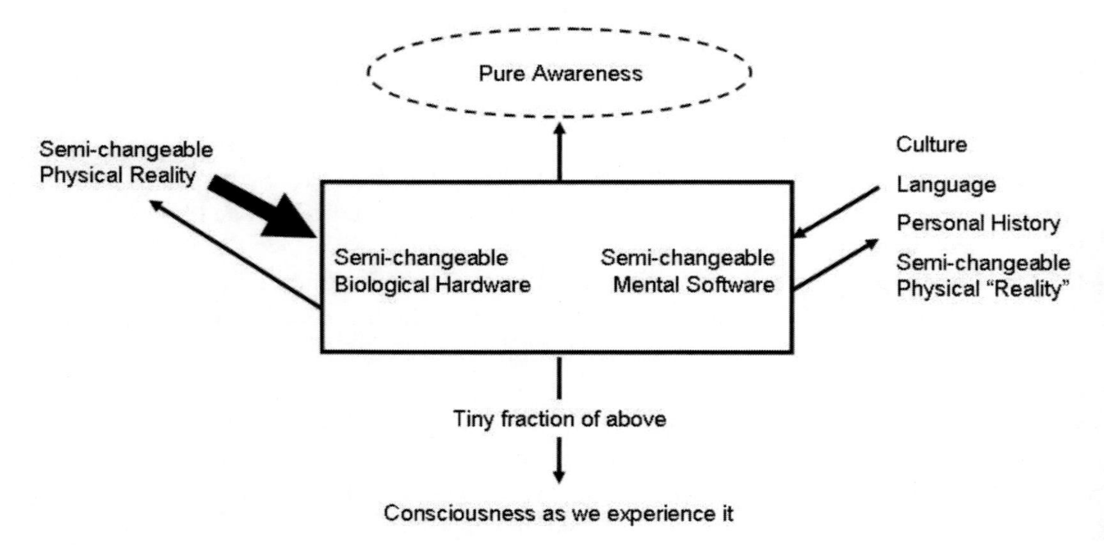

Figure 1. An orthodox, monistic materialist representation of the nature of human consciousness.

[1] Philosophers who are not impressed with this leap of faith refer to this position as promissory materialism, and regard it as inherently unscientific. Proper scientific theories can, in principle, be disproven, whereas you can never disprove that consciousness (or anything) will be explained in terms of X (your favorite explanatory framework) someday... .

The psychoneural identity approach is clearly a useful scientific approach, for it has observable consequences. Again, it predicts, that a physiological correlate of any and every kind of experience can ultimately be found. It further predicts that no experiences can occur *in reality* that violate the basic physical laws and system operation laws that govern the operation of the brain, although the brain may produce *illusory* experiences that *seem* to violate basic physical laws.

COMPLEXITY

In mentioning laws governing systems operation, I am reminded of the other disappointment I had with formal philosophical writings on the nature of consciousness. That was their typical obsession with an absolutistic understanding of simple mental events, when it has always been obvious to me that consciousness represents an exceptionally complex and dynamic *system*, not a simple mechanism. My early experience in working with complex electronic systems – I was a ham radio operator and then a Federally licensed Radio Engineer - where alterations in one component could have many effects on the whole system's operation, effects which were often not at all obviously predictable beforehand, sensitized me to this issue. Modern brain theory now recognizes the complexity of the brain and nervous system. Starting from a simplistic approach that likened each neural junction to a relay and thought the complexity of brain function could be handled by a simple additive operation of all these individual relay operations[2], modern understandings of the brain are increasingly looking to general systems theory ((Miller, 1978)) to provide general laws about emergent properties of brain functioning, properties that are holistic outcomes of total system operation rather than simple linear additions of more basic subsystem elements. The MINDS approach to understanding consciousness that I shall outline here tries to take this complexity, these mutually interacting, emergent system properties of brain functioning (and, as we shall see, of *mind* functioning, the dualistic and distributed, non-local aspects) into account, as well as dealing with the fundamental experience of a dualistic difference between experience and the physical world.

As a final introductory note, I should say that if I had to characterize my philosophical bias it is to be pragmatic. As a scientist, I am committed to the proposition that data, experience, is primary, and our conceptualizations, our theories about the meaning of that data are secondary. If I cannot adequately or logically express my data or experience, that is a shortcoming of my philosophy or grammar, not an invalidation of my date or experience. Theories must always be adjusted to account for the data, and theories must have consequences in terms of observable data. If any theory has no testable consequences, it may be intellectually interesting, but it is not scientifically worthwhile. I believe that the dualistic

[2] In graduate school fifty years ago, there was an assignment in the neurophysiology course I was taking to see how many electrical circuits we could devise from just three neurons, back when a neuron was thought of over-simplistically as simply an on-off device, a kind of relay. With my previous experience in electronics I readily thought of half a dozen circuits I could make with only three neurons. Given a zillion such neurons in the brain, the possibilities were staggering, so I decided not to continue in neurophysiology, it would take too many researchers too many lifetimes to really figure out the brain this way. Now, of course, that simple "relay" has dozens of inputs, computes various functions all by itself, is bathed in a chemical bath of neural messenger substances further modifying its activity, etc.... Thus I applaud all those working to understand neural and brain functioning, but am even more overwhelmed by the complexity of it.

theory of consciousness I shall now present, MINDS, has such testable consequences, and so forms a basis of a scientific set of theories about consciousness.

Another aspect of this pragmatism is that I do not want to get into the kind of absolutism that marks philosophical discourse. I have no way to satisfactorily define concepts like “mind” versus “matter” or “mind” versus “brain” in any kind of absolute fashion. If I say that something is “mental” or “non-material,” what I am saying is that something seems to have observable or experiencable properties which cannot be adequately explained in terms of our current understanding of the physical world, or reasonable extrapolations of that understanding. It is quite possible that future advances at the cutting edge of physics will drastically change our conception of what is and is not "physical." and what can and cannot be handled within a physical explanatory system. Radin’s recent book ((Radin, 2006)) arguing that psi functions are quite sensible within a quantum world view, or Stapp’s recent book ((Stapp, 2007)) arguing that awareness/consciousness is an inherent aspect of the quantum view of the universe, develop these ideas, although I will not pursue them here. Thus, in distinguishing mind and brain, I am doing no more than making distinctions which are pragmatically useful at present, regardless of their absolute validity.

Paraconceptual Phenomena

The basic support for my dualistic approach to understanding consciousness comes from the excellent scientific evidence for the existence of certain "paranormal" phenomena. Given our current understanding of the Newtonian physical world, it is possible to talk about completely isolating or shielding one event from another, so that no known, feasible form of information transfer channel exists between two such shielded, isolated objects or events. If we now make physical observations, either the behaviors of people or the readings of physical instruments, which indicate that an information or energy transfer has nevertheless occurred between two such isolated events. we have a paranormal or, more appropriately, a para*conceptual* event. We have an observation that cannot be satisfactorily explained by our theories. Since the majority of the population in America believe they have experienced some kinds of psi. ((Greeley, 1975)). these events are hardly para*normal*, beyond the norm, but they are certainly paraconceptual to the orthodox, current Newtonian scientific view of how the physical universe works. I shall use “paraconceptual” rather than “paranormal” in the rest of this chapter as a more correctly descriptive term.

There have been many types of observations reported on purported paraconceptual events happening in life, and these led to extensive experimental work on five kinds of experimental phenomena, leading us to postulate the existence of five basic types of paraconceptual events namely telepathy. clairvoyance (CL). precognition. psychokinesis (PK), and psychic healing, collectively referred to as psi events. There are dozens to hundreds of experimental reports supporting the existence of each of these kinds of effects. The studies leading to these existence claims are reviewed under the rubric of the *Big Five*, phenomena for which there is, in my opinion, so much high quality experimental evidence that rationality require us to recognize their reality, in my recent *The End of Materialism: How Evidence of the Paranormal is Bringing Science and Spirit Together* ((C. Tart, 2009)), as well as some other recent books ((Radin, 1997) (Braude, 2007) (Broughton, 1991) (Kelly, 2007) (C. Tart, Puthoff, & Targ, 1979)), so I shall only briefly describe them here.

Typically. we define (a) *telepathy* as mind to mind communication, (b) *clairvoyance* as matter to mind communication or sensing the physical state of affairs directly with the mind, (c) *precognition* as predicting a future state of events that cannot be logically predicted with any amount of knowledge of current conditions (that we might further subdivide into precognitive telepathy or precognitive clairvoyance), (d) *psychokinesis* (PK) as directly effecting a state of physical events simply by wishing for it, and (e) *psychic healing* as a direct effect of mental intention on biological organisms or processes, all of these under conditions which eliminate known physical mechanisms as carriers of information or force.

These conventional types of definitions have an implicit dualism in them, so we could be more formal and distinguish the above five phenomena simply by the kinds of experimental operations by which they have been established. Thus, (A) telepathy becomes a matter of a percipient making a behavioral response that is supposed to relate to what is in someone else's mind (as judged by his behavior), (B) clairvoyance as perception of a physical event that is not in, known to, anyone's mind (as fixed by the experimental situation) at the time, etc. Basically a mental intention ultimately results in a physical observable phenomenon. They require a dualistic view because no known physical properties of brains have been found to be able to produce these results, at least not in a Newtonian universe. This lack of possible physical explanation is expanded on in my recent *The End of Materialism* ((C. Tart, 2009)).

The existence of these paraconceptual or psi phenomena provides a general basis for arguing that a dualistic view of mind and matter is a useful, realistic and required view; that is, that it reflects the differing nature of physical and mental, rather than just being a semantically convenient distinction. The monistic view of mind and matter, the psychoneural identity hypothesis so widely accepted in science, is one result of a world view that totally denies (usually irrationally) the existence of psi phenomena as we experimentally know them. The existence of psi phenomena is a clear-cut scientific demonstration, however, that our understanding of the nature of the world is quite inadequate and will require major revisions. These paraconceptual events demonstrate the incompleteness of the overall conceptual system from which monism is derived. Thus, in a general sense, we can argue that a psychoneural identity position as a total solution to the question of consciousness is far from proven, because it rests on an incomplete and, therefore, faulty conceptual system. Research within a monistic materialistic framework is quite useful, of course, it must simple be understood that this is a convenient simplification for working with material kinds of effects, not a total explanatory system.

My psychological studies of consciousness and states of consciousness, as well as and especially my parapsychological studies, have forced me to go a step further than this and postulate that experience and high quality scientific data basically indicate that mind is of a fundamentally different nature than matter as we know it today, and, more specifically, postulate that *certain psi functions are the mechanisms of mind-brain interaction.* Consciousness, as we experience it, is an emergent property of this mind-brain interaction. This theory is represented in simplified form in Figure 2.

Here the factor I call "pure awareness" is shown as entering into the structure of consciousness from "outside," i.e. that something of a quite different nature than brain matter and processes is added to produce consciousness as we know it. Later we will call that dualistic factor "mind/life."

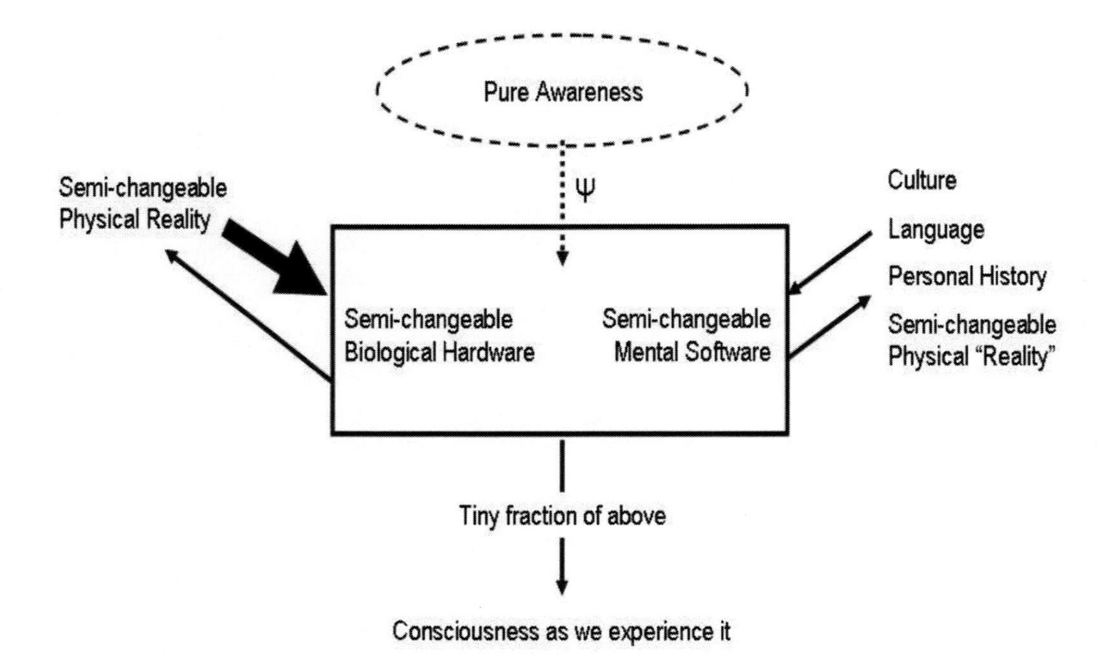

Figure 2. Simplified representation of the MINDS position, in which consciousness, as experienced, is an emergent, system property of two basically different component systems interacting via psi.

Thus consciousness, as we ordinarily experience it, is the higher level emergent of the psi interaction of brain and mind/life. To put it more formally, *experienced consciousness is a system property, an emergent, of the complex interaction of the subsystems of brain on the one hand and the "mind" factor on the other.*

The brain is, of course, the link between consciousness and the physical world about us. Environmental factors are detected through the sense organs and end up represented by electrical/chemical patterns within the brain. Is that all that is needed to put them "in" the "mind?" Monistic physicalism says yes. Actions begin as electrical/chemical patterns within the brain and end up as specific impulses to motor apparatus that create our overt behavior. Is that all there is to decisions to act? Again, monistic physicalism says yes, neural processes either create that decision or merely reflect that decision as an experience even though it has been made by non-conscious brain processes ((Libet, 1985)).

MINDS has a more complex view than physicalistic monism, at least for some perceptions and actions. The brain is an ultra-complex and especially interesting structure, for while many aspects of brain functioning seem completely determined, such as basic reflexes or patterns controlled by sensory input, many other important aspects seem to be under the control of quasi-random or fully random processes. That is, they are controlled by neurons or neural ensembles that are often almost-but-not-quite ready to fire ((Eccles, 1990)). My MINDS approach postulates that:

- *The mind/life factor can cognize important aspects of the state of the brain by means of clairvoyance*, that is, that mind/life uses clairvoyance to 'read" the brain and thus the state of the body and the body's immediate sensory world.

- Further, *the action of the brain is sometimes influenced at critical junctures by PK from mind/ life.* That is, in addition to self-organizational system properties of its own, there are control functions exerted over the brain by mind/life through psychokinetic modification of brain firing.

The holistic emergent of this interaction, the mutual interaction and mutual patterning of brain and mind/life on each other via clairvoyance and PK, leads to an overall pattern of functioning and experience that is consciousness as we experience it. Ordinarily, when we consult our own experience, we do not experience what brain alone is like, or experience what mind alone is like; we experience the emergent from their interaction, for which I use the term consciousness. Thus the mutual interaction (MI) in MINDS, between brain and a non-local (N) mind factors, resulting in systems (S) emergents.

Having sketched the basic postulates of MINDS in terms of "brain," "consciousness," and "mind/life," I must now face the semantic problem that others have used these and related terms in wider, overlapping and sometimes contradictory ways, as I myself have done in the past. While I could request that you listen to these terms only in the way I define them, it is not that easy to drop lifetime associative patterns, so I shall try to lessen semantic problems by adopting more neutral abbreviations, terms with less associational histories, for the remainder of this paper. I shall use the term "B system" to refer to those physical functions of the brain, body, and nervous system that we already understand in physical concepts or expect to understand with straightforward extensions of current physical concepts. I shall use the term "M/L system" for those non-physical, non-local (by current and straightforward extensions of current physical concepts) aspects I have been calling mind/life. I shall retain "consciousness," with a reminder that I restrict it to our usual experience of ourselves, not to more exotic experiences (I'll discuss altered states below). As for the psi interactions, I shall add the prefix *auto* to designate psi in general or clairvoyance or PK in particular that is concerned with a person's M/L system interacting with his own B system: thus auto-psi, auto-clairvoyance (auto-CL in later diagrams), and auto-PK. For those cases where psi reaches outside the physical bounds of the normal B system and M/L system interrelationship, as when we ask a percipient to tell us, e.g., what the order of a sealed deck of cards is or to describe some remote location, I shall add the prefix *allo-*: thus allo-psi in general, or more specifically, allo-clairvoyance, allo-PK, allo-telepathy, etc.

Figure 2 was a very general schematic of B system and M/L system interaction and their emergent properties. A more realistic schematic. using just present knowledge, would be of the sort shown in Figure 3.

This figure helps me introduce a number of further considerations. First, there are various hierarchical levels of organization in the B system alone, without even beginning to bring the M/L system into the picture. The lowest level shown on the left hand side of the figure would be individual neurons. These have properties we are beginning to understand fairly well, although constant new discoveries about their complexity keep coming in. Neurons are organized into basic neural ensembles at the next level, so this next systems level has emergent and system properties, the S in MINDS. That is, simple neuron ensembles can have properties which are not clearly predicted from those of neurons alone. These level two neuron ensembles are influenced by the lower level properties of neurons, and these level two properties in turn downwardly influence level one functioning, thus the arrows representing interaction, the I in MINDS. Similarly, neuron ensembles are organized into more complex ensembles etc., up to very high levels of complexity. System, emergent properties occur at all these various levels, as do numerous and complex interactions. It will not be an easy job to understand the B system, especially since the brain alone, without even bringing in the M/L system, is so many orders of magnitude more complex than any well understood present day system, even our best computers.

Although we know far more about the B system than the M/L system. I have assumed, on the basis of a principle of symmetry that has proven very useful in physics ((C. Tart, 1975a), chapter 18), that the M/L system itself is probably also a system of many hierarchical levels, and have diagramed it accordingly on the right side of Figure 3. I have avoided putting any labeling on that part of the scheme other than distinguishing the most basic life "energies" at the lowest levels versus more "mental" levels higher up in the system hierarchy. This is a matter of being cautious and not pretending to know more than we do know, but it seems very likely that there are fundamental aspects of the M/L system that interact with each other, produce more complex, emergent system properties, and so on, as with B system processes. All interaction within the M/L system is mediated by some kind of auto-psi, which might or might not be the same kind of psi auto-mediating M/L system and B system interactions. Thus we have an emergence of system properties on the M/L side of the diagram as well as the B system side.

I have shown auto-CL and auto-PK interactions between the B and M/L systems as potentially occurring between similar hierarchical levels of B and M/L subsystems, as well as potential cross level auto-psi interactions. There may be no single locus of interaction of auto-clairvoyance and auto-PK between the B and M/L systems, but a variety of interactions occurring at different levels. Lower level auto-psi interactions between B and M/L systems, then, may change the isolated properties of both neural tissue and basic life energies at those lower levels, which in turn are reflected in further interactions and system property emergence in both the B and M/L system levels, further complicating interactions at higher levels.

I regret that this is not the simplistic kind of picture we seem to prefer, but real systems are complex!

If one could separate out B system properties alone, one would observe an emergent that I have labeled as "mechanical brain" at the top of the left hand systems hierarchy in Figure 3. Similarly, if one could separate out M/L system properties and functionings without any interactions with brain ones, one would observe something I have called "pure awareness" at the top of the M/L side of the diagram. I suspect that we actually have some data on both of these relatively pure cases, but not in a form we can clearly recognize and make use of. Some meditative practices, for example. such as those leading to “contentless awareness,” or variants of out-of-the-body experiences (OBEs), lead to experiences which are usually described as "ineffable," that is, they cannot be expressed in terms of the emergent of language which deals with consciousness: these may be instances of isolated M/L system operation. And, more speculatively, some cases of deep coma may represent instances of the M/L system no longer interacting with the B system, although we are probably also dealing with a badly damaged B system.

This has been a basic outline of the MINDS position. Let us now consider a variety of topics from this point of view, starting with the psychological factor of automatization.

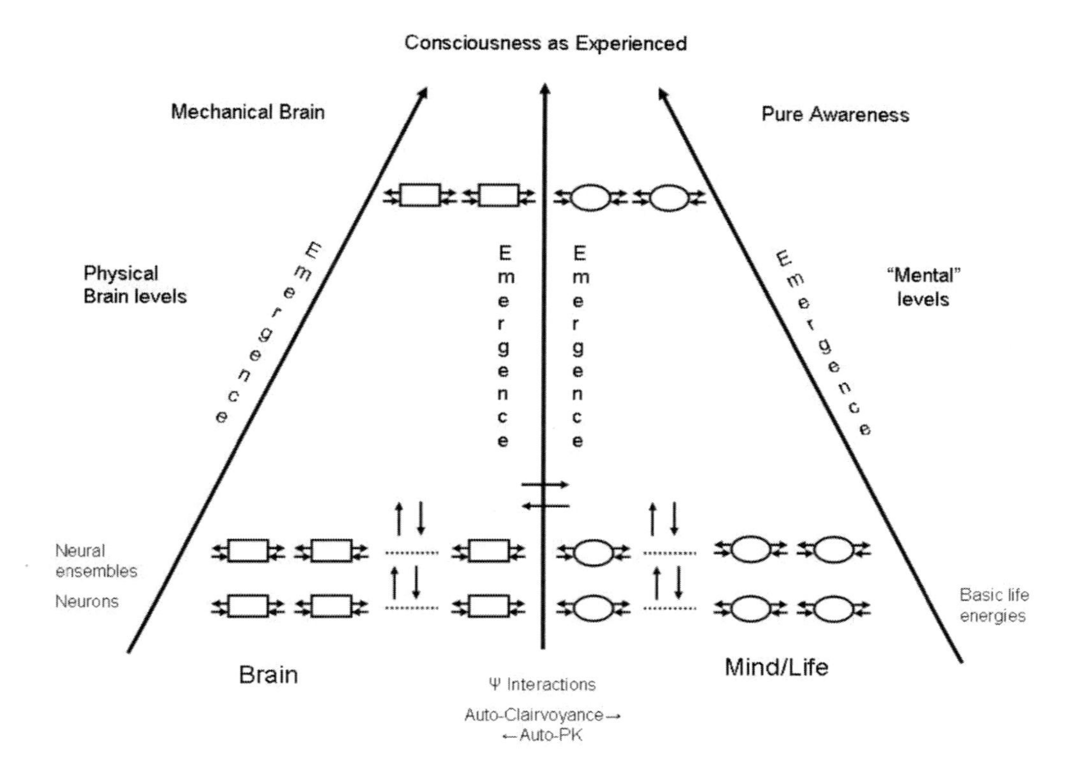

Figure 3. The MINDS representation of consciousness.

AUTOMATIZATION

An important psychological consideration to now introduce into this MINDS approach is that of *automatization.* the habitual, automatic way that consciousness seems to function a great deal of the time. The ability to automatize almost anything is a basic human ability, certainly present in the B-system, possibly also in the M/L system. Much specific automatization results from the socialization process, where various assumptions and habit patterns become implicit. Automatizations lead to semi-permanent physical modifications in the B system which automatically tend to guide B system functioning (and M/L system interaction) along certain lines, lines which simply seem like the "natural" way of doing things, a process I have discussed at length elsewhere ((C. Tart, 1975a)). One important consequence of this in terms of the MINDS approach is that *a great deal of information processing, decision making, "perception" and action may take place in the B system without there necessarily being any auto-psi interaction with the M/L system.* The B system, as it were, can do a good many things "on automatic," without the M/L system being involved. We shall consider aspects of this in more detail later. For now, this point can be illustrated by considering this MINDS point of view as analogous with the operation of a "smart" computer terminal.

Ordinary computer terminals originally consisted essentially of a keyboard or other input device whose sole function was to transmit data and instructions and a monitor to receive data from a remote computer. The remote computer did some kind of processing of the information and sent back output, it sent "decisions" back to the ordinary computer terminal which simply printed them out, unaltered. Such setups are still in use and are known as *dumb* terminals. A *smart* computer terminal, on the other hand, actually has a computer of its own built into the terminal. Certain kinds of data may be inputted to this terminal and, rather than simply transmitting it unaltered to the remote computer, the terminal will carry out some processing on the data right there. The resulting abstractions or transformations of the input data may then be sent to the remote computer when the remote computer is ready to accept them, and/or an output, a decision, may be made right there at the smart computer terminal and activate its output printer or control devices.

Let the B system be analogous with the smart computer terminal and the M/L system be analogous to the remote computer. A good deal of information processing from both sensory input and internal, habitual needs and concerns is carried on by the mechanical processes of the B system alone and outputs (behaviors or ideas) made. For much of this, there may be little or no auto-psi connection with the "remote computer,“ the M/L system, at all, other than a kind of passive, emergent consciousness. Sometimes, however, the remote computer, the M/L system, is consulted and it modifies the action of the B system, the smart computer terminal, in ways which are not predictable from a knowledge of the smart computer terminal alone. The kinds of behaviors Stanford ((Stanford, 1974a) (Stanford, 1974b)) has described as *psi-mediated instrumental responses* (PMIRs) are excellent examples of this. In PMIRs, given the sensory and stored information available to the person and the processing capacities of his B system, he does not have the information necessary to reach a decision to carry out a certain kind of action which will be need relevant, yet he nevertheless behaves appropriately, for the M/ L system has used allo-psi to scan the environment for need-relevant information, gather the useful information and then influenced B system processes by auto-psi to modify the final emergent, the person's ideas or behavior, in ways which are need relevant. In the PMIR process, the auto-psi process need not always modify the emergent of *conscious* experience, however; the person just happens to do the right thing without knowing why.

I believe the tremendous complexity of the B system and the automatization of much of its action in the course of ordinary socialization offers a partial explanation for why allo-psi about external events does not work very well in our ordinary state of consciousness. The information processing activity in the B system has become habitual and continuous, and it ties up most or all of the processing capacity of the B system. In terms of possible allo-psi messages being received or allo-psi outputs being initiated (via auto-psi intermediation), this produces a very high “noise” level that makes it unlikely that auto-psi will be able to influence the B system or vice versa. This view is congruent with various experimental data we have that indicate that allo-psi conducive states (quieting meditation, e.g.) involve cutting down internal noise levels from irrelevant B system processes. In my extended presentation of my theory that immediate feedback will help learning ((C. Tart, 1977d)), I also stress that learning to discriminate the relevant psi signals from internal B system noise is a major requirement of success.

ALTERED STATES OF CONSCIOUSNESS

In my systems approach to understanding altered states of consciousness ((Lee, Ornstein, Galin, Deikman, & Tart, 1975) (C. Tart, 1974a) (C. Tart, 1975a) (C. Tart, 1976) (C. Tart, 1977b) (C. Tart, 1977e) (C. Tart, 1980)) (C. Tart, 1977b) (C. Tart, 1977e) (C. Tart, 1977a) (C. Tart, 1978a) (C. Tart, 1978b) (C. Tart, 1979) (C. Tart, 1981) (C. Tart, 1986) (C. Tart, 1989) (C. Tart, 1990) (C. Tart, 1991b) (C. T. Tart, 1993) (C. T. Tart, 1994) (C. Tart, 1998b) (C. Tart, 2000) (Charles Tart, 2000) (C. Tart, 2004) (C. Tart, 2008a) (C. Tart, 2008b) (C. Tart, 2011), I defined a *discrete altered state of consciousness* (d-ASC) as *a radical pattern change in the functioning of consciousness, a combination both of particular subsystems, processes or aspects of consciousness changing as well as the consequent emergent, system properties of consciousness changing*. I deliberately did not to bring in serious dualistic considerations in these earlier writings intended for mainstream colleagues, in order not to arouse possible prejudices in the psychologist audience the theory was primarily intended for. My systems approach is quite useful for many aspects of materialistic monist approaches. Thus while I talked about "awareness" as constituting a kind of activating energy for affecting the operation of subsystems of consciousness, I was careful to ambiguously use "awareness" this way as primarily a matter of semantic convenience, if one adopted a monistic position. For the dualistic MINDS position I am now proposing, however, some further distinctions about the nature of altered states of consciousness can be made.

Any *discrete state of consciousness* (d-SoC) consists of a particular pattern of functioning, a system emergent functioning within *both* B system and M/L systems. The d-SoC, the experienced consciousness, is the emergent from the interaction of both of these B and M/L levels of organization. A discrete *altered* state of consciousness[3], a radical pattern shift, can be induced by either (1) changing the organization/functioning of subsystems of the B system alone; and/or (2) changing the organization/functioning of subsystems of the M/L system alone; and/or (3) changing the nature of the auto-psi interactions between B and M/L system levels. In terms of this MINDS approach, some d-ASCs will turn out to be explainable strictly in terms of alterations of B system functioning, but others will not be reducible simply to alterations in B system functioning. Becoming able to investigate changes in the M/L system in isolation and/or its auto-psi interaction with the B system will require considerable practical development in the use of allo-psi as an observational tool.

This possibility that some d-ASCs are primarily functions of M/L system changes or auto-psi interaction changes has important implications for parapsychological research. We have a scattering of evidence to suggest that various altered states may be conducive to allo-psi functioning, more so than ordinary consciousness. This may be partially due to the fact that well ingrained B system habits (automatisms) that create the noise that interferes with psi functioning in our ordinary state are no longer functioning as strongly due to changes in B system operation. It may also mean that certain d-ASCs have their balance of functioning shifted more toward the M/L system side, for which psi is a direct mode of expression. Thus we might expect some important breakthroughs for enhancing allo-psi by discovering which particular d-ASCs are most favorable in this way.

[3] Lack of space prevents treatment of many details of d-ASCs here, but my original writings, referenced earlier, have this information.

ORDINARY PSI AND NON-ORDINARY PSI

Given this MINDS view of consciousness, it becomes clear that psi is being used a large amount of the time in everyone's life, but is being used, as it were, "internally". We frequently use auto-clairvoyance to read our own B system and auto-PK to affect our B system. This is ordinary but "invisible" psi, auto-psi. What we observe in parapsychological experiments, however, is non-ordinary psi, it is taking a process ordinarily confined "within" a single organism and pushing it outside, making it allo-psi. I have tried to represent the general situation in an amplified model of consciousness in Figure 4.

When the M/L system learns the state of its own B system, we term this auto-clairvoyance (auto-CL in the figure); when the M/L system influences B system operation, we term it auto-PK. The relatively unusual, non-"normal" use of psi outside of the organism results in allo-clairvoyance (allo-CL) to obtain information about the external environment, and allo-PK to affect the external environment. This is non-B system matter to M/L system information flow, and M/ L system to non-B system information/energy flow. Communication from one distinct M/L system to another, telepathy, can conceptually be subdivided into *receptive* telepathy, picking up information from another M/L system, and *projective* telepathy, sending information to another M/L system, a useful division for maintaining symmetry with the clairvoyance and PK processes. Given our terminological convention, telepathy with other minds is a form of allo-psi. Indeed, a fundamental distinction seems to be with psi that deals with M/L system to M/L system interaction, and psi that deals with matter and M/L system interaction.

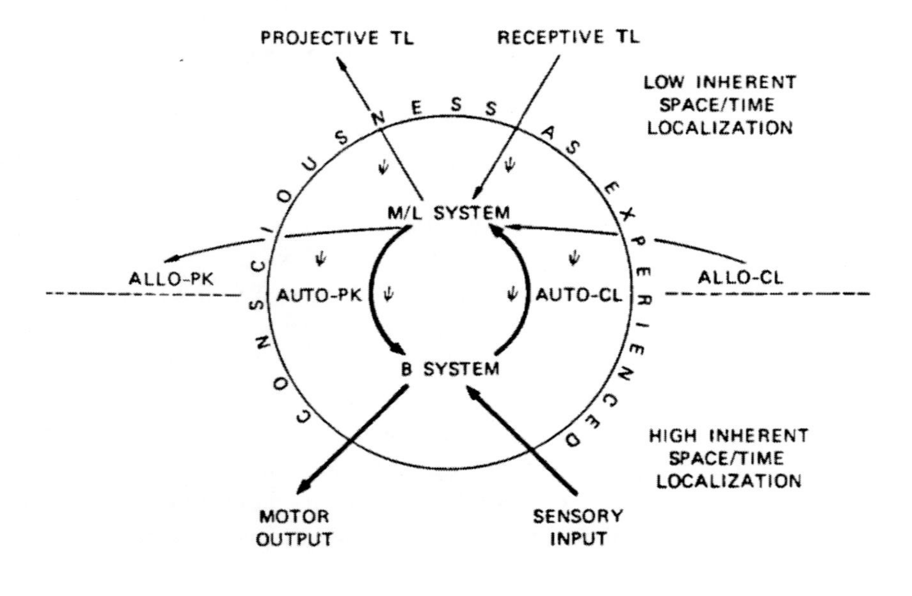

Figure 4. Psi processes within and external to an organism from the MINDS approach.

On the B system side of consciousness, sensory input brings in information about the material world around us, automatically abstracts this information along value lines and creates a continuous simulation of our environment (which we call perceiving the world), what I have elsewhere ((C. Tart, 1991a) (C. T. Tart, 1993) (C. Tart, 1993)) called a *biological-psychological virtual reality* (BPVR). Motor output sends, though our various musculatures, information and energy back out into that matter world. I have shown the sensory input and motor output arrows, and the auto-CL and auto-PK arrows in heavy lines to represent the most prominent information flow channels ordinarily active in an organism. I have also drawn in consciousness-as-experienced as a circle around these other processes, to remind us that it is an emergent of B system and M/L system auto-psi interaction.

Earlier I listed precognition as one of the basic psi phenomena but I am now inclined not to consider the temporal distinction of past, present and future that we make in everyday consciousness as quite so real or basic. In Figure 4 I have indicated that the M/L aspect of consciousness has *a low inherent degree of localization in space and time* (an idea developed further in my discussion of *trans-temporal inhibition* ((C. Tart, 1977c) (C. Tart, 1978c)), while the B system aspect of consciousness is very highly localized in space and time. That is, the B system belongs to an order of reality in which you can specify with great confidence that a particular event is happening at a certain time, at a certain location in space and possesses highly specifiable and predictable matter and physical energy properties. The M/L system, on the other hand, does not seem to be so localized in terms of physical space/time measures[4]. While it usually centers around the here and now of B system space and time for a particular person, I postulate it to be more widely spatially and temporally spread than that (thus the need for *trans-temporal inhibition* in efficient ESP), (C. Tart, 1978c), but it can, to some degree, volitionally focus at different spatial and temporal location than the B system and its associated sensory and motor apparatus is at. By analogy, the B system is a tight flashlight beam focused on a small segment of space and time (the so-called specious present), but the M/L system is a wider (and perhaps dimmer) flashlight beam whose spatial "here" takes in things not present to the B system senses and whose temporal "now" includes some of the past and future as well as the present.

To be more speculative, the very diffuseness or non-localization of the M/L system has something to do with the reason that it is associated with a particular B system, for that B system acts as a stabilizing, influence on the operation of the M/L system, it focuses and anchors that M/ L system to a particular location and moment in space and time for evolutionary reasons[5]. On the one hand, as far as biological survival is concerned, events within the sensory range of the B system are almost always the most important ones for the organism to be concerned with, so the style of M/ L system interaction with the B system would evolve toward maximizing the efficiency of the B system consciousness for biological survival. To the extent to which this becomes habitual and automatized, this would be a reason why allo-psi seems relatively rare: the psi capacity is almost totally, used up in auto-

---

[4] It is tempting to compare this with the inherent non-locality of quantum phenomena before acts of conscious observation, but as I have no expertise in quantum physics I mention this only because it might usefully stimulate others with relevant expertise.

[5] I am influenced by ideas I have come across in various spiritual systems that posit that mind, unassociated with a physical body, is ultra-flexible, but at the same time cannot do some things as it's too loose, too flexible. Being associated with a physical body gives mind leverage or an anchor, a solid base to build from. You can't move things with a level unless part of it is firmly anchored.

psi functioning which is geared to maximizing the functioning of the total organism in its physical environment. On the other hand, insofar as the M/L system has goals of "surviving" and evolving (Murphy, 2011), the stabilization resulting from it's lifetime association with a B system allow for greater progress.

SPECIAL SENSITIVITY OF B SYSTEM TO PSI

The B system, from this MINDS point of view, has two main properties of concern here. First are many self-organizing properties, independent of interaction with the M/L system, that are adaptive in dealing with the needs for maintaining homeostasis within the physical organism and dealing effectively with the physical environment. Second, it must have properties that not only make it receptive to M/L system influences via auto-PK, it should be *efficiently* receptive to these influences in order to maximize survival potential. This means semi-independent associative and decision making properties. "perceptual" properties, with respect to M/L system influences that compensate for possible inefficiencies and deficiencies in the psi interaction. The B system, for example, might automatically "fill in" a message from the M/L system that is a little incomplete and, in cases of doubt, fill it in along lines which are most relevant to biological survival. For example, if an ambiguous pattern seen in some bushes could be a tiger, it is highly adaptive for the B system's simulation of the environment to make you perceive it as a tiger and take fast escape action, rather than ignore it because you aren't *sure* it's a tiger. The receptive function of the B system, then, for auto-PK, is likely to be elaborative as well as efficiently receptive, and, by being elaborative, it can be prone, like any similar communication system, to produce incorrect outputs. In terms of observable data, this "filling in with errors" is commonly seen in remote viewing clairvoyance procedures, where a percipient's impressions of the desired distant target scene may have a number of very accurate, correct elements, along with incorrect ones that my be generated by the B system's attempt at error correction.

This line of reasoning has two important consequences for parapsychological research. First, the B system must be especially sensitive to auto-PK, and, insofar as auto-PK and allo-PK are probably manifestations of the same fundamental psychokinetic process, investigation of what aspects of B system functioning make it especially sensitive to PK should be of great value in designing other physical processes which would be sensitive PK detectors. Second, not only does the M/L system need to use an appropriate allo-psi process to gather the relevant information from some distant target other than the percipient's own B system, this information must be then put into, or influence the B system of the percipient by auto-PK effects on B system functioning, in order to get relevant information into the emergent consciousness of the percipient, information which she can then express behaviorally so we can observe it. Auto-PK at least needs to affect relevant aspects of the B system so we can observe a behavioral or physiological effect that manifests the psi information, even if it does not reach the percipient's consciousness. But, the B system is constantly producing an adaptive simulation of the percipient's immediate sensory environment (modified by his psychological concerns, the BPVR) in a way largely independent of current B and M/L system interactions, and this constitutes a high noise level that the auto-PK information carrying the allo-psi information must compete with. Further, the elaborative aspects of the B system's receptivity to auto-PK means that there is a strong probability that the psi message will tend to be elaborated/distorted in ways which fit the ongoing, survival-oriented simulation of the immediate physical environment being continuously constructed by the B system. The very "efficiency" and partial independence of the B system, then, automatically makes some distortion of allo-psi messages likely. Given this as a basic characteristic of the B system, practical measures to increase the incidence of psi in parapsychological experiments would need to involve some combination of B system noise reduction (as in. e.g. ganzfeld or perhaps quieting meditation techniques), discrimination training (as in immediate feedback training), and enhanced discriminability of the allo-psi targets themselves (distinct remote locations for the remote viewing procedure, e.g. versus similar playing cards differing only in number).

Complexity of Psi Tasks

The B system is an incredibly complex system, so auto-psi interaction with the M/L system must also be of a very complex nature. This leads to an interesting comparison: the kinds of allo-ESP and allo-PK tasks we have given percipients and agents in the laboratory have probably been enormously simple (and perhaps trivial) compared to what is routinely done by auto-clairvoyance and auto-PK. In an earlier modeling of PK along conventional lines ((C. Tart, 1966) (C. Tart, 1977d)). for example. I argued that influencing a tumbling die by PK is quite complex when modeled from a Newtonian perspective, requiring continuous clairvoyant feedback about its three dimensional motion, its mass-energy parameters and the surface characteristics of the surface it would bounce against, so just the right amount of PK force could be applied in just the right places and directions at just the right moments. This is, indeed, a formidable task from the viewpoint of physical mechanics as we currently conceive it, but from the point of view of an M/L system used to constantly reading and influencing enormous numbers of cells in a dynamically changing brain, the task may well be so trivial as to be hardly capable of attracting much attention! Similarly, the circuits of the electronic random number generators which have been influenced by allo-PK to date may also be trivially simple compared to the typical operations of auto-CL and auto-PK. As a comparison, to ask a person to control activity in a single muscle fiber without affecting activity in surrounding fibers is normally impossible – although it can be done with biofeedback from that single fiber.

This leads me to a prediction, namely, that allo-PK should work more successfully when directed toward super-complex systems, such as brains or huge computers with huge numbers of random elements in them, rather than when directed toward simple physical tasks: it's what the M/L system is used to doing, and habit is hard to break. Further, we probably can't reliably detect differences in PK efficacy for simple tasks that involve, say, influencing one versus two or three, or ten random decision-making elements. They are all ridiculously simple; we need to compare PK on single decision making RNGs versus those that employ millions or more of interacting random of semi-random decision-making elements leading to a final output.

A similar line of thinking might be applied to ESP psi tasks; perhaps ESP is more successful at detecting the overall pattern of complex elements than at picking out single elements? Rather spectacular successes in the remote viewing procedure sometimes occur, for example, which may be due to the fact that a target pool consisting of thousands or millions of real-world, complex targets may fit this habit of dealing with complex material, rather than the very simple kinds of ESP tasks like card-guessing or electronic random number generators with only a few possible outputs.

OUT-OF-THE-BODY EXPERIENCES

Out-of-the-body experiences (OBEs) are especially interesting from a MINDS point of view. While there are a wide variety of reported experiences and much looseness in the use of the term OBE by various writers ((C. Tart, 1974b)), the basic, "classical" OBE that we will consider here has two distinguishing elements. First, the experiencer perceives himself as located at some location other than where he knows his physical body is located. Second, and of crucial importance in definition, the experiencer knows *during* the experience that his consciousness is basically functioning in the pattern he recognizes as his ordinary state. He can call upon most or all of his ordinary cognitive abilities during the OBE, typically recognizing, e.g.. the "impossibility" of his ongoing experience according to what he has been taught. “How can I possibly be clearly experiencing myself as being at this distant location when I know that my physical body is at home, in bed?” As far as he can tell, he is perfectly "normal" in all mental ways that matter; it's just that he is obviously located somewhere other than where his physical body is[6].

Although some people manage to retrospectively talk themselves out of the clear experiential reality of their experience, most people who have an OBE become confirmed dualists based on this experience. No matter what "logical" arguments one may make, they *know* that their consciousness is of a different nature than their physical body, because they've experienced them as separated.

As an outsider, listening to someone else's account of his OBE, we can dismiss the implications of his experience and remain convinced monists without much psychological effort. Indeed, the OBE, as defined so far, can be seen as an interesting hallucination. It is like a dream in that a realistic, but hallucinatory environment is present. but obviously certain other parts of the B system responsible for ordinary consciousness, for memory, reasoning, and simulating an environment, are also activated. Indeed, if the experiencer would only call his experience a "lucid dream," instead of an OBE, that is stop insisting that his experience was *real* and agree with our view that it was hallucinatory. even if it *seemed* real, he would not bother a confirmed materialistic monist. It is easy from a monistic point of view to model brain functioning that would create a lucid dream.

As defined so far, OBEs could easily be included within the domain of ordinary psychology, and there have been some recent attempts, quite inadequate in my assessment, to explain OBEs away in terms of pathological brain functioning. Some OBEs may indeed by something like lucid dreams and so explainable on those terms, for I haven't put any psi element into the above definition. Indeed, it is useful to define them in purely psychological terms just to make them legitimate subjects for investigation by psychologists who might shy away from psi phenomena. But we know, of course, that in some OBEs the experiencer accurately describes a distant location that he could not have known about except by psi, as

---

[6] Since the experient’s consciousness feels pretty much like ordinary consciousness, it doesn’t seem correct to call an OBE a d-ASC. This tempts me to propose the acronym ASL, an Altered State of Location, for OBEs, but we have enough acronyms already..... ;-).

when my Miss Z correctly read a five-digit random number on a shelf above her head ((C. Tart, 1968)) as an outstanding example.

Because of the strong psi component of some OBEs, I am inclined, from a MINDS approach. to regard at least some of them as being pretty much what they seem to be to the experiencers, a temporary spatial/functional separation of the M/ L system from the B system. The separation is temporary (otherwise we wouldn't get any report!). It is generally probably only partial, with the M/L system still interacting with the B system to some extent. Several aspects of OBEs support this partial separation view.

First, in most OBEs the person experiences his consciousness as very like ordinary consciousness, yet ordinary consciousness arises as an emergent from B and M/ L system interaction and mutual patterning. This suggests that a great deal of this interaction is still occurring, and/or that the force of habit, the lifetime practice of this patterning is still fairly active in the M/ L system alone.

Second, in cases of prolonged (more than a few minutes apparent duration) OBEs, or people who have had many OBEs, or OBEs associated with severe disruption of physical functioning, as in near-death experiences (NDEs), consciousness, as experienced, tends to drift away from its ordinary patterning into various d-ASCs. The OBE starts to become difficult to describe very well or more of a "mystical experience," even though it retains the basic feeling of separation of B and M/ L systems[7]. This is what we would expect for greatly reduced auto-psi interaction between these two systems; both the B system and the M/ L system would start drifting toward unique patterns of functioning determined by their own inherent characteristics, now manifesting as they are freed from mutual interactive patterning of each other. Indeed, it is these kinds of unusual OBEs that may give us valuable insights into what the M/L system in and of itself may be like, unpatterned by the B system.

Third, the sparse (and largely anecdotal) evidence we have on it suggests that there are few, if any, physiological changes of medical consequence during brief OBEs. The B system functions pretty much as usual. But during prolonged OBEs, larger and potentially fatal physiological changes may begin to occur. Robert Monroe, for example, reports that his physical body has become quite chilled following prolonged OBEs ((Monroe, 1971)). NDEs are usually caused by major B system change involved in almost or temporarily dying, and, to go to an extreme for which we have good physiological data, in the famous Pam Reynolds case ((Sabom, 1998)) her NDE seemed to occur at a time when no possible functioning of her B system was occurring. I see this as showing that life and consciousness, as we know them, arise from the mutual interaction and patterning of the B and M/ L systems, and when the patterning of the M/L system upon the B system begins to break down, the brain by itself cannot adequately run the complex system of the body, and small errors start to cumulate. In principle, this would eventually lead to death.

[7] Indeed I usually distinguish typical OBEs from typical NDEs by noting that NDEs frequently start with an OBE. The experiencer finds herself outside her physical body, observing it and the scene around it, with consciousness functioning pretty much as it does ordinarily, but as the NDE goes on, describing it becomes harder, words like "ineffable" are used, for the functioning of consciousness itself alters to some kind of d-ASC.

SURVIVAL

The MINDS approach allows for some kind(s) of potential survival of bodily death, but it would not necessarily be the kind of postmortem survival we usually conceive of. Our usual conceptions of survival mean survival of the basic pattern of our consciousness, our experience of our mental life, our feelings of personal identity, our personality. But consciousness, as we have seen, is an emergent of the auto-psi interactions of both the B and the M/ L systems, an emergent of constant patterning of each system upon the other. If the B system ceases functioning in death, the patterning influence of the B system upon the M/ L system will cease, so how is ordinary consciousness, as we know it, to survive? What is the emergent to emerge from?

One answer may be that ordinary personal identity, which is so intimately intertwined with ordinary consciousness (see my *States of Consciousness* for a discussion of this, (C. Tart, 1975a)), does not survive death, at least not for very long. The M/L system may survive, with the length of postmortem survival being determined by currently unknown characteristics of M/ L systems in general, but this is survival of some *aspect* of a person, not the "person" *per se*. Indeed, we would expect this aspect to be quite different from the ordinarily embodied person.

This answer should be partially modified by referring back to our discussion of OBEs, where we noted that rather ordinary consciousness is frequently maintained for at least short periods in many OBEs. The customary patterning of the M/L system by the B system is thus capable, at least for short periods, of continuing to pattern the M/L system with reduced or perhaps temporarily eliminated auto-psi interaction. The patterning parameters may be stored in something analogous to ordinary "memory" in the M/L system, or the M/L system may be permanently or semi-permanently modified in its own stable pattern of functioning as a result of prolonged auto-psi interaction with the B system in its developmental history.

If B system patterning and consequent "ordinary" consciousness can manifest in the M/L system alone, at least temporarily in OBEs, then it is possible to conceive of survival of personal identity in at least some people. To the degree that a particular person's sense of identity was not strongly and permanently patterned in the M/ L system *per se*, but was supported largely through environmental, bodily, and social constancies patterned in the B system, then we would expect the emergent of consciousness and personal identity to disintegrate rapidly once the B system ceased functioning. At the other extreme, if basic personal identity and consciousness patterns were strongly and permanently stored at the M/L level, for whatever reasons, such a person might withstand the loss of B system patterning influence and still maintain consciousness and personal identity patterning in the M/ L system after death, thus achieving personal survival. Such intense patterning of the M/ L system might arise for a variety of reasons, such as deliberate practice of meditative techniques or sheer psychological rigidity and fanaticism. Some spiritual systems (Gurdjieff's Fourth Way, for instance – see Ouspensky (Ouspensky, 1949)) postulate survival only for those who have developed sufficiently in life[8].

There is no space here to compare this MINDS approach with the data about mediumistic communications, some of which seems to strongly indicate some kind of survival of

[8] Personally as a person raised in a democratic society and believing in equal human rights, I find this view distasteful, but scientifically deserving of consideration.

relatively full personal identity, as that is an area of complex phenomena strongly affected by social beliefs and experimenter/sitter biases. Eventually comparison should require enrichment or modification of the MINDS approach.

What Can we Learn about the M/L System in Isolation?

As discussed earlier, we ordinarily know almost nothing about what the M/L system *per se* is like, the consciousness we ordinarily experience is an emergent from the extensive interactions and mutual patternings of the B and M/L systems. Yet, as touched on somewhat above, I believe we can learn at least some things about the properties of the M/ L system in and of itself, when it is not patterned, or at least is patterned to a much lesser degree by the B system.

The characteristics of allo-psi processes give us some clue to what the M/L system is like, so that we can generally say that the M/L system is probably capable of gathering information about and affecting at least some aspects of physical reality which are sensorily/energetically remote and shielded from the B system by either spatial shields or distances or temporal distances. That is, the M/L system can exercise allo-psi of the clairvoyant and psychokinetic type, either in real time or precognitively, and possibly postcognitively. Although there is little evidence for "pure" telepathy (where a clairvoyance interpretation of the data is completely excluded). I shall presume that the M/L system can also exercise allo-telepathy in both real time and pre and postcognitively.

As far as ordinary physical limits are concerned, our present knowledge of allo-psi indicates no obvious limits, but we have really only investigated a quite limited range of physical variables. There may well be limits inherent in the nature of psi that are perceptible from the point of view of the M/L system. even if not detectable from physical measures, although the best estimate from current data is that there may well be no spatial and temporal limits – that the M/L system is indeed non-local, the N in MINDS. This last point about the detectability of limits or characteristics of psi being related to the perspective from which it is viewed, leads us to a specific proposal within a more general conceptual framework that I have written about elsewhere ((C. Tart, 1972) (C. Tart, 1975a) (C. Tart, 1975b, 1998a) (Charles Tart, 2000) (C. Tart, 2008a)), namely the development of *state specific sciences* as a means of understanding psi (and consciousness in general).

The state of consciousness in which we ordinarily carry out scientific research is an emergent from auto-psi interaction between the B and M/L systems. It is not "unlimited" consciousness, not "pure awareness" *per se*, but a specific kind of consciousness. Its characteristics and limitations are governed by the inherent properties of the B system, the inherent properties of the M/ L system, the laws which govern auto-psi interaction, and the general laws of emergence which we hope to understand adequately some day through development of general systems theory ((Miller, 1978)). The part of all this to emphasize for our purposes here is that ordinary consciousness has limitations, limitations in the way reality can be perceived, limitations in the kinds of concepts that can be generated about reality, and limitations in the way such concepts can be tested.

While I believe that a great deal can be learned about psi from skillful scientific work in our ordinary state of consciousness, I suspect that important aspects of it will not be comprehensible, will remain *paraconceptual* to ordinary consciousness because of these limitations. The little scientific and anecdotal knowledge we have about the range of functioning available in various d-ASCs, however, suggests that there are alternative modes of consciousness, quite different emergents from B and M/L system interaction, that may yield more useful perceptions of, concepts about, and tests of psi functioning. The paraconceptual aspect of psi is not saying something about any inherent perversity in the universe that restricts understanding; it is saying something about the limitations of ordinary consciousness stemming from its specializations in promoting biological survival and evolution (and perhaps, insofar as the B system provides a stable platform for the M/L system, as mentioned earlier, promoting M/L system evolution).

In discussing OBEs, I suggested that certain OBEs, especially when they become NDEs, show drastic alterations in consciousness because there is greatly reduced B and M/ L interaction, so the consciousness experienced reflects M/ L characteristics *per se* more than ordinary consciousness does. I have also suggested that in general some d-ASCs may come about through reduced or altered B and M/L system interaction. The state-specific sciences that could potentially be developed for these d-ASCs then, including OBEs[9], could lead us to increased experiential and scientific knowledge about the M/L system under conditions of greatly reduced interaction with the B system, from which we could make more accurate extrapolations to what the totally isolated M/ L system would be like.

Our knowledge base is too small to warrant further speculation now about specific d-ASCs and directions of development that will be useful for understanding psi, but this is the direction I ultimately see the field going in.

Summary

I have proposed the beginnings of a dualistic theory of consciousness, a Mutually Interacting Non-local Dualistic Systems approach, MINDS, which is intended to be scientifically useful and that is pragmatic, that has empirical, testable consequences. The existence of psi phenomena, which are paraconceptual for physicalistic monism, is the basic evidence for a pragmatic dualism, a recognition of the need to understand consciousness in terms of two qualitatively different aspects of reality, what I have called the B system, the brain, body, and nervous system and the physical laws which govern it, and the M/L system, the mental and life aspects of reality, which is not so local to ordinary space and time as the B system is. Consciousness is seen as a system property, an emergent from the auto-psi interaction of the B and M/L systems. Ultimate understanding of consciousness, then, while it requires further and extensive development of conventional approaches in the study of brain functioning and physical law, also requires extensive development of our knowledge of psi, as well as development of general systems theory, so principles of emergence in complex systems can be better understood. While this view is complex, it is more adequate to the reality of psi than a physicalistic monism, and exciting discoveries await us!

[9] And NDEs if they could be safely induced, without leading to actual death or lasting damage.

REFERENCES

Braude, S. (2007). *The gold leaf lady and other parapsychological investigations*. Chicago, IL: University of Chicago Press.

Broughton, R. (1991). *Parapsychology: The controversial science*. New York: Ballantine.

Chalmers, D. (1996). Facing up to the problems of consciousness. In S. Hameroff, Kaszniak, A., and A. Scott. (Eds.), *Toward a science of consciousness: The first Tucson discussions and debates*. Cambridge, MA: MIT Press, pp. 5-28.

Eccles, J. (1990). A unitary hypothesis of mind-brain interaction in the cerebral cortex. *Proceedings of the Royal Society of London, Series B, Biology, 240*, 433-451.

Forman, R. (1999). *Mysticism, mind, consciousness.* Albany, NY: SUNY Press.

Greeley, A. (1975). *The sociology of the paranormal: A Reconnaissance*. Beverly Hills, CA: Sage Publications.

Kelly, E., Kelly, E., Crabtree, A., Gauld, A., Grosso, M., and Greyson, B. (2007). *Irreducible mind: Toward a psychology for the 21st century*. Lanham, MD: Rowman and Littlefield.

Lee, P., Ornstein, R., Galin, D., Deikman, A., & Tart, C. (1975). *Symposium on Consciousness*. New York: Viking Press.

Libet, B. (1985). Unconscious cerebral initiative and the role of conscious will in voluntary action. *Behavioral and Brain Sciences, 8*, 529-566.

Miller, J. (1978). *Living systems*. New York: McGraw-Hill.

Monroe, R. A. (1971). *Journeys out of the body*. Garden City, NJ: Anchor Books.

Ouspensky, P. D. (1949). *In search of the miraculous*. New York: Harcourt, Brace and World.

Radin, D. (1997). *The conscious universe: The Scientific truth of psychic phenomena*. San Francisco: HarperOne.

Radin, D. (2006). *Entangled minds: extrasensory experiences in a quantum reality*. New York: Paraview Pocket Books.

Sabom, M. (1998). *Light and death: One doctor's fascinating account of near-death experiences*. Grand Rapids, MI: Zondervan Publishing.

Stanford, R. (1974a). An experimentally testable model for spontaneous psi events. I. Extrasensory events. *Journal of the American Society for Psychical Research, 68*, 34-57.

Stanford, R. (1974b). An experimentally testable model for spontaneous psi events. II. Psychokinetic events. *Journal of the American Society for Psychical Research, 68*, 321-356.

Stapp, H. (2007). *Mindful universe: Quantum mechanics and the participating observer*. New York: Springer.

Tart, C. (1966). Models for explanation of extrasensory perception. *International Journal of Neuropsychiatry, 1*, 488-504.

Tart, C. (1968). A psychophysiological study of out-of-the-body experiences in a selected subject. *Journal of the American Society for Psychical Research, 62*, 3-27.

Tart, C. (1972). States of consciousness and state-specific sciences. *Science, 176*, 1203- 1210.

Tart, C. (1974a). On the nature of altered states of consciousness, with special reference to parapsychological phenomena. In W. Roll, R. Morris, and J. Morris (Eds.), *Research in Parapsychology, 1973* (pp. 163-218). Metuchen, NJ: Scarecrow Press.

Tart, C. (1974b). Some methodological problems in out-of-the-body experiences research. In R. M. W. Roll and J. Morris (Eds.), *Research in parapsychology, 1973.* Metuchen, NJ: Scarecrow Press, pp. 116-120.

Tart, C. (1975a). *States of consciousness*. New York: E. P. Dutton.

Tart, C. (1975b). *Transpersonal psychologies*. New York: Harper and Row.

Tart, C. (1976). The basic nature of altered states of consciousness: A systems approach. *Journal of Transpersonal Psychology, 8*, 45-64.

Tart, C. (1977a). Beyond consensus reality: Psychotherapy, altered states of conscious ness, and the cultivation of awareness. In O. L. McCabe (Ed.), *Psychotherapy and behavior change: Trends, innovations and future directions*. New York: Grune and Stratton, pp. 173-187.

Tart, C. (1977b). Drug-induced states of consciousness. In B. Wolman, L. Dale, G. Schmeidler, and M. Ullman (Eds.), *Handbook of parapsychology*. New York: Van Nostrand Reinhold, pp. 500-525.

Tart, C. (1977c). Improving real time ESP by suppressing the future: Trans-temporal inhibition. Electro/77 Special Session: The State of the Art in Psychic Research, pp. 1-16.

Tart, C. (1977d). *Psi: Scientific studies of the psychic realm*. New York: E. P. Dutton.

Tart, C. (1977e). Putting the pieces together: A conceptual framework for understanding discrete states of consciousness. In N. Zinberg (Ed.), *Alternate states of consciousness.*. New York: Free Press, pp. 158-219.

Tart, C. (1978a). Altered states of consciousness: Putting the pieces together. In A. Sugerman and R. Tarter (Eds.), *Expanding dimensions of consciousness*. New York: Springer Verlag, pp. 58-78.

Tart, C. (1978b). Psi functioning and altered states of consciousness: A perspective. In B. Shapin & L. Coly (Eds.), *Psi and States of Awareness*. New York: Parapsychology Foundation, pp. 180-210.

Tart, C. (1978c). Space, time, and mind. In W. Roll (Ed.), *Research in parapsychology 1977*. Metuchen, NJ: Scarecrow Press, pp. 197-250.

Tart, C. (1979). An emergent-interactionist understanding of human consciousness. In B. Shapin and L. Coly (Eds.), *Brain/mind and parapsychology*. New York: Parapsychology Foundation, pp. 177-200.

Tart, C. (1980). A systems approach to altered states of consciousness. In J. Davidson and R. Davidson (Eds.), *The Psychobiology of consciousness*. New York: Plenum, pp. 243-269.

Tart, C. (1981). Transpersonal realities or neurophysiological illusions? Toward a dualistic theory of consciousness. In R. V. R. v. Eckartsberg (Ed.), *The metaphors of consciousness* (pp. 199-222). New York: Plenum.

Tart, C. (1986). Consciousness, altered states, and worlds of experience. *Journal of Transpersonal Psychology,18*, 159-170.

Tart, C. (1989). Enlightenment, altered states of consciousness and parapsychology. In B. Shapin and L. Coly (Eds.), *Parapsychology and human nature*. New York: Parapsychology Foundation, pp. 150-169.

Tart, C. (1990). Psi-mediated emergent interactionism and the nature of consciousness. In R. K. A. Sheikh (Ed.), *The psychophysiology of mental imagery: Theory, research and application*. Amityville NY: Baywood, pp. 37-63.

Tart, C. (1991a). Multiple personality, altered states and virtual reality: The world simulation process approach. *Dissociation, 3,* 222-233.

Tart, C. (1991b). Multiple personality, altered states and virtual reality: The world simulation process approach. *Dissociation, 3*, 222-233.

Tart, C. (1993). Mind embodied: Computer-generated virtual reality as a new, dualistic-interactive model for transpersonal psychology. In K. R. Rao (Ed.), *Cultivating consciousness: Enhancing human potential, wellness and healing* (pp. 123-137). Westport, CT: Praeger.

Tart, C. (1998a). Investigating altered states of consciousness on their own terms: A proposal for the creation of state-specific sciences. *Ciencia e Cultura, Journal of the Brazilian Association for the Advancement of Science, 50,* (2/3), 103-116.

Tart, C. (1998b). Transpersonal psychology and methodologies for a comprehensive science of consciousness. In S. Hameroff, A. Kaszniak, and A. Scott (Eds.), *Toward a science of consciousness: The First Tucson discussions and debates.* Cambridge, MA: MIT Press, pp. 669-675.

Tart, C. (2000). Investigating altered states of consciousness on their own terms: State-specific sciences. In M. Velmans (Ed.), *Investigating phenomenal consciousness.* Amsterdam, The Netherlands: John Benjamins, pp. 255-278.

Tart, C. (2000). Prelude to Investigating altered states of consciousness on their own terms: A proposal for the creation of state-specific sciences. *International Journal of Parapsychology, 11*, 3-5.

Tart, C. (2004). On the scientific foundations of transpersonal psychology: Contribu tions from parapsychology. *Journal of Transpersonal Psychology, 36*, 66-90.

Tart, C. (2008a). Altered states of consciousness and the spiritual traditions: The pro posal for the creation of state-specific sciences. In K. R. Rao, A. C. Paranjpe, and A. K. Dalal (Ed.), *Handbook of Indian psychology.* New Delhi: Cambridge University Press India Pvt. Ltd, pp. 577-607.

Tart, C. (2008b). Consciousness: A psychological, transpersonal and parapsychological approach. In T. Simon (Ed.), *Measuring the immeasurable: The scientific case for spirituality.* Boulder, CO: Sounds True., pp. 313-326.

Tart, C. (2009). *The end of materialism: How evidence of the paranormal is bringing science and spirit together.* Oakland, CA: New Harbinger.

Tart, C. (2011). Extending our knowledge of consciousness. In E. W. Cardeña (Ed.), *Altering consciousness: Multidisciplinary perspectives: Volume 1: History, culture and the humanities* (pp. ix-xx). Santa Barbara CA: Praeger.

Tart, C., Puthoff, H., and Targ, R. (Eds.). (1979). *Mind at large: Institute of electrical and electronic engineers symposia on the nature of extrasensory perception.* New York: Praeger.

Tart, C. T. (1993). Mind embodied: Computer-generated virtual reality as a new, dualis tic-interactive model for transpersonal psychology. In K. Rao (Ed.), *Cultivating con sciousness: Enhancing human potential, wellness and healing.* Westport, CT: Praeger, pp. 123-137.

Tart, C. T. (1994). The structure and dynamics of waking sleep. *Journal of Transpersonal Psychology, 25*, 141-168.

The Psychological Approaches

In: Consciousness: Its Nature and Functions
Editors: Shulamith Kreitler and Oded Maimon
ISBN 978-1-62081-096-5

Chapter 12

COGNITIVE EFFECTS OF STATES OF CONSCIOUSNESS: DO CHANGES IN STATES OF CONSCIOUSNESS AFFECT JUDGMENTS AND EVALUATIONS?

Yuval Rotstein[1], Oded Maimon[1], and Shulamith Kreitler[2,3]

[1]Department of Industrial Engineering, Tel-Aviv University, Israel
[2]Department of Psychology, Tel-Aviv University, Israel
[3]Psychooncology Research Center, Sheba Medical Center, Tel Hashomer, Israel

ABSTRACT

Evaluation of severity of various indicators, signs and symptoms is a highly important cognitive act, because it often precedes the decision to act or not to act in order to forestall the event or phenomenon implicated by the indicators. We perform many severity evaluations and judgments over the course of our everyday lives: we evaluate the soundness of our computers that sometimes seem too slow, we evaluate the relationships we have with our life partners, we evaluate dangers and we even evaluate the probability for world peace. Evaluations of severity represent one kind of evaluations people often do. But there are many more kinds, such as evaluating honesty (of politicians and others), likelihood of events, and the kind of emotion people's faces express. Evaluations of these kinds seem to be based mainly on the data and knowledge that the evaluator possesses. Nevertheless, they cannot be dissociated from personal influences to which individuals are subjected, such as their emotional states, their changes in mood and even their states of consciousness. This assumption needs support through research. The present chapter describes research that was done at Tel-Aviv University, which examines the effects of states of consciousness on severity judgments and evaluations. The study is based on the theory of meaning (Kreitler and Kreitler, 1990) that defines states of consciousness as the organization of the entire cognitive system at a given time. This theory will be described further on.

WHAT IS A STATE OF CONSCIOUSNESS?

At the beginning of the 20th century, William James (1890, 1904) raised the issue of states of consciousness. He noticed that there is a "normal/rational state of mind", as he called it. But he also pointed out that this is only one specific state of consciousness, and that there are additional states of consciousness. This statement is seemingly obvious and natural, but the problem arises when we try to define and classify the states of consciousness. What is a state of consciousness and what are the differences between the various states? Analyzing several of the common approaches to this problem may serve to highlight the difficulties involved in attempted solutions (see Figure 1).

One of the simplest approaches is to define states of consciousness in line with the processes or behaviors that enabled or promoted "entry into" that state, such as hypnosis, alcohol consumption, sleep, drug ingestion, etc. (Blackmore, 2003). This approach may represent the simplest and most intuitive solution, but also the most problematic one. Its major shortcoming is that it lacks uniformity, since drugs or alcohol, for instance, do not necessarily affect two different individuals in the same way.

Another approach is to identify states of consciousness by means of the individual's physiological and behavioral measures. The physiological measures include, among others, oxygen consumption and heart rate, whereas the behavioral measures include the ability to walk in a straight line. The problem with this method is that very few states of consciousness are related to unique patterns of physiology and behavior, and as such this method is too crude. There are many situations in which the states of consciousness are completely different, without this being manifested in any way in physiological or behavioral measures, and vice versa.

A third common approach assumes that all individuals know their own "normal state of consciousness", and therefore the most adequate and proper way to check if someone is in a "normal" state or a different state is simply to ask the individual. Nevertheless, inasmuch as this method is simple, and in everyday life we usually use it intuitively, it also raises questions. First, there is no precise and agreed upon definition of "the normal state", so that it can hardly serve as anchor for defining other states of consciousness. Secondly, in many states of consciousness the participants cannot be relied upon to attest to their own state. For example, individuals under the impact of drugs, often claim that they are in a completely normal state. In cases such as these, the researchers suggest combining this approach with physiological measures.

A fourth approach to defining states of consciousness is based on using mental function characteristics (Tart, 1972; Farthing, 1992). Accordingly, states of consciousness can be defined by administering to the participants valid tests for several basic mental functions, such as attention, problem solving, imagery, planning and memory (Baars and McGovern, 1996). Although this approach is highly acceptable in psychology, it too has drawbacks. There is still no consensus about the assessment of each of these functions, which could serve as standard criteria.

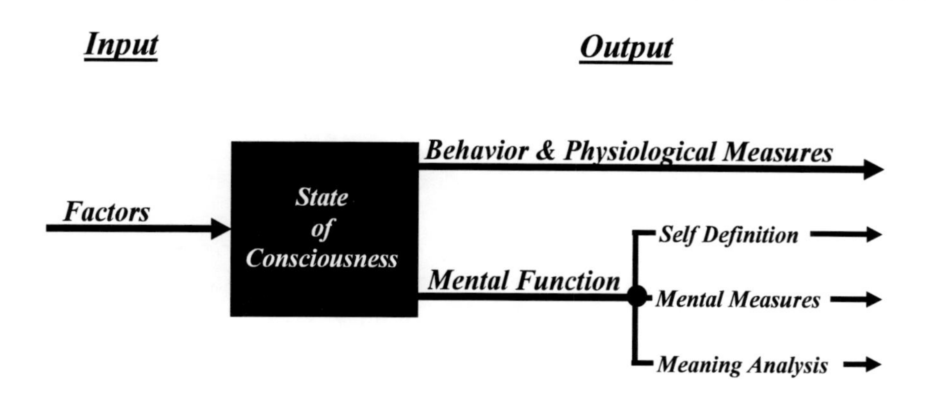

Figure 1. Different approaches to defining states of consciousness.

Finally, the fifth approach is the one followed in our research. This definition is based on the analysis of meaning, in line with the theory of meaning (Kreitler and Kreitler, 1990). This is an innovative approach in that it tries to define the states of consciousness according to the internal state of the cognitive system rather than in terms of external characteristics, as the other approaches do. Previous research provided a lot of information supporting the supposition that the approach based on the analysis of meaning provides an adequate theoretical framework for studying states of consciousness. According to the meaning theory, a "state of consciousness" is defined as a particular state of the cognitive system conditioned by a meaning-based organizational transformation (Kreitler, 1999, p. 194). Hence, at any given time, the cognitive system is in some state of consciousness, not only during specific times or in particular situations. The cognitive state of organization can be characterized in terms of the major components (contents and processes) of the system of meaning. Consequently, there is a great number of existing and potential states of consciousness (Kreitler, 2001, 2002, 2009),

The meaning-based approach has three major advantages. First, assuming that the cognitive system is indeed determined mostly by the meaning system, this approach truly defines states of consciousness in terms of the system's components and not according to external characteristics alone. Second, the tests for determining the state of consciousness are indirect and not related to consciousness in any obvious way, so that the responses are not affected by factors, such as social desirability or specific conceptions of the participants in regard to consciousness.. Third, by grounding the definition of states of consciousness in the meaning-based organization of the cognitive system it is possible to account for and even predict functioning and changes not only in cognition but also in other systems affected by the state of consciousness, mainly emotions, reality perception (including time and place perception), self control, self concept, body awareness, and even manifestation of various personality traits. The reason is that studies showed how cognitive functions, affected by the processes and contents of meaning, are involved in changes in all these different domains (Kreitler and Kreitler, 1985a, 1985b, 1986, 1987, 1990a, 1990b, 1994; Kreitler, 2003, 2010, 2011, 2012).

One important conclusion is that the meaning-based approach to consciousness includes also the characteristics based on the mental function approach. Further, it is related also to the other approaches to consciousness in that it assumes that the described effects of states of consciousness are mediated through the impact the different factors have on cognition. For example, an emotion or a drug or a certain behavior like dancing may affect consciousness by means of specific changes produced in the organization of the cognitive system.

THE CONSCIOUSNESS STATES EXAMINED IN THIS RESEARCH

In accordance with the meaning-based approach, there are a great many states of consciousness. Actually, the cognitive system is always in some state, characterized by the organization of the processes and contents of cognition. However, not all of them have been identified, described, and evaluated. There are probably three major criteria that play a role in determining which organizational state of the cognitive system will be identified as a state of consciousness. One criterion is inherent in the organizational transformations in the meaning system. Changes in organization that affect the whole of the cognitive system are more likely to be recognized as states of consciousness. These are mostly massive changes or changes that affect factors that play a central role in the overall organization of the cognitive system, such as focusing on personal meanings (see below). A second criterion is information and awareness of the factors able to produce changes in the cognitive system that are noticeable by the affected individuals. Thus, changes in cognition following the ingestion of drugs, intoxication, sensory deprivation, fasting, sleep, dancing, meditation, hypnosis and intense emotions are likely to be noticed and accorded a special status by the affected individuals. Moreover, changes due to factors of this kind enjoy a priority in that the factors responsible for the changes are often under the control of the individuals. At present this holds for all the listed factors to a greater extent than for emotions. Finally, a third criterion is the significance and role attributed to a particular state of cognition within a conceptual framework, such as a culture, a religion and even a scientific approach. Thus, states of consciousness discussed, used and promoted by Zen, meditation based on Yoga practices, the Kabbalah or Shamanism are likely to be identified as states of consciousness by the members of the respective cultures or religions and other practitioners (Faber, 1981; Johnston, 1971; Fischer, 1978; Kaplan, 1982). It is however likely that the salience of specific states of cognition is accounted for both by the sanction or legitimacy accorded to specific states by the conceptual values and framework but also by familiarity on the part of the individuals. Familiarity is promoted by the easier accessibility of specific states due to their legitimacy in the culture as well as by the availability of and information about means likely to evoke these states which are increased in regard to states sanctioned by the culture.

Familiarity is an important factor, as demonstrated by the following examples. Thus, people who have gone through the Zen experience became attuned through the training to note differences between four kinds or stages of consciousness: (1) ego loss (koan), (2) deep meditation (sunmay), (3) hallucinatory phase (makyo) and (4) enlightenment (satori or kensho) (Johnston, 1971). For those who are not Zen conoisseurs the distinctions are unclear conceptually and non-existent experientially. Likewise, the training of Yoga may focus on differentiation of states of consciousness that a regular untrained person from Western culture can hardly make sense of (e.g., 'Dharna' - "an uninterrupted concentration to hold the mind on a fixed center (such as a thought, an object..." versus 'Dhyan' - maintaining the former "but at a constant rate of flow" Fischer, 1978, p. 42).

The following three pairs of states of consciousness were selectedfor the set of studies performed within the framework of the presented research:

a. Emotional states of consciousness: the state of anger compared with the state of happiness.
b. Conceptual states of consciousness: the state of concrete perception compared with abstract perception.
c. States of consciousness resulting from the manner in which people relate to their internal world ("personal") and the manner in which they relate to their environment ("interpersonal").

Notably, the choice of the first set corresponds to the second above-mentioned criterion, insofar as emotions represent a factor likely to produce major changes in the state of the cognitive system. The choice of the second set corresponds to the third criterion according to which conceptually-supported states are likely to enjoy the status of states of consciousness. We chose however a set of states corresponding to a neutral scientifically-supported approach so as to avoid the issues of possible attitudes toward a particular culture or religion. Finally, the choice of the third set corresponds to the first mentioned criterion of organizational transformations in the cognitive system itself insofar as focusing on personal or interpersonal types of meaning evokes directly major transformations in the structural organization of the cognitive system (see also chapters 3 and 13,this book).

Emotional States of Consciousness: Anger and Happiness

The Relation between Emotion and Cognition

Emotion is a phenomenon that includes physiological components (in the brain and the body), activity or behavioral components (muscular) and cognitive components. The cognitive system affects the emotional stimuli and is affected by them (Dalgleish and Power, 2000). Emotion can affect various functions of the cognitive system, such as logical thought, short and long-term memory, attention, concentration, problem-solving and creativity (Barrett, 2005; Martin and Clore, 2001) (see Figure 2).

The cognitive system is not only affected by the emotional system but also affects it. In the course of this research we were interested only in the affect of emotion on cognitive function.

Emotional states can be intensified and reinforced by psychological means, for example, by presenting stimuli that evoke emotions (for instance, certain words, films, images, drawings, stories, melodies), through guided imagery (creating stimuli that evoke feelings using personal memories and imaginary situations), by role-playing in emotional contexts and by virtual means. In the framework of this research we tried to take advantage of this property in order to construct methodologies that would enable us to evoke in participants intense states of anger and happiness (see description of the methods in “Is it possible to change states of consciousness proactively?”)

The emotional states examined in this research are anger and happiness. Both are highly familiar to us in our everyday lives, and may be manifested in various degrees of intensity.

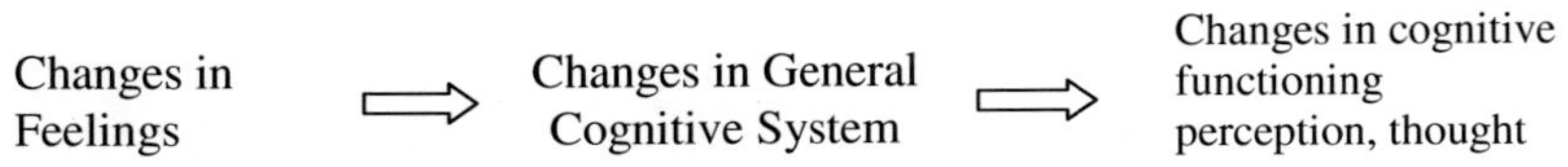

Changes in Feelings ⟹ Changes in General Cognitive System ⟹ Changes in cognitive functioning perception, thought

Figure 2. The effect of emotion on the cognitive system.

Anger and Happiness

Happiness is a positive emotion that produces in the individual an uplifted mood. It usually shows up following events and experiences that are considered positive by that individual. The feeling is characterized by certain cognitive and physical manifestations, such as optimistic thinking, joyful facial expressions, shining eyes, smiles, and an energetic and light body sensation.

Anger is an intense emotion designed mainly to induce an individual to act aggressively and assertively. Anger is expressed in terms of psychological and physiological characteristics. The physical symptoms include, for example, increased blood flow to the muscles, heightened heart and lung activity, expansion of blood vessels, drive to act, restlessness and increased muscle tension. The psychological symptoms include, for example, extreme focus and concentration on the disturbing issue, a decrease in analytical and logical skills, and weakening of the ability to concentrate.

These two states were chosen because they seemed to provide a good basis for the research on severity evaluations. It could be hypothesized that an individual in a state of happiness tends to make milder severity evaluations, whereas an angry individual tends to make harsher severity evaluations. In the study these two expectations were tested for the two states of consciousness in terms of the evaluations themselves as well as the changes in evaluations.

States of Consciousness of Concrete and Abstract Thinking

The definitionsof concrete and abstract states of consciousness are taken from the clinical psychology literature. The distinction between concrete and abstract thought was made for the first time in regard to individuals suffering from various brain injuries and diseases (Goldstein, 1942). The distinctions made by the researchers working with these disorders can be projected on the characteristics of states of consciousness. However, it is evident that in healthy individuals the changes are less extreme, and a person in a state of consciousness tending towards the concrete, will still have the ability for abstract thinking, and vice versa.

Goldstein and Scheerer (1941) define a *concrete approach* as realistic, in the sense of comprehension and direct and immediate experience of the event or situation while they are still being formed. The thoughts and reactions are directed by the facts that we simply come across. The facts need neither to be based on actual experience, nor to be evaluated and checked in terms of general criteria.

In contrast, the *abstract approach* is defined as consisting of a stimulus or perception grasped within an adequate context, detecting general characteristics on the basis of a number of different and particular qualities, and assigning them to general categories or attributing to them an overall significance.

The abstract approach differs from the concrete approach in the following 8 characteristics:

a. Disconnecting the ego from the outside world or from internal experiences – for instance, the ability of individuals to express a reality that is different from the one known to them, for instance, expressing a falsity.
b. Entering a given mental state at will – for instance, the ability of individuals to assume that they are already in a given situation and to continue the task from that point onward.
c. Explaining acts or events around us – for instance, the ability to describe where a certain noise comes from, or to discern which of two objects is more distant.
d. Shifting between different aspects of a given situation – for instance, the ability to perform two tasks that need to be done simultaneously, the ability to simultaneously discern a number of different characteristics of an object, such as color, shape and dimensions.
e. Discriminating between different aspects of the same situation – for instance, the ability to distinguish between mixed up items, and the like.
f. Comprehending the whole in general, or assembling the whole from various details – for instance, the ability to understand a given situation from a number of sequential events.
g. Creating hierarchal concepts – for instance, the ability to organize different items according to their sizes or their quantities.
h. Explaining ideas – for instance, demonstrating an action without the use of objects.

These characteristics are organized in a rising order of complexity, so that the last mentioned ones (i.e., g. and h.) are more complex than those placed in the beginning (i.e., a. and b.). The normal individual combines different kinds of thinking, and passes between the different kinds according to the demands of the situation.

In our experiment, we did not attempt to put individuals into an absolute concrete or abstract state, because this would not have served the purpose of our research which was to simulate normal everyday situations. Our main intent in using manipulations was to try and tip slightly the thinking approach in one or the opposed direction.

Accordingly, in manipulations designed to promote a concrete state of consciousness we asked our participants to focus on very clear and definite details in images or in complex situations. For instance, in an urban landscape image, we asked that they focus on a particular structure, or focus on the shape of a window and on the sensation of touching the depicted building. On the other hand, in manipulations designed to promote an abstract state we asked that they look at the structure of the city as a whole and not at the specific structures that constitute the city.

The Personal and Interpersonal States of Consciousness

The state of personal consciousness is evoked in individuals when they focus on their personal world and think about things that are related to personal experiences and to things personally related to them, in subjective terms. The typical mode of thought in this state is expressed mainly by using examples, illustrative details, exemplifying situations or dramatic scenes, as well as metaphors and symbols, that is, concrete exhibits that illustrate general and abstract ideas, whereby the connection between the exhibit and the general idea is completely personal and not presented in any dictionary. For example, a metaphor for the concept “life” could be “a handkerchief in the hand of a magician”, for “success” - “reaching the top of a mountain where you find yourself alone and cold”, or for “love” – “bright sunlight all around”.

In the personal state of consciousness there is also a tendency to generalities and abstractions, such as “happiness” – “this is the place where you are not at present but have probably been there before” or “love” – “meeting yourself through another person”. This state is further characterized by the tendency or personal reference to issues and problems, with an emphasis on subjective emotions. On the other hand, there is no tendency towards organized and accurate thought, no tendency to methodical arrangement of matters, no focus on one matter, no reference to quantities or numbers, and no mentioning of accurate dates.

The state of interpersonal consciousness is characterized by organized, logical and rational thinking. In this state individuals express ideas in terms that are interpersonally accepted, and that can in principle be inserted in a dictionary. These individuals distance themselves from examples and metaphors, and use instead direct and simple assertions, as well as comparisons (emphasizing similarities, differences and contradictions, as well as reciprocal relations between issues or exhibits). In general they prefer the objective over the subjective, the general over the particular. Individuals in this state can undoubtedly relate to themselves and to their feelings, but these emotions neither overwhelm them nor control them or their thoughts. They will define “love”, for instance, not by examples but rather as “a feeling between two people”.

The interpersonal state is usually noticeable when individuals try to communicate with others, to pass on messages, information or instructions. A personal state is particularly noticeable when individuals delve into their inner worlds, when they deal with their emotions, their experiences and their personal positions. Many comparative experiments found that a great number of effects were created as a result of the transition from one state to another. For instance, under the impact of the personal state individuals got higher scores in visual memory and in number of associations, but lower scores in verbal memory, in comprehension accuracy and in the evaluation of lines and circles (Kreitler, 1999, 2002; Kreitler, Kreitler and Wanounou, 1987-1988).

Is it Possible to Change States of Consciousness Proactively?

In order to carry out the research, we were required to "transport" the participants into the desired states of consciousness. For this purpose, we created manipulations that used a combination of the two following techniques: (1) Guided imagery – inducing the participants to create in fantasy stimuli, personal memories and imaginary situations; and (2) Presentation of visual stimuli – exposing the participants to images, visual illustration, including a short film of a sequence of images. Every manipulation designed to change the state of consciousness lasted about five minutes, during which the participants were asked to concentrate on the method so as to optimize its impact.

Prior to applying the created transformational techniques it was necessary to evaluate them. For this purpose, we did a preliminary study, during which we tested the techniques on 40 participants. After implementing improvements and changes, we succeeded in reaching the position where over 82% of the participants testified that the different manipulations had substantial impact on their states of consciousness.

This validation stage demonstrated that the manipulations we had developed influenced the participants' states of mind, and confirmed the assumption that the various manipulations were adequate for the study and had a high probability of creating the desired changes in the state of consciousness.

The following sections summarize the major components of each of the manipulations we used:

Emotional Manipulations – Happiness/Anger

The participants were shown several images expressing happiness/anger and moments of happiness/anger (see examples of images used in the manipulation of happiness). At the same time, they were asked to concentrate on the images that portrayed happiness/anger to them.

After being shown the sequence of images, the participants were asked to close their eyes and to experience a specific moment in which they had been very happy/angry. At the same time, the participants were instructed to intensify the experience of happiness/anger and to experience it in all parts of their bodies. The duration of the whole manipulation was about five minutes.

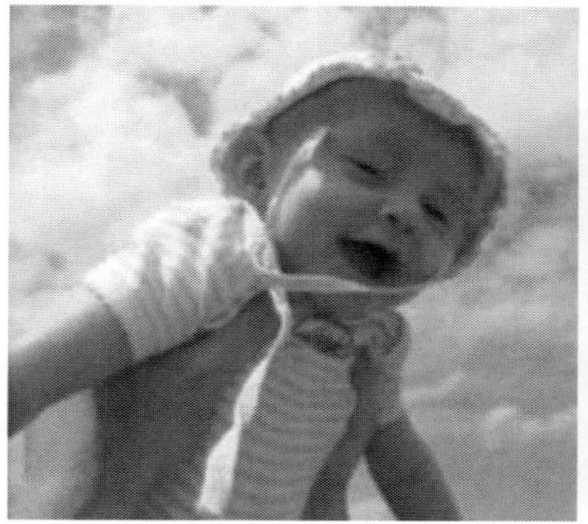

Examples of images used in the manipulation of happiness

Manipulations for Concrete and Abstract States of Consciousness

In order to get the participants into a concrete state of mind, they were shown a sequence of images of objects and various illustrations for example, a cup of coffee, a façade of a building or a street. In regard to each image, the participants were asked to concentrate for 30 seconds on the characteristics of the presented object and on the small details of the image. The purpose was to cause the participants to develop thoughts on minor details rather than more general concepts.

Images from the manipulation of the concrete state of mind

In order to get the participants into an abstract state of mind, they were shown a sequence of images of different objects, different pieces of furniture or landscapes.. In regard to each image, they were asked to concentrate for 30 seconds on the general characteristics of the image, focusing on the connection between the objects presented in the image and the general concept of the image. The aim was to cause the participants to develop general thoughts,rather than focusing on the minor details.

Images from the manipulation of the abstract state of mind

Manipulations for Personal/Interpersonal States of Consciousness

In order to put the participants into a personal state of mind, they were shown a sequence of images, for example, a classroom, a park or a beach. They were asked to concentrate on each image for 30 seconds and to think about what it symbolized for them. They were also asked what each image reminded them of and what kind of personl effects it evoked in them.

In order to put the participants into an interpersonal state of mind, they were shown a sequence of images, for example, a crowded street, a shop or a highway. They were asked to concentrate for 30 seconds, on the familiar details of each image, specifically those details that are likely to be considered similarly by all observers and not only by them personally.

An image from the manipulation of the interpersonal state of mind

DID THE CHANGES IN THE STATES OF CONSCIOUSNESS AFFECT JUDGMENTS? THE EXPERIMENTAL DESIGN

General Description of the Procedure

The experiment was based on the analysis of questionnaires, and over 300 participants from various population groups took part in it. Each participant was asked to fill out three identical questionnaires in a sequence:

- Filling out the questionnaire for the first time: "zero state", before the participants were put through any manipulation.
- Filling out the questionnaire for the second time: after the participants had been through one manipulation aimed at changing their state of consciousness.
- Filling out the questionnaire for the third time: after the participants have been through manipulations aimed at changing their state of consciousness to another state.

Each of the questionnaires included ten questions in which the participants were required to carry out severity evaluations based on textual information data they had been provided, and five severity evaluations based on visual images they were shown.

After completing the data collection stage, a comparison was made of the severity evaluations completed by each participant in each of the three states of consciousness in which the task (i.e., filling out the severity evaluation questionnaire) was done.

The following diagram presents the course of the experiment for each participant:

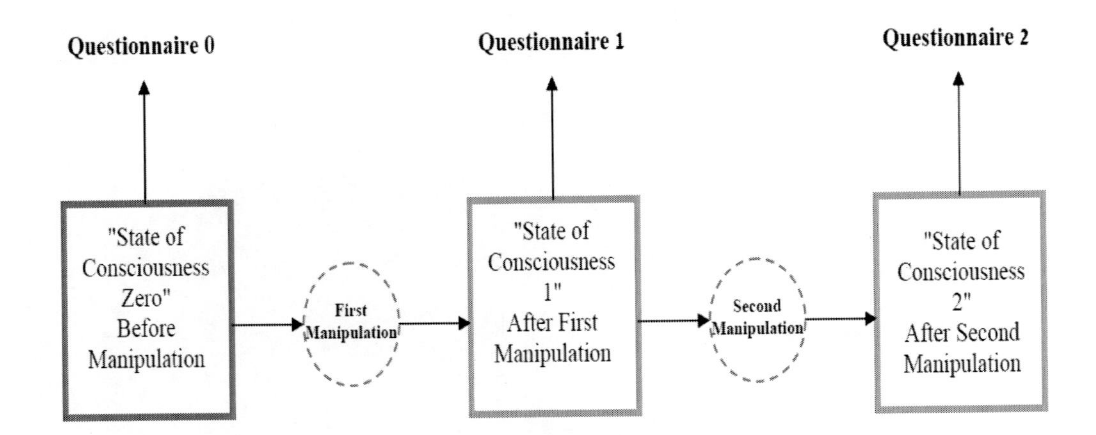

Diagram: Flow diagram of the experimental procedure.

The analysis of the results focused on comparing the changes in severity evaluations carried out by each participant, and analyzing the general change trends of all the participants (expectancy and variance).

THE QUESTIONNAIRE STRUCTURE AND DESCRIPTION OF TASKS

The Severity Evaluation Task

The role of the severity evaluation was to simulate situations in which we are required to perform severity evaluations and judgments of various everyday situations, based on the data and textual information presented to us. The task was designed to create a database that would enable us to compare severity evaluations of participants in each of the states of consciousness. The scenarios presented to the participants were identical in each of the states of consciousness. The order of the questions and the order of the data in the questionnaires were changed randomly in order to minimize and possibly even cancel the effect of memorizing the questions.

The Structure of the Severity Evaluation Task

In each questionnaire there were 10 scenarios (5-6 facts for each scenario). For each scenario, the participants were asked to evaluate the severity of the scenario. The scenarios were divided into 3 types:

Personal scenarios:

- Evaluate your present state of health, if it is known that...
- Evaluate the degree of severity of danger, if it is known that...
- Evaluate the probability that your partner is cheating on you, if it is known that...
- Evaluate the probability that your boss will fire you soon, if it is known that...

Impersonal scenarios (technical):'

- Evaluate the severity of your computer's state of repair, if you know that...
- Evaluate the severity of your car's state of repair, if it is known that...
- Evaluate the severity of the state of health of some person, if it is known that...
- Evaluate the degree of success of a student in an upcoming examination, if it is known that...

Scenarios in regard to opinions:

- Evaluate the probability that within the next 5 years, the state of Israel will sign a new peace agreement, with at least one of its neighbors...
- Evaluate the probability that within the next 5 years, there will be a large military confrontation between the state of Israel and at least one of its neighbors...

The participants were asked to check their evaluation on the severity axis:

Very Grave	Completely Intact

The "Facial Expression Identification" Task

The role of the facial expression identification task was to simulate situations, in which we are required to carry out everyday evaluations, based on the visual information to which we are exposed, such as images, graphs, diagrams, and any other visual stimuli. The use of facial expressions, each different from the other, allowed us to confront the participants with different visual situations, whose nature they were required to evaluate. The aim of this section was to create a database that would enable us to compare the participants' evaluations of facial emotional expressions in each of the states of consciousness, when the objects represent emotional expressions and the evaluation refers to identifying the expressed emotion. The participants were shown the same facial expressions during each of the states of consciousness. The order of the facial expressions in the questionnaires was changed randomly.

The participants were shown sequentially 5 photographs of facial expressions, and were asked to evaluate their nature from the 7 presented possibilities: happiness, sadness, anger, surprise, disgust, natural, fear.

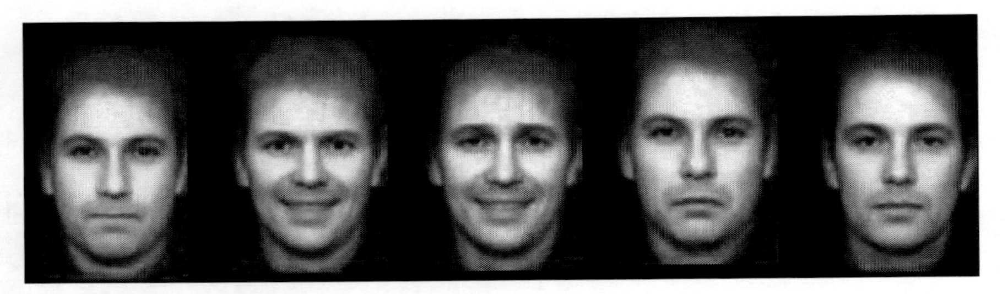

Diagram: the expressions each participant was required to assess

RESEARCH RESULTS

Task Performance without Manipulation -Control Group

During the first stage, we wanted to examine the supposition that participants who had not been through any manipulation, would be consistent so that there would be no significant changes in their severity evaluation results and their facial expression evaluation results. To do this, 40 participants were asked to do the tasks several times, each time after a few minutes break, without going through any manipulation. The following diagram presents the results for this stage:

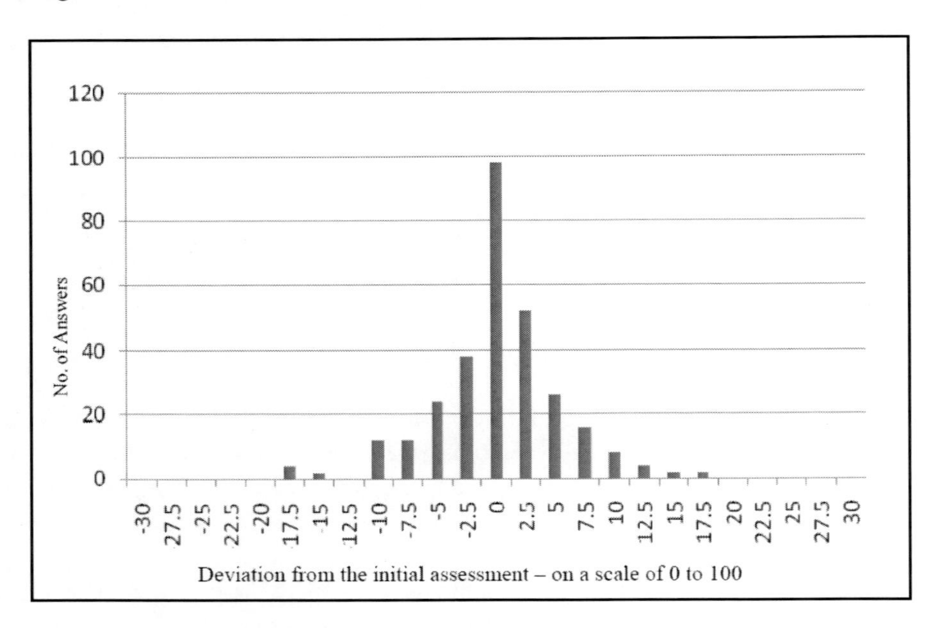

Diagram:. Expectancy assessment changes of severity evaluations in the control group.

The diagram shows that for each of the questions, the test hypothesis does not contradict the hypothesis that the expectancy assessment of the control group will not change significantly without manipulation. The "natural deviation" expectancy (a deviation that does not result from the states of consciousness) of the assessments obtained was only 0.16 units (on a scale from 0 to 100). The conclusion was that when participants underwent no manipulations designed to change their states of consciousness, the expectancy of their severity evaluations was almost identical to the evaluations that they gave a short time before. We also noticed that there was no consistency in the direction in which the severity evaluation changed. Some changes were towards harsher evaluations, whereas others were towards milder evaluations.

Finding the Natural Change in the "Facial Expressions" Task – Control Group

We did a similar experiment (without manipulation) for the facial expressions task, and there too, the results were similar: only 3.3% of the facial expression evaluations changed. That is, the participants had no tendency to change their evaluations in the "facial expressions" task, when they underwent no manipulation of their states of consciousness (a similar result to that received for the severity evaluation task).

Results Regarding Emotional States of Consciousness: Anger and Happiness

Analysis of the results with regard to the effect of emotional states shows that anger and happiness states of consciousness affect severity evaluations in the following manner:

- A happiness state of consciousness caused milder severity evaluations.
- An angry state of consciousness caused harsher severity evaluations.

The extent of change to severity evaluations is not consistent and changes from question to question. Even so, we succeeded in finding a relation between the extent of the impact of emotional states of consciousness on severity evaluations, and the type of subject that the question dealt with, where the questions were classified into 3 topics:

a. Questions dealing with severity evaluations and judgments of personal situations.
b. Questions dealing with severity evaluations and judgments of technical matters.
c. Questions dealing with severity evaluations of general and strategic matters.

The results showed that in situations dealing with personal matters, the state of consciousness of anger had a significant effect (towards harsher evaluations), whereas a state of consciousness of happiness had a less significant effect (towards milder evaluations).

On the other hand, in questions dealing with technical matters, the happiness state actually had a greater impact than that of anger.

The following explanation can be given in regard to the effect of anger and happiness on technical matters in comparison with personal matters: it is possible in everyday life, that when we are in a "regular" state of consciousness, ("neutral" state – neither angry nor happy), we have a natural tendency to make harsher evaluations in technical matters, and milder severity evaluations in regard to our personal matters. Therefore, our being happier affects mainly technical evaluations towards which we routinely have a harsher tendency;, but it does not affect personal evaluations as much because anyway we deal with those in a milder fashion. On the other hand, when we are in a state of anger, the tendency is exactly the opposite. There may certainly be other explanations for the phenomenon we noticed.

In the third type of questions, we dealt with evaluating the probabilities of war and peace and other general strategic topics, where no change at all was discovered in the severity evaluations. It is possible that in these matters, evaluations of individuals are based on previously-held personal opinions and represent a general and deeper perspective. Therefore momentary changes in states of consciousness have no significant impact on the evaluations.

Table 1. Summary of conclusions and findings regarding severity evaluations in two emotional states of consciousness

Questions Regarding **Strategic** Situations	Questions Regarding **Technical** Situations	Questions Regarding **Personal** Situations	Impact Direction Compared to State Zero	State of Consciousness
No Change in Severity Evaluation	Low Level of Significant Change	**High Level of Significant Change**	Less Severe Evaluation (Milder Evaluation)	**Happiness**
No Change in Severity Evaluation	**High Level of Significant Change**	Low Level of Significant Change	More Severe Evaluation (Harsher Evaluation)	**Anger**

Conclusions Regarding the "Facial Expression Task"

The conclusions and findings obtained for judgments and evaluations based on visual information are similar to those obtained for the severity evaluation task. We found that the tendency was identical for both tasks: individuals tend to make harsher evaluations when in a state of anger, whereas, in a state of happiness, the evaluations were generally milder. Nonetheless, in the "facial expression" task, there was less change in evaluations than in the severity evaluation task. It is possible that the discrepancy is a result of the difference in the nature of applied method (the choice between the alternatives given in a finite list, compared with marking the severity evaluation on an axis).

Conclusions Regarding Concrete/Abstract and Personal/Interpersonal States of Consciousness

Conclusions Regarding Concrete and Abstract States of Consciousness

The study shows that there is some impact on severity evaluations. It seems that in most cases, the severity evaluations in concrete and abstract states of consciousness were less severe than in the initial stage of control ("zero state"). Nonetheless, almost all the changes we found were not significant. Yet, we succeeded in detecting a tendency for change in the severity evaluations between the concrete state and the abstract state. Likewise, no impact of these states of consciousness was discovered in regard to the "facial expressions" task.

Summary of the Results for the Concrete and Abstract States of Consciousness

It can be definitely stated that transformations in states of consciousness to the concrete and abstract states caused a change in task responses. Nonetheless, it is difficult to characterize the change and to draw conclusions or make rules that characterize the impact. It may be possible to characterize the impact of these states of consciousness on other cognitive tasks that do not use a linear axis to portray the results as less or more severe.

Conclusions Regarding Personal and Interpersonal States of Consciousness

Transformations of states of consciousness to these states had no impact on the examined tasks, i.e., both severity evaluations and the "facial expression" task. In this case too, it is possible that the impact of these states of consciousness can be characterized using other cognitive tasks.

SUMMARY AND GENERAL CONCLUSIONS

The research results lend support to the central hypothesis that we wanted to test, and show that changes in states of consciousness may definitely affect severity evaluations and judgments of facial expressions, that is, both for judgments based on data and textual information, and for judgments based on visual data.

Notably, not all the states of consciousness examined affected the severity evaluations and judgments. States of anger and happiness had significant impact with clear tendencies in regard to the severity evaluations; the concrete and abstract states of consciousness had little impact; and no impact at all was noticed in the severity evaluations caused by the transformations to personal and interpersonal states of consciousness.

It was also noticed that the impact of states of consciousness on severity evaluations was derived from, among other things, the type of questions and situations that the participants were asked to evaluate.

THE CONTRIBUTION OF THE RESULTS

In the present section we will focus on the conclusions and findings for each of the states of consciousness and the overall contribution of the research findings.

a. We saw that states of consciousness sometimes affect severity evaluations. Regarding states of anger and happiness, we even succeeded in characterizing the states' manner of impact on the judgment evaluation results. This characterization allows us to try and minimize this impact in the future, to take it into consideration and to improve severity evaluation capabilities.

   For example, if individuals carry out a judgment evaluation *of a personal matter,* when it is known to us in advance that they are in *a state of anger*, it will be possible for us to assume that their severity evaluations are *overly harsh*. In cases of evaluations of great importance, it will even be possible to proactively try and "make a correction" to ease the severity of the evaluation. Nonetheless, there are still two problematic issues: 1) the extent of "correction" needs to be determined. Notably, according to the research results, the extent of change in the evaluations was not consistent. 2) We need to identify the individual's state of consciousness. But there does not exist one consensual method for determining an individual's state of consciousness (for example, defining a state of anger). In our research, we presented a number of acceptable methods for doing this.

As to the situation existing today and the results as they are, it is possible to try and decide randomly on a reasonable "degree of correction" and on one method for determining a state of consciousness (for instance having the participants testify for themselves), in order to improve the trustworthiness of the evaluations. Another solution is to take the evaluations as they are, but to also take into account that a deviation resulting from the participants' states of consciousness is possible. In any case, particularly when evaluating answers, it is recommended to ask individuals to perform a number of evaluations at different times, and so to minimize the effects of extreme states of consciousness, such as anger.

b. Within the framework of the research, we developed manipulations designed to change states of consciousness that combined guided imagery (such as personal memories and situation imagery) and presenting visual stimuli (a short film combining photographs, animation and colors). At the different stages of the research, we also did manipulation validation, and observed that they indeed did succeed in creating a change in the state of consciousness of the participants. Future research into states of consciousness will be able to use the manipulations developed in this research, or to develop similar manipulations for other states of consciousness.

WHAT WE LEARNED ABOUT CONSCIOUSNESS

The study showed that it is possible to change states of consciousness and that the changes in states of consciousness affect the output in specific cognitive acts. However, these conclusions apply more to changes in emotional states of consciousness than to the other states and more in regard to severity evaluations than to the evaluation of facial expressions.

To our mind, there are important conclusions also regarding the negative results. First, it appears that not all cognitive tasks are affected to the same extent or at all by all states of consciousness. For example, concrete/abstract states of consciousness do not affect the evaluation of facial expressions. Also, as we saw,the effects differ in their direction (i.e., increased or decreased severity). Similar findings were reported earlier (Kreitler, 1999, 2009).

Second, it is possible that the impact of a state of consciousness may depend on the means or technique used for its evocation. In the present study, the means were characterized by two features: a heavy emphasis on visual stimuli, and a narrowing down of the active role of the participant in the phases of inducing the states of consciousness. In former studies (e.g., Kreitler, 1999) with more salient effects in a range of cognitive tasks the induction procedures were purely verbal and emphasized the active role of the participant. Further research conducted at present is devoted to clarifying the dependence of the depth and effects of states of consciousness on the methods of induction.

EXPANDING THE RESEARCH TO ADDITIONAL TASKS

The tasks examined in the framework of the research were concerned with severity evaluations and judgments. Although we succeeded in detecting a connection between the anger and happiness states and the performance of the tasks, it was still difficult to characterize this connection regarding the other examined states of consciousness. This research can be expanded in the future for the same states of consciousness, while integrating within it other cognitive tasks, such as creativity, logic tests and more. Although it is difficult to characterize the connection between concrete/abstract and personal/interpersonal states of consciousness and severity evaluations, it may actually be possible to identify this connection using other tasks.

EXPANDING THE RESEARCH TO ADDITIONAL STATES OF CONSCIOUSNESS

Within the framework of this research, we chose to focus on 3 pairs of states of consciousness. In the future, the same research can be implemented regarding other states of consciousness, for example, states that involve other emotions, states that are caused by the involvement of external substances, such as alcohol, and additional states. It would be interesting to examine how, for instance, a state of drunkenness or a state of fear, affects severity evaluations and judgments. Likewise, if a researcher has results for different states of consciousness, it would be possible to try and characterize similar features for different groups of states of consciousness.

REFERENCES

Baars, B. J. and McGovern, K. (1996). Cognitive views of consciousness: What are the facts? How can we explain them? In M. Velmans (Ed.), *The science of consciousness: Psychological, neuropsychological, and clinical reviews*. London: Routledge, pp. 63-95.

Barrett, L. F. (Ed.) (2005). *Emotion and cognition*. New York, NY: Guilford.

Blackmore, S. J. (2003). *Consciousness – an introduction*. New York: Oxford University Press.

Dalgleish, T. and Power, M. (Eds.), (2000). *Handbook of cognition and emotion*. Chichester, UK: Wiley and Sons.

Faber, M. D. (1981). *Culture and consciousness: The social meaning of altered_awareness*. New York: Human Sciences Press.

Farthing, G.W (1992). *The psychology of consciousness*. Englewood Cliffs, N.J., Prentice-Hall.

Fischer, R. (1978). Cartography of conscious states: Integration of East and West. In A. A. Sugerman and R. E. Tarter (Eds.), *Expanding dimensions of consciousness*. New York: Springer Publishing, pp. 24-57.

Goldstein, K. and Scheerer, M. (1941). Abstract and concrete behavior: An experimental study with special tests. *Psychological Monographs, 53,* No. 2 (whole No. 239), pp. 1-151.

Goldstein, K. (1942). *After effects of brain injuries in war.* New York: Grune and Stratton.

James, W. (1890). *The principles of psychology (2 Vols.).* London: Macmillan.

James, W. (1904). Does 'consciousness' exist? *Journal of Philosophy, Psychology, and Scientific Methods, 1*, 477-491.

Johnston, W. (1971). *The still point.* New York: Perenial Library, Fordham University Press, Harper and Row

Kaplan, A. (1982). *Meditation and kabbalah.* York Beach, ME: Samuel Weiser.

Kreitler, S. (1999). Consciousness and meaning. In J. Singer and P. Salovey (Eds.), *At play in the fields of consciousness: Essays in honor of Jerome L. Singer* Mahwah, NJ: Erlbaum, pp. 175-206..

Kreitler, S. (2001). Psychological perspective on virtual reality. In A. Riegel, M. F. Peschl, K. Edlinger, G. Fleck and W. Feigl (Eds.), *Virtual reality: Cognitive foundations, technological issues and philosophical implications.* Frankfurt, Germany: Peter Lang, pp. 33-44.

Kreitler, S. (2002). Consciousness and states of consciousness: An evolutionary Perspective. *Evolution and Cognition, 8*, 27-42.

Kreitler, S. (2003). Dynamics of fear and anxiety. In P. L. Gower (Ed.), *Psychology of fear.* Hauppauge, NY: Nova Science Publishers, pp. 1-17.

Kreitler, S. (2009). Altered states of consciousness as structural variations of the cognitive system. In E. Franco (Ed., in collab. with D. Eigner), *Yogic perception, meditation and altered states of consciousness.* Vienna, Austria: Oestrreichische Akademie der Wissenschaften, pp. 407-434.

Kreitler, S. (2010). Meaning correlates of value orientations. In Franziska Deutsch, Mandy Boehnke, Ulrich Kühnen and Klaus Boehnke (Eds.), *Crossing borders: Cross- cultural and cultural psychology as an interdisciplinary, multi-method endeavor.* E-book.

Kreitler, S. (2011). Anger: Cognitive and motivational determinants. In J. P. Welty (Ed.), *Psychology of anger: Symptoms, causes and coping.* Hauppauge, NY: Nova Science Publishers, pp. 179-195.

Kreitler, S. (2012). The psychosemantic approach to logic. In S. Kreitler, L. Ropolyi, G. Fleck and D. Eigner (Eds.), *Systems of logic and the_construction of order.* Bern, New York, Vienna: Peter Lang Publishing Group, pp. 33-60.

Kreitler, S. and Kreitler, H. (1985a).The psychosemantic foundations of comprehension. *Theoretical Linguistics, 12*, 185-195.

Kreitler, S. and Kreitler, H. (1985b). The psychosemantic determinants of anxiety: A cognitive approach. In H. van der Ploeg, R. Schwarzer and C. D. Spielberger (Eds.), *Advances in test anxiety research*, Vol. 4. Lisse: The Netherlands and Hillsdale, NJ: Swets and Zeitlinger and Erlbaum, pp. 117-135.

Kreitler, S. and Kreitler, H. (1986). Individuality in planning: Meaning patterns of planning styles. *International Journal of Psychology,* 21, 565-587.

Kreitler, S. and Kreitler, H. (1987). Psychosemantic aspects of the self. In T. M. Honess and K. M. Yardley (Eds.), *Self and identity: Individual change and development.* London: Routledge and Kegan Paul, pp. 338-358.

Kreitler, S. and Kreitler, H. (1990a). *Cognitive foundations of personality traits.* New York: Plenum.

Kreitler, H. and Kreitler, S. (1990b). The psychosemantic foundations of creativity. In K. J. Gilhooly, M. Keane, R. Logie and G. Erdos (Eds.), *Lines of thought: Reflections on the psychology of thinking* (Vol. 2). Chichester, UK: Wiley, pp. 191-201.

Kreitler, S. and Kreitler, H. (1994). Motivational and cognitive determinants of exploration. In H. Keller, K. Schneider and B. Henderson (Eds.), *Curiosity and exploration.* New York: Springer-Verlag, pp. 259-284.

Kreitler, S., Kreitler, H. and Wanounou, V. (1987-1988). Cognitive modification of test performance in schizophrenics and normals. *Imagination, Cognition, and Personality, 7,* 227-249.

Martin, L.L. and Clore, G. L.(Eds.) (2001). *Theories of mood and cognition: A user's guidebook.* Mahwah, N.J.: Erlbaum.

O'Brien, F. J. Jr. (1987). The Goldstein-Scheerer tests of abstract and concrete thinking. Test Review. ED289891.

Peters, J. and Daum, I. (2008). Differential effects of normal aging on recollection of concrete and abstract words. *Neuropsychology, 22*, 255–261.

Tart, C.T. (1972) States of consciousness and state-specific sciences. *Science, 176*, 1203-1210.

In: Consciousness: Its Nature and Functions
Editors: Shulamith Kreitler and Oded Maimon
ISBN 978-1-62081-096-5

Chapter 13

CONSCIOUSNESS AND KNOWLEDGE: THE PSYCHOSEMANTIC APPROACH

Shulamith Kreitler
Department of Psychology, Tel-Aviv University, Israel

ABSTRACT

Knowledge is used in many definitions of consciousness, which suggests that consciousness is intimately related to knowledge. A survey of the definitions led to the following conclusions that may be relevant for understanding consciousness: there may be different ways of getting knowledge, different states of knowledge, and different contents of knowledge. In order to clarify the relations of consciousness to knowledge a psychosemantic epistemology based on the meaning system by Kreitler and Kreitler was presented. This approach includes the conceptual tools for describing different domains of contents, corresponding to types of knowledge, and cognitive processes, corresponding to ways of knowledge production and presentation. The meaning system provides the contents and processes for the performance of different cognitive acts and is indirectly involved also in personality traits, emotions, attitudes etc. In this framework consciousness is defined as the state of the whole cognitive system at a given time that is determined by the organizational structure of the underlying system of meaning. The organizational structure implies that some meaning variables are dominant and accessible whereas others may be in varying degrees of availability. In each consciousness state only specific types of knowledge and related ways of knowing are available. There are different states of knowledge, namely, the dominant types of knowledge change when the state of consciousness changes. States of knowledge may be produced by a variety of means, all of which affect in the last count the organizational structure of cognition.

INTRODUCTION: RELATIONS OF CONSCIOUSNESS TO KNOWLEDGE

Reference to Knowledge in the Definitions of Consciousness

Knowledge figures as a major construct in many definitions of consciousness. Here are some examples. For lack of a better starting point let us start with quoting the Free Online

Dictionary which defines consciousness in two complementary ways: First, consciousness is "an alert cognitive state in which you are aware of yourself and your situation" and second, "having knowledge of" as in "he had no awareness of his mistakes" or "his sudden consciousness of the problem he faced". In The Blackwell Companion to Consciousness we find the following definition: "Anything that we are aware of at a given moment forms part of our consciousness" (Schneider and Veldman, 2008). In her ground-breaking book about consciousness Susan Blackmore (2004, p. 2) defines 'conscious' as "the equivalent of knowing something or attending to something, as in 'She wasn't conscious of the crimes he'd committed'". The concept most closely affiliated to consciousness is awareness, which itself is defined by the Free Online Dictionary as knowing or being able to know feelings, perceptions and stimuli in general. Not surprisingly, the etymological origin of the English word "conscious" is the Latin word *conscius* (*con-* "together" + *scire* "to know") which meant having common knowledge with someone else and also knowing that one knows (Lewis, 1990, chap. 8) or having knowledge of knowledge (Koriat, 2000). Adam Zeman (2002) who noted that the word 'consciousness' tends to cover all the activities of the mind (ibid, pp. 28-29) nevertheless disentangled out of the multiplicity of connotations the following meanings: being awake and noting one's environment, knowing this or that, experiencing with knowledge this or that, having a concept of oneself. Notably, he found these different senses of consciousness represented in over 30 languages he examined, including some outside the Indo-European cycle, like Chinese (ibid, pp. 32-35).

The close affinity between consciousness and knowledge is evident also in the Indian tradition which defines consciousness as "the direct knowledge" that one gets from one's 'Atman' (which in Hinduism refers to one's *true* self beyond identification with phenomena). Direct knowledge is described as complete , self-evident, valid and satisfying, in contrast to knowledge gained from external sources which is incomplete and requires validation. Accordingly, gaining consciousness consists in acquiring self-knowledge, namely, realising experientially that one's true self is identical with the transcendent self (Harvey, 1995; Rama, 2002).

Also in specific domains like medicine the uses of the term consciousness focus around cognizance, and the abilities to know, to perceive, and share one's knowledge. A medical dictionary emphasizes the following aspects of consciousness: the state of being conscious; being fully alert, aware, oriented, and responsive to the environment; subjective awareness of one's own existence, sensations, thoughts, environment; full activity of the mind and senses as in waking life (Stedman, 2006). Similarly, a textbook of neurology defines consciousness as "the state of awareness of the self and the environment" (Aminoff, 2008, chap. 62). Accordingly, in medicine consciousness is assessed in terms of the patient's arousal and responsiveness, and is considered as a continuum of states ranging from full alertness and comprehension, through disorientation, delirium, and loss of meaningful communication, to absence of movement in response to stimuli causing pain (Güzeldere, 1997). The contrast of 'being conscious' is 'being unconscious', which denotes that the cognitive system is not functioning fully or adequately in the service of knowledge-grounded interaction with the environment.

Finally, in the context of psychoanalysis and the dynamically oriented psychological schools, consciousness is defined as a state of interacting cognitively with external reality, whereas the unconscious is defined as a state or reservoir of emotions, thoughts, drives and memories outside of our conscious awareness, which are mostly unacceptable or unpleasant for the individual (Akhtar, 2009). Freud (1960) himself defined consciousness as the perception of information arising both from the external world and from the internal world. Notably, in this context too consciousness or its absence are bound with knowledge of particular contents.

The brief survey of the definitions of consciousness leads to the following conclusions. First, consciousness is intimately related to knowledge. Secondly, there may be *different kinds or different ways of getting knowledge* that could be relevant for understanding consciousness. For example, there may be knowledge obtained through the experiential way and knowledge acquired through the senses. Third, there may be *different states of knowledge* that could be relevant for understanding consciousness. For example, there may be very clear knowledge, or fuzzy knowledge. Or again, there may be knowledge that exists but is not accessible and knowledge that is in a state of accessibility which is a necessary condition for awareness. Fourth, there may be *different contents of knowledge* that could be relevant for understanding consciousness. For example, according to the psychoanalytic approach personal contents related to pain or anxiety tend to be kept in a state of unconscious knowledge. According to Hinduism, consciousness refers to a particular kind of knowledge of the self.

These conclusions suggest the importance and necessity of clarifying and grounding theoretically the relations of consciousness to knowledge. This is the major goal of the present chapter.

Knowledge and Epistemology

The idea that there are different kinds of knowledge is not completely new. The Greeks differentiated between four kinds of knowledge: *doxa* (=an opinion or knowledge based on experience), *episteme* (=rational knowledge), *gnosis* (=true knowledge in the metaphysical sense), and *sophia* (=wisdom). Also the Hebrew language differentiates between *Hochma* (=wisdom), *Bina* (knowledge based on rational understanding), *Tvuna* (=knowledge based on drawing implications), *Da'at* (knowledge based on the senses, on perception), *Yeda* (=knowledge that consists of validated facts), *Meida* (=knowledge concerning a specific theme or information), and *Haskel* (=knowledge that refers to right and wrong). However, the different constructs referring to knowledge need to be grounded in a common theoretical framework so that their relations to consciousness may be clarified.

When the theme of knowledge enters the scene, we are in fact in the territory of so-called epistemology. As a discipline of philosophy epistemology has dealt with defining and conceptualizing knowledge in terms of philosophy for hundreds of years. Epistemology promoted traditionally different kinds of knowledge, primarily empirical, rational and idealistic (intuitive), or analytic vs synthetic knowledge, and a priori vs a posteriori knowledge. One of the major assumptions of the epistemological study of knowledge is that knowledge should be investigated and evaluated within the framework of contexts defined in terms of particular functions (Zalta, 2011). Hence, the knowledge with which epistemology proper deals is propositional knowledge, namely, knowledge of propositions of the kind '*S* knows that *p*' ('*S*' stands for the subject who has knowledge and '*p*' for the proposition that is known), whose validity or justification or truth value need to be specified. Knowledge in other domains of inquiry which do not fit the schematic propositions is disregarded by traditional epistemology and is referred instead to other frameworks of epistemology, such as virtue epistemology, moral epistemology, religious epistemology, social epistemology, feminist epistemology and natural (or cognitive) epistemology, which deal with epistemic qualities of statements about virtues, religion, social objects and events, feminine themes and natural world issues including cognitive processes, respectively. One of the first precursors of this approach was the philosopher Ernst Cassirer (1944) who identified several major content domains as basic forms of knowledge construction, e.g., language, mythology, science and art.

In line with the approach of so-called multiple epistemologies, it appears justified to apply a specific epistemology for clarifying the relations of consciousness to knowledge. Since consciousness is a characteristic of cognition and concerns a broad variety of domains of contents, the epistemic approach adequate for analyzing the relations of consciousness to knowledge should be cognitively-based and broad-ranging in its applicability. The psychosemantic approach to cognition appears to fulfill both requirements. This approach will be described, with an emphasis on the parts that will be applied for analyzing the relations of knowledge with consciousness.

THE PSYCHOSEMANTIC APPROACH: THE THEORY OF MEANING

The Definition and Variables of Meaning

The psychosemantic approach to cognition is grounded in the theory of meaning (Kreitler and Kreitler, 1990a). The major thesis of this approach is that meaning defines the essential contents and functioning of cognition, enabling the performance of a variety of acts including comprehension, communication and problem solving. Meaning consists of meaning units, which include two components: 'the referent' which is the input, the stimulus, or the subject to which meaning is assigned, and 'the meaning value' which is the cognitive contents designed to express or communicate the meaning of the referent. Hence, meaning may be defined as a pattern of cognitive contents focused on a referent. The following are four examples of meaning units: "table – serves for eating", "bread – is on the table", "milk – is produced by cows", "bottle – is made of glass". In these meaning units, 'table', 'bread', 'milk' and 'bottle' are the referents and 'serves for eating', 'is on the table', 'is produced by cows' and 'is made of glass' are the meaning values. Each meaning unit may be characterized in terms of meaning variables of the five following classes: meaning dimensions – which characterize the contents of the meaning values (e.g., locational qualities, material), types of relation – which characterize the immediacy of the relation between the referent and the meaning value (e.g., attributive, metaphoric-symbolic), forms of relation – which characterize the logical-formal properties of the relation between the referent and the meaning value (e.g., positive, conjunctive, partial), shifts of referent – which characterize the relations of the present referent to the initial input and previous referents (e.g., identical, partial, opposite), and forms of expression – which characterize the media of expression of the referent and/or the meaning value (e.g., verbal, graphic, motional). The meaning system consists of the whole set of the meaning variables (see Table 1).

Table 1. Major Variables of the Meaning System: The Meaning Variables

Meaning dimensions		**Forms of relation**	
Dim. 1	Contextual Allocation	FR 1	Propositional (1a: Positive; 1b: Negative)
Dim. 2	Range of Inclusion (2a: Sub-classes; 2b: Parts)	FR 2	Partial (2a: Positive; 2b: Negative)
Dim. 3	Function, Purpose & Role	FR 3	Universal (3a: Positive; 3b: Negative)
Dim. 4	Actions & Potentialities for Actions (4a: by referent; 4b: to referent)	FR 4	Conjunctive (4a: Positive; 4b: Negative)
Dim. 5	Manner of Occurrence & Operation	FR 5	Disjunctive (5a: Positive; 5b: Negative)
Dim. 6	Antecedents & Causes	FR 6	Normative (6a: Positive; 6b: Negative)
Dim. 7	Consequences & Results	FR 7	Questioning (7a: Positive; 7b: Negative)
Dim. 8	Domain of Application (8a: as subject; 8b: as object)	FR 8	Desired, wished (8a: Positive; 8b: Negative)
Dim. 9	Material	**Shifts in referent**[b]	
Dim. 10	Structure	SR 1	Identical
Dim. 11	State & Possible changes in it	SR 2	Opposite
Dim. 12	Weight & Mass	SR 3	Partial
Dim. 13	Size & Dimensionality	SR 4	Modified by addition
Dim. 14	Quantity & Mass	SR 5	Previous meaning value
Dim. 15	Locational Qualities	SR 6	Association
Dim. 16	Temporal Qualities	SR 7	Unrelated
Dim. 17	Possessions (17a) & Belongingness (17b)	SR 8	Verbal label
Dim. 18	Development	SR 9	Grammatical variation
Dim. 19	Sensory Qualities[c] (19a: of referent; 19b: perceived by referent)	SR 10	Previous meaning values combined
Dim. 20	Feelings & Emotions (20a: evoked by referent; 20b: felt by referent)	SR 11	Superordinate
Dim. 21	Judgments & Evaluations (21a: about referent; 21b: held by referent)	SR 12	Synonym (12a: in original language; 12b: translated in another language; 12c: label in another medium; 12d a different formulation for the same referent on the same level)
Dim. 22	Cognitive Qualities (22a: evoked by referent; 22b: of referent)	SR 13	Replacement by implicit meaning value

Table 1. (Continued)

Types of relation[a]		Forms of expression	
TR 1	Attributive (1a: Qualities to substance; 1b: Actions to agent)	FE 1	Verbal (1a: Actual enactment; 1b: Verbally described; 1c: Using available materials)
TR 2	Comparative (2a: Similarity; 2b: Difference; 2c: Complementariness; 2d: Relationality	FE 2	Graphic (2a: Actual enactment; 2b: Verbally described; 2c: Using available materials)
TR 3	Exemplifying-Illustrative (3a: Exemplifying instance; 3b: Exemplifying situation; 3c: Exemplifying scene)	FE 3	Motoric (3a: Actual enactment; 3b: Verbally described; 3c: Using available materials)
TR 4	Metaphoric-Symbolic (4a: Interpretation; 4b: Conventional metaphor; 4c: Original metaphor; 4d: Symbol)	FE4	Sounds & Tones (4a: Actual enactment; 4b: Verbally described; 4c: Using available materials)
		FE5	Denotative (5a: Actual enactment; 5b: Verbally described; 5c: Using available materials)

[a] Modes of meaning: Lexical mode: TR1+TR2; Personal mode: TR3+TR4.

[b] Close SR: 1+3+9+12 Medium SR: 2+4+5+6+10+11 Distant SR: 7+8+13.

[c] This meaning dimension includes a listing of subcategories of the different senses/sensations: [for special purposes they may also be grouped into "external sensations" and "internal sensations"] e.g., color, form, taste, sound, smell, pain, humidity and various internal sensations.

The Assumptions Underlying the System of Meaning

The following assumptions enabled the conceptualization, the methodology and assessment of meaning as the psychosemantic understructure of cognition. The first assumption was that meaning is essentially even if not completely communicable. The rationale was that most of the meanings are acquired and learned from and through others. The second assumption was that meaning includes both an interpersonally-shared part and a personal-subjective part, so that it can serve not only communication between individuals but also the expression of internally and subjectively meaningful contents. The third assumption was that meaning is a complex structure in that the meaning of any referent mostly includes several meaning units often representing different meaning variables. The fourth assumption was that meaning may be expressed in terms of any verbal and non-verbal forms of expression. And the fifth assumption was that meaning is a dynamic structure, in the sense that the meaning of any referent and the composition and structure of the whole system undergo changes, some of which are due to development whereas others are due to other causes which affect meanings.

Structure of the System of Meaning

As noted, the system of meaning includes variables of five groups. Of the total of 90 meaning variables, 33.3% are meaning dimensions, 17.8% types of relation, 17.8% forms of relation, 14.4% shifts of referent and 16.7% forms of expression. Each of the five sets is complete in itself and independent of the other sets. Thus, characterizing a meaning unit involves using one variable from each of the five sets. Hence, when we have a group of meaning units characterized in terms of meaning variables and we count the frequencies of meaning variables used in characterizing these meaning units, we get in fact five independent groups of frequencies, namely, one for meaning dimensions, one for types of relation, one for forms of relation, one for shifts of referent, and one for forms of expression. Each of these five groups of frequencies amounts to the same total but consists of different meaning variables.

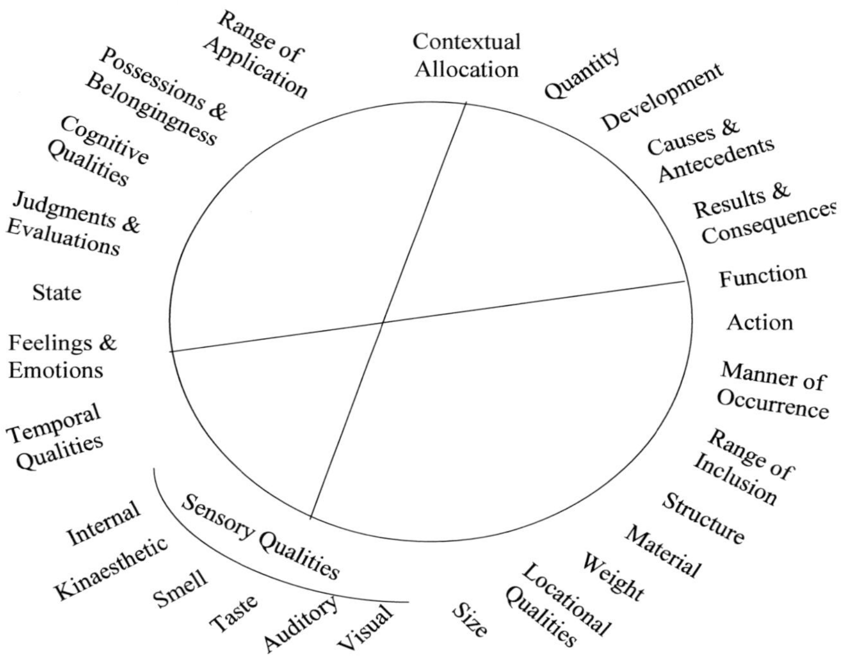

Note. The figure represents schematically the relations between the meaning dimensions in the system of meaning that seem likely on the basis of data available up to date. Some of the relations are still merely hypothesized. The locational position of the meaning dimensions represents their proximity. The closest relations are between adjoining variables, the furthest are between variables placed opposite each other on the circumference of the circle. The two intersecting lines represent factors identified in several studies. The variables at opposite poles represent meaning dimensions with positive and negative loadings on the factors, respectively.

Figure 1. The Circumplex Model of Meaning Dimensions.

Each group of meaning variables has a unique structure. Meaning dimensions form a circumplex structure. Accordingly, they may be ordered in line with the similarity in their contents along the circumference of a circle, so that the more similar the two meaning dimensions are the closer they are placed to each other. The arrangement is based on varied data sources, including studies of multidimensional scaling, characteristics of participants' responses and testing of hypotheses based on contents similarity (Kreitler and Kreitler,1989a). Further, the circular arrangement is defined in terms of two major axes recurring in most studies: an approximately vertical axis of abstractness-concreteness, anchored on the poles of contextual allocation and sensory qualities, and an approximately horizontal axis of action-emotion, anchored on the poles of action and feelings and emotions. In addition, each meaning dimension represents towards the center of the circle more general and global meaning values which become increasingly differentiated the closer they lie to the external circumference of the circle (see Figure 1).

The circumplex structure of meaning dimensions suggests also groupings of the meaning dimensions in line with their proximities, which represent common spheres of contents. One grouping includes the meaning dimensions referring to concrete and sensory qualities of referents (see Fig. 1, meaning dimensions from Range of inclusion to Temporal Qualities). A second grouping includes the meaning dimensions referring to the dynamic and actional qualities of referents (see Fig. 1, meaning dimensions from Function to Manner of Occurrence). A third grouping includes the meaning dimensions referring to contextual aspects of the referents (see Fig. 1, meaning dimensions from Possessions and Belongingness to Results). A fourth grouping includes the meaning dimensions referring to inner life and experiences of the referents (see Fig. 1, meaning dimensions from Feelings to Cognitive Qualities) (Margaliot,2012).

The types of relation also form a circumplex structure. Figure 2 shows the proximity placements on the circumference of the circle as well as the two major axes describing the arrangement. Here too the meaning values placed closer to the external circumference are increasingly more differentiated (Kreitler and Kreitler, 1985b).

The forms of relation form three sets of meaning variables. One set specifies the existential status of the relation between the referent and the meaning value, which may be factual (propositional), normative (required, necessary), questioned (does it exist?), or desired-wished for. A second set specifies the quantitative nature of the relation, which may be universal-absolute or partial. A third set specifies the relation when more than one meaning value is involved, so that it may be conjunctive or disjunctive. Each of the specified variables may be positive or negative.

The referent shifts form three classes of variables. One class includes the referent shifts to referents close to the original one (i.e., identical, partial, the original referent plus an addition, grammatical variation, synonym). A second class includes referent shifts to referents at medium distance from the original one (i.e., opposite, a previous meaning value, an association, a superordinate referent). A third class includes referent shifts to referents relatively far removed from the original one (i.e., unrelated referent, verbal label, a combination of several previous referents, an implicit unstated meaning value of the original referent).

The forms of expression include five sets which differ primarily in the mode of expression: verbal, graphic, movement, sounds, denotation. Each set includes variables which differ in whether the medium is used directly, is verbally described or is applied by means of materials.

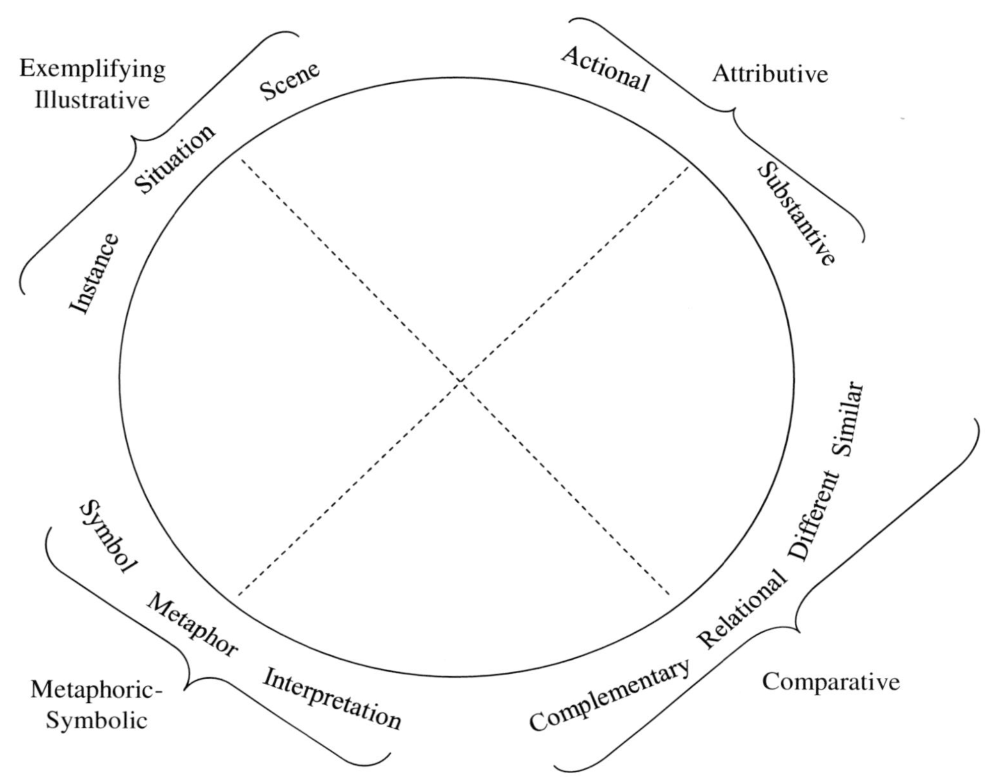

Note. The figure represents in a schematic way the relations between the types of relation in the system of meaning that seem likely on the basis of available data up to date. The locational position of the types of relation represents their proximity. The closest relations are between adjoining variables, the furthest are between variables placed opposite each other on the circumference of the circle. The two intersecting lines represent factors identified in several studies. The variables at opposite poles represent types of relation with positive and negative loadings on the factors, respectively.

Figure 2. The Circumplex Model of Types of Relation.

Static and Dynamic Properties of Meaning

Each of the meaning variables may be considered in static or dynamic forms. The static form refers to the meaning variable as a set of contents or outputs characterized by the feature characterizing the meaning variable, such as comparative contents (Table 1, TR2b), statements of functions (Table 1, Dim 3), or descriptions of emotions (Table 1, Dim 20). The dynamic form refers to the meaning variable as one or more processes involved in setting into operation the domain of outputs designated by the meaning variable, for example, causal thinking (Table 1, Dim. 6), analyzing (Table 1, Dim. 2), or metaphorization (Table 1, TR4c). The static and dynamic aspects of the meaning variables complement each other.

Also the meaning system as a whole may be characterized in terms of static (or structural) aspects and in terms of dynamic properties. The major characteristics reflecting the static or structural aspects of the system are (a) that it is a system, (b) that it is complex, and (c) that its elements are defined in terms of other elements of the system (namely, it is self-embedded and regressive). The properties that describe the dynamic aspects of meaning are: (a) it is a developing system both ontogenetically and phylogenetically; (b) it is a selective system dependent in its structure and functioning on properties of the species, the individual and the input; and (c) it is a dynamic system, whose special characteristics become manifest when it is activated for meaning assignment.

The static and dynamic properties of the meaning system interact within the framework in which meaning functions. One such framework is the individual. Each individual disposes over a certain selected part of the meaning system which represents the kinds of information the individual has or uses and his or her specific tendencies to apply the meaning system in information processing. Thus, each individual tends to use specific meaning variables with higher frequency and other meaning variables with medium or low frequency. The frequencies with which the individual tends to use each meaning variable are assessed by means of a test (The Meaning Test) and constitute the individual's meaning profile.

The interactions between the static and dynamic aspects of the meaning system are manifested also in the social and cultural frameworks. Here too the meaning variables define the major domains of contents relevant in the socio-cultural setup (e.g., functions, possessions) as well as the major processes subserving the maintenance of these outputs (e.g., definition of functions, or legal aspects of possessions).

Functions of Meaning

The major and most essential function of meaning is *input identification.* This function is implemented by providing the contents and processes enabling meaning assignment to inputs.Input identification ranges all the way from limited identification in terms of a stimulus for a particular action to highly complex meaning elaborations necessary for acts involving cognitive, emotional, physiological and behavioral components. Input identification is the prior condition for any further action on any level. If a personality trait is to be enacted or an emotion is to be evoked depends primarily on how the input has been identified. For example, an input that has been identified as related to success or failure is likely to enable enactment of achievement-related traits or emotions such as pride in achievements, or fear of failure (Kreitler and Kreitler, 1984).

A further function of the meaning system is to provide the cognitive contents and processes necessary for *performing cognitive acts*. Studies showed that each meaning variable represents a specific set of contents and processes. For example, the meaning dimension Locational Qualities represents the set of contents denoting location (e.g., special, geographic) and the processes involved in dealing cognitively with locations (e.g., identifying, specifying, recalling, or transforming locations). Further studies showed that each type of cognitive act corresponds to a specific pattern of meaning variables that provide a description of the contents and processes involved in its enactment. For example, meaning variables involved in planning include structure, temporal qualities, and causes and antecedents (Kreitler and Kreitler, 1986, 1987a). If the individual's meaning profile includes a sufficient proportion of the meaning variables included in the pattern corresponding to the particular cognitive act, that individual will be able to perform well the particular cognitive act (Arnon and Kreitler, 1984; Casakin and Kreitler, 2005; Kreitler and Kreitler, 1985a, 1989a, 1990b; Kreitler, 2012).

A third function of the meaning system consists in providing the cognitive contents and processes underlying the *formation of personality traits*. A body of research showed that each of over 350 personality traits corresponds to a specific pattern of meaning variables. Again, as in the case of cognitive acts, the pattern of meaning variables may be considered as providing a description of the contents and processes involved in the enactment of the specific trait. For example, the meaning variables in the pattern corresponding to extraversion include high salience of the meaning dimensions of action, sensory qualities, temporal qualities and belongingness of objects, as well as low salience of the meaning dimensions of internal sensations and cognitive qualities (Kreitler and Kreitler, 1990a, 1997). If the individual's meaning profile contains a sufficient proportion of the meaning variables included in the pattern corresponding to the particular personality trait, it is highly likely that the individual scores high on that personality trait.

The same holds in regard to further tendencies in the domain of personality, such as *personality dispositions, defense mechanisms* (Kreitler and Kreitler, 1993a), *attitudes and values* (Kreitler, 2010) and even *the self*. An individual's self concept was shown to consist of contents representing the major meaning variables in the individual's meaning profile (Kreitler and Kreitler, 1987b).

The fourth function of the meaning system is in regard to *emotions*. Here too studies showed that particular patterns of meaning variables correspond to specific emotions, such as anger, anxiety or fear (Kreitler, 2003, 2011). Further, changing experimentally specific components of the pattern corresponding to anxiety decreased the level of anxiety and its effects in the individual (Kreitler and Kreitler, 1985b, 1987c).

In sum, the four functions of the meaning system that have been identified indicate that the meaning system provides the understructure - that is, the raw materials in terms of contents and processes – for input identification, cognitive functioning, personality tendencies and emotions. All four functions depend on meaning assignment and reflect the central role of meaning for and within cognition. This has given rise to the psychosemantic conceptualization of cognition as a meaning-processing and meaning-processed system.

MEANING AND KNOWLEDGE: THE BLUEPRINT FOR A PSYCHOSEMANTIC EPISTEMOLOGY

Meaning Variables and Domains of Knowledge

In line with the psychosemantic approach, knowledge corresponds in terms of contents to the meaning dimensions and in terms of form of presentation to the types of relation. The forms of relation function as modulators of the presented contents. The shifts of referent refer to the extent to which the presented contents are related to the initial input or point of departure, or the extent to which far-ranging domains of content are brought into play, so that they represent the connectedness of the knowledge to the original issue. The forms of expression describe the means of expressing or communicating the knowledge. Thus, the meaning system provides a framework for defining kinds of knowledge that differ in contents, form of presentation, modulatory character, connectedness with the original focus, and means of expression.

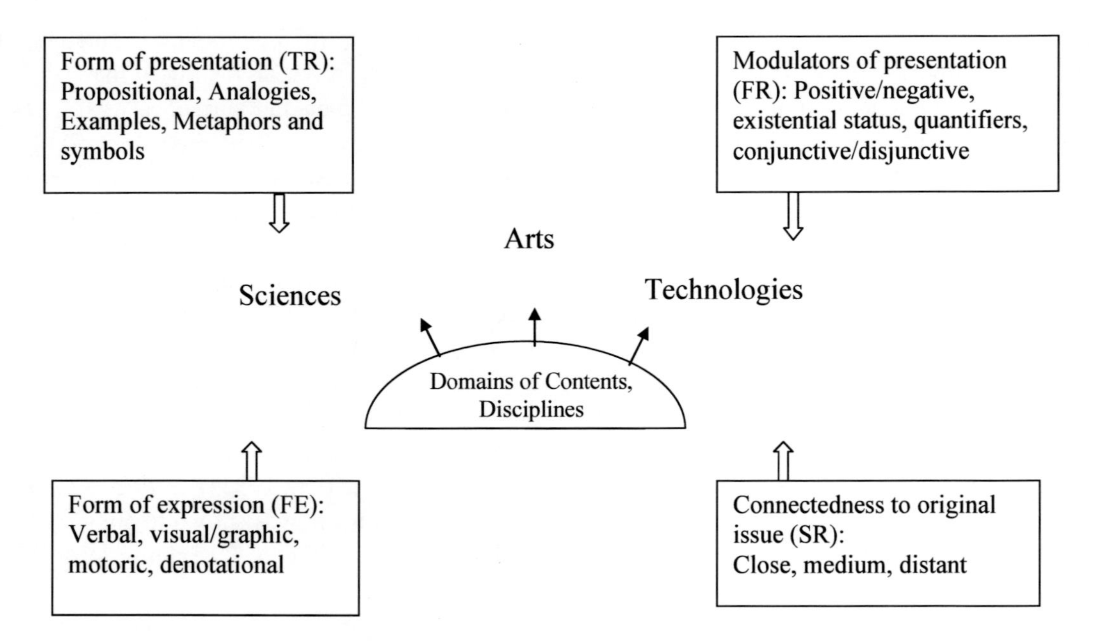

Figure 3. The psychosemantic presentation of knowledge. Fig. 3 should be placed on page 12

The most salient part of the psychosemantic epistemology are the domains of contents. Table 2 presents some examples of domains of contents or disciplines that correspond to meaning dimensions, for example, the domain of contents of chemistry corresponds to the meaning dimension of materials, that of psychology to the meaning dimensions of emotions and cognitive qualities. This presentation is provisional and schematic. It is evident that in most if not all cases more than one meaning dimension is involved in the particular discipline. For example, architecture deals in the very least with contents that belong to the meaning dimensions of structure, materials and functions, all of which are of prime importance when constructing buildings. Further, each discipline may apply any one of a set of meaning dimensions, depending on the explored issues, for example, physics may be focused on causes, materials, or manner of operation.

The form of presentation of the domains of contents, based on types of relation, contributes to extending the range of represented sectors of knowledge (Kreitler, 1965, 2011). Most traditional disciplines apply mainly or exclusively the propositional form of presentation, that is, the form of statement with a definite clearly identified subject and a predicate attributed to it (e.g., S is/does P). Another common form of presentation applies examples. This has become customary in the domains of law in which precedents are often cited and in medicine which often accumulates and presents knowledge in the form of case studies. Analogies and comparisons are commonly used in diverse domains, such as mathematics, law and linguistics. Finally, metaphors and symbols have traditionally been applied in the different domains of art, ranging from literature and poetry to the visual arts, drama, and dance, as well as in mythology, religion and the different kinds of hermeneutics and mystical knowledge.

The modulators of knowledge, based on forms of relation, become evident in specific domains of knowledge. For example, the quantifiers fulfill an important role in specific branches of mathematics (e.g., arithmetics) and in classical logic. The modifiers that define the existential status of knowledge (e.g., factual, doubted/skeptical, wished for, normative) characterize scientific knowledge (i.e., factual), ethical and legislative knowledge (i.e., normative), philosophical knowledge (i.e., doubtd/skeptical), and idealistic or utopian knowledge (i.e., wished for).

The connectedness of knowledge to the initial focus or issue characterizes the style of presentation in specific disciplines. For example, in news reports it is tolerated much less than in scientific presentations, in which in turn it is tolerated less than in the arts.

Finally, forms of expression extend knowledge beyond the classical verbal form (Kreitler and Kreitler, 1972). Through the arts different forms of expression came to play a role in knowledge presentation. The basic forms of expression are the visual media (static and in motion), sound (instrumental, singing), music, dance, literature and poetry and the more complex forms that use several media, such as theater and opera. Whether art can have anything to do with knowledge is the theme of a debate that has been going on in philosophy for many hundreds of years. Arts are discarded as a source of knowledge by those who hold that only propositional knowledge counts (e.g., Stolnitz, 1922; Carroll, 2002). However, those who accept the epistemological contributions of the arts (e.g., Harrison, 1998) can count as their supporter Plato who in the *Republic* warned against the potentially dangerous effects of knowledge communicated through the arts, because it is emotionally transmitted. It is evident that the knowledge presented by the arts is of a special kind, because the means of expression used for its communication is adequate for presentation of knowledge meaningful for the individual, subjective, often value-oriented, mostly not verifiable.

The five sets of meaning variables can thus be used for creating a psychosemantically-grounded epistemology. In the paragraphs above each set of meaning variables was used separately for demonstrating its specific contribution to the psychosemantic epistemology. In practice, however, all sets of meaning variables have to be applied together for the full characterization of the psychosemantic structure.

Meaning Variables and Ways of Knowing

All the examples up to now referred to the static aspects of meaning, namely, to contents that already exist in some form. However, a full-fledged psychosemantic epistemology requires also consideration of the dynamic aspect that concerns the generation or acquisition of knowledge. Already on the level of children it is apparent that the applied manner of getting knowledge, e.g., by observing, manipulating objects motorically, reflecting or asking others, determines the kind of knowledge one obtains (Kreitler and Kreitler, 1994). In terms of knowledge, the difference between the static and dynamic aspects translates to the difference commonly discussed in philosophy between 'knowing that' and 'knowing how' (Polanyi, 1962; Ryle, 2002).

Consideration of ways of acquiring knowledge has always been a part of epistemology, mostly with an emphasis on the source of the knowledge (Zalta, 2011). This is reflected, for example, in the traditional distinctions between a priori (intuitive, originating in the mind) and a posteriori (empirically based) knowledge; between analytic (based on definitions of the involved terms) and synthetic (based on learning, experience) knowledge; between empiricism (based on experimental evidence), rationalism (based on reason, theorizing) and idealism (based on intuition); as well as in the more recent distinction (Russell, 1910-1911) between knowing by acquaintance (based on direct experience) and knowing by description (based on information provided indirectly by others). An extreme position has been adopted by the constructivist approach which claims that the means of knowledge production determine the kind of knowledge that one gets (Glasersfeld, 1997).

Notably, the ways of knowledge generation discussed by philosophers are neither comprehensive nor based on any theoretically sound framework. The meaning system offers a theoretical framework of this kind, which moreover is economical in the sense that it does not require any additional constructs than those already introduced for dealing with the static aspects of knowledge. As stated above, each meaning variable represents meaning values (the static aspect) and a process involved in generating this kind of meaning values (the dynamic aspect). Thus, one and the same construct supports the static contents of a particular kind as well as its generation, production and applications. For example, the meaning dimension Locational Qualities fulfills a role in characterizing the set of contents referring to sites, locations and space as well as in regard to the cognitive acts, such as learning, processing, thinking, memorizing, imagining, and problem solving concerning locations. Similarly, the meaning dimension of actions represents the way of knowing that is based on manipulating dynamically objects and constructs, while the meaning variables of verbal and nonverbal expressions correspond to obtaining knowledge by verbal means, such as questions (Kreitler and Kreitelr, 1990c) or the various nonverbal media.

Table 2 presents examples of the ways of knowing based on each of the meaning variables. Some of these ways of knowing have been identified before and have been applied in diverse contexts. For example, learning by doing (based on meaning dimension 4) has been placed by Piaget (1937/1954) in the center of the sensori-motor stage of cognitive development; the functional approach (based on meaning dimension 3) has been emphasized by the pragmatic trend in philosophy and education; learning by experiencing (based on meaning dimension 20) forms the focus of experiential education popularized by David Kolb and others (Itin, 1999). Each one of these and similar examples focuses on one particular way of knowing but does not examine it in the comprehensive theoretically and empirically grounded framework representing the various available ways of knowing. This is the particular contribution of the psychosemanticaly-based approach.

Table 2. Domains of knowledge and ways of knowing represesnted by the meaning variables of the five sets

Meaning Variables	Examples of domains of knowledge (sciences, arts, etc.)	Ways of knowing
Meaning Dimensions		
Contextual. Allocation	Biology – taxonomy	By categorizing, by embedding the object/construct in different contexts
Range of inclusion Dim. 2a: Sub-classes Dim. 2b: Parts, components	Chemistry, botany, linguistics	By analyzing and forming sub-classes or sub-categories (Dim. 2a); by identifying parts, constituents, or components
Functions	Sociology, biology	By examining the functions attained by the different phenomena

Meaning Variables	Examples of domains of knowledge (sciences, arts, etc.)	Ways of knowing
Meaning Dimensions		
Actions Dim. 4a: Done by agent Dim. 4b: Done to/with agent	The natural sciences: Physics/biology, sports, the art of dancing	By actions, by doing, by motoric manipulation: Performing active actions of the object (Dim. 4a), or passive actions with/to the object (Dim. 4b).
Manner of operation	Technology, engineering, linguistics	By examining how it operates and functions, by making it operate
Causes	The natural sciences: Physics/biology	By investigating what causes it
Results	The natural sciences: Physics/biology, logic, mathematics	By manipulating in order to check results and effects
Range of application Dim. 8a: Subjects, agents Dim. 8b: Objects	Criminology, the natural sciences: Physics/biology	By noting who uses it, for whom it is relevant (Dim. 8a); by examining on what it has effects, what or who is affected by it (Dim. 8b)
Materials	Chemistry, technology, engineering	By testing materials
Structure	Mathematics, architecture	By characterizing structures
State	Chemistry, climatology, medicine	By observing changes in state
Weight	Chemistry	By measuring or evaluating weight and mass
Size	Engineering	By measuring or evaluating size and dimensions
Quantity	Mathematics	By counting, measuring, assessing quantities
Place	Geography, astronomy	By noting, examining or recording sites and locations
Time	Archeology, history, geology	By recording time, age, periods
Possession Dim. 17a: by subject Dim. 17b: belongingness	Social policy, law, economy	By checking relations of possessing and belongingness: Examining possessions by the object (Dim. 17a); by examining belongingness of the object (Dim. 17b).
Development	History, evolution	By investigating how it developed and projecting development in the future
Sensory qualities: 19a: observed 19n: experienced	The arts, cooking, visual arts, music, cosmetics, aromatics and perfumes	By observing different sensory qualities (Dim. 19a); or by experiencing different sensory modes (visual, auditory, olfactory, tactile etc.) (Dim. 19b)
Emotions: Dim. 20a: evoked Dim. 20b: experienced	Psychology, rhetoric, media and communication	By observing the emotional impact on others (Dim. 20a); by experiencing emotionally (Dim. 20b)
Judgments and evaluations Dim. 21a: applied, evoked in others Dim. 21b: held	Law, philosophy, religion	By checking judgments, beliefs and evaluations evoked in others (Dim. 21a); by judging, evaluating, criticizing (Dim. 21b)

Table 2. (Continued)

Meaning Variables	Examples of domains of knowledge (sciences, arts, etc.)	Ways of knowing
Meaning Dimensions		
Cognitive actions Dim. 22a: evoked in others Dim. 22b: experienced	Logic, cognition, psychology, ethics	By checking cognitive effects on others (Dim. 22a); by reflection, thinking about, imagining
Types of Relation		
Attributional TR1a: substance to qualities TR1b: agent to actions	Law, sciences	By testing various attributions of the referent – qualities (TR1a), and actions (TR1b)
Comparative TR 2a: SimilarityTR 2b: Dissimilarity, difference, contradiction TR 2c: Complementariness TR 2d: Relationality	Poetry, creativity in technology (inventiveness)	By creating analogies and looking for similarities between objects (TR 2a); by checking contradictions and dissimilarities (TR 2b); by defining and checking interactions and complementary relations (TR 2c); by defining and checking relativity and relationality (TR 2d).
Exemplifying-illustrative TR 3a: Instance TR 3b: Situation TR 3c: Scene	Teaching, pedagogy	By producing and studying examples of instances (TR3a), situations (TR 3b), and dynamic scenes (TR 3c)
Metaphoric-Symbolic TR 4a: Interpretation TR 4b: Conventiaonal metaphor TR 4c: Original metaphor TR 4d: Symbol	Different forms of art, religion, literature (myths, legends, folktales etc.), and science	By studying and creating metaphors and symbols of different kinds: abstract interprettive (TR 4a), conventional metaphors (TR 4b), original metaphors (TR 4c), and symbols (TR 4d)
Forms of Relation		
Existential nature of the relation (propositional, normative etc.)	Logic	By formulating and examining statements that differ in their existential status, i.e., are propositional (FR 1), normative (FR 6), questioning (FR 7) or express desired/wished relations (FR 8)
Quantitative nature of the relation (universal, partial)	Sciences, logic	By formulating and examining statements that differ in the quantifying relation, e.g., are based on universal (FR 2) or partial relations (FR 3) Do not type this in bold
Conjunctive or disjunctive relation	Logic	By formulating and examining statements based on conjunctive (FR 4) or disjunctive relations (FR 5)
Shifts of Referent		
Meaning Variables	Examples of domains of knowledge (sciences, arts, etc.)	Ways of knowing
Close to input	Manuals of instructions for operating any system or instrument	By formulating, using and examining various shifts of referent close to the original input: identical (SR 1), representing parts of the input (SR 2), grammatical variations (SR 9), and synonyms (SR 12)

Meaning Variables	Examples of domains of knowledge (sciences, arts, etc.)	Ways of knowing
Shifts of Referent		
Medium distance	Sciences	By formulating, using and examining various shifts of referent at medium distance from the original input: opposite (SR 2), modified by additions (SR 4), previous (SR 5), related by association (SR 6), representinga combination of several previous referents (SR 10), superordinate (SR 11)
Far- removed	Philosophy	By formulating, using and examining various shifts of referent far-removed from the original input: unrelated (SR 7), focused on the verbal label (SR 8), representing some implicitlyr elated input (SR 13)
Forms of Expression		
Verbal	Sciences, literature	By formulating, using and examining various verbal forms of expression (FE 1)
Sensory, e.g. graphic, movement, sounds, odors etc.	Arts, drama	By formulating, using and examining various sensory and motoric forms of expression (FE 2, FE 3, FE 4)
Objects	Teaching, pedagogy	By formulating, using and examining various kinds of denotation to objects (FE 5)

STATES OF CONSCIOUSENESS AND KNOWLEDGE

Selectivity as a Characteristic of the Meaning System

One of the major characteristics of the meaning system is selectivity, which means that the system never functions in its totality, just as no meaning of any stimulus is manifested in its totality (Kreitler and Kreitler, 1993). There are several levels on which selectivity can be detected and characterized. One level is defined by socio-cultural and demographic categories. There is evidence that groups characterized in terms of basic categories, such as gender, age, ethnic background, or culture tend to use specific meaning variables with specific frequencies, which suggests that there exist meaning profiles of groups (Kreitler and Kreitler, 1988, 1989b). These meaning profiles may differ not only in the constitution of the meaning variables but also in their relative strength. For example, in certain cultures a meaning dimension like possession and belongingness may be given priority over a meaning dimension such as feelings and emotions.

Further, selectivity is manifested also on the individual level. Each individual has at his or her disposal only a certain section of the total system. This section is assessed by means of the Meaning Test which shows which meaning variables the person tends to use and with which frequencies, i.e., how strong or typical of the person they are. The meaning variables and the frequencies constitute the individual's meaning profile. The selectivity on this level is determined by a variety of factors, such as the individual's background in terms of education, family, socioeconomic status, culture and ethnicity, as well as his or her profession, age, gender, and even state of health.

A third important context of selectivity is the psychological construct that is activated, be it a cognitive act, a personality trait, an emotion, or some other behavioral tendency. Each such construct requires the activation of only certain meaning variables with particular strength which constitute the meaning profile of that construct.

Thus, when an individual is engaged in some act, cognitive or other, the meaning system gets activated selectively in several senses: there is the selectivity dictated by the kind of input in the given situation and by the act that is to be performed; and there is in addition the selectivity in terms of the individual's meaning profile, which is determined in turn by factors shaped by the individual's family background, culture and ethnic background, religion, gender and age, to mention just a few of the selectivity-determining factors.

States of Consciousness as Factors Determining Selectivity

As noted, in each activity only those cognitive processes and contents that are relevant for the task as well as accessible to the individual are involved. One major factor that defines and modifies the accessibility of the adequate cognitive processes in the individual is the state of the cognitive system in the course of performing the task. The state of the cognitive system is defined in terms of the kind and number of meaning variables that are in a focal position and salient at the time, namely, they have an organizational primacy and a functional advantage for elicitation and involvement in the act, whereas the other meaning variables are in the background in different states of inactivation (Kreitler and Kreitler, 1999, 2001, 2002, 2009).

A great many changes occur in the cognitive system for a variety of reasons. These include actions due to some externally presented task, such as solving a problem; handling some task arising from the needs of the cognitive system itself, e.g., organizing material; or performing a cognitive or other act in response to other needs of the organisms, such as social or emotional. Some of the changes are relatively small, for example, changes in contents within one meaning variable; other changes may be larger in the sense that more processes are involved, or more complex, in the sense that the changes are interdependent and more enduring. However, regardless of how encompassing or how long they last, changes of this kind do not affect the cognitive system as a whole.

Changes that affect the whole cognitive system may be brought about by means of organizational transformations in the meaning system. These kinds of transformations take place because of the needs and dynamics of the meaning system itself or in response to the needs of the organism, for example, reorganizing when a mass of new contents has become available, developing structural complexity, complementing a rudimentary or fragmentary view of reality, etc. Changes motivated by the dynamics of the meaning system typically consist of placing in the focal position one or more specific meaning variables or even merely one or more meaning values and changing accordingly the whole structure of the meaning system. Therefore changes of this kind may be considered as involving an organizational transformation of the system.

Changes motivated by the meaning system include, for example, placing in a focal position the meaning dimensions 'Contextual Allocation', 'Results and Consequences', and 'Causes and Antecedents', which manifest the so-called 'abstract approach'; or placing in a focal position the meaning dimensions 'Sensory Qualities', 'Size and Dimensions', 'Weight and Mass', and 'Locational Qualities',which manifest the so-called 'concrete approach' or 'concrete thinking'. In fact, almost any of the meaning variables and quite a number of sets of meaning variables could serve as foci for the meaning system and be the carriers of an organizational transformation. For example, the central positioning of the meaning dimension 'Feelings and Emotions', or one or more of the comparative types of relation, or any one of the nonverbal forms of expression (e.g., gestural, graphic) could promote the 'emotional approach', the comparative kind of thinking or the nonverbal style, respectively.

Promoting one or more meaning variables into a central position in the meaning system at least temporarily can be attained by various means. One validated way is experimental training or priming of the relevant meaning variables according to a procedure developed in the framework of the meaning theory. It consists essentially in eliciting meaning values in the desired meaning variables by using adequate stimuli (e.g. verbal, visual, musical) and reinforcing the desired meaning values that are evoked by the individual (Kreitler, 1999, 2001, 2009). Other ways that have been used for hundreds of years in different cultures consist of repetitive movements, monotonous singing tones, hypnosis in different variants, food or fluid deprivation for some time, and even ingestion of drugs. It is likely that regardless of the induction manner, the effect of the procedure on the individuals is mediated by changes in the meaning system. It remains to be studied which effects in the meaning system are produced by specific induction procedures which are essentially not cognitive in nature.

Regardless of the induction procedure, the organizational changes in the meaning system produce a range of effects. For example, the induction of personal meaning (based on promoting the exemplifying-illustrative and metaphoric-symbolic types of relation) as compared with the induction of interpersonally-shared meaning (based on promoting the attributive and comparative types of relation) resulted in better performance on visual memory tasks, identifying embedded figures, recounting of bizarre experiences, creativity tests assessing fluency, flexibility and originality, and the production of more associations; but worse performance on judging the validity of logical syllogisms as well as reality testing and emotional control in the Rorschach test (Kreitler, Kreitler and Wanounou, 1987-1988). The major conclusions based on findings of this kind is first, that it is possible to produce cognitive changes by manipulating meaning variables, and second, that the level of performance of specific cognitive tasks depends on the organizational state of the cognitive system.

The changes brought about by the placement of different meaning constituents in a focal position include changes in the nature, salience, and interconnectedness of contents and cognitive processes that affect cognitive functioning. But the changes are not limited to the cognitive sphere. Since, as noted, the meaning system is also involved in personality traits, the self-concept and emotions, it is likely that the organizational transformations of the meaning system affect these spheres too, directly or indirectly. Hence, one may expect the organizational transformations of the meaning system to be manifested in the form of changes in cognitive functioning (e.g., changes in attention, memory, creativity, the difficulty of solving different types of problems, styles of decision making, fluency and flexibility of associations, etc.), in the self concept (e.g., thoughts about oneself, self-esteem, one's biographical narrative, the experiential atmosphere of the self, etc.), in personality traits (e.g. changes in the strength and salience of different traits and other personality dispositions), and in emotions (e.g., changes in the strength and salience of different emotions and moods). These changes in turn may bring about further changes in the affected domains as well as in other domains, including overt behavior and physiological reactions.

It is evident that the meaning system is always at any given time in a specific state. The state is based on a particular organization of the meaning variables, whereby some are in a focal and dominant position and the others in a weaker and more subsidiary position. The organization of the meaning system determines which meaning variables are available and potentially accessible on the cognitive level for any act. Hence, the state of the meaning system is a selectivity-determining factor that affects the cognitive system directly and through it indirectly other systems and outputs of the organism, including emotions, personality, the self concept and the body image.

In view of these considerations, it seems justified to suggest that consciousness is the property that characterizes the whole of the cognitive system at any given time, including the contents and processes that are potentially available at this given time. The potential availability is dependent upon the organizational structure of the underlying meaning system. Hence, the state of the cognitive system at any given time can be considered as manifesting a state of consciousness. Notably, it is not consciousness per se but a state of consciousness because the cognitive system cannot have consciousness or be consciousness; it can only be in a state of consciousness.

States of Consciousness and Knowledge

As defined above, a state of consciousness (SOC) is a state of the whole cognitive system at a given time that is determined by the organizational structure of the underlying system of meaning. The organizational structure implies that some meaning variables are dominant and clearly accessible whereas others may be in varying degrees of availability.

The meaning variables that are dominant and available in a given SOC determine which kinds of contents and processes are accessible and dominate the individual's cognition. When a meaning variable is dominant it means that the individual is aware of contents that relate to this meaning variable, perceives the relevant contents, can readily retrieve from memory and store in memory contents that relate to it, has a great number of associations in regard to these contents, and can easily apply it in thinking, problem solving and further elaborations. In short, what all this means is that meaning variables in a dominant position in the system determine the kind of knowledge that is considered, that is relevant, and that commands attention at a given time. In other word, it is knowledge that one has, that one handles and uses as knowledge, that one knows one has and knows how to get, produce and evaluate it. However, this is not a constant state. The kind of relevant knowledge that dominates the scene may change when the SOC changes. Knowledge that dominates the scene in one SOC may be overlooked when the system is under the sway of another SOC.

Just as there are some meaning variables that are dominant at a given time, there are others that are less dominant. In each SOC there are contents or processes that are not available and may hence be considered as unavailable or even so-called "unconscious". The difference between the SOCs consists not only in the contents and processes that constitute knowledge but also in those contents and processes that do not constitute knowledge, that are not known or are unconscious. Thus, every SOC has an unconscious but its contents differs across the SOCs. For example. in one SOC the information accessible through the external sense organs constitutes knowledge whereas the information accessible in principle through the internal sense organs is barely perceived, noticed or processed. In another complementary

SOC, the situation may be reversed: information from the external world is barely accessible or blocked whereas information from the internal world is freely available. The former SOC corresponds to what is usually called the regular common SOC of the adult in the Western world, who is awake and alert, neither fatigued nor drunk; whereas the latter SOC corresponds rather to the state of a person who is hypnotized or engaged in guided imagery.

The differentiation in dominance or salience of the different meaning variables is not of an all-or-nothing nature but rather gradual. The meaning variables that are dominant in a given SOC have a focal position and enjoy the status of being handled by the individual as "deserving consideration as an important/focal domain". Meaning variables with weaker levels of dominance may be described in graduated terms such as "deserving study" or some consideration, "deserving exploration" in order to decide whether they deserve attention, "deserving minimal attention", at the utmost merely "taking notice of", and "not deserving any attention" (see Figure 4). These descriptions are designed merely to illustrate decreasing levels of attention accorded to domains of contents representing meaning variables in increasingly lower states of dominance or situated further from the central focal position of dominance. The further from the core position that the domains of contents are placed, the less clear, less well-defined, less perceivable, and less relevant and noticeable they become for the individual.

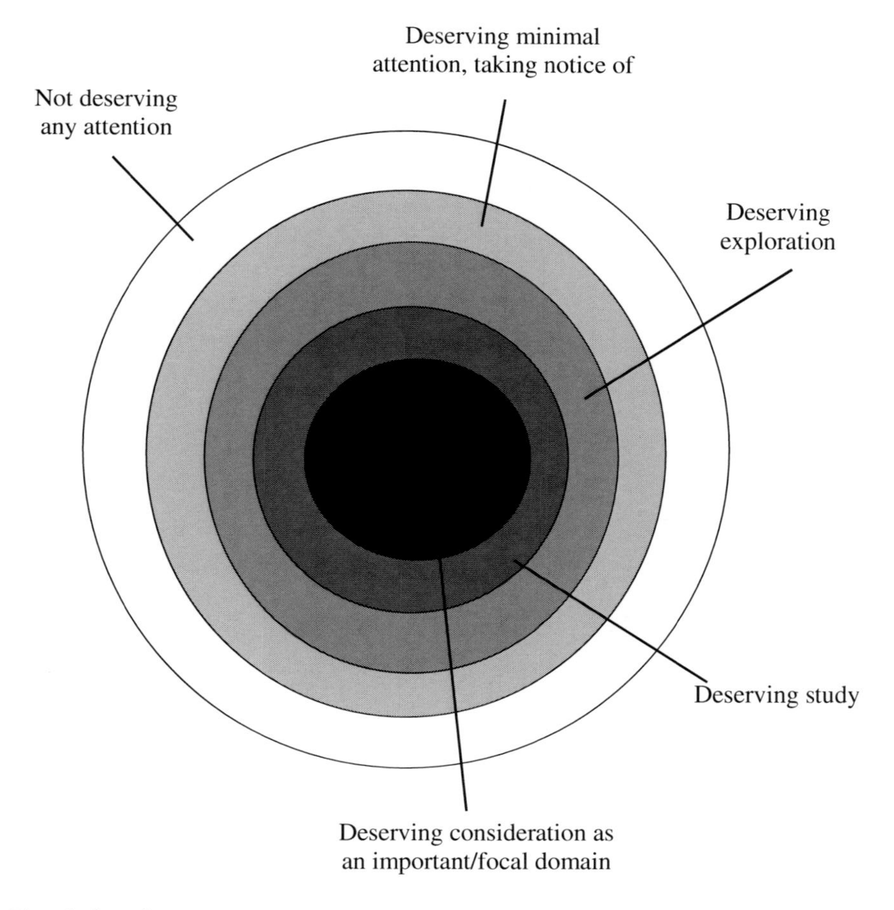

Figure 4. The circles of awareness.

Studies about attention, learning and memory support the conclusion that contents or information removed from the focus of attention may still be retrieved and evoked under special circumstances, such as change of context, free associations, or special effort at retrieval (Eysenck and Keane, 2010). Hence, it seems that the degree to which the out-of-focus contents may be revoked depends to some extent on the amount of effort that is invested in noticing these contents and drawing them even into a position where attention will be accorded to them. This type of attention is called endogenous attention and is considered to be under the control of the individual's goals and other top-down processes, in contrast to exogenous attention which is regulated by the external qualities of stimuli and other bottom-up processes (Posner and Petersen, 1990; Posner and Rothbart, 1998).

The above considerations indicate that the contents that is accorded the status of knowledge within the framework of a SOC depends to some extent also on goals, beliefs, values and other factors that are determined both by the personality of the individual and his or her culture. For example, introverted persons who are focused on their internal world, and cultures that put a premium on the internal world of individuals are likely to reinforce attention to domains of contents such as emotions, internal images, imagination and thoughts, so that even when these domains of contents are outside a given SOC they are endowed with more salience than otherwise, that is, in the case of individuals with other personalities and cultures with emphases on other contents.

The suggested meaning-based approach to consciousness contrasts with the bulk of definitions that assume the existence of 'the' consciousness (presumably denoting ordinary consciousness) and so-called altered SOCs. Rather, it implies that there are an infinite number of potential SOCs and all are evaluated as of equal potential importance and status. Indeed, any one of them can become dominant for any duration and can even come to characterize a given culture. It is possible that some of the possible SOCs are not yet known or described. Moreover, it is likely that it is even possible to invent new SOCs.

It is likely that some changes in SOCs become noticeable because they are sanctioned by the culture to which the individual belongs, or are bound to a specific technique that is salient in a particular culture (Faber, 1981). Thus, the training of Yoga may focus on differentiation of SOCs that a regular untrained person from Western culture can hardly make sense of. For example, the differences between the two following SOCs, that form part of Buddhist meditation, *Dhāranā* and *Dhyāna*, seem to refer to associations and dissociations between the observer, the object or thought, and the act of observing or thinking, that are barely comprehensible to the untrained.

Production of SOCs and Knowledge

Being in a SOC entails considering certain types of knowledge as focal and dominant while viewing other forms of knowledge as secondary or peripheral in various degrees, and completely overlooking some forms of knowledge. Personality and cultural tendencies may modulate these effects to some degree. In short, being in a SOC – and one is always in some kind of SOC, regardless of whether one is aware of it or not – means that one has available specific kinds of knowledge and does not have available other kinds of knowledge, or at least not to the same degree.

Having available a certain kind of knowledge entails having accessibility to the contents of that type of knowledge and in addition, to the characteristic modes of production of that kind of knowledge. For example, one may focus on judgments and evaluations as contents and on the manner in which evaluations are produced and maintained. The contents and the production procedures interact so closely, complementing and enhancing each other, that the result resembles what Knorr-Cetina (1999, p. 1) defines as "*epistemic cultures*", namely, an "amalgam of arrangements and mechanisms - bonded through affinity, necessity and historical coincidence - which in a given field, make up how we know what we know", what is adequate knowledge and how it is acquired. What we know enhances how we know and vice versa.

An important implication of the suggested definition of SOC is that SOC depends upon and is characterized by changes occurring in the cognitive system (through an organizational transformation in the meaning system), regardless of the nature of the agent or conditions that brought about the changes. Even when the changes are induced by behavioral, emotional, physiological, technological (e.g., virtual reality) and other conditions external to meaning and cognition, the changes that form the basis for SOC occur on the level of cognition. In fact, there may be a whole array of factors that produce SOC which have inherently nothing to do with SOC or for that matter with cognition.

This raises an interesting possibility. As has often been reiterated, SOC affects the kinds of knowledge available to the individual. It is most likely that changing the kinds of knowledge available to the individual may in itself be a means of changing SOC, and if it already exists, enhancing and stabilizing it. This is probably what actually happens in the regular course of events. An individual in a given SOC deals with knowledge that the SOC makes available, and this in turn enhances the SOC. The effect of the available kinds of knowledge on the existing SOC is so much greater in the case of SOC on the level of a group, a society or even a culture. The effect is greater because the means of promoting kinds of knowledge that are at the disposal of a larger group of people are stronger and more varied than on the level of the individual. Moreover, in the case of a SOC on the level of the individual, enhancement through available kinds of knowledge depends on active efforts and directing of attention on the part of the individual. In contrast, in the case of a SOC on the level of a society or culture, enhancement through available kinds of knowledge depends to a greater extent on exposure to sources and presentations of knowledge that surround the individual almost continuously and require much less directed attention on his or her part. This is probably the way in which cultures become associated with particular SOCs and the bond between specific SOCs and the corresponding kinds of knowledge becomes strengthened to the point where it seems so natural that no other kinds of knowledge appear as relevant or for that matter even possible.

Some Conclusions

The relations of consciousness with knowledge are complex. Consciousness can be defined in terms of the kinds of knowledge that it evokes and which constitute it. Thus, consciousness is both determined by knowledge and determines knowledge. The apparent magic cycle can be resolved by considering that the relations of consciousness to knowledge are not the same as those of knowledge to consciousness. It is consciousness, or the organizational structure of cognition, that makes the evocation of knowledge possible and defines what may potentially be considered as knowledge. However, it is knowledge or rather the different types of knowledge that constitute consciousness and render it a state that can affect different systems in the organism, including cognition but also personality, emotions, and attitudes. The complexity can be resolved by emphasizing that consciousness determines knowledge on the theoretical level, but is determined by knowledge on the experiential and operative levels.

Thus, it is through knowledge or rather different kinds of knowledge available to the individual or to a culture that consciousness becomes manifest, accessible and an active agent in the lives of individuals and cultures. Further, it is through the different available kinds of knowledge that consciousness is established, enhanced, stabilized and further elaborated. In other words, consciousness promotes different kinds of knowledge, and these in turn promote consciousness.

Being aware of these multi-levelled interactions may serve as basis for expanding and elaborating different kinds of knowledge whereby consciousness will undergo changes which may be expected to result in the development of the individual and of society and cultures too.

References

Akhtar, S. (2009). *Comprehensive dictionary of psychoanalysis*. London, UK: Karnac Books.

Aminoff, M. J. (2008). *Neurology and general medicine* (4th ed.). New York: Elsevier.

Arnon, R. and Kreitler, S. (1984). Effects of meaning training on overcoming functional fixedness. *Current Psychological Research and Reviews,* 3, 11-24.

Blackmore, S. (2004). *Consciousness—an introduction***.** New York: Oxford University Press.

Carroll, N. (2002). The wheel of virtue: Art, literature, and moral knowledge. *Journal of Aesthetics and Art Criticism* 60 (1), 3-26.

Casakin, H., & Kreitler, S. (2005). The determinants of creativity: flexibility in design. In P. Rodgers, L. Brodhurst , & D. Hepburn (Eds.). *Engineering and product design education conference: Crossing design boundaries*. Edinburgh, U.K. Napier University, pp. 303-308.

Cassirer, E. (1944). *An essay on man*. New Haven, CT: Yale University Press

Eysenck, M. and Keane, M. T. (2010). *Cognitive psychology: A student's handbook* (6th ed.). New York, NY: Psychology Press.

Freud, S. (1960). The ego and the id. In J. Strachey (Ed.), *The standard edition of the complete psychological works of Sigmund Freud*. New York, NY: W. W. Norton.

Glasersfeld, E. von (1997). *Wege des Wissens (Ways of knowing: Constructivist explorations of thinking).* Heidelberg, Germany: Carl Auer.

Güzeldere, G. (1997). The many faces of consciousness: A field guide. In N. Block, O. Flanagan and G. Güzeldere (Eds.), *The nature of consciousness: Philosophical debates.* Cambridge, MA: MIT Press, pp. 1–67.

Harrison, B. (1998). Literature and cognition. In M. Kelly (Ed.), *Encyclopedia of aesthetics*. Oxford: Oxford University Press.

Harvey, P. (1995). *The selfless mind*. Richmond, UK: Curzon Press.

Itin, C. M. (1999). Reasserting the philosophy of experiential education as a vehicle for change in the 21st Century. *The Journal of Experiential Education, 22*(2), 91-98.

Kahneman, D. (1973). *Attention and effort*. Englewood Cliffs, NJ: Prentice-Hall.

Knorr-Cetina, K. (1999). *Epistemic cultures: How the sciences make knowledge*. Cambridge, MA: Harvard University Press.

Koriat, A. (2000). The feeling of knowing: Some metatheoretical implications for consciousness and control. *Consciousness and Cognition, 9*, 149-171.

Kreitler, S. (1965). *Symbolschoepfung und Symbolerfassung: Eine Experimental-psychologische Studie (The formation and perception of symbols: Experimental studies)*. Munich-Basel: Reinhardt Verlag.

Kreitler, S. (1999). Consciousness and meaning. In J. Singer and P. Salovey (Eds.), *At play in the fields of consciousness: Essays in honor of Jerome L. Singer* Mahwah, NJ: Erlbaum, pp. 175-206..

Kreitler, S. (2001). Psychological perspective on virtual reality. In A. Riegel, M. F. Peschl, K. Edlinger, G. Fleck and W. Feigl (Eds.), *Virtual reality: Cognitive foundations, technological issues and philosophical implications.* Frankfurt, Germany: Peter Lang, pp. 33-44.

Kreitler, S. (2002). Consciousness and states of consciousness: An evolutionary Perspective. *Evolution and Cognition, 8*, 27-42.

Kreitler, S. (2003). Dynamics of fear and anxiety. In P. L. Gower (Ed.), *Psychology of fear.* Hauppauge, NY: Nova Science Publishers, pp. 1-17.

Kreitler, S. (2009). Altered states of consciousness as structural variations of the cognitive system. In E. Franco (Ed., in collab. with D. Eigner), *Yogic perception, meditation and altered states of consciousness.* Vienna, Austria: Oestrreichische Akademie der Wissenschaften, pp. 407-434.

Kreitler, S. (2010). Meaning correlates of value orientations. In Franziska Deutsch, Mandy Boehnke, Ulrich Kühnen and Klaus Boehnke (Eds.), *Crossing borders: Cross- cultural and cultural psychology as an interdisciplinary, multi-method endeavor.* E- book

Kreitler, S. (2011). Anger: Cognitive and motivational determinants. In J. P. Welty (Ed.) Psychology of *anger: Symptoms, causes and coping*. Hauppauge, NY: Nova Science Publishers, , pp. 179-195.

Kreitler, S. (2012). The psychosemantic approach to logic. In S. Kreitler, L. Ropolyi, G. Fleck and D. Eigner (Eds.) *Systems of logic and the construction of order*. Bern, New York, Vienna: Peter Lang Publishing Group, pp. 33-60.

Kreitler, H. and Kreitler, S. (1972). *Psychology of the arts*. Durham, NC: Duke University Press.

Kreitler, S. and Kreitler, H. (1984). Meaning assignment in perception. In W. D. Froehlich, G. J. W. Smith, J. G. Draguns and U. Hentschel (Eds.), *Psychological processes in cognition and personality*. Washington: Hemisphere Publishing Corporation/McGraw-Hill, pp. 173-191.

Kreitler, S. and Kreitler, H. (1985a).The psychosemantic foundations of comprehension. *Theoretical Linguistics, 12*, 185-195.

Kreitler, S. and Kreitler, H. (1985b). The psychosemantic determinants of anxiety: A cognitive approach. In H. van der Ploeg, R. Schwarzer and C. D. Spielberger (Eds.) *Advances in test anxiety research,* Vol. 4. isse, The Netherlands and Hillsdale, NJ: Swets & Zeitlinger and Erlbaum, pp. 117-135.

Kreitler, S. and Kreitler, H. (1986). Individuality in planning: Meaning patterns of planning styles. *International Journal of Psychology,* 21, 565-587.

Kreitler, S. and Kreitler, H. (1987a). The motivational and cognitive determinants of individual planning. *Genetic, Social and General Psychology Monographs, 113*, 81-107.

Kreitler, S. and Kreitler, H. (1987b). Psychosemantic aspects of the self. In T. M. Honess and K. M. Yardley (Eds.), *Self and identity: Individual change and development*. London: Routledge & Kegan Paul, pp. 338-358.

Kreitler, S. and Kreitler, H. (1987c). Modifying anxiety by cognitive means.In R. Schwarzer, H. M. van der Ploeg and C. D. Spielberger (Eds.), *Advances in test_anxiety research*, Vol. 5,. Lisse, The Netherlands and Hillsdale, NJ:Swets & Zeitlinger and Erlbaum, pp. 195-206.

Kreitler, S. and. Kreitler, H. (1988). Meanings, culture and communication. *Journal of Pragmatics, 12*,135-152.

Kreitler, S. and Kreitler, H. (1989a). Horizontal decalage: A problem and its resolution. *Cognitive Development*, 4, 89-119.

Kreitler, S. and Kreitler, H. (1989b). Meanings, culture, and communication. In A. Kasher (Ed.), *Cognitive aspects of language use*. Amsterdam, The Netherlands: Elsevier Science Publishers, pp. 221-238.

Kreitler, S. and Kreitler, H. (1990a). *Cognitive foundations of personality traits*. New York: Plenum.

Kreitler, H. and Kreitler, S. (1990b). The psychosemantic foundations of creativity. In K. J. Gilhooly, M. Keane, R. Logie nd G. Erdos (Eds.), *Lines of thought: Reflections on the psychology of thinking* (Vol. 2). Chichester, UK: Wiley, pp. 191-201.

Kreitler, H. and Kreitler, S. (1990c). The psychosemantics of responses to questions. In K. J. Gilhooly, M. Keane, R. Logie and G. Erdos (Eds.) *Lines of thought: Reflections on the psychology of thinking,* Vol. 1. Chichester, UK: Wiley,pp. 15-28.

Kreitler, S., & Kreitler, H. (1993a). The cognitive determinants of defense mechanisms. In U. Hentschel, G. Smith, W. Ehlers and J. G. Draguns (Eds.), *The concept of defense mechanisms in contemporary psychology: Theoretical, research and clinical perspectives..* New York: Springer-Verlag, pp. 152-183.

Kreitler, S.and Kreitler, H. (1993b). Meaning effects of context. *Discourse Processes, 16,* 423-449.

Kreitler, S. and Kreitler, H. (1994). Motivational and cognitive determinants of exploration. In H. Keller, K. Schneider and B. Henderson (Eds.), *Curiosity and_exploration.* New York: Springer-Verlag, pp. 259-284. Kreitler, S. and Kreitler, H. (1997). The paranoid person: Cognitive motivations and personality traits. *European Journal of Personality, 11*, 101-132.

Kreitler, S., Kreitler, H. and Wanounou, V. (187-1988). Cognitive modification of test performance in schizophrenics and normals. *Imagination, Cognition, and Personality, 7*, 227-249.

Lewis, C. S. (1990). *Studies in words*. Cambridge University Press.

Margaliot, A. (2012). *What is not meaning?* Tel-Aviv: Thema publications, Mofet Institute [in Hebrew]

Piaget, J. (1937/1954). *La construction du réel chez l'enfant (The construction of reality in the child)*. New York: Basic Books.

Polanyi, M. (1962). *Personal knowledge: Towards a post-critical philosophy*. Chicago, IL: University of Chicago Press.

Posner, M. I. and Petersen, S. E. (1990) The attention system of the human brain. *Annual Review of Neuroscience, 13*, 25-42. Posner, M.I. and Rothbart, M.K. (1998). Attention, self-regulation and consciousness. *Philosophical Transactions of the Royal Society B: Biological Sciences, 353*, 1915-1927.

Rama, S. (2002). *The essence of spiritual life*. Twin Lakes, WI: Lotus Press, Himalayan Institute Hospital Trust, Swami Rama Foundation.

Russell, B. (1910-1911). Knowledge by acquaintance and knowledge by description. *Proceedings of the Aristotelian Society (New Series),11*, pp.108-128. Ryle, G. (2002). *The concept of mind*. Chicago: University of Chicago Press. Schneider, S. and Velmans,

M. (2008). Introduction. In M. Velmans and S. Schneider (Eds.), *The Blackwell companion to consciousness*. Malden MA: Wiley.

Stedman, T. L. (2006). *Stedman's medical dictionary* (28th ed). Baltimore, Maryland: Lippincott, Williams & Wilkins.

Stolnitz, J. (1992). On the cognitive triviality of art. *British Journal of Aesthetics,* 32, 191-200.

Young, J. O. (2001). *Art and knowledge*. London: Routledge.

Zalta, E. N. (2011) (Ed.). *Stanford encyclopedia of philosophy*. Stanford, CA: Stanford University, Metaphysics Research Lab, Center for the Study of Language and Information.

Zeman, A. (2002). *Consciousness – a user's guide*. New Haven, CT: Yale University Press.

In: Consciousness: Its Nature and Functions
Editors: Shulamith Kreitler and Oded Maimon
ISBN 978-1-62081-096-5

Chapter 14

NOVELTY, NOT INTEGRATION: FINDING THE FUNCTION OF CONSCIOUS AWARENESS

Liad Mudrik[1], Leon Y. Deouell[2,3], and Dominique Lamy[1]
[1]Department of Psychology, Tel Aviv University;
[2]Department of Psychology, The Hebrew University of Jerusalem;
[3]Edmond and Lily Safra Center for Brain Sciences,
The Hebrew University of Jerusalem, Israel

ABSTRACT

The possible functions of conscious awareness have been a matter of controversy among scientists and philosophers for centuries. While some view consciousness as an epiphenomenon with no causal role in human behavior, others stipulate that high-level cognitive functions cannot be performed without consciousness. More specifically, they view consciousness as an integrating mechanism that conjoins different types of information into a unified percept or idea. In this chapter we review the different theories in the field, and present empirical evidence for integration between object and background in the absence of conscious perception. Our data suggest that conscious awareness is not necessary for integration but is called upon when facing a conceptually novel situation or scene.

Keywords: consciousness, functions, epiphenomenalism, novelty, integration, continuous flash suppression

Conscious awareness constitutes one of the biggest mysteries of science; for centuries, attempts have been made by scholars of different disciplines, from neuroscience to philosophy, to reach a better understanding of consciousness and its place in nature. No less of an interest was drawn to the question of what conscious awareness is good for. In this chapter, we set out to examine the leading suggestions regarding the possible functions of consciousness, ranging from a firm denial of such functions to assigning consciousness with a crucial role in thought and behaviour. We do so by presenting our own findings, which shed new light on this long-lasting question and challenge some of these prominent theories.

WHAT IS CONSCIOUSNESS?

In many ways, our consciousness makes us who we are. It is, by definition, the way we experience the world, our singular perspective on everything that surrounds us. It is this perspective that forms our sense of self, the basis for agency and personhood. But, try to imagine our lives without consciousness – what would be missed in our behavior? Surprisingly enough, it seems that we could achieve quite a bit without conscious awareness. We would most likely be able to differentiate between the yellow of a banana and the yellow of a school bus – computers easily make such judgements, and they are not assumed to have a conscious mind. We could still make tactical and even strategic decisions, perhaps even better than the ones we normally make: most of us would be embarrassingly beaten at chess by our cellular phone, which would turn out be the better strategist. We could even engage in relatively complex verbal exchanges, as the Turing test for intelligence requires: current-day robots can astonish their listeners with human-like responses and sentences and can also learn and adjust their behavior according to what they have learned.

Clearly, robots, cell phones and computers do not have conscious awareness (at least for now). They perform high-level functions (i.e., learning, communicating, decision making or forming perceptual judgements) despite having no ability to perceive, think or feel. Does this imply that conscious awareness does not, in actual truth, endow us with a cognitive or functional advantage, and that contrary to our strong intuitions, it plays no role in information processing and behaviour?

THE FUNCTIONS OF CONSCIOUSNESS

The possible functions of consciousness (or lack thereof) have been a matter of ongoing debate for hundreds of years. Much like other controversies in human sciences and philosophy,[1] its roots go back to ancient Greece, where Socrates defended the causal efficacy of the soul[2] in an argument with Simmias (see the dialogue "Phaedo"; Plato, 1966). The argument revolves around the immortality of the soul, but it also reveals Socrates' conceptions about its causal role. There, Simmias compared the soul to the harmony produced by the lyre (i.e., the body), hereby stating that the body leads the soul and not vice versa. Socrates, on the other hand, maintained that the soul governs the body and can oppose its appetites. According to him, it has many functions and is crucial for human existence.

[1] This seems like yet another example of Alfred North Whitehead's famous claim about all philosophy being no more than a footnote to Plato. Evidently, this applies also to cognitive psychology.

[2] Socrates speaks about the soul, the *psuchê*, and not about consciousness. In fact, the term consciousness did not even exist in his time, but appeared only hundreds of years later, in 1620 a sermon given by Archbishop Usher. However, his *psuchê* is akin in many ways to current ideas about consciousness. It serves as the subject/agent of moral judgment, choice, and action, the locus of the self, or the "I" of consciousness and personality, and the part of us that engages in reasoning and other intellectual activity.

In many ways, Simmias and Socrates represent the two ends of the continuum of modern views about the function of consciousness. In claiming that the soul (i.e., consciousness) is nothing but a mere by-product of the body's (i.e., brain) activity, Simmias may be considered as an early epiphenomenalist. Epiphenomenalism emerged in the nineteen century when Thomas H. Huxley first spoke about "conscious automata": he claimed that humans are conscious, but nevertheless respond in an automated fashion that is completely independent of conscious processes. To him, mental states are to behavior what a steam whistle is to a locomotive: it accompanies it yet contributes nothing. Accordingly, the term "epiphenomena", which was introduced by William James five years later, describes phenomena that lack any causal efficacy. In its modern version, epiphenomenalism holds that there is only one-way psychophysical causation – from the physical to the phenomenal. Consciousness itself is a real phenomenon (and not an illusion or a collection of behavioral dispositions, as suggested by Daniel Dennett and others), but it is no more than a by-product of neural activity, emerging from the complexity of the brain without having a function of any kind (e.g., Jackson, 1982). This view is not only shared by philosophers – it was also put forward by scientists who study consciousness experimentally, like Max Velmans, who wrote in his book "Understanding Consciousness" (Velmans, 2009):

> "It is only when we experience entities, events and processes for ourselves that they become subjectively real. It is through consciousness that we real-ise the world. That, and that alone, is its function" (p. 260).

On the other hand, Socrates joins, or forms, the school of thinkers who ascribe cognitive functions to conscious awareness (e.g., Baars, 2005; Searle, 1992). This school includes a wide range of views as to the functions of awareness. Some endow awareness with a crucial role in our existence. The philosopher John Searle, for example, suggests that consciousness mediates the causal relations between input stimuli and output behaviors, so that the conscious mind is the entity that governs behavior (Searle, 1992). This claim should not be interpreted in dualistic terms, as Searle does not, in any way, describe consciousness along the lines of Descartes' immaterial soul. Rather, he views it as a biological phenomenon, a high-level property of the brain that exerts its effects on behavior.

A leading advocate of the functionality of consciousness is Benjamin Baars. In his Global Workspace Theory[3] (for a recent formulation, see Baars, 2005), he ascribes several functions to conscious awareness: (1) it enables access to widespread brain sources, (2) it serves as a necessary condition for both working memory and learning (episodic and explicit learning, but also skill learning), (3) it allows the organism to voluntarily control its motor functions, (4) it evokes and influences selective mechanisms of attention, and (5) it enables access to the ‘‘observing self’: the first-person perspective we have of our surroundings. Note that Baars' last function is not necessarily different from Velmans' 'epiphenomenalist' approach. However, this comparison raises the crucial issue of whether this first-person perspective affects our behavior. In other words, while it seems clear that the self-person perspective is

---

[3] In a nut shell, the global workspace theory maintains that consciousness serves as a primary agent that facilitates widespread access between massively distributed sets of specialized networks in the brain. Accordingly, unconscious processing is held to take place in discrete nodes or small networks in parallel, up until the point where consciousness comes in, enabling the integration of the otherwise disjointed information, to form a unified percept of representation.

the defining characteristic of conscious awareness, the critical question is whether or not the other functions depend on it.

In the same vein, the psychologist George Mandler states that consciousness is a necessary condition for decision making, action selection and self-deliberation (Mandler, 2002). He describes consciousness as a scratch-pad for the choice of action, which enables the organism to predict the possible outcomes of a certain behavior, compare it with the desired outcomes and the subject's set of personal goals, and assess its appropriateness in the relevant situation. It thus serves as the basis for long- and short-range planning (see Crick and Koch, 2003, for a similar proposal). It is also the mechanism that prompts the retrieval and use of long-term memories, and the processing of upcoming information.

The "Consciousness as Integration" Hypothesis: Putting Things Together

Common to all the theories that assign functional significance to awareness is the notion that information integration is one of its fundamental features. The main function of awareness is held to be the rapid conjunction of signals from a great variety of modalities and sub-modalities in order to create a unified percept or idea (Tononi & Edelman, 1998). Terms used to describe the neural correlates of consciousness, like Baars' 'global workspace' (Baars, 2005) or Crick and Koch's 'coalitions of neurons' principle[4] (Crick & Koch, 2003) clearly convey the same notion: the brain is viewed as a massive parallel set of specialized processors, each handling a different dimension of a stimulus or a thought. Consciousness functions as a central information exchange system, allowing some processors to distribute information to the system as a whole (Baars, 2005).

Claims about the functions of conscious awareness have direct implications for the functions one assigns to unconscious processes. By definition, according to the epiphenomenal view, since consciousness has no causal influence on the organism, all functional processes (e.g., information processing, decision making and action initiating), including information integration, can be performed without consciousness. By contrast, "functional-consciousness" theories usually view unconscious processes as being unsophisticated and stereotypic. Accordingly, they predict that establishing a conceptual relationship between objects does not occur, and may be impossible, without conscious perception. In the words of Tononi and Edelman (1998):

> "Categorizations of causally unconnected parts of the world can be correlated and bound flexibly and dynamically together inside consciousness but not outside it" (p. 247).

[4] This principle is one of ten put forward by Crick and Koch as a theoretical framework for the neuroscientific study of conscious awareness. "Coalitions of neurons" refer to the notion that assemblies of neurons rival for gaining access to awareness. The winning coalition is then somewhat sustained, and embodies the content of awareness.

PUTTING THE "CONSCIOUSNESS AS INTEGRATION" HYPOTHESIS TO TEST

This contention was tested in an experiment recently conducted by our group (Mudrik, Lamy, & Deouell, 2011). We asked whether or not unconscious processing can support semantic integration of the different constituents that make up a visual scene. Previous attempts at demonstrating semantic unconscious processing have been limited to the analysis of a single word or object. Thus, before our study, there had been little evidence for integration between multiple components of a visual scene that is not consciously perceived (for review, see Kouider & Dehaene, 2007).

We used the Continuous Flash Suppression method (CFS; Tsuchiya & Koch, 2005) in order to prevent conscious perception of a visual scene. In CFS, distinct color images ("Mondrians") flashed successively at approximately 10 Hz into one eye can reliably suppress conscious awareness of an image presented at the same time to the other eye, for relatively long durations. Nevertheless, at some point the suppressed image gains dominance over the Mondrian suppressor, and thereby becomes visible to the observer. This point in time can serve as a measure of the level of unconscious processing during suppression: when images that differ on some dimension break suppression systematically at different times, one must infer that this dimension was processed while the images were suppressed, so to allow one image to break suppression sooner.

For example, Jiang and his colleagues (Jiang, Costello, & He, 2007) used CFS to suppress upright and inverted faces, and showed that the former break suppression faster than the latter. Similarly, recognizable words (i.e., words in the observer's own language) were found to break suppression faster than unrecognizable ones (i.e., words in a foreign language that the observer cannot read). The authors concluded that high-level processing of stimulus meaning takes place during suppression, allowing upright faces and meaningful words to emerge earlier into awareness.

Relying on the same logic, we suppressed awareness of scenes in which a critical object was either congruent or incongruent with the overall context (e.g., a man drinking from a glass vs. a man "drinking" from a hairbrush), and compared the exposure time required to detect each type of scenes during CFS. As a measure of detection, the scenes could be presented at either the right or left side of the visual field, and subjects had to report the side of the screen in which the scene appeared, as quickly as possible (cf. Jiang et al., 2007). Recognition of the scene or its semantic congruency was not requested. We reasoned that if the integration between an object and its background scene can be achieved when neither is consciously perceived, suppression durations should differ between congruent and incongruent stimuli. Notably, all low-level differences between the two types of images (e.g., brightness and contrast, spatial frequencies, chromaticity) were controlled for or equated.

Our results showed that incongruent scenes broke suppression faster than congruent scenes in the CFS condition. A control condition showed that this was not a result of 'partial awareness' of the stimuli (Kouider & Dupoux, 2004). Thus, while subjects did not consciously perceive the congruent and incongruent scenes they were viewing, not only did they process the displayed object and its background, but they also attempted to integrate the two into a meaningful and coherent scene. This finding provides a clear demonstration of conceptual integration between an object and its background without conscious awareness of either. What, then, can be learned from this finding about the function of consciousness?

The Functions of Consciousness Revisited

By showing that awareness is not a necessary condition for grasping the semantic relations between the constituents of a visual scene and for establishing structured mental representations our experiment widens the realm of unconscious processing to territories previously assigned exclusively to conscious awareness. In that respect, they are consistent with other demonstrations of unconscious high-level processes, commonly attributed to conscious awareness.

For example, in an fMRI study (Lau & Passingham, 2007), prefrontal activity was recorded while subjects were instructed to perform either a lexical or a phonological decision about a word. A task cue (either a diamond or a rectangle) was briefly presented before the target word and instructed the subject about the appropriate task for the upcoming target. Another task cue immediately followed. It was presented subliminally and was either congruent or incongruent with the visible task cue. Relative to congruent ones, incongruent invisible task cues impeded performance and elicited increased prefrontal activity. According to the authors, these findings show that a conflicting unconscious information can activate the conflict-related prefrontal control system, which has typically been associated with consciously perceived conflicting stimuli.

This notion challenges traditional views which hold that a gulf separates conscious and unconscious processes: while the former are intentional, complex and flexible, the latter are automatic, involuntary, simple and stereotypic (Baars, 2005; Searle, 1992). More specifically with regard to visual perception, the new findings of high-level processing during unconscious perception are inconsistent with classical versions of feature-binding theories (Treisman & Gelade, 1980) which claim that scenes' and objects' features cannot be bound without awareness and attention. The findings are more consistent with the Reverse Hierarchy theory[5] (Hochstein & Ahissar, 2002) as well as more recent revisions of the feature integration models (Evans & Treisman, 2005), and the model suggested by Rousselet, Thorpe, and Fabre-Thorpe (2004), which posit parallel automatic processing of objects without the need for conscious awareness.

Rousselet and his colleagues review studies that investigated the behavior of neurons in the inferior temporal cortex (ITC; some studies refer specifically to area TE within the ITC) in monkeys. ITC neurons typically fire in response to whole complex objects, as opposed to relatively simple features such as local contours and colors. Arguably, parallel processing of objects takes place within a few tens of milliseconds, in which significant amounts of information about these objects are transferred to the ITC. This allows high-level object representations to be established even before they enter into competition for attentional resources needed for further processing. Such rapid binding stands at the core of the "vision at a glance" feedforward processing suggested by the Reverse Hierarchy theory (Hochstein &

[5] According to which perception involves two stages; the first is a feedforward sweep ("vision at a glance"), in which spread attention enables perception of the scene's gist. The second stage of scene processing ("vision with scrutiny") is led by focused attention and allows the perception of the scene's finer details. See also the description in the following paragraph.

Ahissar, 2002). During this feedforward sweep, spread attention crudely “detects” objects and binds their features together, sometime erroneously. Then, during "vision with scrutiny", incorrectly bound features are unbound by focused attention. Accordingly and contrary to more traditional views of scene processing, feature binding and initial integration *can* be rapidly achieved in the ventral visual pathway, while attention and awareness are only needed for a more detailed inspection. Critically, our CFS research (Mudrik, et al., 2011) shows that not only objects parts and features get bound together at the early pre-attentive\awareness stage, and not only does scenes' gist get extracted, but our perceptual system also integrates those representations together. This finding seems to strip from consciousness one if its most cardinal putative functions, perhaps its *raison d'être*: integration (Tononi & Edelman, 1998).

The reallocation of such a cardinal function to unconscious processing inevitably casts doubts on the possible functionality of consciousness, thereby seemingly tipping the scale in favor of epiphenomenalism. However, this same finding may actually highlight a crucial role of awareness in perception and scene interpretation: the preferential access of incongruent scenes to awareness could have resulted from the conceptual difficulty to integrate (pre-consciously) the meaning of the incongruent object with its background. Possibly then, while awareness is not needed for object-background integration *per se*, it may become necessary when this integration yields a conceptual conflict. In other words, perceptual flexibility in the face of novelty may call for awareness.

A Possible Function for Consciousness: Dealing with the Unknown

Awareness was previously assigned with a special role in handling novel, unusual situations that require flexibility and reevaluation of expectations (Baars, 2005; Dehaene & Naccache, 2001; Searle, 1992). In fact, awareness was even suggested to be an "anti-habit" mechanism or a "novelty detector" (Gray, 1995), that comes into play when the organism is confronted with an unpredicted situation that requires the overruling of routine behaviors and automatic, unconscious processes. This can be easily illustrated by an everyday situation, familiar to many of us (especially those who have had their share of academic life). Imagine sitting in class, typing what your lecturer is saying onto your laptop. Alas, this particular lecturer is extremely boring, and you soon find yourself daydreaming. In most cases, you would have no difficulty to continue typing every word the lecturer is saying: you have been typing for so long that it has become something you can perform almost automatically. It is almost as if your ear sends signals directly to your fingers. But what happens when your lecturer decides to mention some famous warrior from the third century, whose name you have never heard before, and have no idea how it should be spelled? To make matters worse, let's say that the name was in a foreign language that has different pronunciation rules. Then, most probably, you will either lose track of what is being said and miss this warrior's name, or unwillingly abandon your daydreaming in order to become fully aware of what the lecturer is saying, so that you could employ the less familiar pronunciation rules and type the name correctly. Your unconscious mechanisms were extremely effective when dealing with habitual typing. But when new or less familiar rules had to be applied, conscious awareness was needed.

More formally, Merikle and Cheesman (Merikle & Cheesman, 1987) elegantly demonstrated the need for consciousness in dealing with novel situations by using a version of the classical Stroop paradigm: two color words (GREEN/RED) were used to prime responses to two target colors: green and red. Replicating the Stroop effect,[6] subjects' classification of the target color was faster when the prime word and color were congruent than when they were incongruent. However, when the proportion of incongruent trials exceeded the proportion of congruent trials (75% of the trials were incongruent), the effect reversed: subjects became faster on incongruent trials than on congruent trials. Arguably, subjects could strategically take advantage of the predictability of the target from the prime: given that there were only two possible colours, the best strategy would be to expect the target colour to be the one that is not written in the prime word. Crucially, this strategic inversion only occurred when the prime *was consciously perceived*. When the primes were presented below subjects' subjective threshold (i.e., when subjects reported not seeing the prime), or outside the focus of attention, performance for congruent trials was still better than for incongruent trials, despite their low predictive value. These findings imply that awareness is necessary for inhibiting an automatic stream of processes in order to deploy an updated strategy based on the reevaluation of a novel situation (Crick & Koch, 2003).

The potential for conscious, but not unconscious, inhibition is the premise upon which the Process Dissociation Procedure rests (PDP; Debner & Jacoby, 1994). The procedure was originally devised to separate the contributions of automatic and intentional uses of memory, and later applied to the field of consciousness studies. A target word is presented subliminally or at threshold level (so that subjects consciously perceive it at about 50% of the trials), and is followed by a word stem completion test. Subjects are instructed to fill the stems in one of two ways: in the *inclusion* condition, they are required to complete it using the target word they have seen before, and if unable to do so – to provide the first word that comes to mind. In the *exclusion* condition, they must not use the target word they have seen before, but instead complete the stem with another suitable word. If they have not perceived any word, they are instructed to again provide the first word that comes to mind. To illustrate, suppose the subject has been presented with the word FLOWER. In the inclusion condition, if the subject recreates the target word it is either because the subject consciously perceived it or because even though conscious perception failed, the effects of unconscious perception were sufficient for the word FLOWER to be the first to come to mind. Thus, conscious and unconscious processes act in concert here. In the exclusion condition, on the other hand, the subject may provide the target word only if the word was NOT consciously perceived but the effect of unconscious perception was sufficient for the target word to be the first to come to mind[7]. In terms of the current discussion, conscious and unconscious processes act in opposition in the exclusion condition, thereby leading to two possible scenarios. If subjects consciously perceive the word, they can exert volitional inhibition of it, and provide a different word. If, on the other hand, the word fails to reach their awareness, such inhibition is impossible, and subjects tend to provide the target word despite the instruction not to do so. Thus, based on subjects' responses, processes relying on conscious and unconscious processing can be dissociated. Note however that the logic of the PDP relies entirely on the

[6] The Stroop effect occurs in tasks that involve a conflict between two dimensions of a stimulus: classically, such a conflict is achieved by presenting the word "RED" written in blue color. When asked to specify the color of the word (rather than read its content), subjects perform slower for incongruent words (e.g., the word RED written in blue) than for congruent ones (e.g., the word RED written in red).

[7] The chance of simple guessing is kept very low by using words that are unlikely to be used, unprimed, as completions to the given stems.

idea that the flexible suppression of primed responses can only be accomplished if the prime was consciously perceived.

The dependence of flexible behavior on the presence of conscious awareness can also be gleaned from the behavior of neurological patients. Searle (Searle, 1992) pointed out the behavior of epileptic patients, who Penfield (Penfield, 1975) described as exhibiting seemingly normal, goal-directed behavior (like walking, playing the piano or even driving) during a seizure, despite the fact that the epileptic seizure rendered them completely unconscious to their surroundings[8]. Searle maintained that these behaviors were possible despite lack of awareness because they were all routine, previously learned and rehearsed by the patients. Novel behaviors, or coping with unexpected states of affairs, on the other hand, were not observed during seizures. Searle concluded that one of the major evolutionary advantages of awareness is the ability to interpret new situations, reach better discriminations and conceptualizations, and act accordingly.

Presumably then, the fact that incongruent scenes emerged earlier than the more conceptually familiar scenes in our CFS experiment (Mudrik, et al., 2011), was due to subjects' difficulty to reconcile the incongruent images with their expectations, which in turn made conscious inspection necessary. Only with conscious awareness could subjects overcome the conflict of experience-based rules regarding the appropriate relations between the object and its background, and compose a new interpretation of the scene. Overall, the findings suggest that consciousness is more about flexibility in face of novelty than about integration: while integration can be unconsciously performed, when this integration leads to a conceptually unexpected result, awareness is needed[9]. And so, it seems that the "zombie within" (Crick & Koch, 2003), that is, the unconscious processes underlying perception, behavior and cognition, may be much more sophisticated than was previously thought, but its sophistication may not be all-encompassing.

The suggestion that flexible response to novel situations requires conscious awareness does not mean that novelty detection per se requires attention or awareness. On the contrary, evidence from a variety of paradigms suggests that detecting novelty is one of the most basic operations, which can be done without focused attention, and possibly without awareness. In previous studies (Berns, Cohen, & Mintun, 1997; Ursu, et al., 2009), subjects performed a simple reaction-time task in which all stimuli were said to be equally likely to appear. Unbeknownst to the subjects, the targets actually followed a fixed complex sequence. Subjects' performance indicated that they learned the sequences even though they were unaware of their existence, much like other demonstrations of implicit learning. Once the subjects were trained, a different target word was presented, violating its *implicitly learned* context (i.e., the sequence). Although subjects were unaware of the violation, their reaction

[8] Searle refers to the patients as suffering from "petit mal" seizures, although the behavior of these patients seems more typical of temporal lobe epilepsy. In Penfield's book, three patients are described: the first is indeed said to have petit mal seizures, the second suffers from temporal lobe epilepsy, and no specifics are given as to the source of epilepsy of the third.

[9] One might claim that integration and novelty are intrinsically related: because individual objects may be combined in an infinite number of ways, each scene inevitably leads to a novel configuration, and calls for awareness. However, such novelty is perceptual in nature, while the novelty in our stimuli is semantic or conceptual: seeing a picture of an unfamiliar person playing tennis is novel in the sense that we have never seen this person or this particular game before, but it is not conceptually novel (we have seen tennis players in the past). In that respect, our findings actually show that not all configurations are equally novel. Thus, it seems that the association of conscious awareness with the need to address conceptually novel percepts is more accurate.

times (RTs) were slower, and activity was found in the ventral striatum, held to be involved in novelty detection (Berns, et al., 1997), and in the anterior cingulate cortex (Ursu, et al., 2009), which was correlated with conflict monitoring. This unconscious novelty detection seems to undermine the claim that awareness is necessary for processing novel events. However, detecting a novel event and processing novelty in terms of reevaluation and strategy selection are two different things. Our results also speak to the fact that the mere novelty of the scenes was detected even when subjects were not aware of the scenes. It is this detection that promoted awareness of the scenes, arguably for re-evaluation.

PHILOSOPHICAL CONSIDERATIONS

Although these results suggest a function for conscious awareness, it could be argued on philosophical grounds that the *metaphysical* dispute over mental causation (or lack thereof) is not a matter that can be empirically settled: any attempt to refute epiphenomenalism by empirical evidence is doomed to fail. Here, for example, the claim that awareness was needed in order to resolve the conceptual difficulty evoked by incongruent images can be converted to accommodate epiphenomenal views. To do this, it could be stated that awareness is a mere by-product of the brain's activity, one that appears when highly complicated computations take place. It is those complicated computations, rather than the phenomenal experience that accompanies them (and has no function), that are needed for the processing of incongruent scenes. And so, epiphenomenalism holds.

To resolve this conundrum, Block (1995) proposes that it applies only to phenomenal consciousness. According to him, scientists and philosophers have mistakenly conflated two types of consciousness: phenomenal consciousness (P-consciousness), which can be described as mere experience, the "what it is like" to feel something,[10] and access consciousness (A-consciousness), which is the ability to report, perform judgments or rationally control a specific mental content. While the former has no functions whatsoever, the latter is functional: it is required for flexibility, planning, integration etc. In other words, Block argues that empirical findings that allegedly demonstrate the functions of awareness actually pertain to A-consciousness rather than P-consciousness. In the CFS Experiment, then, the processes that *enabled subjects' report* or judgment about an incongruent scene, rather than *the actual phenomenal experience* of perceiving the scene, would be the ones endowed with a functional role (namely, processing the novelty of the images and reaching a suitable interpretation for them).

CONCLUSIONS

The goal of the current chapter was to examine the possible functions of conscious awareness. More specifically, we aimed at putting the "Consciousness as integration" hypothesis to empirical test. Combining our findings with the philosophical concerns raised

[10] Block himself, much like Searle, admits that P-consciousness cannot be defined in any non-circular way. However, they both contend that this problem in defining P-consciousness does not imply that it does not exist, or cannot be investigated (although Block is much more skeptical about the prospects of such an investigation, if it is led by science).

above, we might conclude on a more cautious note: we suggest that either phenomenal awareness, or the processes which it accompanies (i.e., the processes that make the information accessible to the observer, A-Consciousness), are required when facing a novel, unexpected situation. Then, these serve as an executive mechanism that reinterprets the new state of affairs, evaluates it, deploys an appropriate strategy, and controls its execution. Which of these aspects of conscious awareness is the more cardinal, or whether this question is at all tractable, remains for future empirical research as well as philosophical theorization. However, our findings clearly show that integration is not the cardinal feature of either P-consciousness or A-consciousness. In fact, integration may even not be a sufficient condition for conscious awareness. If that were the case, congruent scenes should have emerged into awareness earlier (being easier to integrate). The fact that it was incongruent scenes that were consciously perceived first suggests that integration did not require conscious awareness nor did it pave the way for consciousness. Rather, it was novelty that prompted the transition from unconscious to conscious perception.

REFERENCES

Baars, B. J. (2005). Global workspace theory of consciousness: toward a cognitive neuroscience of human experience. *Progress in brain research, 150*, 45-53.

Berns, G., Cohen, J., & Mintun, M. (1997). Brain regions responsive to novelty in the absence of awareness. *Science, 276*(5316), 1272-1275.

Block, N. (1995). On a confusion about a function of consciousness. *Behavioral and Brain Sciences, 18*(2), 227-287.

Crick, F., & Koch, C. (2003). A framework for consciousness. *Nature Neuroscience, 6*(2), 119-126.

Debner, J., & Jacoby, L. (1994). Unconscious perception: attention, awareness, and control. *J Exp Psychol Learn Mem Cogn, 20*(2), 304-317.

Dehaene, S., & Naccache, L. (2001). Towards a cognitive neuroscience of consciousness: basic evidence and a workspace framework. *Cognition, 79*(1-2), 1-37.

Gray, J. A. (1995). The Contents of Consciousness - a Neuropsychological Conjecture. *Behavioral and Brain Sciences, 18*(4), 659-676.

Hochstein, S., & Ahissar, M. (2002). View from the top: Hierarchies and reverse hierarchies in the visual system. *Neuron, 36*(5), 791-804.

Jackson, F. (1982). Epiphenomenal Qualia. *Philosophical Quarterly, 32*, 127-136.

Jiang, Y., Costello, P., & He, S. (2007). Processing of invisible stimuli: Advantage of upright faces and recognizable words in overcoming interocular suppression. *Psychological Science, 18*(4), 349-355.

Kouider, S., & Dehaene, S. (2007). Levels of processing during non-conscious perception: a critical review of visual masking. *Philosophical Transactions of the Royal Society of London. Series B: Biological Sciences, 362*(1481), 857-875.

Lau, H. C., & Passingham, R. E. (2007). Unconscious activation of the cognitive control system in the human prefrontal cortex. *Journal of Neuroscience, 27*(21), 5805-5811.

Mandler, G. (2002). *Consciousness recovered: psychological functions and origins of conscious thought* Amsterdam ; Philadelphia, PA John Benjamins Publishers.

Merikle, P. M., & Cheesman, J. (1987). Current Status of Research on Subliminal Perception. *Advances in Consumer Research, 14*, 298-302.

Mudrik, L., Lamy, D., & Deouell, L. Y. (2011). Integration without awareness: expanding the limits of unconscious processing. *Psychological Science, 22*(6), 764-770.

Penfield, W. (1975). *Mystery of the Mind: A Critical Study of Consciousness and the Human Brain*. Princeton: Princeton University Press.

Plato (1966). Translated by Harold North Fowler; Introduction by W.R.M. Lamb. Cambridge, MA: Harvard University Press.

Rousselet, G. A., Thorpe, S. J., & Fabre-Thorpe, M. (2004). How parallel is visual processing in the ventral pathway? *Trends in Cognitive Sciences, 8*(8), 363-370.

Searle, J. R. (1992). *The Rediscovery of Mind*. Cambridge, Massachusetts: MIT Press.

Tononi, G., & Edelman, G. M. (1998). Consciousness and the integration of information in the brain. In H. H. Jasper, L. Descarries, V. F. Castellucci & S. Rosignol (Eds.), *Consciousness: At the frontiers of Neuroscience, Advances in Neurology*, Philadelphia: Lippincott-Raven Publishers, pp. 245-279.

Treisman, A. M., & Gelade, G. (1980). Feature-Integration Theory of Attention. *Cognitive Psychology, 12*(1), 97-136.

Tsuchiya, N., & Koch, C. (2005). Continuous flash suppression reduces negative afterimages. *Nature Neuroscience, 8*(8), 1096-1101.

Ursu, S., Clark, K. A., Aizenstein, H. J., Stenger, V. A., & Carter, C. S. (2009). Conflict-related activity in the caudal anterior cingulate cortex in the absence of awareness. *Biological Psychology, 80*(3), 279-286.

Velmans, M. (2009). *Understanding consciousness* (2 ed.). London: Routledge.

In: Consciousness: Its Nature and Functions
Editors: Shulamith Kreitler and Oded Maimon
ISBN 978-1-62081-096-5

Chapter 15

Why the Mind Works: The Emergence of Consciousness from Mental Dynamics

Robin R. Vallacher and Jay L. Michaels
Florida Atlantic University, Boca Raton, FL, US

Abstract

Consciousness is a defining feature of human experience, yet it's nature and functions remain poorly understood. We provide a framework for reframing issues concerning consciousness that borrows from recent advances in complexity science and nonlinear dynamical systems. In this view, conscious awareness emerges from the self-organization of lower-level, largely non-conscious mental processes. It has a reciprocal causal relation with action that is captured by the principles of action identification theory (Vallacher and Wegner, 2010). The conscious representation and control of action can optimize performance but it can also promote performance impairment under specified conditions. Consciousness provides a sense of personal agency and free will, but this sense is largely illusory, reflecting the limitations of consciousness in achieving awareness of its own antecedents, constraints, and dynamics.

Keywords: reductionism, neuroscience, dynamical systems, complexity science, self-organization, emergence, fixed-point attractor, mental control, action identification, optimality, performance impairment, self-awareness, free will

Why the Mind Works: The Emergence of Consciousness from Mental Dynamics

Consciousness is one of psychology's persistent gremlins. The idea that people are aware of their thoughts, feelings, and actions is self-evident to lay people and is acknowledged—albeit, often begrudgingly—by psychologists from all quarters. And most psychologists largely agree what conscious experience is in broad, subjective terms. Behaviorists, psychoanalysts, cognitive psychologists, and neuroscientists concur that consciousness is a phenomenon whereby personal experience is accessible for verbal report, reasoning, and evaluation (cf. Damasio, 1999; Dennett, 2004). But there is little consensus when attempts are made to unpack this subjective experience into its objective properties and processes. Where does consciousness fit in models of cognition? What brain regions and neural mechanisms are responsible for conscious awareness? What role, if any, does consciousness play in other psychological processes such as decision-making, social judgment, and morality? Is consciousness a causal force in thought and action, or is it merely an epiphenomenon that exists in parallel with mental operations that shape our interactions with the world?

Issues concerning the nature and function of consciousness have proven remarkably resistant to resolution for centuries and we do not expect to settle them in this chapter. Our goal instead is to provide a new perspective within which the issues can be examined in a fresh light. Rather than starting with the assumption that consciousness is a uniquely human phenomenon that must be understood in terms of correspondingly unique principles, we approach the topic from the opposite vantage point—that consciousness can be understood in terms of basic principles that underlie phenomena throughout nature. A central goal of science is the identification of fundamental processes that are common to different topics. In recent years, progress in service of this goal has been achieved by framing diverse phenomena in all areas of science as complex systems that function in accordance with a remarkably small set of rules (cf. Miller and Page, 2007). These rules are proving to be relevant to topics and issues in psychology as well—including those that revolve around subjective experience (cf. Guastello et al., 2009; Vallacher and Nowak, 2007). Against this backdrop, we suggest that consciousness can be conceptualized and investigated in terms of dynamic processes that generate both complexity and coherence in systems of all kinds in the physical and phenomenological world.

THE ELUSIVE NATURE OF CONSCIOUSNESS

For a phenomenon that has been continually revisited since the beginnings of psychology, consciousness has proven to be stubbornly elusive, both conceptually and empirically. The central problem, as the behaviorist school admonished, is the inherently subjective nature of consciousness, which renders it difficult to investigate with objective measures. To get around the subjectivity of conscious experience, researchers tend to rely on either reductionism or experimentation contingent upon certain preconceptions. Although both approaches can contribute new insights and empirical evidence to the science of conscious experience, each has certain limitations.

The reductionist approach is represented in contemporary psychology by many research approaches in neuroscience (cf. Damasio, 1999). Indeed, there is considerable optimism about the capacity of the field to explain a majority—if not all—psychological phenomena based on their reduction to specific neural pathways, structures, or activity patterns. The field has in fact been quite successful to date in introducing impressive findings about topics ranging from learning and memory to patterns of drug addiction. With respect to consciousness, some have suggested that it originates from a constructive process in which interactions among neural networks within certain brain regions form dynamic cores of high complexity (cf. Lamme, 2006; Tononi and Edelman, 1998).

It remains to be seen, however, whether higher order psychological phenomena can be reduced to neural function or brain activity patterns. A central concern is that reductionism is poorly equipped to work with the influence of time. Even if a system is progressively deconstructed to finer and finer components, specific events at specific points of time that influenced the given system cannot be easily isolated without knowing that the events took place. Time-specific events can have important consequences. For example, Sharma et al. (2000) demonstrated that brain connectivity and neural patterns depend on the influence of sensory input over time early in development. The direct relationship between neurons and time-specific events reveals that the brain is a biological system with sensitivity to specific initial conditions. As is recognized by contemporary biology, chemistry, and physics, a system that exhibits sensitivity to initial conditions can have infinite pathways to its current state and is thereby irreducible to simple cause and effect relationships. So although neural pathways and patterns doubtlessly are associated with the experience of consciousness, it is doubtful that reducing consciousness to simple neural properties will ever be able to fully explain this phenomenon.

Experimental approaches to the study of consciousness also encounter some difficulties.

Foremost, experimental studies typically rely on paradigms that examine conscious experience using methods from attention, perception, and automaticity research. Reliance on such paradigms conflates consciousness with specific cognitive functions based on preconceptions about what conscious experience means in light of the paradigm itself (Lamme, 2006). For example, is a seemingly automatic process one that is not conscious? In research that aligns conscious experience with a sense of free will, it is asserted that automatic processes are beyond conscious experience (Bargh, 1997). However, recent integrative work that tries to disentangle attention from consciousness indicates that automatic processes may operate as conscious experiences. When presented with very brief (30-ms) visual stimuli in different attention tasks, for example, people show remarkable capacity to not only discriminate, but also experience frustration at task failure (Koch and Tsuchiya, 2006), even when the visual information is presented for too brief a duration to allow for much processing to transpire.

Although some researchers ascribe these types of effects to unconscious processes (Bargh, 1997), others consider them evidence of conscious experience outside of attentional awareness (Koch and Tsuchiya, 2006). The difference between the two perspectives has less to do with conscious experience than it does with how the researchers frame consciousness in light of the *a priori* assumptions concerning the relationship between consciousness and attention. As a result, there is little consensus about the source of conscious experience in terms of mental processes. Some researchers argue that people live predominantly without conscious experience, behaving and deciding in a largely automatic way (e.g., Bargh, 1997), whereas others posit that consciousness acts as a disruptive force that allows people to shift from habitual behaviors or thought to engage in adaptive, active decision-making (e.g., Baumeister and Sommer, 1997; Vohs, 2010).

There are no simple solutions to these difficulties inherent in the examination of conscious experience. Progressive reductionism and experimental work relying on different paradigms introduce unique problems when it comes to arriving at a truly empirical understanding of consciousness. Instead of assuming that current approaches can fully unravel the complexities of conscious experience, it may be beneficial to reframe how consciousness is understood. Rather than a benign experience that is somehow dichotomous (e.g., people do or do not have consciousness), perhaps consciousness can be recast as an experiential phenomenon that emerges from specific psychological processes. Linking consciousness to specific process-based antecedents can help integrate seemingly disparate perspectives on this subject by providing a framework that clearly articulates how specific cognitive processes (e.g., attention) and experiences interact to generate a sense of consciousness. Within this perspective, consciousness becomes a dynamic part of a process-driven model that can guide research and integrate theory across many domains of inquiry, from neural signaling research to whole brain imaging to experimental work.

DYNAMICAL FOUNDATIONS OF CONSCIOUSNESS

We suggest that insights from complexity science and nonlinear dynamical systems are useful in understanding the interactions among processes that give rise to conscious experience. Broadly defined, a dynamical system is a set of interconnected elements that undergo change (cf. Miller and Page, 2007). The ability to evolve in time is the most fundamental property of a dynamical system. The state of each element at a given moment in time may be described by values of one or more variables. Because these values change in time depending on the state of the system and in response to external influences, the variables describing elements are called *dynamical variables*. Dynamical systems theory characterizes the relations among the dynamical variables, and hence the mutual influences among elements, and the macro-level properties of the system to which these relations give rise.

In psychology, systems and their associated elements can be defined on different levels. Thus, the coordination of motor movements produces a system of action, the interplay of cognitive and affective elements forms a system of attitudes and judgments, and the influence among people creates a system at the level of social groups. In each case, the interactions among elements (i.e., among limbs, thoughts, or individuals) promote the emergence of macro-level properties that characterize the system as a whole. Consequently, systems of action, judgment, and group behavior are each associated with phenomena that can be discussed independently of the elements comprising the system. Action may ultimately derive from the coordination of limb movements, for example, but it can be characterized in terms of such dimensions as goal-directedness and appropriateness that are not inherent in specific movements.

Attractors, Emergence, and Self-Organization

Over time, a system's dynamics tend to cohere into reliable patterns, referred to as attractors. Psychological systems are commonly characterized by a *fixed-point attractor*, which is a state of coherence among the individual elements (e.g., thoughts or actions) comprising the system. Attractors are evident in a wide variety of psychological contexts. On a microscopic level, neurons exhibit attractors by having specific resting potentials to which the neurons' electrical states return after perturbation from an action potential. On a macroscopic level, attractors appear as a person's habits, values, and goals, or as a society's customs and norms.

The fact that systems converge on attractors does not mean that systems are unresponsive to external influences. Environmental obstacles obviously affect action, new information influences judgments, and external threats transform group behavior. The perturbing influence of such factors tends to be short-lived, however, so that the system quickly returns to its attractor. An individual with high self-esteem, for example, might feel bad about him or herself in response to negative social feedback, but over time his or her sense of self-worth will be restored. Attractors are psychological forces that generate particular patterns that constrain the evolution of thoughts, emotions, and behaviors. As such, attractors are a "ghost in the system" (Michaels and Vallacher, 2009)—they influence how people think, feel, and act without being explicitly held in conscious awareness.

As with any force, though, attractors do not simply conjure something out of nothing to miraculously appear within a psychological system. Attractors are the natural consequence of self-organization. *Self-organization* refers to the tendency for multiple interacting elements (e.g., thoughts, emotions, or behaviors) to adjust to one-another over time. With repeated mutual adjustments, interacting elements tend to lose degrees of freedom, eventually forming attractors. For example, during opinion formation of a novel target, a person initially can have nearly any potential opinion. The person's thoughts and emotions about the target are ambiguous and disorganized. As the individual is provided more information about the target, however, his or her thoughts and emotions become constrained. The press for coherence in the mental system drives the separate and disorganized thoughts and emotions into a unified whole as each new judgment of the target is updated in light of previous judgments, while being simultaneously influenced by new, incoming information.

Consciousness in the Mental System

The properties of dynamical systems have implications for the nature of consciousness. Research in cognitive and social psychology has revealed that the vast majority of cognitive operations occur non-consciously, often in parallel, and without placing a drain on mental resources (cf. Bargh, 1997; Nisbett and Wilson, 1977; Schneider and Shiffrin, 1977). In Freud's famous metaphor, conscious awareness is the tip of the iceberg, the output of operations that proceed autonomously and automatically in accordance with a host of processing rules.

From the perspective of dynamical systems, the entry into conscious awareness of what has been accomplished at the non-conscious level represents the emergence of a global product from the integration of various lower-level processes. In neuroscience, for example, the sudden synchronization of distinct neural pathways to produce a coherent neural network promotes the emergence of a higher-order mental state (e.g., perception, thought, judgment) that presumably is experienced consciously (e.g., Tononi and Edelman, 1998). In social judgment, specific thoughts and considerations can achieve integration in the mental background and suddenly pop into conscious awareness as a new insight or global evaluation (Vallacher and Nowak, 2007). In both cases, there is a loss in the system's degrees of freedom as the elements (neurons, thoughts) become linked in a coordinative structure that is experienced as a new macro-level state. From this perspective, consciousness has an "Aha!"

aspect to it, as new ideas, insights, and judgments emerge from cognitive processes operating outside of awareness.

The emergent products that become the object of consciousness tend to converge on common patterns and themes that function as fixed-point attractors in the mental system. For example, a person may form distinct impressions of specific individuals, but the impressions are all likely to reflect the chronic concerns and personal constructs of the person. These dimensions of personal meaning operate as tacit constraints on the person's social judgment tendencies, but they are rarely themselves the object of conscious attention. In effect, the person looks "though" these dimensions, not "at" them, in the same way that people typically look through window at the outside world rather than at the window itself. Psychological attractors, in other words, constrain the content of consciousness but are not themselves the focus of conscious attention (Michaels and Vallacher, 2009).

DYNAMICS OF CONSCIOUSNESS IN ACTION

Consciousness is a relatively small part of the mental system, but it is a part nonetheless. This raises a fundamental question: what is the point of consciousness? What role, if any, does this subjective phenomenon play in psychological systems? Is conscious awareness of what one is thinking or doing simply a consequence of system dynamics, or does it have something to do? There is reason to think that consciousness is more than an epiphenomenon. We suggest, in fact, that consciousness affects thought and behavior in several ways—some of them constructive and adaptive, others less so.

Mental Control of Action

When people do things, they could in principle recognize everything about the action—from its mechanical details to its personal and social consequences. But this is not the way it works. Consciousness is an expensive resource that is used in limited fashion (Schneider and Shiffrin, 1977; Vohs, 2010). So at any one point in time, people tend to be conscious of their action with respect to a particular representation (Vallacher and Wegner, 1985). In driving a car, for example, a person could think about turning the steering wheel, making a turn, getting to a destination, or saving time. Focusing on one of these identities removes the others from conscious attention, at least for the moment. Making a turn may occupy conscious attention for a brief period, for example, with getting to a destination supplanting the turning identity once the turn is made.

Action identification theory (Vallacher and Wegner, 1985) specifies the principles that determine which representation out of the set of potential representations occupies conscious awareness, as well as the transitions between the potential representations. The theory begins with the recognition that there are many ways of identifying what one is doing, has done, or intends to do, and that these act identities sort themselves out hierarchically in an overall act identity structure. Lower-level act identities in the hierarchy convey the details or specifics of the action and thus indicate *how* the action is done. Higher-level act identities convey a more general understanding of the action, indicating *why* the action is done or what its effects and implications are. Identification level is a relative concept—whether a particular act identity is considered a means or an end, a detail or an implication, depends on the identity with which it is compared. “Making a turn” is a high-level identity with respect to “turning the steering wheel,” for example, but a low-level identity with respect to “getting to a destination.”

The interplay of three simple principles determines the level at which an action is consciously represented. Principle 1 states that action in maintained with respect to its prepotent identity. This principle acknowledges the mental control of action that is reflected in many theoretical traditions (e.g., Carver and Scheier, 2002; Miller et al., 1960). Because act identities exist at different levels, the principle implies that people can be conscious of what they are doing in very different ways. Thus, a person may concentrate on a pattern of physical movements and monitor his or her subsequent action to see whether this intention was fulfilled, or he or she could focus instead on the consequences of the action and monitor the action with this higher-level identity in mind.

Principle 2 holds that when both a lower-level and a higher-level act identity are available, there is a tendency for the higher-level identity to become prepotent. The idea is that people prefer to think about their action in terms of its larger meaning, effects, and consequences. A person could think about his or her behavior as “using eating utensils,” for example, but is likely to gloss over such details and identify the action in higher-level terms as “eating dinner,” or in even higher-level terms as “satisfying my appetite” or “putting on weight.” In social psychology, this principle is reflected in the emphasis on values, goals, and other global constructs that are said to motivate personal and interpersonal behavior. But the essence of this principle can be found in other theoretical traditions, including learning under reinforcement contingencies, pattern recognition and formation in Gestalt psychology, inductive reasoning in cognitive psychology, the development of action mastery, and the search for meaning in existential psychology. In each case, higher-level meaning (perception, judgment, skill, value) emerges from the lower-level components comprising the phenomenon (features, facts, movements, life experiences).

Principle 3 holds that when an action cannot be performed in terms of its current identity, there is a tendency for a lower-level identity to become prepotent. If the second principle were the only basis for action identification, people’s minds would be populated by hopes, fears, abstractions, and fantasies that charge even the most rudimentary acts with high-level significance. In the real world, of course, there are obstacles to enacting goals and achieving desires, and these obstacles can derail an action consciously represented in high-level terms. A person might set out to “demonstrate racquetball skill,” for example, but proceed to lose game after game against an opponent. To regain control of the action, the person is inclined to become consciously concerned with lower-level aspects of the action, such as “gripping the racquet” or “following through on the swing.” So whereas the second principle can lead to progressively higher levels of identification, the third principle pulls people in the opposite direction, with consciousness devoted to increasingly molecular features of what they are doing.

Considered together, the interplay among the three principles imparts a dynamic pattern to the conscious representation of action. Each time one’s attention is diverted from a higher-level representation to lower-level identities (Principle 3), there is a press to reassemble these elements into a higher-level identity and reestablish consciousness of the action in more comprehensive terms (Principle 2). This movement from low- to higher-level representation captures the essence of emergence in dynamical systems (Vallacher and Wegner, 2012). The emergent high-level representation, moreover, may provide a qualitatively different integration of the lower-level identities than that provided by the initial high-level representation. With each enactment of the disruption-repair scenario, then, there is potential for the creation of a new higher order system of mental control. From this perspective, the content of consciousness is open-ended and ever changing, representing a constructive process that fosters adaptation to changing task demands.

Performance Impairment

The tension between preference (Principle 2) and necessity (Principle 3) is experienced as oscillation over time between comprehensive and detailed modes of conscious representation. This oscillation eventually dissipates, with the mental system converging on a restricted range of action representations that provides a balance between the two tendencies. This range represents the actions' *optimal level of identification*. At a party, for example, a person may be consciously concerned with "demonstrating wit and charm," but the unimpressed reactions of the other partygoers may promote a conscious concern with enacting decidedly lower-level identities such as "think of funny comments," "smile and maintain eye contact," or "look for gullible people." Once control is established with respect to these identities, the person may reconnect with somewhat higher-level identities (e.g., "show interest in what others are saying") that come close to his or her original goal of impressing the partygoers. The optimal level, in other words, represents a compromise between comprehensive understanding and effective performance. As such, it signifies the individual's level of action mastery, either generally (e.g., social skill) or in specific performance contexts (e.g., self-presentation at parties).

Despite the tendency toward optimality in action identification, people's conscious representations are not always in resonance with their overt behavior (Vallacher, 1993). Indeed, the fascination on the part of lay people with psychology is attributable in large part to problems in the feedback between mind and action expressed in the optimality hypothesis. People routinely fail to do what they intend, and often fail to profit from their mistakes in subsequent planning and behavior. To be sure, poor performance sometimes merely reflects a lack of skill or experience with respect to the action in question. The more interesting cases, however, represent lapses in mental control, with people making mistakes despite having the requisite skill, experience, and motivation to perform the action effectively.

The potential for non-optimal conscious representation exists because an action's available identities are constrained by the context surrounding the action. Most contexts in daily life are stacked in favor of relatively high-level identities, with salient cues to an action's causal effects, social implications, and potential for self-evaluation. The offer of reward, the threat of punishment, and other pressures to do well may keep the person mindful of higher-level identities at the expense of the action's more molecular representations. If the action is unfamiliar or personally difficult, a conscious concern with fulfilling these higher-level identities lacks sufficient detail and coordination of the action components necessary for optimal performance.

By the same token, some contexts can induce a level of conscious representation that is too *low* for effective performance. This can occur when there are obstacles, distractions, or other sources of disruption that sensitize the person to the action's details. An unfamiliar keyboard, for example, can change a person's conscious representation from "preparing a chapter" to "hitting the right keys." Lower-level identities also occupy the mind in novel settings lacking familiar cues to higher-level meaning or when attention is drawn to the details of the person's behavior. If a higher-level representation is optimal for the action (e.g., because the action is familiar and easy to perform), the prepotence of lower-level identities resulting from disruption, novelty, or directed attention can undermine the quality of the person's performance. Beyond interfering with an action's performance, a non-optimal representation can promote anxiety and self-consciousness (Vallacher et al., 1989). In dynamical terms, these aversive phenomenal states signal incoherence in the mental system.

CONSCIOUSNESS AND FREE WILL

It is tempting to equate consciousness with choice, personal control, and even free will (e.g., Baumeister and Sommer, 1997). After all, if people are aware of what they are thinking or doing, they are in a position to change course as they see fit. Rather than being automatons whose thoughts are constrained by automatic processes and whose actions are merely reactions to external events and social influence, people have a sense of personal agency that empowers them to chart their own course of thought and behavior.

At least that's what we like to think. From a scientific perspective, the feeling of personal agency, choice, and free will is itself another psychological process that conforms to specified rules and that varies in strength under different sets of conditions (cf. Bargh and Chartrand, 1999; Baumeister, 2008; Wegner, 2002). The emergence of any conscious experience is dependent on neuronal synchronization (e.g., Tononi and Edelman, 1998) and the capacity limits of attention and cognition restrict conscious experience to a small subset of mental states (e.g., Schneider and Shiffrin, 1977; Vohs, 2010). Principles of action identification (Vallacher and Wegner, 2012), meanwhile, specify which subset of conscious mental states is associated with the feeling of personal agency. So why do people feel free and insist that they are autonomous agents who can think and do whatever they like?

The illusion of free will can be traced, somewhat ironically, to the very nature of consciousness. In the dynamical account, the mental system is characterized by a press for coherence that forges reliable patterns of thought and judgment. These patterns function as fixed-point attractors to which mental content tends to converge over time and that resist change. The attractor itself, however, is rarely the object of conscious attention, but rather provides the tacit boundary conditions within which people experience a sequence of specific thoughts and feelings. On a moment-to-moment basis, where people spend the majority of their time mentally, the mind seems to generate thoughts that arise out of nowhere in an almost whimsical fashion. People therefore feel in charge of their conscious thoughts, even though the flow of thought is directed and constrained by the mental systems' attractors. Because the mechanisms in service of an attractor operate outside of awareness, people attribute their moment-to-moment thoughts and feelings to their conscious will.

Of course, consciousness can expand to consider the larger patterns and internal forces shaping the content of thought. People can reflect on themselves in a detached existential sense, for example, or they can ruminate on their entrenched patterns of thinking that pose problems and seem impervious to change. Introspection is notoriously unreliable, however, as noted by cognitive and social psychologists (e.g., Nisbett and Wilson, 1977). More fundamentally, reflection on one's pattern of thinking is itself shaped by forces that ultimately derive from non-conscious processes that are tacit in their operation. People can then reflect on *those* processes, but this again transforms the flow of conscious awareness and involves yet other tacit processes, and so on, in an infinite regress that is as counter-productive as chasing one's own shadow. Consciousness, in short, involves looking *through* mental mechanisms, not *at* them.

Conscious awareness can occur without being constrained by a strong attractor. In an unfamiliar setting or when faced with a novel task, the person may lack a coherent higher-order state that constrains the flow of conscious thought (Vallacher and Wegner, 2012). In the absence of an attractor, the trajectory of conscious thought can conform to one of two very different scenarios. In one, the flow of thought is under the control of the physical or social context surrounding the person. This scenario corresponds to the "power of the situation" perspective that underlies much of the theorizing in social psychology. Sometimes, though, the context offers weak, ambiguous, or conflicting guides for thought and behavior. In this scenario, the person's thoughts are unconstrained, capable of evolving in quite different directions as a function of slight variations in initial conditions (e.g., a random thought, a spontaneous lower-level action, or a trivial event).

The lack of constraint in the second scenario would seem to capture the essence of free will. However, this is not the sort of free will that people find desirable, let alone adaptive. To function effectively in their personal and interpersonal lives, people need a frame of reference that provides a coherent and stable platform for decision-making and action. When confronted with a surplus of choices, people's sense of personal agency and free will is undermined rather than enhanced. People have a greater sense of freedom and control, for example, when they choose from six as opposed to twelve alternatives (Harvey and Jellison, 1974). In effect, too much freedom (i.e., too little constraint) reduces a person's freedom to a sense of uncertainty and indecision. When faced with this prospect, people become willing to embrace cues to higher-order meaning that provide a coherent basis for thought and action (Vallacher and Wegner, 1985).

Conclusion: The Double-Edged Sword of Consciousness

Consciousness is both a blessing and a curse. The ability to report on one's thoughts, feelings, and actions is central to humans' unique standing in the animal kingdom. We do more than interact with the world, we *know* that we are doing so, and we can reflect on the quality of this interaction. Consciousness deepens the quality of mental experience, provides the foundation for shared reality among people, and creates the potential for close relationships, business transactions, and political alliances. Consciousness is also implicated in action, generating new and increasingly comprehensive courses of conduct and repairing existing actions that have run into trouble. And consciousness gives us a sense of personal agency, the feeling that we are masters of our own destiny, able to withstand the power of pressures and pleasures in our daily lives.

But consciousness has downsides as well. Although it can generate, guide, and correct action, conscious awareness of what we are doing can also paralyze and otherwise derail actions we have the skill to perform effectively. An action that might proceed smoothly and effectively if left to automatic processes can become fragmented and hesitant when the action's details are consciously monitored and controlled. Conversely, a difficult or unfamiliar action that is consciously represented in terms of its consequences and self-evaluative implications can lead people to choke under pressure, experience self-doubt, and engage in self-defeating behavior. The ability to reflect on one's experiences and mental processes can be informative and adaptive, but it can also promote endless cycles of rumination and remorse. And there is no guarantee that introspection and meta-cognition will provide an accurate portrayal of what's really going on in one's mind, since these manifestations of consciousness are themselves the product of other mental processes. Personal agency is desirable, but it can be built upon a misunderstanding of an action's true causes, and thus foster a sense of free will that is illusory rather than real.

For all its drawbacks, and despite its uniqueness, consciousness is ultimately a manifestation of largely adaptive processes that generate order out of complexity in dynamical systems throughout nature. The experience of consciousness emerges from the coordination of lower-level, largely non-conscious, mental processes. Some conscious experiences are ephemeral, dissipating as quickly as they arise, but others exert control over subsequent mental experiences, thereby providing coherence to the ongoing stream of thought. The emergence of mental order and control is uniquely human, but it is also testament to our intimate connection with systems throughout the natural world—none of which share with us the recognition of their own existence.

REFERENCES

Bargh, J. A. (1997). The automaticity of everyday life. In R. S. Wyer, Jr. (Ed.), *The automaticity of everyday Life*. Mahwah, NJ: Lawrence Erlbaum, pp. 1-62.

Bargh, J. A. and Chartrand, T. L. (1999). The unbearable automaticity of being. *American Psychologist*, 54, 462-479.

Baumeiser, R. F. (2008). Free will in scientific psychology. *Perspectives on Psychological Science, 3*, 14-19.

Baumeister, R. F. and Sommer, K. L. (1997). Consciousness, free-choice, and automaticity. In R. S. Wyer, Jr. (Ed.), *The automaticity of everyday life*. Mahwah, NJ: Lawrence-Erlbaum, 75-81.

Carver, C. S. and Scheier, M. F. (2002). Control processes and self-organization as complementary principles underlying behavior. *Personality and Social Psychology Review, 6*, 304-315.

Damasio, A. (1999). *The feeling of what happens: Body and emotion in the making of consciousness*. New York: Harcourt Brace.

Dennett, D. (2004). *Consciousness explained*. Middlesex, UK: The Penguin Press.

Guastello, S., Koopmans, M., and Pincus, D. (Eds.) (2009). *Chaos and complexity in psychology: Theory of nonlinear dynamical systems*. Boston: Cambridge University Press.

Harvey, J. H. and Jellison, J. M. (1974). Determinants of perceived choice, number of options, and perceived time in making a selection. *Memory and Cognition, 2*, 539-544.

Koch, C. and Tsuchiya, N. (2006). Attention and consciousness: Two distinct brain processes. *Trends in Cognitive Sciences, 11*, 16-22.

Lamme, V. A. F. (2006). Towards a true neural stance on consciousness. *Trends in Cognitive Sciences, 10*, 494-501.

Michaels, J. and Vallacher, R. R. (2009). The ghost in the system: Where free will lurks in the human mind. *In-Mind Magazine*. http://www.in-mind.org.

Miller, G. A., Galanter, E., and Pribram, K. H. (1960). *Plans and the structure of behavior*. New York: Holt.

Miller, J. H. and Page, S. E. (2007). *Complex adaptive systems: An introduction to computational models of social life*. Princeton, NJ: Princeton University Press.

Nisbett, R. E. and Wilson, T. D. (1977). Telling more than we can know: Verbal reports on mental processes. *Psychological Review, 84*, 231-259.

Schneider, W. and Shiffrin, R. M. (1977). Controlled versus automatic human information processing: I. Detection, search, and attention. *Psychological Review, 84*, 1-66.

Sharma, J., Angelucci, A., and Sur, M. (2000). Induction of visual orientation modules in auditory cortext. *Nature, 404*, 841-847.

Tononi, G. and Edelman, G. M. (1998). Consciousness and complexity. *Science, 282*, 1846-1851.

Vallacher, R. R. (1993). Mental calibration: Forging a working relationship between mind and action. In D. M. Wegner and J. W. Pennebaker (Eds.), *Handbook of mental control*. Englewood Cliffs, NJ: Prentice-Hall, pp. 443-472.

Vallacher, R. R. and Nowak, A. (2007). Dynamical social psychology: Finding order in the flow of human experience. In A. W. Kruglanski and E. T. Higgins (Eds.), *Social psychology: Handbook of basic principles*. New York: Guilford, pp. 734-758.

Vallacher, R. R., and Wegner, D. M. (1985). *A theory of action identification*. Lawrence Erlbaum Associates, Hillsdale.

Vallacher, R. R. and Wegner, D. M. (2012). Action identification theory. In P. Van Lange, A. W. Kruglanski, and E. T. Higgins (Eds.). *Handbook of theories in social psychology*. London: Sage, pp. 327-348.

Vallacher, R. R., Wegner, D. M., and Somoza, M. P. (1989). That's easy for you to say: Action identification and speech fluency. *Journal of Personality and Social Psychology, 56*, 199-208.

Vohs, K. D. (2010). Free will is costly: Action control, making choices, mental time travel, and impression management use precious volitional resources. In R. F. Baumeister, A. R. Mele, and K. D. Vohs (Eds.), *Free will and consciousness: How might they work?* New York: Oxford, pp. 66-81.

Wegner, D. M. (2002). *The illusion of conscious will*. Cambridge, MA: MIT Press.

THE NEUROPSYCHOLOGICAL APPROACHES

In: Consciousness: Its Nature and Functions
Editors: Shulamith Kreitler and Oded Maimon
ISBN 978-1-62081-096-5

Chapter 16

CONSCIOUSNESS AND PARALYSIS

***C. A. Ruf*[1,*] *and N. Birbaumer*[1,2,†]**
[1]Institute of Medical Psychology and Behavioural Neurobiology, University of Tübingen, Tübingen, Germany
[2]Ospedale San Camillo IRCCS, Venezia-Lido, Italy

ABSTRACT

This chapter reviews measurement of and interventions by brain-computer interface (BCI) in different states of consciousness associated with severe neurological diseases. The two dimensions of consciousness are awareness and wakefulness. Patients in a vegetative state (VS) and with locked-in syndrome (LIS) do not differ in their wakefulness but with regard to awareness, they are polarized on two extremes of the continuum.

Event-related potentials (ERPs) and electric brain oscillations in electroencephalography (EEG), magnetoencephalography (MEG), and electrocorticography (ECoG) reflect information processing mechanisms in the brain. These cortical responses help to describe disorders of states of consciousness such as vegetative state, coma and locked-in syndrome. They can also be used for controlling a brain-computer interface (BCI). BCIs translate brain signals into computer commands such as cursor control or selection of letters for communication and control of external devices, e.g. robots or movement prosthesis. Paralyzed people and patients with LIS can use BCIs for communication and environmental control. By applying operant and classical conditioning the users learn to regulate their brain activity in order to operate BCIs even with advanced LIS and probably with complete locked-in syndrome (CLIS).

Current research investigates the potential of BCIs for differential diagnosis and assessment of non-responsive patients. In the approach described here, a hierarchical paradigm including passive and active stimulation, instructions, volitional tasks and decision making with a BCI indicates preserved or absent cognitive abilities.

Keywords: vegetative state (VS), locked-in syndrome (LIS), brain-computer interface (BCI), minimally conscious state (MCS), communication, event-related potentials (ERPs),

* Corresponding author. E-mail: carolin.ruf@medizin.uni-tuebingen.de
† E-mail: niels.birbaumer@uni-tuebingen.de

cognitive processing, P300, electroencephalogram (EEG), functional magnetic resonance imaging (fMRI), near-infrared spectroscopy (NIRS), disorders of consciousness (DOC), conditioning

Concepts of Consciousness

There is no generally accepted definition of consciousness. In psychology and neuroscience consciousness is often described as awareness related to the self and the environment. Zeman (2005) contrasted consciousness (different states like sleep or coma) with awareness (content of consciousness), but both can also be seen as two different components of consciousness: the level of consciousness or wakefulness and the content of consciousness or awareness (Laureys, 2005).

Wakefulness is a tonic state on a continuum between sleep and conscious wakefulness, awareness comprises thoughts and emotions of a subject, awareness of self and the environment. A person is considered to be conscious by showing high activity on both dimensions (which are usually correlated). The exploration of diseases in which there is a separation of both dimensions of consciousness will be described here (see also Figure 1), for example there exists a high level on wakefulness but low level of awareness in vegetative state (VS).

Consciousness is subject to spontaneous and induced fluctuations. Altered states of consciousness (ASC) can be classified into different groups by origin and method of induction (Vaitl et al., 2005). Spontaneously occurring ASC are drowsiness, daydreaming, hypnagogic states, sleep and dreaming as well as near-death experiences. They are transient and depend on the arousal level of the central nervous system. Physically and physiologically induced ASC are caused by extreme environmental conditions, starvation and diet, sexual activity and orgasm and respiratory maneuvers (for example Yoga practice). Psychologically induced ASC can be a consequence of sensory deprivation, homogenization and overload, rhythm induced trance, relaxation, meditation, hypnosis and biofeedback. The heterogeneous group of disease-induced ASC includes states of consciousness changed by psychotic disorders, coma and vegetative state or epilepsy.

Non-responsive patients need to be carefully investigated regarding their conscious state, but due to the lack of overt behavioural reactions, neurophysiological recording methods such as electroencephalography and functional magnetic resonance imaging are applied for assessment of cognitive functions and for localising these patients on a presumed continuum of consciousness. The following paragraph provides a short overview about neuroimaging techniques for detection of conscious states and awareness in humans.

Recording of Brain Activity

Most of the patients with disorders of consciousness (DOC) are by definition non-responders, and do not show any goal-directed behavioural reaction. The problem of correct assessment of consciousness in these patients can be approached by applying neurophysiological methods that measure brain activity and indicate information processing in the brain.

Electroencephalography (EEG)

Electroencephalography measures the electrical activity of the brain with electrodes on the scalp. The neural activity created by synchronously firing neurons summates to electrical potential changes. Most of the electrical activity recorded on the surface of the scalp is generated by excitatory activity of the dendrites of pyramidal cells of the cortex.

Event-related potentials (ERPs) are changes in amplitude of the EEG signal time-locked to external or internal stimulation (perception, thoughts, cognitive processing). Exogenously induced ERPs, defined as changes within the first 100 ms, depend upon physical stimulus characteristics and early non-conscious attentional changes. Amplitude changes after more than 100 ms, termed as endogenous ERPs, are related to cognitive states of a person. A well-known subtype of endogenous ERP is the P300, a positive deflection approximately 300 ms after stimulation. Usually it is elicited in the so called "oddball paradigm" in which the appearance of a target stimulus has a lower probability compared to the appearance of a non-target. The P300 is related to orienting to novel stimuli and is considered to be a neural correlate of awareness of the stimulus.

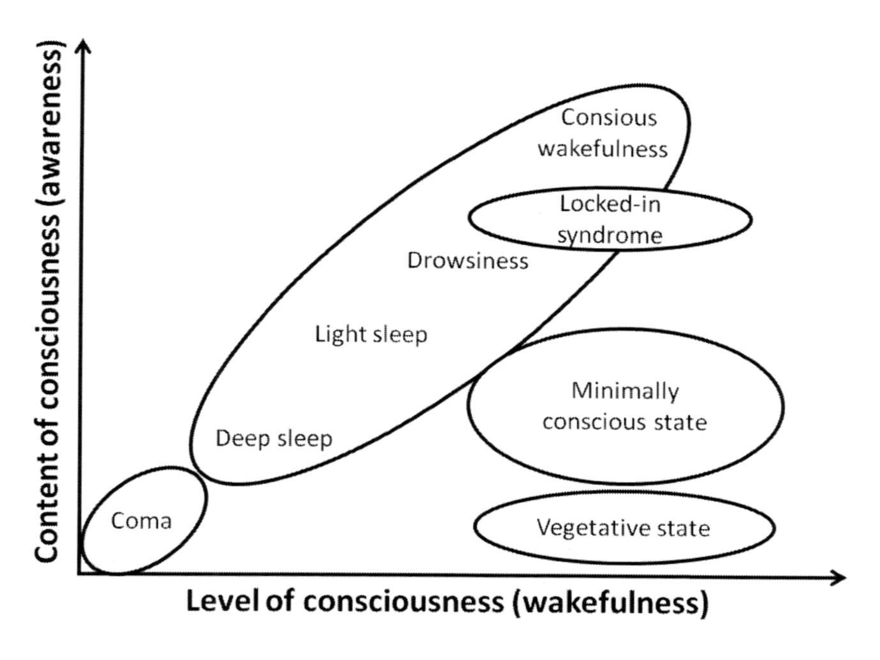

Figure 1. The two dimensions of consciousness, awareness and wakefulness, and the localization of different states of consciousness within these dimensions. (The figure is adapted from Laureys, 2005a).

Magnetoencephalography (MEG)

In contrast to EEG, the magnetoencephalography records magnetic fields of electrical potential changes of the brain tangential to the cortical surface, mainly from the gyri. The MEG system consists of superconducting quantum interference devices (SQUID) arranged in a helmet with up to 300 sensors. Compared to EEG recording, the MEG signal has higher spatial and frequency resolutions, but needs a magnetically shielded room and cannot be applied at the patient's bedside.

Functional Magnetic Resonance Imaging (fMRI)

Functional magnetic resonance imaging measures aspects of the hemodynamic response in the brain caused by neural activity. The paramagnetic load of deoxygenated haemoglobin in the blood changes through activation of the brain, especially in relation to the apical dendrites. The blood oxygenation level-dependent (BOLD) signal is an indirect measurement of brain activity, but seems to correlate with synaptic electrical brain activity (Logothetis et al., 2001). It provides excellent spatial resolution but poor temporal resolution compared to electrophysiological measures. However, the same as for MEG, it is bound to a laboratory environment and not portable.

Near-infrared Spectroscopy (NIRS)

Similar to fMRI, near-infrared spectroscopy relies on the oxygenation of brain regions. Near-infrared light passes through the brain and is partially absorbed by haemoglobin. The reflected light is collected by detectors and provides information about the hemodynamic activity of different cortical areas. The system for NIRS is an affordable device with a lower spatial resolution than that achieved by fMRI, but has the same poor temporal resolution.

Invasive Recordings

Invasive recordings of brain activity require surgery and craniotomy (opening of the skull). For electrocorticogram (ECoG) electrodes are implanted epidurally or subdurally on the brain.

ECoG measures local field potentials (as the EEG does); its advantages include higher spatial and frequency resolution, improved signal-to-noise-ratio, and a broader bandwidth and higher amplitudes. It is mostly used in epileptic patients to locate the sources (foci) of seizure.

Intracortical measurements with microelectrode arrays or electrodes record local field potentials (originating from synapses) or action potentials of single neurons (originating from axon hillocks and axons). They are usually applied in animal research and only a few cases involving humans have been published (Hochberg et al., 2006).

Disorders of Consciousness

Progress in emergency medicine resulted in an increased number of severely brain damaged patients who survive and remain in different stages of consciousness. Awaking from coma, a state in which wakefulness and awareness are both absent, patients may enter or pass through the vegetative state (VS) and the minimally conscious state (MCS). It is possible for them to remain in one of the states or to recover and retrieve the ability to communicate.

Some patients develop a locked-in syndrome (LIS) in which cognitive functioning of the brain remains intact and the affected patients are fully conscious inside a paralyzed body.

Coma

In coma, patients do not respond to any kind of stimulation, not even to highly intensive or painful stimuli. No circadian sleep-wake-cycle is present in these patients. The EEG shows continuous slow delta activity. When classifying coma in comparison to syncope and other transient states of consciousness, coma is defined as a state remaining for more than one hour. However, recordings of sleep-wake cycles are necessary to validate the diagnosis. Coma results from brain damage, e.g. large bihemispheric damage, lesion of the brainstem or thalamus, lack of oxygen supply or intoxication.

The level of consciousness while in a coma is measured subjectively with the Glasgow Coma Scale assessing motor, verbal and eye responses. A minimum of three points (no reaction in all three dimensions) indicates deep coma. The brains of coma patients show 50-70 % of the normal metabolic activity of healthy controls.

Although by definition a patient in a coma is deficient on both dimensions of consciousness, studies report cortical processing in isolated cases.

In a sample of 42 coma patients semantic processing and detection of deviants in an oddball paradigm was investigated (Daltrozzo et al., 2009). Two different paradigms used congruent and incongruent word pairs or sentences to elicit N400 negativity in the ERP after semantic incongruence. In the first paradigm three patients showed indications of semantic processing, while in the second paradigm seven patients showed such a response, but it remains unclear whether or not this difference in ERPs is related to conscious perception of semantic incongruence. The ERP response in the oddball paradigm (P300) correlated with recovery from coma and can serve as a positive outcome predictor.

Vegetative State (VS)

Vegetative state patients have a close to normal or a normal level of consciousness (wakefulness, sleep-wake cycle) but deficiencies lie in their awareness (content of consciousness). The Multi-Society Task Force defined vegetative state as a condition with complete absence of awareness of self or of the environment, an existing sleep-wake cycle and completely or partially preserved hypothalamic and brain-stem function (The Multi-Society Task Force on PVS, 1994).

There is no evidence for consistent verbal comprehension or expression, no voluntary, purposeful, reproducible or sustained behavioral responses to any kind of sensory or noxious stimulation in vegetative state patients. Cranial-nerve and spinal reflexes may be preserved. Responses to stimuli in these patients are "automatic" (implicit) and not conscious; they show only reflexive behavior, no conscious and consistent perception and no interaction with the social or physical environment around them.

Acute traumatic and nontraumatic injuries, as well as degenerative and metabolic disorders and developmental malformations, can cause a vegetative state.

According to the criteria of the Task Force the vegetative state is defined as persistent after being maintained for one month. By contrast, a "permanent vegetative state" is defined as irreversible, and is concluded with a high degree of certainty after three months of being in the vegetative state for patients with a non-traumatic and after one year for patients with traumatic brain injury.

The term permanent vegetative state (PVS) is based on probabilities; even if partial recovery after more than one year is unlikely, certainly exceptions exist in which patients recover from VS after this time period.

The correct diagnosis of VS appears to be difficult because, by definition, the patients are non-responders and can only be diagnosed using exclusion criteria (no changes in behavioural response or in facial expression based on commands given). Up to 40 % of all VS patients were misdiagnosed and their level of awareness was inadequately assessed (Schnakers et al., 2009). For this reason (functional) neuroimaging plays an important role in the diagnostic process. FRMI and positron emission tomography (PET, a nuclear imaging technique for brain metabolism using radioactive tracers) examinations of patients in VS often show a brain metabolism which is 50 to 60 % lower than in healthy controls. In addition, severe lesions in the neocortex of these patients indicate that awareness is highly improbable. The reduced activity is found mainly in bifrontal areas, pariotemporal and posterior parietal regions, mesiofrontal and posterior cingulum and preacuneal cortex. These areas are involved in executive functions, planning, decision making, attention, language and memory.

Several studies attempted to detect awareness and learning in VS patients. The results from auditory habituation in 33 VS patients showed a decrease of the N1 amplitude, an early ERP response located in frontal regions and in the superior temporal gyrus (Kotchoubey et al., 2006). The results indicate for the first time elementary cortical learning in a large sample of VS patients. The demonstration of cortical learning is an indicator that patients with VS may have islands of preserved functioning and it underlines that the VS is an extremely heterogeneous diagnostic category.

Minimally Conscious State (MCS)

The diagnosis "minimally conscious state" is given to patients who show partial signs of awareness and do not meet the criteria for coma or VS. In those patients inconsistent but reproducible behavioral evidence of awareness of self or the environment is observable (Giacino et al., 2002). MCS is diagnosed by at least one of the following behavioral criterion: the patient follows simple commands and/or shows behavioral or simple verbal response to yes/no questions, meaningful (simple) verbalization or any kind of goal-directed responding or non-reflexive behavior in a contingent relation to environmental stimuli, e.g. purposeful motor activity, smiling or crying after perceiving emotional stimuli.

An important question in MCS is the uncertainty of whether patients feel pain in reaction to noxious stimuli. A PET study involving MCS patients showed brain activation patterns to pain which are comparable to healthy controls (Boly et al., 2008). This indicates processing of pain in these patients and the need for analgesic treatment. However, lack of pain should not be assumed in any human patient, and analgesic treatment should be applied in all situations in which pain perception is even slightly probable.

The MCS can be a transient or permanent state and it is common for the patients' state to fluctuate in terms of wakefulness and awareness (Giacino et al., 2002). No definite prognosis can be made, but as in VS, the probability of recovery is likely to decrease over time.

The main neurophysiological difference between VS and MCS patients seems to be higher activations of the praecuneus and the posterior gyrus cinguli in MCS patients.

MCS can be a consequence of traumatic or nontraumatic brain injury, stroke, tumors, neurometabolic diseases, progressive degenerative disorders, and congenital or developmental malformations.

The major problem in diagnosing VS and MCS is to define the borders of consciousness. The diagnostic process in MCS is highly difficult. Due to the fluctuations of the assumed mental functions, the diagnostic test procedures should be repeated at different times of day and night as changes in the circadian rhythm are frequent in all disorders of consciousness.

Recovery from MCS is characterized by functional use of objects or by functional communication. The demonstration of the ability to discriminate among different objects, the proper use of at least two objects or appropriate yes/no responses to questions indicates emergence from MCS.

A striking number of recent studies investigated information processing and neural correlates in order to differentiate between MCS and VS patients.

An fMRI study described BOLD responses during imagination in 54 VS and MCS patients (Monti et al., 2010). The patients were instructed to execute motor imagery (playing tennis), and spatial navigation (navigate through own apartment or known city streets). Five of the patients (four of them in VS) were able to modulate their brain activity according to commands and one of these even used the imagination tasks to answer yes/no questions.

Another study used an active cognitive paradigm for exploring differences in MCS and VS patients (Schnakers et al., 2008). They were instructed to count their own name which was presented in a sequence of other names. MCS patients showed higher P300 amplitudes compared to VS patients both in listening to and counting their own name. The authors explain this finding as a better ability of MCS patients to focus attention on the instruction and the experimental stimuli.

Trace conditioning of the blink reflex as an indicator of recovery was investigated in a study with 13 VS and 7 MCS subjects (Bekinschtein et al., 2009). An air puff as the unconditioned stimulus (US) was paired with a tone (conditioned stimulus, CS); a long inter stimulus interval of 500 ms prevented overlapping of the CS and US. Trace conditioning, which requires explicit knowledge of temporal contingency and might therefore be an indicator for consciousness, predicted recovery with 86 % accuracy in both the MCS and VS patients.

Locked-In Syndrome (LIS)

Plum and Posner were the first to use the term locked-in syndrome (LIS) to describe people who are fully conscious but completely paralyzed and unable to communicate with the outside world using speech or any kind of muscular activity besides eye movement (Plum and Posner, 1983).

The LIS can be a consequence of a bilateral ventral pontine lesion, basilar artery thrombosis, brain stem insult, brain tumor, traumatic brain injury, polyneuropathy, end-stage amyotrophic lateral sclerosis (ALS), chronic Guillain-Barré syndrome and end-stage multiple sclerosis.

The neocortex of LIS patients is largely undamaged, leading to a “locked-in” mind with preserved cognitive functions (language, memory, semantic and auditory perception and processing) in a paralyzed body (Laureys et al., 2005).

The “classical” LIS is defined as complete paralysis of the body except for voluntary eye movements and blinking whereas the incomplete LIS describes as state with residual voluntary movement, e.g. face muscles, head movement (Bauer, Gerstenbrand, and Rumpl, 1979). The total or complete LIS (CLIS) represents complete paralysis with no voluntary control of eye movements, blinking or external sphincter control (Murguialday et al., 2010).

The results of neuroimaging studies in people with LIS show circumscribed lesions of the brain stem (after stroke) and a brain activity significantly higher than in VS patients, which is comparable to healthy controls.

A study using cognitive ERPs with ALS patients in CLIS showed preserved cognitive abilities in two out of three patients (Kotchoubey et al., 2003). As people in CLIS suffer from complete lack of speech or motor expression, the authors used different ERP paradigms for assessment of cognitive processing. Different auditory stimuli were presented to the CLIS patients to investigate auditory processing, associative learning, semantic and emotional processing and imagining movement preparation. Based on the results the authors conclude that non-motor perceptual and cognitive processing seems to be independent of motor functions.

Patients in the incomplete or classical LIS are able to communicate through the use of assistive technology, e.g. an eye-tracking system, blinking for answering yes/no questions or selecting letters from a letter board. Patients in LIS are awake and aware (high on both dimensions of consciousness), but due to failure of motor execution they have a similar appearance as patients in VS and are often misdiagnosed.

Residual Consciousness in otherwise Non-conscious Patients?

Disorders of consciousness can be conceptually defined on a continuum of non-awareness and non-wakefulness (as in a state of coma), non-awareness with wakefulness (VS), partial awareness (MCS), and full awareness (LIS and CLIS).

But all of the disorders are characterized by non-responsiveness and an inability of the patients to interact with their environment.

One study investigated the P300 in response to one’s own name with VS, MCS and LIS patients (Perrin et al., 2006). Significant differences in the P300 latency were observed between the VS, LIS and MCS patients in comparison to the healthy controls such that the patients showed a delayed reaction. Three out of five VS patients did not show a different response to their own name compared to other names. While performance differs from healthy controls, the results suggest the existence of semantic processing of highly salient stimuli in patients with DOC, even in some VS patients. However, the P300 response in this study was not suitable for differentiating between patients with (LIS, MCS) and without (VS) rudimentary conscious perception.

This study and some of the above mentioned studies raise the question whether or not ERPs can be used for more reliable diagnosis of DOC and whether they can provide evidence for cognitive processing in these patients. The relation between the degree of complexity of cognitive tasks and the presence or absence of an ERP response is inconsistent (Kotchoubey, 2005) and most of the results in the literature indicate cognitive processing in some of the VS patients but rarely in all. Therefore, it is impossible to draw a general conclusion about VS patients as a whole. There might be residual consciousness in some VS patients, but some of the connectivity within and between higher-order cortical areas is lost. Subcortical lesions in the ascending reticular-activation system (Hobson, 2009) are usually excluding conscious awareness but exceptions have been reported (Faran et al., 2006).

Brain-Computer Interfaces (BCIs)

Lack of communication is one of the major problems for people with disorders of consciousness or locked-in syndrome. Above all, locked-in patients often cannot use assistive technology for communication with the outside world as all such devices rely on muscular activity. The development of brain-computer interfaces (BCIs) allows muscle-independent communication for people who have lost the ability to communicate by means of speech or muscles. BCIs use recordings of brain activity such as EEG, MEG and recordings of metabolic changes (BOLD in fMRI and NIRS) as input signal to control external devices, e.g. computer and language programs, wheelchairs or prosthetic devices. Control over BCIs can be achieved by learning self-regulation of the brain (operant-instrumental conditioning), classical conditioning of brain changes or cognitive stimulation paradigms and attentional manipulations.

Electroencephalography Brain-computer Interfaces (EEG- BCIs)

In most BCI applications EEG is the preferred method which is justified by its high temporal resolution and the advantage of being portable and relatively inexpensive.

In the last two decades three main input signals in the EEG were used for control of BCIs: Slow cortical potentials (Birbaumer et al., 1999), sensorimotor rhythms (Wolpaw et al., 2002), and the P300 event-related potential (Farwell and Donchin, 1988).

Slow Cortical Potentials (SCPs)

SCPs are slow shifts in the EEG (below 1 Hz) lasting from 300 ms to several seconds generated in the dendrites of pyramidal cells of the cortex. The negativity results from an increased depolarization while positive amplitudes represent reduced cortical activation. Humans can learn to voluntarily regulate their own SCPs by operant conditioning (Birbaumer et al., 1990). Visual feedback representing SCP changes is often provided by a cursor moving toward targets at the top or bottom of a computer screen. It is designed such that a negative amplitude shift (compared to baseline) moves the cursor up, while a positive amplitude shift leads to downward movement. Reinforcement of correct response trials (operant feedback) is critical for self-regulation of the SCPs. Paralyzed users (e.g. ALS patients) need about one to five months of training (one to several training sessions per week) to learn to control a BCI using SCPs in order to attain an accuracy rate of about 70 % (Birbaumer et al., 2000).

Sensorimotor Rhythm (SMR)

Sensorimotor rhythm, also termed mu-rhythm, is recorded over primary sensorimotor areas of the brain with frequencies of 8 - 12 Hz and 18 - 26 Hz. It desynchronizes with actual movement, preparation for movement or imagination of movement (event-related desynchronization, ERD) and synchronizes again after movements or during muscular relaxation (event-related synchronization, ERS). The imagination of movement (kinesthetically) is sufficient to cause an ERD which is often used as an input signal for BCI. Applying the principles of operant conditioning, humans learn to regulate their SMR activity. Feedback is presented on a screen via movement of a cursor or variation of tones. A vertical bar on the top or bottom of a screen needs to be contacted by the cursor which moves from left to right. Different combinations of imagined movements (for example left hand versus right hand or left hand versus foot) cause the vertical cursor movement which provides feedback of the ERD. In several studies healthy subjects have been shown to achieve mean performance accuracies between 60 and 100 %, while severely impaired ALS patients have been shown to successfully learn to control their SMR activity with an average accuracy over 75 % after 20 training sessions (Kübler et al., 2005).

SMR-BCIs were shown to be successfully controlled by patients with ALS (Kübler et al., 2005), cerebral palsy (Neuper et al., 2003), spinal cord injury (Pfurtscheller et al., 2003) and chronic stroke (Buch et al., 2008).

Event-related Potential P300

The P300 is a positive deflection in the EEG which occurs approximately 300 ms after a rare deviant stimulus (target) and shows the largest amplitude over centro-parietal regions. A speller application based on visual P300-BCI presents letters and symbols in the form of a 6 x 6 matrix on a screen (Farwell and Donchin, 1988). The rows and columns of the matrix flash in a random order. For selection of a letter the subjects are instructed to focus their attention on the target letter every time the flash occurs. The low probability of the flashing of the target row or column in comparison to all the non-target rows and columns elicits a P300 reaction. The intersection of the row and column responses results in the selection of the target letter.

P300-BCIs have the advantage of not requiring training sessions over several days or weeks. Most people are able to use it after short screening periods of about 10-15 min.

In a study with 100 healthy participants, Guger and colleagues showed that about 73 % of all participants were able to control P300-BCI with 100 % accuracy and 89 % of the participants achieved between 80 and 100 % accuracy (Guger et al., 2009). Several studies including patients with ALS, Guillan-Barré, muscular dystrophy and spinal cord injury prove the applicability of visual P300-BCI in clinical populations (Kübler and Birbaumer, 2008). Considering the target patient groups for BCI use, in some diseases visual abilities are impaired and patients are no longer able to control a visual BCI. Auditory P300-based BCIs aim to overcome these problems as they replace visual feedback with auditory stimuli (Furdea et al., 2009).

Functional Magnetic Resonance Imaging Brain-computer Interface (fMRI-BCI)

Hemodynamic brain responses have only recently been used as input signals for BCIs. The high spatial resolution of fMRI-based BCIs allows selective self-regulation of circumscribed brain areas and their connectivity whereas the portability of NIRS may offer broader applications in different populations.

The first studies involving fMRI-based BCIs were reported only a few years ago. Participants learned to control the BOLD response of the supplementary motor area and parahippocampal place area with online feedback of local BOLD activity (Weiskopf et al., 2003). Several studies have also shown the ability of users to learn to control BOLD activity in brain areas related to conscious pain perception (deCharms et al., 2005).

Up- and down-regulation of the anterior insula was found to correlate with changes in emotional perception of negative emotional pictures (Caria et al., 2007). Clinical application of fMRI-BCI might help patients with anxiety disorders, schizophrenia, psychopathy, and pain disorders to modify their brain activity and behavior. Ruiz and colleagues demonstrated successful control of the insula in chronic schizophrenics after 12 sessions resulting in improved recognition of negative emotional faces (Ruiz et al., submitted).

Near-infrared Spectroscopy Brain-computer Interface (NIRS-BCI)

Investigation of a multichannel NIRS-BCI demonstrated the feasibility of decoding motor execution and imagination of finger tapping in both hands (Sitaram et al., 2007). An offline classification accuracy of more than 80 % was achieved by applying different machine learning algorithms for pattern recognition. The authors propose a word speller application based on hemodynamic responses recorded via NIRS. A cursor controlled by motor imagery could be used for the selection of letters on a screen, but to date no application of such NIRS-based movement imaginary BCI has been reported.

Communication in CLIS Using BCIs

One of the main aims in the development of BCIs is to provide a communication channel for people in the CLIS. A meta-analysis reviewed BCIs based on SCP, SMR and P300 and drew the conclusion that none of the involved seven CLIS patients could control a BCI (Kübler and Birbaumer, 2008). The investigated CLIS subjects were unable to self-regulate their brain activity consistently or use the P300 as an input signal; however none of them had participated in BCI training before entering into the CLIS. Several reasons could be responsible for the lack of BCI control in CLIS: First, a drop in cognitive abilities, arousal or attention might be responsible for the inability to concentrate on the task. Inadequate artificial ventilation in the CLIS subjects could lead to impairments of cognitive abilities. Likewise, there is uncertainty about the state of wakefulness of the CLIS participant during BCI training (Murguialday et al., 2010). As a second explanation, the authors propose that the perception of contingency between a patient's own behavior and the consequences of the behavior is lost in CLIS. As a consequence, operant (voluntary) learning (which is necessary for learning to control a SMR- or SCP- based BCI) is not possible presumably because extinction of goal-directed thinking occurs (Birbaumer and Cohen, 2007).

While none of the existing invasive or non-invasive BCIs have been successfully used by CLIS patients, in the last years of research promising developments using both brain activity and peripheral measurements have appeared. One peripheral approach involved the use of changes in the pH-level of saliva for communication (Wilhelm, Jordan, and Birbaumer, 2006). A CLIS patient was able to decrease or increase her salivary pH value with up to 100 % accuracy by imagining the taste of milk or lemon. The patient used this method to answer questions with known and unknown answers and finally gave consent to a surgical procedure.

Sniffing was tested as an input signal for the control of external devices (wheelchairs or computers for communication) in three LIS subjects (Plotkin et al., 2010). A sniff controller device measured the nasal pressure that is manipulated through opening and closing the soft palate by volitional nasal inspiration. Two of the LIS subjects were able to gain control over the device and used it for the selection of letters on a computer screen with an average speed of one to two letters per minute. The authors claim that the device would be working even in respirated patients but none of the participating subjects were artificially respirated. A second problem of applying the sniff controlling device in CLIS may appear in the required control over the soft palate. Based on the hypothesis of extinction of goal-directed thinking (Kübler and Birbaumer, 2008) the CLIS patients might not be able to exert volitional control of their soft palate even if they would have been able to do so before entering the CLIS state.

A NIRS study investigated volitional changes in blood volume of 17 totally locked-in ALS patients (Naito et al., 2007). The authors defined the totally locked-in state vaguely as a state in which "they cannot move any part of their bodies", but it remains unclear whether the totally locked-in state in this study is identical to the above described definition of CLIS. Naito and colleagues do not make any statements about life prolonging treatments such as artificial ventilation (which is necessary in patients with CLIS as a consequence of ALS) or any other clinical criterion for CLIS. Also the patients' CLIS state was not documented with measurements of muscle activity (electromyography, EMG) or eve movement recordings (Murguialday et al., 2010). The participants activated the frontal lobe using mental calculations and imagery to answer yes/no questions. In seven of the totally locked-in patients correct answers could be detected with accuracy over 75 %. In the other 10 participants in whom the paradigm failed to achieve adequate accuracy, it is possible that the patients did not comply with the instructions. The authors also referred to potentially lower brain activation in the totally locked-in patients as a possible explanation of the results.

All of the cited studies involving CLIS patients suffer from the problem of non-responding in many or all of the patients.

To overcome these problems a novel approach developed in our laboratory is based on studies with curarized rats that failed to replicate operant learning of autonomic functions shown in earlier studies of the sixties and seventies (Dworkin and Miller, 1986). To explain the failure of operant learning in curarized rats Dworkin hypothesized that unlike for intact animals for which the positive consequences of reward mediate the homeostatic stabilization characteristic of reward, this was not so in curarized rats with artificially maintained body functions. In a later study, Dworkin demonstrated learning of vasoconstriction in curarized rats based on classical Pavlovian conditioning (Dworkin and Dworkin, 1990). In this regard, the curarized rats may resemble the paralyzed CLIS patients who are artificially ventilated and fed. Extending Dworkin's findings to the CLIS problems, classical conditioning appears to be a promising paradigm for learning and BCI control in CLIS. Classical conditioning is expected to overcome the problems of operant learning and voluntary modulation and the assumed extinction of goal-directed thinking because classical conditioning is based on associative learning only and a drop in arousal or attention should not influence the learning process.

In a study with four ALS patients in LIS the subjects were classically conditioned to the meaning of "yes" and "no". In a first experimental phase they listened to true and false statements ("My name is Peter") and were instructed to think "yes" or "no" respectively after each statement. Every true statement was paired with a strong unconditioned stimulus (US 1) (e.g. a loud noise) and every false statement with another unconditioned stimulus (US 2). It was previously established that the US 1 and US 2 elicit a change in brain activity (unconditioned response, UR). Pairing the conditioned stimulus (CS; true and false statements, respectively "yes" and "no" thinking) with the US, the brain learns to associate the US with the meaning of the statements. As a consequence the brain responds with specific brain changes to the two classes of statements. In a second phase the association between the "yes" and "no" answers and the respective changes in brain activity were tested: statements were presented without pairing them with the US. The thinking of "yes" and "no" in brain activity could be classified by single-trial classification (differentiation between "yes" and "no" based on the brain activity of a single EEG segment of 1 second) with up to 80 % accuracy and a mean accuracy of 65 %. The paradigm is now being tested with patients in CLIS, and in order to further validate the paradigm for CLIS, LIS subjects who are already conditioned will be followed and documented throughout their transition to CLIS. Significant classification of semantically affirmative or negative statements would offer the first experimentally valid documentation of socially and semantically meaningful communication in CLIS.

Brain-computer Interfaces for Patients with Disorders of Consciousness

Non-responsive patients are classified along a continuum of consciousness, but the diagnosis is frequently false and subject to underestimation of partially preserved consciousness. Most BCI applications developed for communication require controlled attention, comprehension of instructions and active (goal-directed) learning, but the new paradigms for CLIS patients described above make lower demands on the users. BCI paradigms for improved diagnosis of consciousness in non-responsive patients and for providing a communication channel in (partially or fully) conscious, but paralyzed patients were developed (Kübler, 2008).

In a hierarchical paradigm, cognitive processing of a spectrum of cognitive stimuli ranging from simple to complex is investigated (Kübler and Kotchoubey, 2007). Assessment of resting EEG and auditory evoked potentials allows for the exclusion of patients with hearing loss or a pathological EEG pattern, meaning a flat EEG or with the absence of theta and higher frequencies (Kotchoubey et al., 2005). In a second step several passive neuropsychological paradigms can be applied to investigate the level of awareness. In a passive paradigm external stimulation, mostly auditory and visual, is used to explore perception and processing of the stimuli. Visual stimuli are not feasible for CLIS patients or for many LIS, VS and MCS patients in whom eye movements are impaired and the possibility of visual attention with focusing on a screen cannot be assumed. Auditory oddball paradigms examine exogenous potentials (mismatch negativity) and the endogenous P300 as a potential marker of higher cognitive processes. Semantic processing is investigated via auditory presentation of congruent and incongruent word pairs or sentences.

The presence of the described exogenous and endogenous potentials indicates the existence of higher processes but do not provide information about conscious awareness. Instructing the patients to focus attention on the stimuli is expected to lead to higher ERP amplitudes in those understanding the instructions. In the last step volitional paradigms are applied to those patients showing positive results in the previous tasks. Similar to the study of Monti and colleagues (2010) mental imaginary such as movement imagination and spatial navigation tasks can be used to investigate volitional changes in brain activity. Successful classification of these changes would prove awareness in the participating patient and the use of mental imaginary for free control of a BCI can be tested. Compared to the volitional paradigm, in free BCI communication the patients need to decide which mental task to perform before carrying out the selected task in a second step.

The previously described classical conditioning paradigm is another promising approach which can presumably be applied even after the extinction of goal-directed thinking. Instead of presenting tasks requiring voluntary attention, the classical conditioning avoids the problem of decision making, for which intact attention is necessary.

The hierarchical model provides a new approach for strategic investigation of residual conscious awareness in patients with DOC. However, some patients might be misdiagnosed even with this procedure. In a study with 98 patients having a diagnosis of VS and MCS the hypothesis of hierarchical processing was investigated and could not be confirmed (Kotchoubey et al., 2005). In some patients complex cognitive but not simple physical stimuli lead to brain responses, and for this reason patients should not be excluded from further testing due to missing responses in simple tasks.

Ethical Considerations

A main ethical issue is related to the ongoing debate concerning survival and conscious awareness in the PVS. Caregivers face the problem of respecting patients' autonomy regarding survival or end of life, but VS patients are not able to refuse or accept life prolonging treatment and most of them did not make an advance directive. The hope for recovery and residual awareness in these patients is opposed to the negative prognosis of PVS and the consensus in most caregivers that survival in PVS is not a benefit for the patients (Jennett, 2002). On the other hand, misdiagnosis of awareness in these patients is frequent and only at present are neuroimaging and neurophysiological tests beginning to play a role in the diagnosis.

With the application of BCIs the ethical question arises as to whether BCI research evokes false hope in patients and family members about the possibilities and benefits of BCIs (Haselager et al., 2009). Nonrealistic expectations about the progress in research might have an influence on decisions about life prolonging treatments. The lack of alternative assistive technology for communication in CLIS patients might motivate them to decide for participating in BCI studies or other scientific experimentation which may be seen as an ethical problem (Clausen, 2008). Likewise, informed consent in CLIS, VS and MCS patients can be achieved only through the legal representatives. As a consequence of the missing interaction with the patient, the will of the patient can only be assumed by family members and caregivers who may or may not agree with the patient themselves.

ALS patients in whom the ability to communicate deteriorates with progress of the disease describe effective communication as a main factor for their quality of life (Bach, 1993). Even with impaired communication abilities LIS patients rate their subjective quality of life as comparable to healthy controls (Kübler et al., 2005) and report a low incidence of major depression (Lulé et al. , 2008). BCIs for fully conscious but paralyzed patients are crucial for their interaction with the surrounding world and the maintenance of their quality of life.

CONCLUSION

In this chapter we described disorders of consciousness and tried to define the limits of conscious responses in patients with severe brain disease. Studies with patients in coma, VS and MCS indicate learning and cognitive processing in single cases, but these results can not be generalized to whole groups with these disorders. The relationship between the presence or absence of a particular ERP component and the complexity of a cognitive task was found to be inconsistent. ERPs (e.g. P300, N400) are indicators of the information processing necessary but not sufficient for consciousness. The borders between conscious and non-conscious processing as represented by ERP results are not sharp but fluent. In general, ERPs may rather underestimate the cognitive ability, as fluctuations in attention of the participants will result in a negative outcome of ERP recordings. The absence of an ERP component in patients with DOC is no proof for the absence of cognitive abilities as they are sometimes also absent in healthy people (Kotchoubey et al., 2002).

We presented several studies which have demonstrated indicators for conscious processing in participants with VS and MCS in whom, by definition, no conscious awareness is assumed.

BCIs have proven to be suitable for communication in paralyzed people with preserved cognitive functions such as LIS patients but to date state-of the-art BCIs were not successful in patients with CLIS or DOC. A new approach described here is currently under investigation for its applicability for assessment of consciousness and potential restoration of basic communication in VS and MCS patients.

Future research should identify the amount of awareness necessary for binary BCI control needed for answering yes/no questions in order to evaluate the actual feasibility of BCI communication in VS and MCS. Only the failure to communicate after longer time periods using non-voluntary paradigms, such as that described for semantic classical conditioning, would justify negative outcome predictions in single patients with CLIS or DOC.

ACKNOWLEDGEMENTS

This work is supported by Deutsche Forschungsgemeinschaft (DFG), European Research Council (ERC), Deutscher Akademischer Auslandsdienst (DAAD), European Union (Marie Curie Host Fellowship for Early Stage Researchers Training), and Bundesministerium für

Bildung und Forschung (BMBF) Bernstein Center for Computational Neuroscience, Focus Neurotechnology. This manuscript reflects the authors' views only and funding agencies are not liable for any use that may be made of the information contained herein. We also thank Angela Straub, Andrea Kübler, Sonja Kleih, Ivo Käthner and Tamara Matuz for their contributions to the manuscript and Colleen Dockery for language revisions.

REFERENCES

Bach, J. R. (1993). Amyotrophic lateral sclerosis - communication status and survival with ventilatory support. *American Journal of Physical Medicine & Rehabilitation, 72*(6), 343-349.

Bauer, G., Gerstenbrand, F., and Rumpl, E. (1979). Variables of the locked-in syndrome. *Journal of Neurology, 221*, 77-91.

Bekinschtein, T. A., Shalom, D. E., Forcato, C., Herrera, M., Coleman, M. R., Manes, F. F., et al. (2009). Classical conditioning in the vegetative and minimally conscious state. *Nat Neurosci, 12*(10), 1343-1349.

Birbaumer, N., and Cohen, L. G. (2007). Brain-computer interfaces: communication and restoration of movement in paralysis. *J Physiol, 579*(3), 621-636.

Birbaumer, N., Elbert, T., Canavan, A. G. M., and Rockstroh, B. (1990). Slow potentials of the cerebral cortex and behavior. *Physiological Reviews, 70*(1), 1-41.

Birbaumer, N., Ghanayim, N., Hinterberger, T., Iversen, I., Kotchoubey, B., Kübler, A., et al. (1999). A spelling device for the paralysed. *Nature, 398*(6725), 297-298.

Birbaumer, N., Kubler, A., Ghanayim, N., Hinterberger, T., Perelmouter, J., Kaiser, J., et al. (2000). The thought translation device (TTD) for completely paralyzed patients. *IEEE Trans Rehabil Eng, 8*(2), 190-193.

Boly, M., Faymonville, M.-E., Schnakers, C., Peigneux, P., Lambermont, B., Phillips, C., et al. (2008). Perception of pain in the minimally conscious state with PET activation: an observational study. *The Lancet Neurology, 7*(11), 1013-1020.

Buch, E., Weber, C., Cohen, L. G., Braun, C., Dimyan, M. A., Ard, T., et al. (2008). Think to move: a neuromagnetic brain-computer interface (BCI) system for chronic stroke. *Stroke, 39*(3), 910-917.

Caria, A., Veit, R., Sitaram, R., Lotze, M., Weiskopf, N., Grodd, W., et al. (2007). Regulation of anterior insular cortex activity using real-time fMRI. *Neuroimage, 35*(3), 1238-1246.

Clausen, J. (2008). Moving minds: ethical aspects of neural motor prostheses. *Biotechnol J, 3*(12), 1493-1501.

Daltrozzo, J., Wioland, N., Mutschler, V., Lutun, P., Calon, B., Meyer, A., et al. (2009). Cortical information processing in coma. *Cogn Behav Neurol, 22*(1), 53-62.

deCharms, R. C., Maeda, F., Glover, G. H., Ludlow, D., Pauly, J. M., Soneji, D., et al. (2005). Control over brain activation and pain learned by using real-time functional MRI. *Proc Natl Acad Sci U S A, 102*(51), 18626-18631.

Dworkin, B. R., and Dworkin, S. (1990). Learning of physiological responses: I. Habituation, sensitization, and classical conditioning. *Behav Neurosci, 104*(2), 298-319.

Dworkin, B. R., and Miller, N. E. (1986). Failure to replicate visceral learning in the acute curarized rat preparation. *Behav Neurosci, 100*(3), 299-314.

Faran, S., Vatine, J. J., Lazary, A., Ohry, A., Birbaumer, N., and Kotchoubey, B. (2006). Late recovery from permanent traumatic vegetative state heralded by event-related potentials. *J Neurol Neurosurg Psychiatry, 77*(8), 998-1000.

Farwell, L. A., and Donchin, E. (1988). Talking off the top of your head: toward a mental prosthesis utilizing event-related brain potentials. *Electroencephalogr Clin Neurophysiol, 70*(6), 510-523.

Furdea, A., Halder, S., Krusienski, D. J., Bross, D., Nijboer, F., Birbaumer, N., et al. (2009). An auditory oddball (P300) spelling system for brain-computer interfaces. *Psychophysiology, 46*(3), 617-625.

Giacino, J. T., Ashwal, S., Childs, N., Cranford, R., Jennett, B., Katz, D. I., et al. (2002). The minimally conscious state: definition and diagnostic criteria. *Neurology, 58*(3), 349-353.

Guger, C., Daban, S., Sellers, E., Holzner, C., Krausz, G., Carabalona, R., et al. (2009). How many people are able to control a P300-based brain-computer interface (BCI)? *Neurosci Lett, 462*(1), 94-98.

Haselager, P., Vlek, R., Hill, J., and Nijboer, F. (2009). A note on ethical aspects of BCI. *Neural Netw, 22*(9), 1352-1357.

Hobson, J. A. (2009). REM sleep and dreaming: towards a theory of protoconsciousness. *Nat Rev Neurosci, 10*(11), 803-813.

Hochberg, L. R., Serruya, M. D., Friehs, G. M., Mukand, J. A., Saleh, M., Caplan, A. H., et al. (2006). Neuronal ensemble control of prosthetic devices by a human with tetraplegia. *Nature, 442*(7099), 164-171.

Jennett, B. (2002). The vegetative state. *J Neurol Neurosurg Psychiatry, 73*(4), 355-357.

Kotchoubey, B. (2005). Event-related potential measures of consciousness: two equations with three unknowns. *Prog Brain Res, 150*, 427-444.

Kotchoubey, B., Jetter, U., Lang, S., Semmler, A., Mezger, G., Schmalohr, D., et al. (2006). Evidence of cortical learning in vegetative state. *Journal of Neurology, 253*(10), 1374-1376.

Kotchoubey, B., Lang, S., Bostanov, V., and Birbaumer, N. (2002). Is there a mind? Electrophysiology of unconscious patients. *News Physiol Sci, 17*, 38-42.

Kotchoubey, B., Lang, S., Mezger, G., Schmalohr, D., Schneck, M., Semmler, A., et al. (2005). Information processing in severe disorders of consciousness: vegetative state and minimally conscious state. *Clin Neurophysiol, 116*(10), 2441-2453.

Kotchoubey, B., Lang, S., Winter, S., and Birbaumer, N. (2003). Cognitive processing in completely paralyzed patients with amyotrophic lateral sclerosis. *Eur J Neurol, 10*(5), 551-558.

Kübler, A. (2008). Brain-computer interfaces for communication in paralysed patients and implications for disorders of consciousness. In G. Tononi and S. Laureys (Eds.), *The neurology of consciousness—Cognitive neuroscience and neuropathology*. Amsterdam: Elsevier, pp. 217–233

Kübler, A., and Birbaumer, N. (2008). Brain-computer interfaces and communication in paralysis: extinction of goal directed thinking in completely paralysed patients? *Clin Neurophysiol, 119*(11), 2658-2666.

Kübler, A., and Kotchoubey, B. (2007). Brain-computer interfaces in the continuum of consciousness. *Curr Opin Neurol, 20*(6), 643-649.

Kübler, A., Nijboer, F., Mellinger, J., Vaughan, T. M., Pawelzik, H., Schalk, G., et al. (2005). Patients with ALS can use sensorimotor rhythms to operate a brain-computer interface. *Neurology, 64*(10), 1775-1777.

Kübler, A., Winter, S., Ludolph, A. C., Hautzinger, M., and Birbaumer, N. (2005). Severity of depressive symptoms and quality of life in patients with amyotrophic lateral sclerosis. *Neurorehabil Neural Repair, 19*(3), 182-193.

Laureys, S. (2005). The neural correlate of (un)awareness: lessons from the vegetative state. *TRENDS in Cognitive Science, 9*(12), 556-559.

Laureys, S., Pellas, F., Van Eeckhout, P., Ghorbel, S., Schnakers, C., Perrin, F., et al. (2005). The locked-in syndrome : what is it like to be conscious but paralyzed and voiceless? *Prog Brain Res, 150*, 495-511.

Logothetis, N. K., Pauls, J., Augath, M., Trinath, T., and Oeltermann, A. (2001). Neurophysiological investigation of the basis of the fMRI signal. *Nature, 412*(6843), 150-157.

Lulé, D., Häcker, S., Ludolph, A., Birbaumer, N., and Kübler, A. (2008). Depression and quality of life in patients with amyotrophic lateral sclerosis. *Dtsch Arztebl Int, 105*(23), 397-403.

Monti, M. M., Vanhaudenhuyse, A., Coleman, M. R., Boly, M., Pickard, J. D., Tshibanda, L., et al. (2010). Willful Modulation of Brain Activity in Disorders of Consciousness. *New England Journal of Medicine, 362*(7), 579-589.

Murguialday, A. R., Hill, J., Bensch, M., Martens, S., Halder, S., Nijboer, F., et al. (2010). Transition from the locked in to the completely locked-in state: A physiological analysis. *Clin Neurophysiol., 122*(5), 925-933.

Naito, M., Michioka, Y., Ozawa, K., Ito, Y., Kiguchi, M., and Kanazawa, T. (2007). A communication means for totally locked-in ALS patients based on changes in cerebral blood volume measured with near-infrared light. *IEICE Trans Inf Syst E90-D (7),* 1028-1037.

Neuper, C., Müller, G. R., Kübler, A., Birbaumer, N., and Pfurtscheller, G. (2003). Clinical application of an EEG-based brain-computer interface: a case study in a patient with severe motor impairment. *Clin Neurophysiol, 114*(3), 399-409.

Perrin, F., Schnakers, C., Schabus, M., Degueldre, C., Goldman, S., Bredart, S., et al. (2006). Brain response to one's own name in vegetative state, minimally conscious state, and locked-in syndrome. *Arch Neurol, 63*(4), 562-569.

Pfurtscheller, G., Müller, G. R., Pfurtscheller, J., Gerner, H. J., and Rupp, R. (2003). 'Thought' - control of functional electrical stimulation to restore hand grasp in a patient with tetraplegia. *Neurosci Lett, 351*(1), 33-36.

Plotkin, A., Sela, L., Weissbrod, A., Kahana, R., Haviv, L., Yeshurun, Y., et al. (2010). Sniffing enables communication and environmental control for the severely disabled. *Proceedings of the National Academy of Sciences, 107*(32), 14413-14418.

Plum, F., and Posner, J. B. (1983). *The diagnosis of stupor and coma.* Philadelphia, PA: Davis, F.A.

Ruiz, S., Lee, S., Soekadar, S., Caria, A., Veit, R., Kircher, T., et al. (submitted). Acquired control of insula cortex modulates emotion recognition in schizophrenia.

The Multi-Society Task Force on PVS. (1994). Medical aspects of the persistent vegetative state (1). *N Engl J Med, 330*(21), 1499-1508.

Schnakers, C., Perrin, F., Schabus, M., Majerus, S., Ledoux, D., Damas, P., et al. (2008). Voluntary brain processing in disorders of consciousness. *Neurology, 71*(20), 1614-1620.

Schnakers, C., Vanhaudenhuyse, A., Giacino, J., Ventura, M., Boly, M., Majerus, S., et al. (2009). Diagnostic accuracy of the vegetative and minimally conscious state: clinical consensus versus standardized neurobehavioral assessment. *BMC Neurol, 9*, 35.

Sitaram, R., Zhang, H., Guan, C., Thulasidas, M., Hoshi, Y., Ishikawa, A., et al. (2007). Temporal classification of multichannel near-infrared spectroscopy signals of motor imagery for developing a brain-computer interface. *Neuroimage, 34*(4), 1416-1427.

Vaitl, D., Birbaumer, N., Gruzelier, J., Jamieson, G. A., Kotchoubey, B., Kubler, A., et al. (2005). Psychobiology of altered states of consciousness. *Psychological Bulletin, 131*(1), 98-127.

Weiskopf, N., Veit, R., Erb, M., Mathiak, K., Grodd, W., Goebel, R., et al. (2003). Physiological self-regulation of regional brain activity using real-time functional magnetic resonance imaging (fMRI): methodology and exemplary data. *Neuroimage, 19*(3), 577-586.

Wilhelm, B., Jordan, M., and Birbaumer, N. (2006). Communication in locked-in syndrome: effects of imagery on salivary pH. *Neurology, 67*, 534-535.

Wolpaw, J. R., Birbaumer, N., McFarland, D. J., Pfurtscheller, G., and Vaughan, T. M. (2002). Brain-computer interfaces for communication and control. *Clin Neurophysiol, 113*(6), 767-791.

Zeman, A. (2005). What in the world is consciousness? *Prog Brain Res, 150*, 1-10.

In: Consciousness: Its Nature and Functions
Editors: Shulamith Kreitler and Oded Maimon ISBN 978-1-62081-096-5

Chapter 17

WHAT HAS TMS TAUGHT US ABOUT THE ROLE OF V1 IN CONSCIOUS AND UNCONSCIOUS PROCESSING?

Dominique Lamy*[*] *and Ziv Peremen
Tel Aviv University, Ramat Aviv, Tel Aviv, Israel

ABSTRACT

There exist different theories of the neural mechanisms underlying visual awareness about the role of the primary visual or striate cortex (V1) in conscious and unconscious vision. Transcranial magnetic stimulation (TMS), a technique capable of disrupting cortical activity transiently and reversibly, has been increasingly used to investigate whether V1 activity is necessary for conscious and unconscious perception. The chapter provides a critical review of the different experimental procedures that have been applied in exploring this issue using TMS, summarizes the findings that they have generated and highlight directions for future research.

Keywords: visual awareness, conscious perception, unconscious perception, transcranial magnetic stimulation (TMS), striate cortex, V1, occipital cortex

Despite the subjective impression that we consciously perceive most of the details composing a visual scene in front of our eyes, only a very small portion of the multitude of stimuli that impinge on our retina at a given moment actually reaches our awareness. Thus, only some - but not the whole of the neural activity that is triggered by visual stimulation correlates with conscious experience. While much is known about how visual information is transferred from the retina to different regions of the visual cortex, the neural basis of visual awareness remains controversial. One important question concerns the role of the primary visual or striate cortex (V1) in visual awareness. In the present chapter we review the theoretical approaches to this issue and attempt to determine to what extent recent research

[*] Address correspondence to: Prof. Dominique Lamy, Department of Psychology, Tel Aviv University, Ramat Aviv, POB 39040. Tel Aviv 69978 Israel. Phone: 972 3 6409291. Email: domi@post.tau.ac.il

using Transcranial Magnetic Stimulation (TMS) has yielded significant advances in testing these contrasting theoretical views.

THEORETICAL APPROACHES ABOUT THE ROLE OF V1 IN CONSCIOUS VISION

The primary visual cortex provides the visual input for most of the information that reaches other regions of the cerebral cortex. However, there is a lot of evidence which shows that there exist additional pathways which bypass V1 and project directly to extrastriate visual areas (e.g., Milner and Goodale, 2006). In this context, the debate concerning the contribution of V1 to conscious vision focuses on the question whether it is *only* information that reaches higher visual cortical areas *through V1* that can access visual awareness.

Some models suggest that V1 does not hold any special status in regard to conscious vision. For instance, Zeki and colleagues (e.g., Zeki and and Bartels, 1999) proposed a "micro-consciousness" theory of visual awareness, according to which neural activity in extrastriate areas specialized in the representation of specific features, such as motion or color is necessary and sufficient for visual awareness of these features. A different view is based on the premise that all conscious information must be reportable and available for planning of voluntary action. Accordingly, Crick and Koch (1995) suggested that V1 cannot contribute directly to visual awareness because neurons in V1 have no direct projections in the prefrontal cortex. Therefore, models of this kind predict that conscious vision is possible when extra-striate areas – and not necessarily V1 - operate.

In contrast, other models argue that V1 is necessarily involved in conscious vision. Milner and Goodale (e.g., 2006) suggest a distinction between two pathways serving different functions in visual processing. One is the ventral pathway which mediates "vision for perception" and is important for object recognition and conscious visual perception. The other is the dorsal pathway which mediates "vision for action" and controls visually guided actions. The latter does not require conscious perception and is considered to be responsible for unconscious action priming. Thus, according to this model, conscious perception cannot take place in the absence of primary visual cortex, because without its involvement visual information cannot reach extrastriate areas of the ventral pathway. However, under these conditions visually guided unconscious action priming can still take place because the visual pathways that bypass V1 reach the dorsal cortex.

The models that emphasize most the role of activity in V1 in regard to conscious perception and visual awareness highlight the distinction between feed-forward and recurrent processing in cerebral cortex. After stimulus onset, visual information propagates from lower to higher areas of the visual hierarchy. As soon as a region has been activated, recurrent interactions are initiated between neurons within that region and neurons that have been activated earlier in lower-level regions. Several authors claim that while feedforward processing suffices for unconscious processes, conscious vision is contingent on recurrent connections between posterior extrastriate areas and V1 (e.g., Lamme and Roefselma, 2000). Accordingly, V1 is assumed to play different roles at different stages of visual processing. Early on, information flows through V1 but does not elicit conscious perception, whereas information that reaches V1 from higher areas during recurrent processing is consciously perceived.

MEASURES OF CONSCIOUS AND UNCONSCIOUS PERCEPTION

In order to assess TMS evidence about the role of V1 in conscious processing, it is important to distinguish between different measures of conscious and unconscious processing. The question of what measure should be used to determine whether a subject is aware or unaware of a stimulus has been the focus of an intense yet unresolved debate for decades (e.g., Eriksen, 1960).

Objective measures of awareness are based on the observers' forced-choice decisions regarding different stimulus states. The subjects are required to respond in each trial, and when they do not see the stimulus they are instructed to guess. Conscious perception is considered as having failed when the observer's forced-choice performance is at chance level. Unconscious processing is said to occur when indirect measures of visual processing indicate that the unseen stimulus nonetheless affects behavior, for instance, when response to a target (e.g., an arrow pointing either to the left or to the right) is influenced by its similarity to an invisible prime (e.g., an arrow pointing to the same vs. a different direction). According to the objective-measure approach, subjective reports are likely to be contaminated by response biases. Thus, in the case of ambiguous signals, some subjects may be ready to report having seen the critical stimulus, whereas others may be reluctant to do so, although their conscious experience of the stimulus may in fact be similar.

Subjective measures of awareness are based on the subjects' self-reports of their conscious experiences. The subjects are typically required to press a key when they are perceptually aware of a given target and when they do not see the stimulus they are instructed not to guess. In order to avoid the problems associated with forcing the subjects into a binary classification of their experience, more sensitive scales have been devised in recent years, for example, the subject may be required to rate the visibility of the stimulus from 1 to 4. According to the subjective-measure approach, only the subjects themselves have access to their inner states through introspection, and their subjective report is the only valid marker of their conscious experience. Unconscious processing is said to occur either when objective forced-choice performance is above chance or when indirect measures of visual processing indicate that the stimulus reported as unseen affects behavior.

Accordingly, there are two different measures of unconscious perception: action priming in the absence of objective awareness and above-chance objective performance in the absence of subjective awareness. These two kinds of measures map onto two different approaches: the first suggests that identification and discrimination are necessarily conscious, whereas only perception for action can be unconscious (e.g., Milner and Goodale, 2006); the second proposes that the boundary between conscious and unconscious perception is not related to the goals of perception but rather to the stages of processing at which they occur, so that information is processed unconsciously at early stages, while a small portion of it is further processed and becomes perceived consciously at later stages.

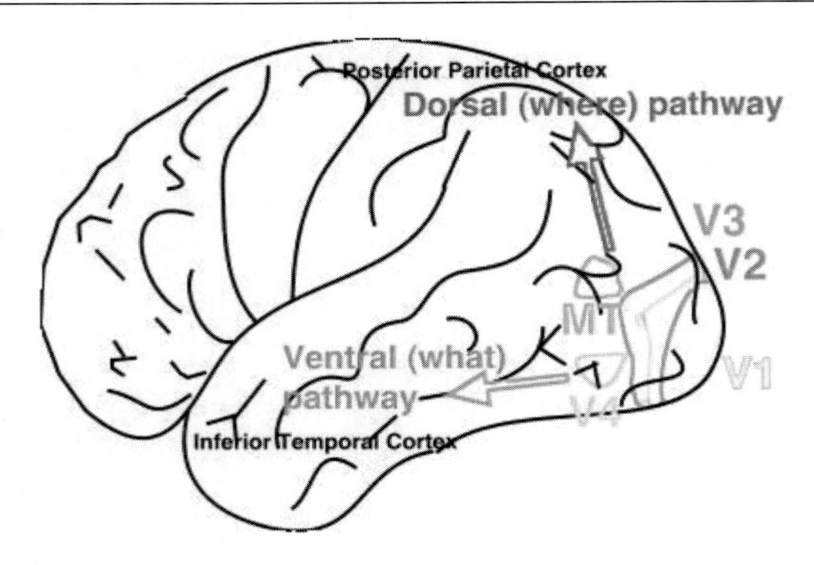

Figure 1.

The issue of the involvement of V1 in conscious perception cannot be directly addressed by brain imaging methods because these are intrinsically correlational. In contrast, the "blindsight" phenomenon, which refers to dissociations between conscious and unconscious vision observed in neuropsychological patients with occipital cortical damage, is often assumed to reflect the necessity of V1 for conscious vision. Patients with blindsight report no phenomenological awareness of stimuli presented in the region of their visual field corresponding to the damaged part of V1. They verbally answer "no" when asked whether they see a visual stimulus. Yet, they show consistent and sometimes impressive ability to discriminate and localize the stimuli to which they report being blind. When required to guess using forced-choice procedures, they are able to make eye movements and manual responses to the stimuli they deny seeing as well as discriminate relatively fine differences in orientation, wavelength or motion direction (see Cowey, 2010, for a review). Further, when tested on subsequent probes related to the unseen stimuli, they show priming effects (e.g., Kentridge, Heywood and Weizkrantz, 1999). These findings are interpreted as indicating that V1 is necessary for conscious vision but not for unconscious processing.

Transcranial Magnetic Stimulation

Criticism has been leveled against the interpretation of blindsight as a deficit of conscious vision as well as against the validity of inferences drawn from this condition for normal vision (e.g., see Cowey, 2010 for a review). For instance, in view of the observation that unconscious processing performance improves remarkably over the years after brain damage, it is possible to argue that such performance results from reorganization of the brain and does not serve as evidence that perceptual processing is possible without V1.

In recent years, Transcranial Magnetic Stimulation (TMS) has proved increasingly to be a useful tool for studying the role of various brain regions in information processing. The application of single-pulse (or double-pulse) TMS to the scalp can transiently disrupt the activity in the underlying cortex, thereby creating short-lived virtual lesions. Applying TMS at varying times during a specific task, makes it possible to learn where cortical activity is necessary for some function and in what specific time windows. This method seems to have the potential to overcome most of the problems associated with blindsight research.

Since the pioneering work of Amassian et al. (1989), there has been a bulk of research using TMS for studying the role of V1 in conscious and unconscious perception. In Amassian et al.'s study, participants viewed non-word trigrams and were required to identify the presented letters. A strong TMS pulse was applied over the occipital cortex with a stimulus onset asynchrony (SOA), that is, an interval between visual stimulus onset and TMS pulse, varying from 0 ms to 200 ms after letter presentation. The average number of identified letters was dramatically reduced for SOAs in the range of 60 and 120 ms.

From that point on, TMS studies of the role of V1 can be broadly divided into two categories. One line of research, which directly follows on the work of Amassian et al. (1989), has investigated the time course of the effect of applying TMS to the occipital cortex. These studies typically tested the issue of whether disruption of V1 activity interferes with visual performance at one or at several time windows within the tested range of stimulus-onset to TMS pulse intervals (henceforth, SOAs). A different line of research used only a narrow range of SOAs, known to be associated with maximal impairments of visual discrimination performance, and investigated which types of unconscious processes are affected and which are left intact under these conditions.

INVESTIGATIONS OF THE TIME-COURSE OF TMS EFFECTS

In studies investigating the time course of TMS effects a TMS pulse is given on the occipital cortex at various times following stimulus onset. The precise timing at which TMS impairs visual performance indicates at what stage(s) of visual processing the activity of the striate cortex is necessary for the task at hand. Early impairment is taken to indicate that visual processing in V1 during the initial volley of neural activity (of feed-forward sweep, e.g., Lamme and Roefselma, 2000) is necessary for conscious perception. Impairment in later stages is taken to indicate that recurrent processing from higher cortical regions to V1 is necessary for conscious perception.

Time-course of TMS Effects with Forced-choice Discrimination (Objective Awareness) Measures

All studies except one (Silvanto, Lavie and Walsh, 2005) measured visual performance with a forced-choice discrimination task, which could involve letter identification (e.g., Corthout, Hallett and Cowey, 2003), motion direction (Koivisto, Mantyl and Silvanto, 2010), arrow direction (Sack, van der Mark et al., 2009; Koivisto, Raikiand and Salminen-Vaparanta, 2010), line orientation (Koivisto et al., 2010b), color discrimination in chromatic and achromatic stimuli (Paulus et al.,1999), Vernier displacement (Kammer, Scharnowski and Herzog, 2003) or bird vs. mammal discrimination (Camprodon, Zohary, Brodbeckand and Pascual-Leone, 2010).

A TMS-induced impairment (or "dip") was typically reported when the TMS pulse was applied around 80 to100 ms after stimulus onset (but see Koivisto et al., 2010a for an exception). In some studies also a second dip was reported, but the SOA at which it occurred varied considerably between studies. In Corthout et al. (2003) the second dip occurred earlier, at a 20ms SOA. However, this finding was replicated in some but not all, of this group's other work, (e.g., Corthout et al., 2000). A similar finding was reported by Paulus et al. (1999) who found an early dip at 0-45 ms SOAs, but only for discrimination of achromatic stimuli, and by Kammer et al. (2003) who found an early dip at 30-ms SOA for one out of the four participants tested. By contrast, Camprodon et al. (2010) reported only a later dip at 220 ms.

These findings were interpreted as showing that the first dip reflects early processing of the stimulus in V1 during what is referred to as the feed-forward sweep, whereas the second dip reflects recurrent processing from higher cortical areas to V1. Accordingly, the studies in which an early dip was reported, interpreted the massively replicated 100-ms dip as reflecting reentrant processing, whereas in a study (Camprodon et al., 2010), in which a later dip was reported, it was taken to index initial processing in V1.

In attempting to make sense of these findings, several considerations are in order. First, while the 20-40 ms SOA was tested in most of the above cited studies, it was observed only in a small minority of them. By contrast, Camprodon et al.'s (2010) study is the only one in which SOAs longer than 200 ms were tested. It will therefore be important that further research should include a larger range of SOAs.

Second, the type of stimuli used seems to have a profound impact on the timing of TMS effects. The dip at the latest SOAs was reported when the task involved complex veridical stimuli, whereas dips at the earlier SOAs were reported with much simpler tasks (e.g., color discrimination of achromatic stimuli or single letter discrimination). Further, systematic investigations of the effect of stimulus contrast on TMS suppression (e.g., Kammer et al., 2005) showed strong modulations, with earlier suppression for lower contrast levels. Similar variability has been reported in studies investigating the latency of the activation of primary cortex using evoked potentials in humans or single-cell recordings in monkeys. The exact speed at which V1 is first activated was shown to depend on stimulus variables, such as contrast and motion (Lamme and Roefselma, 2000). Finally, a similar influence of TMS intensities was reported, with suppression occurring earlier as TMS intensity is increased (e.g., Kammer et al., 2005).

Taken together, these findings suggest that subtle procedural differences can dramatically affect the latency of TMS effects, thereby precluding meaningful between-experiments comparisons of the exact timing of TMS-induced impairment of forced-choice performance. It may thus well be the case that the 100-ms dip reflects recurrent processing in Corthout et al.'s (2003) study and feedforward processing in Camprodon et al.'s study (2010).

Therefore, instead of focusing on precise timing, it may be more fruitful to determine whether one or two dips are observed. The answer to this question awaits further replications with similar task parameters of the few findings pointing to the existence of two dips.

Time-course of TMS Effects: Measuring Conscious vs. Unconscious Perception

In order to investigate the role of V1 in *conscious* vision, one has to ensure (a) that TMS affects measures of visual processing associated with conscious vision and (b) that such effects are specific to conscious vision, that is, are different from the effects of the same TMS stimulation on measures associated with unconscious perception.

As was reviewed in the previous section, most of the studies used forced-choice discrimination performance in order to measure the effects of TMS applied to occipital cortex. However, the status of such objective measures of perception is controversial: while some authors stipulate that forced-choice objective identification performance is a measure of conscious perception (e.g., Sack et al., 2009), others claim that above-chance objective performance in the absence of awareness is a valid marker of unconscious perception (e.g., Koivisto et al., 2010a, 2010b). Obviously, the interpretation of the existing findings is heavily dependent on the theoretical stance adopted in regard to the appropriate measure of conscious perception.

Time-course of TMS Effects: Subjective vs. Objective Measures of Perception

Very few studies have compared subjective and objective measures of visual perception in order to determine whether feedback activity in V1 is necessary specifically for conscious vision or characterizes both conscious and unconscious vision. For instance, Koivisto et al. (2010a) required subjects to rate their experience of the motion of briefly presented stimuli (subjective measure) and in addition, to make forced-choice discrimination judgments regarding the direction of the same moving stimuli (objective measure). They reasoned that "if the 'late' period of V1 activity is unique to visual awareness, the application of TMS during this stage should weaken subjective experience of motion but leave forced-choice discrimination performance on unaware trials unaffected. In contrast, if the 'late' V1 activity reflects a general principle of visual cortical information processing, TMS should disrupt performance in both tasks" (p. 829). They followed a similar rationale in a later study (Koivisto et al., 2010b) in which participants were requested to judge either the orientation of a line or the direction of an arrow.

Their results did not support the predictions of the recurrent processing model of V1 involvement in conscious vision: they found both objective and subjective measures of perception to be affected by TMS for late SOAs. Koivisto et al. (2010a, 2010b) concluded that recurrent activation of early visual cortex is not specific to visual awareness but may be needed in both conscious and unconscious visual processes. Notably, these findings are not entirely surprising in view of the results of studies in which only objective performance was measured (see the previous section of this chapter entitled "*Time-course of TMS effects with forced-choice discrimination measures*"). Indeed, some studies reported both an early and a late dip with forced-choice identification measures (e.g., Corthout et al., 2003) and others reported one dip at an SOA which overlapped the time of recurrent processing to V1 (e.g., Amassian, 1989). In other words, these studies also showed that forced-choice identification accuracy is impaired in the 'late' period of V1.

Do these findings provide compelling evidence that V1 is required for both unconscious and conscious processing? We think not. First, this conclusion is contingent on accepting the premise that objective measures index only unconscious perception, which as has been stated before, remains controversial. However, even if this premise is accepted, several points should be taken into consideration. First, although objective performance can be above chance when subjects deny any conscious perception of the stimulus, it is typically much poorer than when participants report being aware of the stimulus (e.g, Lamy, Salti and Bar-Haim, 2009). It is therefore not surprising to find impaired objective performance when it is measured irrespective of the subjects' reported awareness (Koivisto et al., 2010b), because conscious perception contributes to forced-choice identification performance.

Second, when this problem was avoided and objective performance on only "unaware" trials was considered (Koivisto et al., 2010a), only very small decrements in performance were reported (about 5-6%) which is very remote from the typically reported decrements (e.g., from 16% in Sack et al., 2009 to 40-70% in Corthout et al., 2003). Thus, given the scarcity of studies using this rationale and the weakness of the results, it will be important to await further investigations prior to reaching a conclusion concerning this issue.

Time-course of TMS Effects: Perception for Identification vs. Perception for Action

Only one study (Sack et al., 2009) investigated the time course of TMS effects by contrasting direct measures of perception (subjective or objective) with indirect measures (unconscious priming). An arrow pointing either to the right of to the left was briefly presented and was followed by a probe (the target), also an arrow pointing to the right or left. A preliminary experiment showed that as the time interval between prime onset and probe onset increased, compatibility effects, that is, the RT advantage when the probe pointed in the same vs. a different direction relative to the prime, increased, while direction judgment accuracy remained stable and close to ceiling. Then, single-pulse TMS was applied to the occipital cortex in two separate experiments. Forced-choice identification performance (considered as indexing conscious performance) was measured in one experiment and compatibility effects (considered as indexing unconscious processing) in the other experiment, as a function of stimulus-TMS SOAs, which varied between 20 and 120 ms.

The results showed that TMS over occipital cortex caused a disruption of conscious prime recognition only when applied at 80 ms after prime stimulus onset. Compatibility effects were also reduced with TMS, in the same – albeit wider – time window (60-100 ms). The authors concluded that primary visual cortex is functionally relevant for both conscious perception and perception for action (or behavioral priming) at approximately the same temporal stage during visual information processing.

However, these findings do not indicate whether V1 is *necessary* for action priming. Models that distinguish between perception for action and perception for identification (e.g., Milner and Goodale, 1995) stipulate that processing of visual information for action takes place in the dorsal pathway. Yet, this pathway involves both a route that goes through V1 and a route that bypasses it. Therefore, the findings by Sack et al. (2009) do not challenge the model. In order to provide a valid test for the model, one would have to examine compatibility effects when subjects reported not being aware of the prime: if such effects were completely wiped out, then one could conclude that V1 is necessary for action priming. Unfortunately, as conscious perception and action priming were measured in different experiments, it was not possible to examine priming effects separately for trials in which the subjects were aware of the prime and in trials in which they were unaware of it.

In addition, if the impairment measured in Sack et al.'s (2009) study occurred during the feed forward sweep (as assumed by the authors), the findings also would not challenge the feed-back to V1 hypothesis: this model predicts that both conscious and unconscious processing should be impaired during feed forward processing, which is what Sack et al. reported. The model would also predict that only the conscious measure should be affected at a later SOA, but the study included only SOAs up to 120 ms.

To conclude, in order to determine at what stages of processing V1 is necessary for conscious or for unconscious perception, if at all, one should use different measures of perception (objective, subjective and indirect measures) concomitantly. However, the few studies that have adopted this approach so far did not yield conclusive findings.

Time-course of TMS Effects in Higher Areas

The studies reviewed in the previous section sought to determine whether recurrent processing is necessary for conscious perception by drawing inferences from the time at which disruption of V1 activity maximally impairs conscious perception. A different approach consists in applying TMS to both V1 and higher cortical areas (usually V5/MT) in order to determine whether it is specifically recurrent processing to V1 that is necessary for conscious vision.

Pascual-Leone and Walsh (2001) induced conscious perception of moving flashes of light by stimulating V5/MT using TMS, thereby creating what is referred to as “phosphenes”. This procedure ensured that perception did not result from retinal stimulation and did not involve feed-forward processing through V1. When a subsequent TMS pulse, that was not intense enough to produce a phosphene on its own, was delivered over V1 5-45 ms after V5/MTstimulation, stationary rather than moving phosphenes were perceived. The authors concluded that experiencing moving phosphenes depends not only on activation of MT/V5 but also specifically on a recurrent feedback loop to V1.

In a later study, Silvanto et al. (2005) attempted to make a similar point, using real motion rather than artificially induced perception of motion. They argued that Pascual-Leone and Walsh’s (2001) findings might not hold when external stimulation followed by feed-forward processing through V1 are involved in the sequence of neural events that eventually leads to conscious perception. They administered TMS to V1 or V5/MT in various time intervals from stimulus offset in the course of a simple motion detection task. They reported two time windows during which V1 stimulation impaired conscious perception. The early V1 critical period predated the time window during which V5/MT stimulation impaired conscious perception and the later V1 critical period postdated it. Thus, these findings demonstrated the importance of back-projections from V5/MT to V1 in awareness of real motion stimuli (see also Koivisto et al., 2010a).

Time-course of TMS Effects: Studies Involving Scotoma Descriptions

The typical method used in the studies reviewed above involves briefly flashing a very small stimulus in an empty field, using a variety of procedures in order to ensure that the stimulated area of the occipital cortex corresponds to the area of the visual field where the stimulus is presented. Performance in making perceptual judgments regarding the stimulus when TMS is applied is compared to a control condition (no TMS, sham TMS or TMS in a different region). It follows that, using this method, when the visual stimulus location and TMS do not overlap due to small displacement of the TMS coil, the TMS effects will be missed.

A different method that circumvents this problem and provides a more detailed description of the causal effects of TMS on visual phenomenology consists in using a large stimulus that extends considerably beyond the receptive field of the TMS stimulated site (e.g., Kamitani and Shimojo, 1999). The subjects typically report a scotoma, that is, a region missing from the presented visual pattern, and can be asked to specify its location and extent. Thus, such studies pertain to TMS effects on subjective perception. The results characteristically show scotomas, the specific locations of which correspond to the stimulated site, with peak effects between 80 and 120 ms from stimulus onset. However, unlike target identification studies in which relatively narrow suppression intervals were reported, results from the scotoma literature indicate that phenomenal vision is affected by TMS for extended SOAs (from 58 to 222 ms, Murd, Luiga, Kreegipuu and Bachmann, 2010). One may speculate that while this methodology is sensitive enough to detect subtle changes in the appearance of the stimulus following TMS application, such changes may not be strong enough to impair stimulus identification. Further studies combining the two approaches are needed to test this hypothesis and resolve the contradictions between the two lines of research.

Testing the Role of V1 in Unconscious Perception within a Narrow Time Window

By contrast with studies in which the time interval between stimulus onset and TMS pulse was systematically varied, several studies used only a narrow range of SOAs known to maximally impair visual discrimination performance. They investigated which types of unconscious processes are affected and which are left intact (e.g., Boyer, Harrison & Ro, 2005; Christensen, Kristiansen, Rowe and Nielsen, 2008; Ro, 2008; Ro, Shelton, Lee and Chang, 2004).

For instance, Boyer et al. (2005) used stimuli which could vary either in color (Exp.1) or in orientation (Exp.2). Single-pulse TMS was applied to occipital cortex approximately 100 ms after stimulus onset (100, 114 or 128 ms). Participants had to first report whether or not they were aware of the target's critical property (i.e., either its color or its orientation); then they made a forced-choice discrimination response, and finally rated their confidence in their forced-choice response on a continuous scale (0 to 9). On trials in which the subjects reported being unaware of the target, their forced-choice discrimination performance regarding the target color and orientation were nonetheless above chance. Similar findings were reported with saccade (Ro et al., 2004) and reaching (Ro, 2008) responses as well as with motion correction (Christensen et al., 2008). Boyer et al. (2005) concluded that "these finding suggest that a visual pathway that bypasses V1 must be involved with unconscious vision... direct retinogeniculate–extrastiate pathway to V4 is most likely responsible for our findings of residual unconscious visual processing seen in the absence of normal primary visual cortex functioning " (p.168-178).

This conclusion [in Boyer et al.'s (2005) study as well as in the other studies following the same rationale] is unwarranted for several reasons. First and foremost, as V1 activity was disrupted only during a narrow time window, above-chance discrimination performance may have been mediated by earlier activity in V1 - that was left undisturbed for as long as 100 ms after stimulus presentation! Thus, the findings do not indicate in any way that a "visual pathway that bypasses V1 must be involved with unconscious vision". In addition, as TMS application does not eliminate activity in V1, the level of disruption achieved in these studies may have been sufficient for subjects to report not seeing the target but insufficient to completely eliminate correct forced-choice identification, a measure that is known to be more sensitive than binary yes/no reports of conscious awareness. Finally, and in line with the latter claim, Boyer et al. (2005) reported that forced-choice performance accuracy was significantly above chance for confidence ratings of 4–9, but not for ratings between1 and 3.Confidence judgments are typically taken to index subjective awareness. Thus, if anything, Boyer et al.'s findings demonstrated that there was no unconscious processing when conscious awareness was completely eliminated.

In conclusion, findings of above-chance performance on an objective measure when subjects are rendered subjectively unaware of a target by applying TMS in a narrow (and relatively late) time window cannot be taken as evidence that such unconscious processing is mediated by a neural pathway that does not involve V1. One can only infer that unconscious processing is possible – and in this case, TMS should be considered only as a means to eliminate conscious awareness, irrespective of where or when it was applied.

CONCLUSIONS

Transcranial Magnetic Stimulation is a unique tool because, unlike brain imaging methods, such as fMRI (functional Magnetic Resonance Imagery) or ERPs (Event-Related Potentials), it allows inferences about the necessity of specific regions at a specific time for a specific task. The foregoing review points to two main conclusions.

From a theoretical standpoint, TMS research has as yet only moderately advanced our understanding of the neural processes underlying conscious vision. On the one hand, the findings suggest that recurrent processing from extrastriate to striate cortex seems to play a role in conscious vision, in line with the view advocated by Lamme and colleagues (e.g., Lamme and Roelfsema, 1999). However, whether this role is specific to conscious vision or also pertains to unconscious vision remains unclear. On the other hand, most of the TMS research has aimed at testing the recurrent-processing to V1 hypothesis. Little effort has so far been devoted to testing competing accounts (e.g., Zeki and Bartels' micro-consciousness account, 1999; Crick and Koch's, 1995, account linking consciousness with connectivity with prefrontal cortex). Further, TMS tests of Milner and Goodale's (2006) vision for perception vs. vision for action with respect to the role of V1 is characterized by serious methodological flaws.

From a methodological viewpoint, the gaps in the literature that preclude at this stage of TMS research clear-cut conclusions about conscious vision are due to two main reasons: first, conflicting operational definitions of conscious and unconscious perception used in the different studies, and second, the great sensitivity of TMS effects to stimulus and TMS stimulation parameters. In order to take the best advantage of the research opportunities that

TMS affords, it is advisable that future research pays special consideration to the following four conditions: (a) stimulus to TMS SOAs is to be systematically manipulated at the finest possible temporal resolution; (b) different measures of visual performance (subjective, objective and indirect) are to be used in the same experiment and with the same participants; (c) numerous replications are to be conducted with special attention to stimulus parameters, such as contrast and luminance as well as TMS intensity; and (d) tasks involving target processing and scotoma description are to be combined within the same protocol.

REFERENCES

Amassian, V. E., Cracco, R. Q., Maccabee, P. J., Cracco, J. B., Rudell, A., and Eberle, L. (1989). Suppression of visual perception by magnetic coil stimulation of human occipital cortex.*Electroencephalography and clinical neurophysiology*, *74*, 458-462.

Boyer, J. L., Harrison, S., and Ro, T. (2005). Unconscious processing of orientation and color without primary visual cortex. *Proceedings of the National Academy of Sciences of the United States of America, 102*, 16875–16879.

Camprodon, J. a, Zohary, E., Brodbeck, V., and Pascual-Leone, A. (2010).Two phases of V1 activity for visual recognition of natural images. *Journal of cognitive neuroscience*, *22*, 1262-1269.

Christensen, M. S., Kristiansen, L., Rowe, J. B., and Nielsen, J. B. (2008). Action blindsight in healthy subjects after transcranial magnetic stimulation. *Proceedings of the National Academy of Sciences of the United States of America, 105*, 1353–1357.

Corthout, E., Hallett, M., and Cowey, A. (2003). Interference with vision by TMS over the occipital pole: a fourth period. *Neuroreport*, *14*, 651-655.

Corthout, E., Uttl, B., Juan, C. H., Hallett, M., and Cowey, A. (2000). Suppression of vision by transcranial magnetic stimulation: a third mechanism. *Neuroreport*, *11,* 2345-2349.

Cowey, A. (2010). The blindsight saga.*Experimental brain research*, *200*), 3-24.

Crick, F. and Koch, C. (1995). Are we aware of neural activity in primary visual cortex? *Nature*, *375*(6527), 121–123.

Eriksen, C. W. (1960). Discrimination and learning without awareness: A methodological survey and evaluation. *Psychological Review*, *67,* 279- 300.

Goodale, M. A. and Milner, A. D. (1992). Separate visual pathways for perception and action. *Trends in neurosciences*, *15*, 20-25.

Kamitani, Y. and Shimojo, S., (1999). Manifestation of scotomas created by transcranialmagnetic stimulation of human visual cortex. *Nature Neuroscience, 2,* 767–771.

Kammer, T., Puls, K., Strasburger, H., Hill, N.J., and Wichmann, F.A. (2005). Transcranial magnetic stimulation in the visual system – I. The psychophysics of visual suppression. *Experimental Brain Research, 160*, 118-128.

Kammer, T., Scharnowski, F., and Herzog, M. H. (2003). Combining backward masking and transcranial magnetic stimulation in human observers. *Neuroscience Letters*, *343*, 171-174.

Kentridge, R. W., Heywood, C. A., and Weiskrantz, L. (1999). Attention without awareness in blindsight. *Proceedings. Biological sciences / The Royal Society*, *266*(1430), 1805-1811.

Koivisto, M., Mäntylä, T., and Silvanto, J. (2010a).The role of early visual cortex (V1/V2) in conscious and unconscious visual perception. *NeuroImage*, *51*(2), 828-834.

Koivisto, M., Railo, H., and Salminen-Vaparanta, N. (2010b).Transcranial magnetic stimulation of early visual cortex interferes with subjective visual awareness and objective forced-choice performance. *Consciousness and Cognition*, 1-11.

Lamme, V. A. and Roelfsema, P. R. (2000). The distinct modes of vision offered by feedforward and recurrent processing. *Trends in neurosciences*, *23*, 571-579.

Lamy, D., Salti, M., and Bar-Haim, Y. (2009). The neural correlates of conscious and unconscious perception: an ERP study. *Journal of Cognitive Neuroscience, 21*, 1435–1446.

Milner, A.D. and Goodale, M.A. (2006). *The visual brain in action*. New York: Oxford University Press.

Murd, C., Luiga, I., Kreegipuu K., and Bachmann, T. (2010). Scotomas induced by multiple, spatially invariant TMS pulses have stable size andsubjective contrast. *International Journal of Psychophysiology, 77,*157–165

Paulus, W, Korinth, S, Wischer, S, and Tergau, F. (1999). Differential inhibition of chromatic and achromatic perception by transcranial magnetic stimulation of the human visual cortex.*Neuroreport, 10*, 1245–1248.

Ro, T. (2008). Unconscious vision in action.*Neuropsychologia, 46*, 379–383.

Ro, T., Shelton, D. J. M., Lee, O. L., and Chang, E. (2004). Extrageniculate mediation of unconscious vision in transcranial magnetic stimulation inducedblindsight. *Proceedings of the National Academy of Sciences, 101*, 9933–9935.

Sack, A. T., van der Mark, S., Schuhmann, T., Schwarzbach, J., and Goebel, R. (2009). Symbolic action priming relies on intact neural transmission along the retino-geniculo-striate pathway. *NeuroImage*, *44*(1), 284-293.

Silvanto, J., Lavie, N., and Walsh, V. (2005).Double dissociation of V1 and V5/MT activity in visual awareness.*Cerebral cortex*, *15*, 1736-1741.

Zeki, S., and Bartels, A. (1999).Toward a theory of visual consciousness. *Consciousness and Cognition*, *8*, 225-259.

In: Consciousness: Its Nature and Functions
Editors: Shulamith Kreitler and Oded Maimon
ISBN 978-1-62081-096-5

Chapter 18

NEUROIMAGING APPROACHES TO THE STREAM OF CONSCIOUSNESS: PROBLEMS LOST AND FOUND

M. Gruberger[1,2], E. Ben-Simon[1,3] and T. Hendler[1,2,3]
[1] Functional Brain Center, Wohl Institute for Advanced Imaging, Tel-Aviv Sourasky Medical Center, Tel Aviv, Israel
[2] Department of Psychology and
[3] Faculty of Medicine, Tel-Aviv University, Tel-Aviv, Israel

"Thinking: the talking of the soul with itself"
Plato

ABSTRACT

The experience of mind-wandering, a fundamental element of the "stream of consciousness", is among the most basic and permanent expressions of conscious awareness. It forms the narrative aspect of our self awareness and occupies nearly a half of our waking mental lives. Though the phenomenon of mind-wandering has intrigued philosophers and psychologists for centuries, its unique fleeting and subjective character did not allow for a direct scientific examination. However, recent evidence in neuroscience may possibly put forward the neural basis of this important phenomenon and open the door to promising research in the field of consciousness in general.

A decade ago the neuroscientific discovery of rest related brain activity was formulated. This newly discovered default-mode neural network (DMN) exerts higher activity levels during rest than during task performance and thus serves as a compelling candidate in the search for the neural basis of mind-wandering. In accordance, several studies have found correlations of DMN activity and subjective reports of mind-wandering. Nevertheless, accepted modus operandi in mind-wandering research is scarce, generating a bottle neck for associating it with DMN rest-related activation patterns and pinpointing its neural correlates.

The current chapter overviews existing literature from the fields of philosophy, psychology and neuroscience in order to define five methodological strategies for studying mind-wandering within a functional neuroimaging paradigm. Each strategy is further discussed in terms of its suitability for various neuroimaging and data analysis paradigms. These strategies could serve as the building blocks of mind-wandering research and advance our ability to scientifically explore this unique feature of our daily conscious experience.

I. STREAM OF CONSCIOUSNESS: THE WANDERING MIND

Conscious experience is fluid. Its dynamic nature is constantly present in our awake lives, anchored in our sense of self and in the ongoing spontaneous mentation referred to as mind-wandering (MW). Conceptualized as a core element of what James defined as the 'stream of consciousness' (James 1892), this elementary aspect of human consciousness has gained considerable attention in ancient and modern philosophy as well as in theoretical psychology.

MW is claimed to be essential for any reflective awareness to become possible (Gallagher 2000, Northoff et al., 2006). An example to this essential relationship can be found in the words of Helen Keller after she was taught sign language:

> "I did not know that I am. I lived in a world that was a no world... *My inner life, then, was a blank without past, present, or future, without hope or anticipation...* When I learned the meaning of 'I' and 'me' and found that I was something, I began to think. Then consciousness first existed for me". (Hellen Keller, 1908: quoted by Daniel Dennett 1991, pg 227)

Be it in the sense of James's 'spiritual self' (James 1892), Gallagher's 'narrative self' (Gallagher 2000), Dennett's 'non minimal self' (Dennett 1991) or Damasio's 'autobiographical self'(Damasio 1998), to state a few, the actual flow-ness of consciousness is often represented, within models depicting the human experience of self consciousness, in a module of its own, distinct both from 'lower', more basic, senses of consciousness as well as from 'higher' self-related executive functioning.

This experience of the wandering mind can be attributed certain typical characteristics: as its name implies, MW often occurs in the absence of a demanding task, i.e. mostly during rest; it is spontaneous in nature and hard to restrict or direct by the thinker; it is largely autonomous - sometimes even unaware; and it occupies a great deal of our waking mental activity. Indeed, it has been claimed that by average, we spend a nearly a half of our waking lives engaged in thoughts unrelated to the present task (Kane, Brown et al. 2007; Killingsworth and Gilbert 2010).

Alternative names to the term 'mind-wandering' (Smallwood and Schooler 2006; Mason, Norton et al. 2007) in past and recent literature include 'day-dreaming' (Giambra 1979), 'task unrelated images and thought' (Giambra and Grodsky 1989), 'stimulus independent thought' (Teasdale, Dritschel et al. 1995), 'task unrelated thought' (Smallwood, Baracaia et al. 2003), 'incidental self-processing' (Gilbert, Frith et al. 2005), 'Inner speech' (Morin 2009) and 'spontaneous thought' (Christoff, Gordon et al. 2008).

The robustness, autonomous and continual nature of this psychological process has led writers to suggest that rather than being an undesired lapse of attention to the external world (William James remarked, when he was accused of being absent-minded, that he was really just present-minded to his own thoughts (Barzun 1983)), MW must have an important adaptive value for healthy cognition (Christoff, Gordon et al. 2008; Baars 2010). Yet like the neural basis of MW, its adaptivity and the nature of its interaction with other cognitive processes remain a scientific blind spot.

II. Mind-Wandering: Past and Present Theoretical Conceptualizations

While MW is relatively new to neuroscience, it has, fortunately, gained considerable interest in philosophy and psychology over the years. Though substantial controversy may be found among the immense body of philosophical and psychological literature regarding whether physiology can fully account for the mind and the self, this literature nevertheless encompasses an invaluable resource for conceptualizing and characterizing MW. This characterization can later on facilitate the justification of operational definitions of mind-wandering within experimental paradigms.

Mind-wandering as a Form of Self-related Functioning

The concept of self is debated from the early days of psychology to today's functional neuroimaging studies. Though some writers advocate towards the existence of a core mental sense of self (Beer, 2007, Damasio, 1998, Gusnard, 2005), others, like James (1892), argue that no fixed entity exists behind the continuously ongoing self-referential processing which he called the "stream of thought". While this philosophical debate continues beyond the scope of this paper, contemporary writers (Damasio 1998; Gusnard 2005; Beer 2007) tend to agree that the 'stream of consciousness', be it based on a basal, fixed 'self' or not, is inseparable from a healthy, constant sense of self. According to this notion, mind-wandering, be its content straightforwardly related to the thinker or not, is and a self-related, self-generated, self-sustaining function (Baars 2010), an integral part of self-awareness which is a pre-requisite to healthy psychological functioning.

The conceptualization of mind-wandering as one of many forms of self-related functions produces a hypothesis for an overlap between the neural basis for self-related tasks and the neural basis of mind-wandering. This hypothesis has been translated in some studies into a rational for comparing neural activations during self-related tasks to neural activations occurring when mind-wandering is assumed to occur, as will be demonstrated in the next section.

Mind-wandering as Inverse to External Task Load

That the mind and brain can perform complex cognitive tasks has been demonstrated and investigated for nearly a century. Nevertheless, that mental functioning can be decoupled from task-related functioning, has been recognized nearly solely by theoretical psychology and philosophy. Gallagher (Gallagher 2000) and Dennett (Dennett 1991) recognize the importance of off-task self related functioning in the generation of a continuous sense of self from the actual fragments of everyday experience. According to Dennett's account, when relieved of external demands, humans use language to continuously 'tell stories', and in these stories they create what is called their selves, and extend their biological boundaries to encompass a life of meaningful experience. What is of critical notice is Dennett's recognition of the task-unrelatedness, and in fact, the automaticity with which such 'story telling' occurs. In his words: "for the most part we don't spin them (the stories, M.G.); they spin us" (p. 418).

Already familiar with current thinking of task-deactivated and task-activated neural networks, Northoff et al (Northoff, Heinzel et al. 2006) conceptualize mind-wandering as a 'psychological baseline', a form of continuous self-referential processing which is evident during non-task conditions and which ultimately forms our "continuous stream of subjective experience" or "phenomenal time" where past, present, and future are no longer divided but integrated.

Indeed, mind-wandering occurs in the absence of any external cue, is often unintended and even unaware, takes its own course – probably driven by internally generated cues, and is hard to trace back, replicate or report. These characteristics probably account for the scarceness of explicit measures of MW developed over the years. Nevertheless, the very conceptualization of MW as an inverse, or perhaps complementing, process to cognitive task performance directly magnifies its adequacy for quantified measurements and thus for scientific examination, as will be discussed in the next section.

III. Introduction to the Default Mode Network

The term "resting state" in functional neuroimaging can be defined behaviorally as a state with no demand for task performance usually achieved by eyes closed or by passively looking at a fixation point; and physiologically as the brain activity which accompanies the rest state (Raichle 2010).

The "default-mode network" (DMN) relates to a functionally meaningful rest-related neuronal circuitry, which includes the medial pre frontal cortex, the Precuneus, the posterior Cingulate cortex and the inferior Parietal and lateral Temporal cortices (these areas are portrayed in Figure 1). This network has unique patterns of activation in comparison to other functional neural networks: activation levels in this network, both in terms of energy consumption and in terms of the blood oxygen-level dependent (BOLD) signal, were shown to descend below baseline during cognitively-demanding tasks. Moreover, this network shows high activation levels at rest (non-task conditions) compared to task (Gusnard, Akbudak et al. 2001; Raichle, MacLeod et al. 2001).

The discovery of the DMN nearly a decade ago and the related findings of additional rest-related functionally-connected networks in the brain (Fox, Snyder et al. 2005) have provoked a historical paradigm shift in functional neuroimaging (Raichle 2009): from mostly task oriented exploration, which established many of cognitive neuroscience's milestones, to an ever increasing interest in the exploration of the resting state. Thus, contrary to the earlier presumption, brain activity during rest does not represent random 'white noise' acquired from the brain, but rather orderly, spatially and temporally confined, neural activity which can be claimed to be functionally meaningful. In other words, the brain is never physiologically at rest: when the brain is un-occupied by a demanding task, the DMN activation takes over.

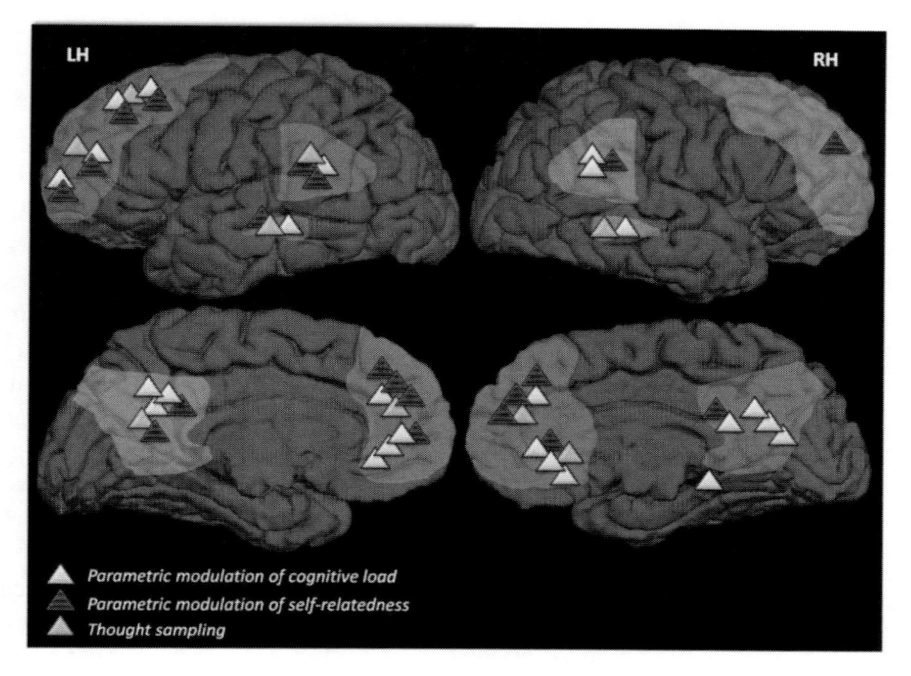

Figure 1. Results of overviewed studies in relation to DMN regions.
Results of overviewed studies, categorized by strategy, superimposed on a template brain. Yellow markings denote DMN areas (in accordance with Buckner et al., 2008) which include: dorsal and ventral medial pre frontal cortex, Lateral temporal cortices, Precuneus and posterior Cingulate cortex.

Originally proposed to be a default baseline system, a physiological rest state of the brain, DMN activity during rest has soon been suggested as the neural basis of the experience of MW. This proposition is based on three main DMN features: First of all, MW and DMN activity both co-occur during rest and show a reverse correlation to cognitive load (Mason, Norton et al. 2007); Secondly, DMN activity is spontaneous and ongoing, only attenuated by goal directed tasks, a fact which is consistent with a unified percept of self which is stable and continuous; And lastly, *task-related* activity in medial prefrontal and parietal areas, which comprise substantial elements of the DMN, have been shown to be associated with self-related functions (Northoff and Bermpohl 2004; Spreng, Mar et al. 2009), making *rest-related* activations in these areas a natural candidate for subserving MW, a spontaneous process of self-related mentation (Baars 2010).

IV. Quantification Strategies for Studying the DMN - MW Relation

MW in its various names has gained considerable attention in ancient and modern philosophy and in theoretical psychology. However, in the relatively short history of cognitive neuroscience, which has inherited much of its models, paradigms, and findings from behavioral and cognitive psychology research, MW is virtually absent (Smallwood and Schooler 2006) as a subject of research. The reluctance in the scientific arena to study MW can be accounted for by its unique, non-behavioral characteristics as discussed earlier.

However, several pioneering attempts have been made to establish a relation between DMN activity and MW, yielding striking results. The current section overviews methodologies and results of this recent literature.

Based on this body of research, we define five methodological strategies for studying MW within a functional neuroimaging paradigm. Two of these strategies rely on theoretical assumptions regarding MW and involve measuring related cognitive functioning which may imply the occurrence of MW. Another two, on the other hand attempt to directly quantify the occurrence of MW, whether in real time - during rest or task performance, or retrospectively. The final strategy employs paradigm-free techniques and derives consequences from the known functionality of networks emerging from connectivity analysis performed on the functional data. Through the prism of these strategies, we review existing literature and findings published mainly in the recent decade. Each strategy will be presented in light of its advantages and disadvantages as well as the degree of its fitting to various paradigms and data analysis techniques in experimental neuroimaging. A broader focus of these strategies in the context of the functional neuroimaging field can be found in (Gruberger, Simon et al. 2011).

The inclusion criterion for studies presented in this chapter as representing examples for each strategy was bringing forward the question of the relation between functionality of rest-related DMN activity on the one hand and rest-related phenomenological experience on the other. Importantly, studies of self-related functions were only included if they state a specific hypothesis regarding rest-related neural and psychological functioning. An illustration of the results obtained by the studies described in this chapter, categorized by strategy, is presented in Figure 1. The below differentiation between *indirect* and *direct* strategies to study whether rest-related DMN activity serves as the neural basis of MW could also be described in terms of determining the dependent and the independent variables within a functional neuroimaging setup: in the case of the *indirect* strategies, cognitive load or self-relatedness are being experimentally manipulated (i.e. independent variable) and are expected to cause differences in the measured neural signal of the DMN (i.e. dependent variable); in the case of the *direct* strategies, typically the degree of DMN neural activation is manipulated (i.e. independent variable) by altering rest and task while scanning, and consequent change in the degree of MW is being assessed (i.e. dependent variable).

A. Indirect Strategies – Studying the Relation between DMN and MW without Direct Measurement of MW

Indirect strategies - strategies in which MW is not directly measured – are typically based on the conceptualization of MW as self-related and as dependent on low cognitive demand as discussed in section II. The hypothesis could be framed as follows: if DMN neural activity during rest is the neural basis of MW, then DMN activations during *rest* and during a *task* should be found to be more similar when the given task is most similar to when MW occurs, i.e. characterized by low cognitive load and high self-relevance.

The advantage of such strategies is straightforward: they overcome the lack of validated behavioral measures of MW by using accepted task related (mostly validated or previously published) behavioral measures to study it.

Strategy 1: Parametric Modulation of Self-relatedness

The first paper to specifically associate MW with DMN activity, though not the first to suggest a relation between rest related neural activity and MW, was published by Gusnard and colleagues (Gusnard, Akbudak et al. 2001), as part of a series of publications in 2001 (Raichle, MacLeod et al. 2001) introducing the concept of the DMN. In this fMRI study, neural activations during rest were compared both to an internally cued condition and to an externally cued condition in a judgment task of affectively normed pictures. In accordance with the predictions, the neural activations during the internally cued condition were found to be more similar to the activations at rest than those during the externally cued condition. A strategy similar in contrasting a self-related task with a similar non-self related task can also be found in fMRI studies such as Johnson et al (Johnson, Baxter et al. 2002), Goldberg et al (Goldberg, Harel et al. 2006) and Andrews-Hanna et al (Andrews-Hanna, Reidler et al. 2010), and in a PET study conducted by D'Argembau et al (D'Argembeau, Collette et al. 2005). The results in all of these studies indicate greater activations in brain areas associated with the DMN, mostly MPFC areas, during self–related tasks than during non self related tasks. Except for Johnson et al (2002), in all of these examples writers discuss their findings in light of the fact that the neural constructs identified by them as 'self-related' are part of the DMN, and suggest that their results might imply a possible functional role of rest-related activations in these areas in self-related spontaneous mental activity or in other words in mind-wandering.

In separate studies as well as in convergence, this is a useful strategy for investigating the functional role of areas within the DMN while staying within the boundaries of accepted neuroimaging paradigms. One drawback of this strategy is the potential of over-stretching the concept of self. This may, in turn, cause confounding the self-relatedness of a task with other characteristics like its emotional valence, as might be claimed to be the case in Gusdnard et al (2001). Therefore, one should pay special attention that the parameter modulated between study conditions is as specific as possible.

Strategy 2: Parametric Modulation of Cognitive Load

Classically in functional neuroimaging, stronger activations in a neural region or network during a task provide support for a possible causal relation between the two. Initial accounts of the DMN demonstrate that areas in this network show higher activations during rest than during task, leading writers to suggest that this network is involved in rest – related mental activity – a non-task related behavior.

When the strategy of parametric modulation of cognitive load, or task difficulty, is used in the context of studying the functionality of the DMN, the contrast of interest in the data analysis is not the commonly used 'task minus rest' contrast, but rather the contrast of 'rest minus task'. Researchers try to demonstrate that the lower the cognitive load in a given task condition, the higher the activations in DMN areas during this condition and thus the smaller the difference between activations in these areas during the task and during rest. This was indeed found to be the case in fMRI studies such as McKiernan et al. (McKiernan, D'Angelo et al. 2006), Christoff et al. (Christoff, Ream et al. 2004) and Mason et al. (Mason, Norton et al. 2007), and in Wicker et al.'s meta-analysis of PET studies (Wicker, Ruby et al. 2003). Moreover, In the case of Mckiernan et al and Mason et al, behavioral measures (described later on in the 'thought sampling' strategy) were added to the study in order to demonstrate a more direct association between high DMN activations, low cognitive demand and mind-wandering.

This strategy yields results which correspond well with the theory concerning mind-wandering as well as with the lay intuition that MW occurs when the mind is free to engage in it. In addition to its intuitiveness, and thus its' simplicity, the advantage of this strategy is in the robustness and replicability of its results across virtually any behavioral task tested, which make it accessible and easy to implement.

It should be taken into account, however, that executive functioning and MW are probably not as anticorrelated as these studies depict. MW may involve executive processes like memory, planning, computing etc., as is indeed reflected by findings of executive networks co-activated with DMN during MW (Christoff, Gordon et al. 2009). Thus, rather than assuming mutual exclusiveness, the degree and direction of the association between neural activity of DMN and of executive networks during MW should be studied in greater experimental resolution.

B. Direct Strategies – Direct Quantification of the Degree of Mind-wandering

Directly quantifying the degree of MW overcomes its non-behavioral nature, and essentially makes conventional experimental methods applicable for studying it: the degree of MW can be used to categorize sessions or groups before analyzing, the correlation between it and degree of activation in selected brain regions of interest can be assessed and compared, it can be regressed into the whole brain within or across subjects to obtain areas which correlate with it in degrees of activity, and so on. The greatest challenge, however, is that in contrast to most behavioral task, the actual tracking, or even mere verbalizing in real time, of MW by subjects tampers with MW itself: an individual busy with reporting her own MW is less free to engage in spontaneous MW comparing to when left to rest quietly. This can probably account for the very few studies, relatively, which have attempted to quantify MW in the history of cognitive neuroscience, and possibly, for the even fewer methods developed to do so. Several quantifications techniques have, however, been employed, some attempting at real-time assessment of the degree of MW while others focusing on post-hoc questioning of subjects.

Strategy 3: Thought Sampling

MW can occur with or without awareness of its occurrence (‘meta-awareness’; (Christoff, Gordon et al. 2009)). Nevertheless, one can normally report if a thought was occurring in their mind or not if interrupted and asked at a given time point. This is the rational underlying the thought sampling (also known as thought probing) technique, applied successfully in studies such as McKiernan et al (McKiernan, D'Angelo et al. 2006) and Mason et al (Mason, Norton et al. 2007)) mentioned earlier, as well as Christoff et al (Christoff, Gordon et al. 2009). Several versions exist for thought sampling, but typically, during a rest or a task scan, a probing tone is presented in even or uneven intervals and subjects are instructed to indicate 'yes' or 'no' according to whether they were experiencing a thought at the time the tone was presented (or, in a similar version, since the previous probe (Giambra 1995)). Each scan session then receives a rating according to the rate of 'yes' answers given in it out of the overall number of tones presented – essentially, a quantified measure of the degree to which MW occurred.

In yet another, less used, version of thought sampling, subjects are requested to press a button each time a thought comes into mind (Giambra 1989). This version seems to be less favorable and can hardly be found in neuroimaging studies, probably because it imposes greater meta-awareness and concentration from subjects and thus interferes with the natural occurrence of MW.

The strategy of thought sampling presents clear advantages of being a real time, direct and quantified measurement of MW occurrence. One should bear in mind, though, that (to the best of our knowledge) it has never been systemically tested for validity and reliability, and thus is only justified by its straight-forwardness and intuitiveness.

Strategy 4: Retrospective Questioning

MW requires a peace of mind; disturbances tend to interrupt its natural flow. In other words, acquiring a concurrent report from subjects regarding MW interferes with its actual occurrence (Filler and Giambra 1973). An informative report regarding MW at a given time period is thus better collected retrospectively, after the session has ended (notably, even then, the contents of MW is not always straightforwardly accessible to memory). Surprisingly, though, in the wide ocean of psychological questionnaires developed in the last century, designated structured psychological questionnaires for explicitly assessing MW in healthy individuals are scarce. The very few examples which can be found in the literature (Giambra et al, 1997; Klinger et al., 1987) did not seem to survive the transition from psychological research to neuroscience, to the best of our knowledge. Therefore unfortunately, there isn't an accumulated body of literature regarding the neural basis of MW, and virtually no experience in the field obtained by retrospective questioning of subjects regarding their MW using validated experimental instruments designated for this matter.

Fortunately, some signs for change in this respect are emerging in recent years. One example can be found in the inspiring PET study by D'argembau et al mentioned earlier (D'Argembeau, Collette et al. 2005), who used an in-house developed questionnaire in which subjects had to rate items like the total amount of thoughts experienced, whatever their content (a similar approach is found earlier in (Mcguire, Paulsau et al. 1996)).

An alternative to using domestically developed questionnaires is to use established questionnaires of experiences which according to theoretical and clinical literature are related to MW, such as self awareness and degree of dissociation, and to infer from these questionnaires on the degree of MW experienced by subjects. Evidence for the potency of this option has been presented by the authors of this chapter in a workshop and a plenary lecture at the Towards a Science of Consciousness 2010 conference (Gruebrger, Ben-Simon et al. 2010) and is submitted for publication. A third example worthy of noting is the Resting State Questionnaire (ReSQ) developed recently by Delamillieure et al (2010) explicitly for usage in a functional neuroimaging setup. The ReSQ consists of 62 items organized by five main types of mental activity: visual mental imagery, inner language, somatosensory awareness, inner musical experience and mental manipulation of numbers. Using a 0–100% scale, the participant retrospectively quantitatively rates the proportion of time spent in each mental activity during the resting state fMRI acquisition. Whether this tool will or will not eventually gain the confidence of the researching community, its great importance lies in that it represents a pioneering effort to encompass the richness and individual nature of MW into a standardized questionnaire.

C. Data Analysis of Neuronal Dynamics

Brain activity is combined of activations of neurons which comprise anatomical and functional networks. Recent advances in functional and computational neuroimaging have provided new tools to examine functional interactions between time series of signals obtained from different brain regions, catalyzing the examination of functional connectivity in the human brain. In fMRI, analysis methods of the resting state signal can typically be placed into model dependent and model free methods (van den Heuvel and Hulshoff Pol 2010), both resulting in connectivity maps - whether correlational or anticorrelational (Uddin, Clare Kelly et al. 2009) - demonstrating anatomical networks which, interestingly, greatly overlap with known functional neural networks. The DMN is one of those emerging networks and it relation to MW can be further characterized in terms of functional connectivity.

Strategy 5: Paradigm-free Techniques

The neuronal basis of spontaneous resting-state fMRI signals was initially regarded as problematic giving rise to unknown parameters of noise as well as known physiological ones. However recent observations validate the potential of these signals allowing for its increasing support (van den Heuvel and Hulshoff Pol 2010) : The first and probably most compelling evidence is that most of the resting-state patterns tend to occur between brain regions overlapping in known functional and neuroanatomical regions (Salvador, Suckling et al. 2005; Damoiseaux, Rombouts et al. 2006; Van Den Heuvel, Mandl et al. 2008). The second observation relates to the frequency of rest related signals revealing that the observed spontaneous BOLD signals are mainly dominated by lower frequencies (b 0.1 Hz) with only a minimal contribution of higher frequent cardiac and respiratory oscillations (N 0.3 Hz) (Cordes, Haughton et al. 2000; Cordes, Haughton et al. 2001). Furthermore an (indirect) association exists between the frequency profiles of slow spontaneous resting-state fMRI and electrophysiological recordings of neuronal firing (Nir, Mukamel et al. 2008) and between spontaneous BOLD fluctuations and simultaneous measured fluctuations in neuronal spiking (Shmuel, Yacoub et al. 2002; Shmuel and Leopold 2008). Valuable anatomical and functional information about the DMN can thus be derived merely by acquiring resting-state functional neuroimaging sessions.

Two studies are brought here to exemplify the usages of a paradigm free strategy to further characterize the relation between MW and spatio-temporal dynamics of the DMN. Horowitz et al (Horovitz, Fukunaga et al. 2008) utilized this strategy to determine whether DMN activity can be de-coupled from conscious awareness. In this study, functional connectivity analysis showed that the level of connectivity within the DMN persisted both during the resting state and during light sleep. The authors conclude that DMN connectivity "does not require or reflect the level of consciousness that is typical for wakefulness" (p. 679), which seems to undermine the idea of a functional involvement of DMN activity in MW. Nevertheless, two alternative explanations are offered by the authors: the first is that these results only decouple wakeful awareness and the degree of *connectivity* within the DMN, but not the amplitude of its *activity*; the other is that light sleep is sometimes characterized by the existence of dream-like reverie activity (a mental activity similar to MW) which like MW might also be a functional product of DMN activity. Another study by the same group (Horovitz, Braun et al. 2009) demonstrated altered correlations between DMN network components during different states of consciousness, most notably a reduced involvement of the medial pre-frontal cortex during sleep. The authors suggest that from among the DMN components, the frontal cortex may play an important role in the sustenance of conscious awareness.

It can be claimed that as some indication exists of the effect of previous task performance on neural activity at subsequent rest (Northoff, Qin et al. 2010), a study design which consist of rest alone will produce results which are more unbiased. In any case, studies of this strategy call attention to the fact that beyond relative degree of neural activity, more holistic parameters of neural dynamics need to be explored to truly characterize the DMN-MW relation, such as temporal and spatial patterns of DMN activity.

V. Discussion and Future Directions

In this chapter we portrayed the evolvement of the neuroscience of MW, while aspiring to lay the grounds for additional research to come. Undoubtfully, studies like the ones overviewed here serve to narrow the gap between theoretical understanding of MW and its scientific exploration. Nevertheless, MW is still by large a mystery, and much work remains to complete the puzzle.

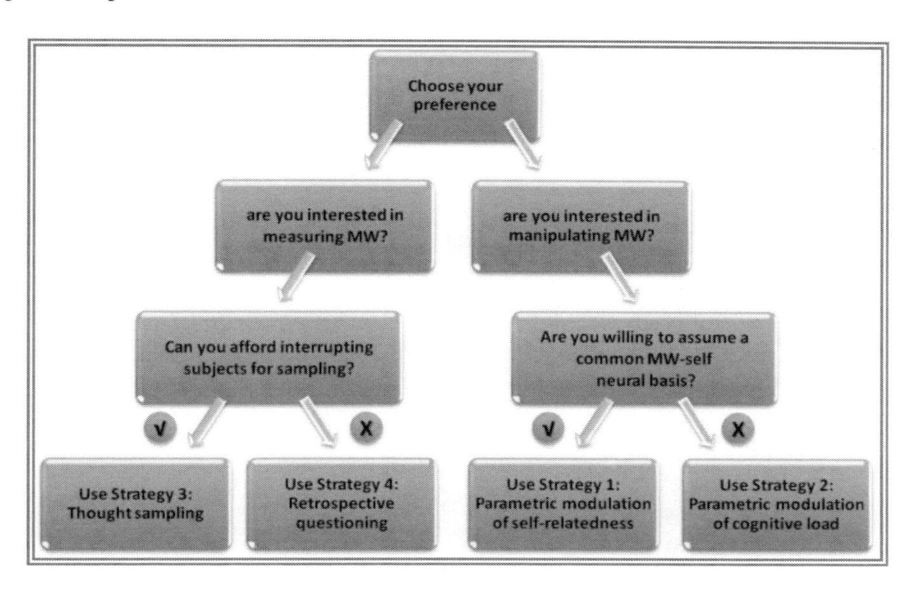

Figure 2. Flowchart of goodness of fit of different strategies according to study aims.
A flowchart which may assist researchers wishing to explore mind-wandering using functional neuroimaging paradigms. According to study aims one should decide on the appropriate strategy taking into consideration advantages and disadvantages of each strategy as discussed in the chapter.

MW can be studied under different contexts involving a wide array of experimental questions. Accordingly, as we tried to exemplify in this chapter, there is no absolute optimal way to study it, but rather it is important to make an informed, educated choice when studying it within a neuroimaging paradigm. Figure 2 depicts a flow chart of relevant considerations.

A. DMN-MW Relation – A Boarder Perspective

Converging results from studies to date suggest that DMN activity is critical for MW to occur (See Figure 1). In light of the infancy of this field, this in itself is a remarkable achievement which provides a rather solid working hypothesis for future research. Nevertheless, this association is not exclusive, as evidence exist both for DMN activity without conscious awareness (Horovitz, Fukunaga et al. 2008), and for additional functional networks contributing to MW (Christoff, Gordon et al. 2009). Though a dichotomy between rest-related and executive related neural networks proved instrumental for the beginning of MW research, a more mature approach might suggest studying the interplay between MW and executive processes and their underlying mechanisms. A better understanding of this interplay, which can be achieved by further implementation of the five strategies defined above, may contribute greatly to the understanding of the adaptive value of MW and its contribution to human cognition.

B. Mind-wandering: The Neural Basis of its Integration and Segregation

Portraying evidence on the brain according to the strategy by which it was obtained as illustrated in Figure 1, suggests that studying MW in different contexts, or with different strategies, indicates both sub-areas within the DMN which are common to MW in any context and sub-areas which are more typically unique to a specific context. For example, and on an impressionist level only, it could be suggested that parametric modulation of cognitive load affected more the medial than lateral cortical wall; that parametric modulation of the self modulated more left than right areas of the lateral frontal cortex; and that thought sampling points towards more left and more lateral areas within the DMN. It is of no doubt that such conclusion requires a careful meta-analysis which is beyond the scope of this essay. Nevertheless, such a neuro-functional differentiation implies that each strategy might reveal, in addition to the network underlying MW, the neural basis of a specific aspect within the large construct of MW. This in turn may serve to re-define MW both theoretically and operationally, as distinct neural processes which underlie its different aspects become clearer.

C. Functionality of MW

The robustness of MW experience across ages, cultures, and individuals, indeed suggests a vital role for it in the functioning human brain. In a meta analysis of task-dependent self-related processing in the brain, Northoff et al (2006) postulate that "below this (self related information processing; M.G.) processing may be a more fundamental…representation of internal and external stimuli including their subjective (or phenomenal) experience as such that is essential for any reflective awareness to become possible" (Northoff, Heinzel et al. 2006, pp.441).

As to the specific adaptive role of MW, we suggest several ideas based on current literature which may be incorporated into future research (also summarized in Box 1):

- *MW serves 'self' functions.* There are theoretical (Gallagher 2000), neuroanatomical (Gusnard 2005; Northoff, Heinzel et al. 2006) and intuitive grounds to claim that MW serves to create and maintain an integrated, meaningful sense of self out of various aspects of self-related information and cognition.
- *MW enables the projection of a 'self' to past and future events.* The idea that MW serves processes of future planning and simulation is based on theory and clinical experience (Wheeler, Stuss et al. 1997), and is strongly supported by the fact that the DMN includes areas such as the pre-frontal cortex, the posterior cingulated cortex, the Precuneus and the Hippocampus, which are known to take part in such mental processes (Buckner, Andrews-Hanna et al. 2008).
- *MW serves as a learning and consolidation mechanism by augmenting the associative abilities of the brain.* This relatively recent idea is presented by contemporary writers (Christoff, Gordon et al. 2008; Baars 2010) and already takes into account what is known about DMN activation patterns as well.

D. Mind-wandering – Questions for Future Research

Understanding MW using brain imaging techniques holds a lot of promise for this new field of research. Listed here are a few lines of thought that could posit an initial framework for future MW studies:

- *Spatial and temporal patterns of MW:* What are the temporal and spatial dynamics of MW in the brain? And what is the effect of MW on brain connectivity?
- *Control of MW:* what is MW's locus of control? Do internal and external abruptions of MW result in similar outcome?
- *MW and pathologies*: Which functions does MW serve and how are they disrupted when MW does not occur. Both the very mechanism and the contents of MW are of great interest to clinical psychology and psychiatry. Clinical and neuronal pathologies associated with pathological functioning of MW may shed light both on the understanding of the symptoms and on the role of MW and its neural basis in healthy functioning.
- *The contents of MW:* In this chapter we put little emphasis on the ever changing contents of MW. This is not to say they are of no importance, only that the studies described here were interested in the common mechanism underlying this changing flow of contents. Future research might very well attempt to segregate neural patterns during MW which are responsible for the experience of different contents or even different time directions (e.g. future or past) as explored by Smallwood (Smallwood, Nind et al. 2009).

To begin answering such questions, society needs agreed upon theoretical definitions as well as normalized, standardized behavioral measures of MW. In the functional neuroimaging field one also needs advanced validated computational methods for studying the temporal dynamics of neural activations in long functional imaging sequences such as present in rest.

E. What Can Be Learned about the Neural Correlates of Consciousness from the Neuroscience of Mind-wandering?

What is the nature of the relationship between consciousness and MW? Is MW simply an expression of conscious experience much like an actor on a stage (Baars, Ramsoy et al. 2003) or is it a substantial part of consciousness giving rise to the stage itself?

The field of the neural correlates of consciousness extends beyond MW and into other states of alert, altered or diminished consciousness. Nevertheless, consciousness can be regarded as the operating system required for mind-wandering to occur. Thus, when capturing brain activity during MW, the results may be conceptualized as representing the sum of a multi-layer process in which consciousness (much like Damasio's hierarchical model (Damasio 1998)), MW and perhaps other related processes take place. It could, for example, be suggested that the formation of the DMN of both 'higher' and 'lower' neural mechanisms subserves this multi-layer phenomenological mechanism. The question of which neural mechanisms are specific to MW and which are involved in mediating further aspects of conscious experience is just one of the scientific mysteries neuroscience and theory will need to unveil.

Nevertheless, as MW is formulating as a stand-alone, legitimate scientific topic, its position within the context of the scientific study of consciousness holds a great potential for promoting both the understanding of MW and of consciousness in general.

Concluding Remarks

MW is a universal phenomenon which accompanies much of our daily lives from childhood to adulthood. Its exploration has a vast potential in leading us to a better and more profound understanding of our ongoing mental selves. Furthermore, its foundations could shed light on the basic properties of conscious experience.

The study of MW is at an exciting position of forming into a field of research in its own. Its relevance to a wide array of disciplines, from neuroscience to philosophy to the clinical world ensures that it will draw a growing number of researchers in the near future. We hope that this chapter serves to set the milestones for a better scientific understanding of the neural basis of this remarkable, unique human quality.

References

Andrews-Hanna, J. R., J. S. Reidler, et al. (2010). "Functional-Anatomic Fractionation of the Brain's Default Network." *Neuron* 65(4): 550.

Baars, B. J. (2010). "Spontaneous Repetitive Thoughts Can Be Adaptive: Postscript on "Mind-wandering"." *Psychological Bulletin* 136(2): 208-210.

Baars, B. J., T. Z. Ramsoy, et al. (2003). "Brain, conscious experience and the observing self." *Trends Neurosci* 26(12): 671-675.

Barzun, J. (1983). *A stroll with William James*, University of Chicago Press

Beer, J. S. (2007). "The default self: feeling good or being right?" *Trends Cogn Sci* 11(5): 187-189.

Buckner, R., J. Andrews-Hanna, et al. (2008). "The brain's default network." *Ann NY Acad Sci* 1124: 1-38.

Christoff, K., A. Gordon, et al. (2008). The role of spontaneous thought in human cognition. *Neuroscience of Decision Making*. O. a. M. Vartanian, R. , Psychology Press.

Christoff, K., A. M. Gordon, et al. (2009). "Experience sampling during fMRI reveals default network and executive system contributions to mind-wandering." *Proceedings of the National Academy of Sciences* 106(21): 8719-8724.

Christoff, K., J. M. Ream, et al. (2004). "Neural basis of spontaneous thought processes." *Cortex* 40(4-5): 623-630.

Cordes, D., V. M. Haughton, et al. (2001). "Frequencies contributing to functional connectivity in the cerebral cortex in" resting-state" data." *American Journal of Neuroradiology* 22(7): 1326.

Cordes, D., V. M. Haughton, et al. (2000). "Mapping functionally related regions of brain with functional connectivity MR imaging." *American Journal of Neuroradiology* 21(9): 1636.

D'Argembeau, A., F. Collette, et al. (2005). "Self-referential reflective activity and its relationship with rest: a PET study." *Neuroimage* 25(2): 616-624.

Damasio, A. R. (1998). "Investigating the biology of consciousness." *Philosophical Transactions of the Royal Society* B: Biological Sciences(353(1377)): 1879-1882.

Damoiseaux, J. S., S. Rombouts, et al. (2006). "Consistent resting-state networks across healthy subjects." *Proceedings of the National Academy of Sciences* 103(37): 13848.

Dennett, D. C. (1991). *Consciousness Explained*. Boston MA, Little Brown & Co.

Filler, M. S. and L. M. Giambra (1973). "Daydreaming as a function of cueing and task difficulty." *Perceptual and motor skills* 37(2): 503.

Fox, M. D., A. Z. Snyder, et al. (2005). "The human brain is intrinsically organized into dynamic, anticorrelated functional networks." *Proceedings of the National Academy of Sciences of the United States of America* 102(27): 9673-9678.

Gallagher, S. (2000). "Philosophical conceptions of the self: Implications for cognitive science." *Trends in Cognitive Sciences* 4(1): 14-21.

Giambra, L. M. (1979). "Sex differences in daydreaming and related mental activity from the late teens to the early nineties." *The International Journal of Aging and Human Development* 10(1): 1-34.

Giambra, L. M. (1989). "Task-Unrelated-Thought Frequency as a Function of Age: A Laboratory Study." *Psychology and Aging* 4(2): 136-143.

Giambra, L. M. (1995). "A Laboratory Method for Investigating Influences on Switching Attention to Task-Unrelated Imagery and Thought." *Consciousness and Cognition* 4(1): 1-21.

Giambra, L. M. and A. Grodsky (1989). "Task-unrelated images and thoughts while reading." *Imagery: Current perspectives*: 26–31.

Gilbert, S. J., C. D. Frith, et al. (2005). "Involvement of rostral prefrontal cortex in selection between stimulus-oriented and stimulus-independent thought." *European Journal of Neuroscience* 21(5): 1423-1431.

Goldberg, II, M. Harel, et al. (2006). "When the brain loses its self: prefrontal inactivation during sensorimotor processing." *Neuron* 50(2): 329-339.

Gruberger, M., E. B. Simon, et al. (2011). "Towards a neuroscience of mind-wandering." *Frontiers in Human Neuroscience* 5.

Gruebrger, M., E. Ben-Simon, et al. (2010). *Experimenting with experience: fMRI, EEG and TMS in service of exploring the functionality of the default-network activity*. Towards a Science of Consciousness, Tucson, Arizona, Center for Consciousness studies.

Gusnard, D. A. (2005). "Being a self: considerations from functional imaging." *Consciousness and Cognition* 14(4): 679-697.

Gusnard, D. A., E. Akbudak, et al. (2001). "Medial prefrontal cortex and self-referential mental activity: relation to a default mode of brain function." *Proc Natl Acad Sci U S A* 98(7): 4259-4264.

Horovitz, S. G., A. R. Braun, et al. (2009). "Decoupling of the brain's default mode network during deep sleep." *Proceedings of the National Academy of Sciences* 106(27): 11376.

Horovitz, S. G., M. Fukunaga, et al. (2008). "Low frequency BOLD fluctuations during resting wakefulness and light sleep: A simultaneous EEG fMRI study." *Human brain mapping* 29(6): 671-682.

James, W. (1892). *Psychology*, Cleveland & New York: World.

Johnson, S. C., L. C. Baxter, et al. (2002). "Neural correlates of self reflection." *Brain* 125(8): 1808-1814.

Kane, M. J., L. H. Brown, et al. (2007). "For whom the mind wanders, and when." *Psychological Science* 18(7): 614.

Killingsworth, M. A. and D. T. Gilbert (2010). "A Wandering Mind Is an Unhappy Mind." *Science* 330(6006): 932.

Mason, M. F., M. I. Norton, et al. (2007). "Wandering minds: the default network and stimulus-independent thought." *Science* 315(5810): 393-395.

Mcguire, P. K., E. Paulsau, et al. (1996). "Brain activity during stimulus independent thought." *Neuroreport* 7(13): 2095-2099.

McKiernan, K. A., B. R. D'Angelo, et al. (2006). "Interrupting the "stream of consciousness": an fMRI investigation." *Neuroimage* 29(4): 1185-1191.

Morin, A. (2009). Inner speech and consciousness. *The Encyclopedia of Consciousness*. B. W, Elsevier.

Nir, Y., R. Mukamel, et al. (2008). "Interhemispheric correlations of slow spontaneous neuronal fluctuations revealed in human sensory cortex." *Nature neuroscience* 11(9): 1100-1108.

Northoff, G. and F. Bermpohl (2004). "Cortical midline structures and the self." *Trends in Cognitive Sciences* 8(3): 102.

Northoff, G., A. Heinzel, et al. (2006). "Self-referential processing in our brain--a meta-analysis of imaging studies on the self." *Neuroimage* 31(1): 440-457.

Northoff, G., P. Qin, et al. (2010). "Rest-stimulus interaction in the brain: a review." *Trends in Neurosciences* In Press, Corrected Proof.

Raichle, M. E. (2009). "A Paradigm Shift in Functional Brain Imaging." *J. Neurosci.* 29(41): 12729-12734.

Raichle, M. E. (2010). "Two views of brain function." *Trends in Cognitive Sciences* 14(4): 180-190.

Raichle, M. E., A. M. MacLeod, et al. (2001). "A default mode of brain function." *Proc Natl Acad Sci U S A* 98(2): 676-682.

Salvador, R., J. Suckling, et al. (2005). "Undirected graphs of frequency-dependent functional connectivity in whole brain networks." *Philosophical Transactions of the Royal Society B: Biological Sciences* 360(1457): 937.

Shmuel, A. and D. A. Leopold (2008). "Neuronal correlates of spontaneous fluctuations in fMRI signals in monkey visual cortex: Implications for functional connectivity at rest." *Human brain mapping* 29(7): 751-761.

Shmuel, A., E. Yacoub, et al. (2002). "Sustained negative BOLD, blood flow and oxygen consumption response and its coupling to the positive response in the human brain." *Neuron* 36(6): 1195-1210.

Smallwood, J., L. Nind, et al. (2009). "When is your head at? An exploration of the factors associated with the temporal focus of the wandering mind." *Consciousness and cognition* 18(1): 118-125.

Smallwood, J. and J. W. Schooler (2006). "The restless mind." *Psychological Bulletin* 132(6): 946.

Smallwood, J. M., S. F. Baracaia, et al. (2003). "Task unrelated thought whilst encoding information." *Consciousness and Cognition* 12(3): 452-484.

Spreng, R. N., R. A. Mar, et al. (2009). "The Common Neural Basis of Autobiographical Memory, Prospection, Navigation, Theory of Mind, and the Default Mode: A Quantitative Meta-analysis." *Journal of Cognitive Neuroscience* 21(3): 489-510.

Teasdale, J., B. Dritschel, et al. (1995). "Stimulus-independent thought depends on central executive resources." *Memory & Cognition* 23(5): 551-559.

Uddin, L. Q., A. Clare Kelly, et al. (2009). "Functional connectivity of default mode network components: Correlation, anticorrelation, and causality." *Human Brain Mapping* 30(2): 625.

Van Den Heuvel, M., R. Mandl, et al. (2008). "Normalized cut group clustering of resting-state FMRI data." *PLoS One* 3(4): 2001.

van den Heuvel, M. P. and H. E. Hulshoff Pol (2010). "Exploring the brain network: A review on resting-state fMRI functional connectivity." *European Neuropsychopharmacology*.

Wheeler, M. A., D. T. Stuss, et al. (1997). "Toward a theory of episodic memory: The frontal lobes and autonoetic consciousness." *Psychological Bulletin* 121: 331-354.

Wicker, B., P. Ruby, et al. (2003). "A relation between rest and the self in the brain?" *Brain Research Reviews* 43(2): 224.

In: Consciousness: Its Nature and Functions
Editors: Shulamith Kreitler and Oded Maimon
ISBN 978-1-62081-096-5

Chapter 19

NEURONAL REFLECTIONS

Rafael Malach[*]
The Department of Neurobiology
Weizmann Institute of Science Rehovot, Israel

ABSTRACT

The search for the link between brain function and conscious awareness poses a profound challenge to Neuroscience research. While a complete explanation of such link may be unfeasible- a more achievable goal could be the discovery of an elegant theory: the identification of a fundamental neuronal principle that may unify the rich and complex world of subjective experiences. Here I present such hypothetical principle termed neuronal reflections. It is proposed that a consensus state achieved through rapid exchange of information in local groups of nerve cells is the fundamental dynamic underlying all subjective experiences. In the visual domain, the binding of visual elements into a holistic template is achieved, subconsciously, via a hierarchical sequence of integration steps. Conscious subjective awareness emerges when the reciprocal exchange of signals between neighboring neurons in high order visual areas unites them into a unique entity. This local reciprocal integration resolves the ambiguity inherent in the activity of isolated neurons- and is the critical event leading to a meaningful conscious image in the mind of the observer. Thus, the original meaning of the term consciousness- collective knowledge- appears to surprisingly capture an essential aspect of the neuronal dynamics leading to phenomenal experience.

INTRODUCTION

The aim of this chapter is to propose a framework for a neuronal theory of conscious awareness. Such an endeavor is tentative since the problem of conscious awareness quite likely poses the most challenging puzzle in neuroscience research. On the other hand, subjective awareness is so immediate, personal and direct that one can not but be irresistibly drawn to understand how it emerges.

[*] E-mail: rafi.malach@weizmann.ac.il Tel: 972-8-934 2758/or -2441; Fax:972-8-934 4131.

I emphasize the obvious- that at this stage any attempt to formulate a brain theory of consciousness must be very speculative due to the scarcity of the relevant experimental information available, particularly from human brain research. On the other hand proposing alternative theories, as has been done by a number of prominent thinkers in the field has its own value. First, it helps in demonstrating what kind of neuronal theories can even be considered in approaching this deeply mysterious problem. Furthermore, such frameworks can guide a research program for neuroscientists- i.e. to outline experiments that will attempt to refute a theory's predictions or improve its details.

This chapter is written with a broad audience in mind, so I made an effort to avoid any technical jargon requiring prior knowledge in neuroscience. I want also to emphasize that while I present here my personal perspective on the subject (which likely will raise many objections) much of it is not original and reflects an integration of a large body of previous ideas generated by a long list of superb scientists, philosophers and psychologists involved in this rapidly expanding field (for a partial sample of informative reviews see (ENREF 5 Zeki, 2001; Baars, Ramsoy and Laureys, 2003; Rees, Kreiman and Koch, 2002; Baars; Block, 2005; Lamme, 2006; Block, 2007; Rosenthal, 2005, Tononi and Koch, 2008; Haynes 2009; Dehaene and Changeux, 2011, Searle et. al. 1997).

CAN CONSCIOUSNESS BE EXPLAINED?

Before I embark on outlining my own hypothesis, I would like to briefly address the recurring arguments that in fact scientific research is unable to contribute anything of value to consciousness research i.e. the mind-body problem. It is argued that this question is so mysterious and unapproachable that it can not be envisioned, even in principle, how scientific tools can "explain" consciousness. I therefore would like to begin by discussing what can science contribute to this deep question, and furthermore, how can we even know that a certain scientific theory is successful in making progress towards the understanding of such seemingly intractable problem.

My personal opinion is that the pessimistic outlook concerning the ability of science to deal with the mind-body problem is unwarranted. I think this pessimism stems from an erroneous conception of the aim of scientific theories in general. A scientific theory does not attempt to completely "explain" any natural phenomenon. If we take the theory of evolution as an example, it is clear that Darwin did not succeed and in fact has not even attempted to fully explain how new species emerge. In Darwin's time the structure of DNA was unknown so obviously major gaps still existed in Darwin's original hypothesis. The reason we consider Darwin's theory as a historical breakthrough is that his theory of evolution was *elegant*- it was able to show that a large number of seemingly unrelated observations - the physical similarity between different species, the appearance of fossils etc. could all be derived from two simple principles. Remarkably, Darwin was even able to correctly estimate, using these principles, the age of the sun!

Our strong sense of understanding when facing Darwin's intellectual triumph is due, then, to the theory's beautiful simplicity. More generally, a successful scientific theory is therefore not judged by its ability to fully explain a certain phenomenon, but rather by its ability to introduce elegance to a disjointed set of observations. A theory is successful if it can show how a large set of phenomena, that superficially appear unrelated, are in fact a reflection of a deeper underlying principle, provided of course that this principle is compatible with all relevant observations. This powerful aesthetic drive is a decisive factor in the success or failure of scientific theories.

How does this view, concerning the nature of scientific theories, apply to the feasibility of a neuronal theory of consciousness? Similar to all other scientific theories, I will argue that in this case as well, the aim of a theory of consciousness is not to explain, or "solve" the incomprehensible link between mind and brain. Rather it is to show how the amazingly complex world of seemingly unrelated conscious mental phenomena may be derived from a single underlying principle. This principle will have to show that the smell of chocolate, the beauty of the starry night and even our ability to reflect upon our own conscious experience- all stem from a single, simple principle. This principle will also have to define the border between conscious and subconscious mental phenomena. To distinguish, for example, the stable percept we experience despite the erratic motions that our eyes make- from the vivid sense of motion we experience even when presented, paradoxically, with physically static images e.g. during a movie.

If a scientific theory could account for all these, we would certainly consider this a major and even dramatic advance in our understanding of the link between the mind and brain. This will be true even if we will not be able to explain why it is the case that this unifying principle works.

This situation is reminiscent of another fundamental triumph of modern science- the theory of quantum-mechanics which describes the physics of phenomena at minute scales. Here the situation is far more advanced- the discovery of a set of elegant principles - described in precise mathematical equations- has in fact already been made and is justifiably considered a major triumph of science. However, we are still in the dark about the explanation (also termed the "interpretation") of these equations – what is the meaning of the laws of quantum mechanics and how come they work in such an incomprehensible manner. In the case of neuroscience – we are still far behind- we are not even close to a successful theory linking subjective experiences and the brain. The hypothesis I will present here both lacks in detail and in sufficient experimental support- and hence should be viewed as a tentative framework. The basis of this framework is a principle that if elaborated more precisely and proven correct, could indeed constitute an important advancement towards a scientific theory of this most mysterious and ancient puzzle- how is subjective experience related to the workings of the physical brain.

The Word

I will begin from a vantage point that will appear at first completely misplaced. About 4000 years ago, one of the most profound cultural developments in human history occurred in the Sinai desert. This revolution was the invention of the written alphabet. It is thrilling to see how we, Hebrew readers, can still identify the letters in the first alphabetic word that has been found. This invention was so powerful and transforming because the alphabet is an example of a deep aspect of nature- the power of combinatorics. It reflected the realization that through different combinations of a few simple elements it is possible to generate a vast, practically unlimited, number of unique entities. In the case of writing- this is of course the combination of different letters into words. What defines the word is the unique assembly of different letters. The individual letter, presented in isolation is not informative- but its contribution becomes critical when it is combined with other letters into a specific word. In written language each letter participates in constructing millions of different words and the number- the "space"- of potential words is astronomical. An important aspect to note is that the mere collection of letters that constitute a word is completely meaningless unless there is some process, some mechanism that "binds" the letters into a meaningful whole. In the case of letters this is of course the reader- without a reader that understands the words- the mere fact that the letters are placed near each other on paper has no meaning. In the following I would like to show how this deep phenomenon, the power inherent in binding few elements into a vast space of complex entities may in fact be the fundamental process underlying all of conscious experiences.

VISION AS A CREATIVE PROCESS

While my overall aim is to link brain processes to any subjective psychological experience, I will focus in this chapter on a specific subset of such experiences- visual percepts. I will ask the deceptively simple question- what needs to happen in our brain so that a subjective experience of a visual image will emerge in our mind.

If we examine how the visual process begins, that is - the very moment when light particles- photons, strike the eye, we can see that at this stage there are clear parallels between the eye and a photographic camera. In both cases, photons of light are reflected from an external physical object. These photons travel through a series of optical focusing layers, both in our eyes and in the camera. They create a pattern of light- an optical pattern- displayed on a thin layer of neural tissue in the eye termed the "retina". This tissue is actually a mosaic made of many thousands of tiny light detectors- also termed-"photoreceptors" and nerve cells that connect them. Without getting into an elaborate description of the light absorbing mechanisms of the retina, which is a fascinating topic in its own right, we can consider the photoreceptor basically as a mechanism whose role is to translate light intensity (rate of photon absorption) into an electrical signal. A critical point to understand is that at any given moment a single photoreceptor is not absorbing photons from the entire visual world but, (if the picture is sharp), it can absorb light from a tiny point in the external world. In other words, each individual photoreceptor can report to the rest of the brain only about a single "pixel" in the external object we look at. We call this tiny pixel the "receptive field" of the specific photoreceptor. Neighboring photoreceptors have neighboring receptive fields that together tile up the entire external world- also called the "visual field"- in front of us.

This simple fact, that the photoreceptors transmit to the brain a collection of isolated points has deep implications to the seeing process. To illustrate this, imagine that you are in a museum, examining the beautiful "Birth of Venus" painting by Boticelli (Figure 1). If we now ask, what information your brain actually gets from the eyes, that is- with what "raw material" the brain begins the seeing process; it is evident that the brain, in fact, does not receive any information that even remotely looks like the picture of Venus as we see it. All the brain receives is a sort of gigantic table of numbers (Figure 1) which are the hundreds of thousands of isolated signals representing the individual pixels in the painting reaching the brain from each eye. These signals reflect the light intensities (more precisely contrast), translated into electrical pulses in each isolated point in the painting (we will not deal here with additional complexities such as wavelength information, low light conditions and the fact that the information is typically delivered from both eyes). These many thousands of isolated signals are received by the brain, and in some mysterious process, that we call "seeing" our brain uses these signals to create the visual image that we experience as the breathtaking picture of Venus.

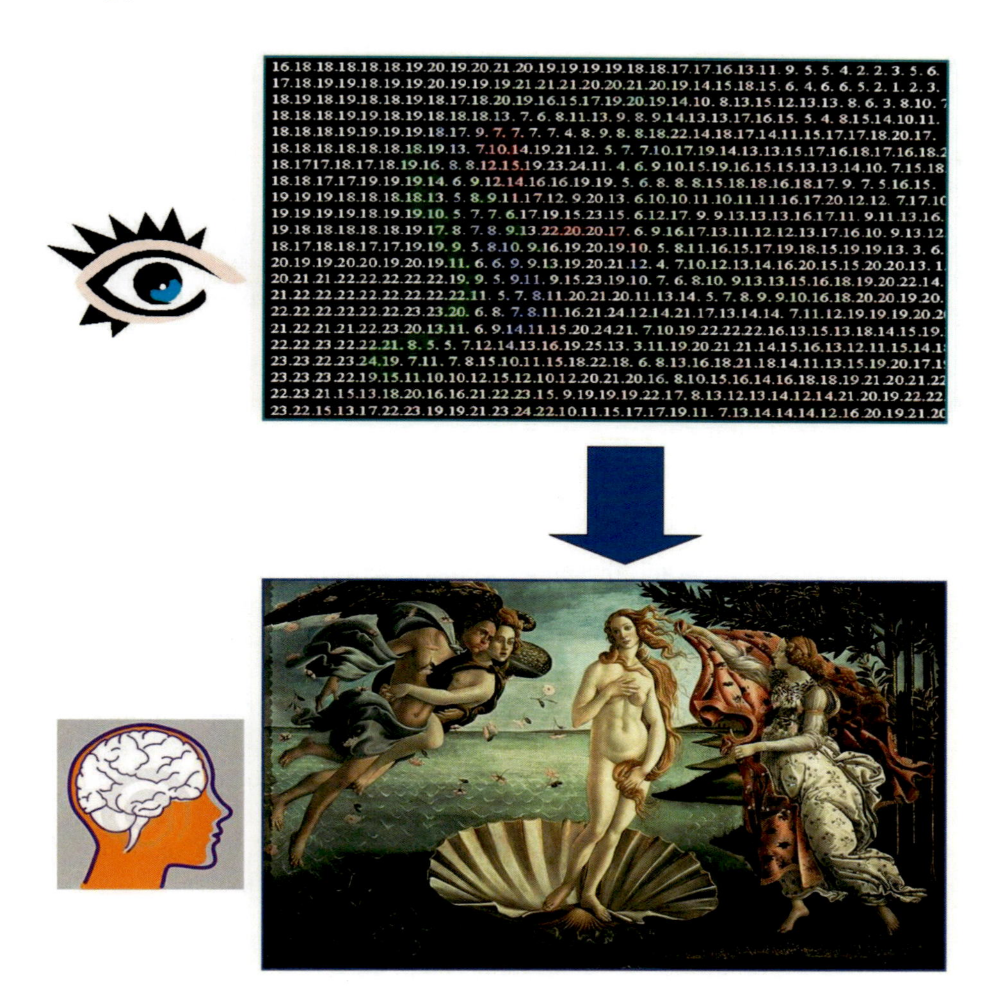

The pictures that we see are not present in the eye but are created by our brains

Figure 1.

We can summarize this as follows: the pictures that we see can not be found in the eye. In fact, they don't even resemble anything that comes from the eyes- and consequently it is wrong to say that what we see is some kind of copy or "representation" of the information that we receive from the external world. The visual picture that we see is the outcome of an internal, creative, process that occurs within our brains, and its relationship to the optical information we derive from the world is complex. Thus, it is meaningless to ask how the external world looks like "on its own". All sensory percepts and visual images are no exception, are the products of an internal image generation process and can be understood only in the context of consciously perceiving brains.

You may protest- that this is just a philosophical word-playing and in fact we have a strong and immediate intuition that the external world looks and feels precisely as we see it. This indeed appears to be the case most of the time, and fortunately so. The fact that the visual images we "invent" are good predictors of the behavior of the physical world is what allowed us, human beings, as well as other animals, to effectively use our vision to compete and survive in the world. However, a large number of visual phenomena- popularly called "visual illusions" repeatedly demonstrate that the coherence between what we create visually and what is actually out there in the world is often loose and sometimes completely incorrect.

Figure 2 provides two striking illustrations of the dissociation between what we see and what is physically there. If we look closely at the triangle shown in Figure 2(top panel) we can see that this is not an ordinary triangle. In fact it is not a triangle at all but a collection of three separated corners. The faint edges of the triangle we perceive- do not exist neither in the world out there, nor in our retina- this is an illusion- a fabrication of our visual brain. It is important to understand the deep significance of this simple illustration- we have here a situation in which the real physical world is made of three separate corners with no edges between them. The photoreceptors in the retina, rather faithfully, inform the brain that between the three corners there is a white homogenous surface. All this is real - nonetheless, our visual brain decides (for reasons that are not important to the point I am trying to make) to generate an image of a triangle that does not exist. In other words, in this case, its not that the brain merely fails to "represents" the external object- we see here that the brain generates an image that actually contradicts the physical reality!

An even more paradoxical illustration is depicted in the picture of the sculpture (Figure 2, bottom panel). At first sight the triangular image we perceive appears clear and quite ordinary, and we may accept it as a perfectly valid physical object. However, closer inspection of this object (the photograph is actually of a real sculpture stationed at the Weizmann Institute Science park in Rehovot, Israel) reveals that such a triangle- having three right angles -is of course physically impossible. Even though we fully realize that this is an object which can not possibly exist in physical reality- this impossibility does not affect in any way the perceptual clarity and ease by which we generate its visual image in our mind. Thus, we are perfectly able to create visual images of objects that *cannot possibly exist.*

The profound and deeply counterintuitive implications of this state of affairs, as has long been contemplated by philosophers from Plato and Kant onward, is that it is simply meaningless to say that the world as such looks like anything or has any visual properties- these are aspects created by each individual's brain and mind.

It is interesting to note the parallel of these conclusions with the picture of reality as is drawn by the physicists. Paradoxically, as modern physics advances, it becomes more unintuitive and unimaginable. Thus, one can not even begin to consider the visualization of true physical reality- as for example formulated in quantum mechanics- an incomprehensible entity whose derivative is a probability cloud spread in space and time.

In summary-despite the fact that our visual percepts of the world are so practically successful- the images we see are our own creation and they are of a fundamentally different nature and quality from what is actually out there in the physical world.

The images we see are not "copies" of real world objects

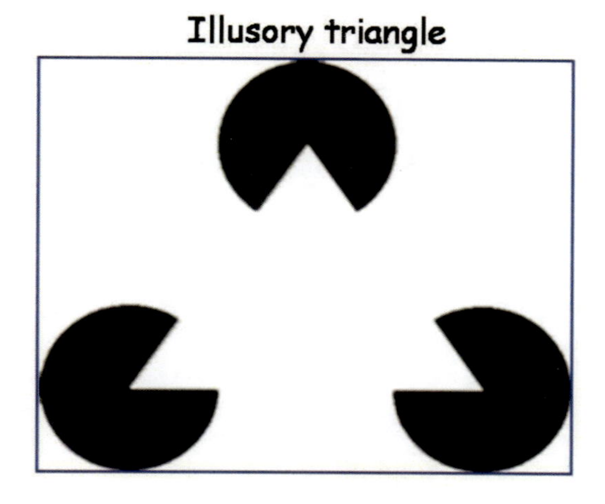

Figure 2.

A Hierarchical Production Line

The brain, then, is the organ that creates the pictures that we see. But how are these pictures actually generated? To understand that, we need to first consider the organization of the visual brain. Visual information, which often triggers the image formation in the brain, flows from the eyes through nerve bundles into a gate-station located roughly at the center of the brain termed the "thalamus" and from there into the outer mantle of the brain- called the "cerebral cortex". The cerebral cortex is a sheet of tissue of about 2.5 mm thickness that covers the entire brain. In humans it is folded in a characteristic pattern (see bottom right corner of figure 2). The "visual" part of the cortex, i.e. those cortical centers that are responsible for generating the visual images, are actually located at the back part of the brain. If, due to an unfortunate accident or lesion, these parts are damaged, a person may become "cortically" blind- i.e. will lose the ability to generate a visual image even though the eyes are perfectly intact- again demonstrating that the visual cortex is crucial for the creation of perceptual images.

The Cortical Hierarchy

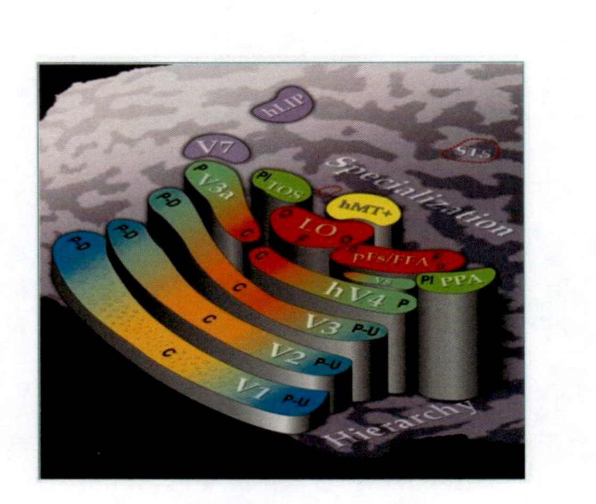

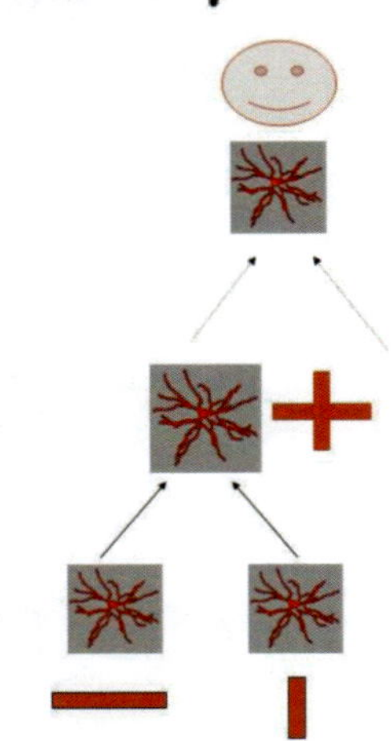

The hierarchy principle

The elements of the visual image are integrated into a holistic template trough a sequential process along the cortical hierarchy of visual areas

Figure 3.

It is important to understand that the visual cortex is not a single, homogenous, entity but is built up of a large number of specialized centers, termed also "Visual areas". A substantial number (about 20) of different visual areas have been identified in the human visual cortex (Figure 3 left panel). However, these areas are not randomly scattered but are organized along clear principles. An important principle is the hierarchical layout of the visual areas. Metaphorically, one can envision this hierarchical principle as if the visual system constructed the perceptual image in a series of steps, along a production line. Thus, in the visual cortex, the visual signals are transferred as a sequence of messages from one visual "station" to another, and at each area the information becomes more integrated and holistic. More specifically- visual signals arriving from the eyes through the thalamus are received in the first cortical area- termed, not surprisingly, area Visual 1 (V1) or primary visual cortex, from there the information is sent to area V2, and then continues along an entire sequence of visual areas until it reaches the highest levels in this sequential "hierarchy" (see Figure 3 left panel).

What precisely happens to the visual information as it moves from one visual area to the next? In order to understand this we first need to get to know a bit better the "work horses" that are actually responsible for the visual process- these are the nerve cells -also called "neurons".

Gossip Machines

When examining a neuron in detail, we find a remarkably complex structure. However, for the purpose of our discussion we can greatly simplify the function of a neuron to its most essential aspect. At this basic level- the neuron can be metaphorically viewed as a specialized "social center". The essence of a neuron's function is the gathering and transmitting of information derived from the community of other neurons. This information transfer is carried by electrical pulses that travel along the neuronal branches- a bit similar to transmitting messages along telephone wires. The signals are actually tiny "explosions" of electrical currents, which are termed "spikes" and when amplified through a speaker system actually sound like machine gun firing-hence when emitting such pulses a neuron is said to "fire spikes". Each neuron receives such pulses from many (often thousands) of other nerve cells in a manner that can be either arousing (excitatory) or depressing (inhibitory).

A crucial aspect of a neuron's function is the "threshold" it is the level of excitation of the neuron above which it begins to fire spikes. Thus, when the signals that impinge on a neuron bring it to a state of sufficient arousal, the neuron's threshold will be crossed and the neuron will send a burst of spikes to its fellow neurons- the more excited the neuron will be, the faster will be the rate of its firing. Importantly, all spikes are roughly equal to each other- so the information transmitted between neurons is embedded in the timing of the spikes- mainly how fast they follow each other- i.e. in the rate by which a neuron will emit such spikes.

Metaphorically we can illustrate the essential role of a neuron as being the ultimate "gossiper". The neuron is specialized for gathering information from its "community". When the information is sufficiently important, the neuron will get excited and share this information with many of its neighbors. Some of these neighbors- and this will turn out to be very important, will directly or indirectly be the same neurons that sent the significant information to the neuron in the first place.

Line Drawing

Now that we understand a bit of the function of a nerve cell- lets consider how this "social" creature plays its role in generating a visual image. We start with the first step of cortical visual integration, which occurs in area V1. An important discovery in this field has been made by two scientists -D. Hubel and T. Wiesel-who received the Nobel Prize for their findings. Hubel and Wiesel inserted tiny micro-wires that can capture the electrical pulses (spikes) into area V1 (in anesthetized animals) and examined what caused the neurons in this area to fire spikes. In contrast to the photoreceptors in the retina that respond to tiny dots of light, the neurons in V1 do not get excited by such dots. Instead, what was found to be particularly effective in making these neurons fire were elongated lines having a specific orientation and location in space. A neuron of this kind could be, for example, sensitive to a horizontal line in the top right corner - and will be excited every time such a line appears on the retina. But if this retinal pattern will rotate and the line will become vertical, this neuron will shut off, and another neuron, which is "specialized" for detecting a vertical line in this location, will begin firing. As a population, there are neurons in V1 that are sensitive to all orientations and all places in the retinal pattern.

How did the neurons in V1 become sensitive to oriented lines instead of tiny points? Recalling how neurons work it should not be too difficult to understand. Imagine that a neuron in V1 is connected in such a way that it receives (indirectly) information that comes only from a group of photoreceptors arranged along a horizontal line; we will call this neuron a "horizontal-line" neuron. If we could color this specific set of photoreceptors we will see that a "template" in the shape of a horizontal line will appear on the retina. Now let's consider how the V1 neuron that receives signals from such a template will react when exposed to bright lines of different orientations. If the presented line is horizontal and its location matches perfectly the template on the retina- then all the photoreceptors that connect to the "horizontal-line" neuron will be simultaneously excited and therefore will emit signals which together will bombard the V1 neuron at high intensity. This in turn will excite the recipient neuron sufficiently to cross its threshold and hence it will start firing. However, if the orientation of the presented bright line will change, so that it will not match the retinal template, the line will now miss most of the connected photoreceptors and only a small fraction of them will be activated and send signals. This small fraction will reach the V1 neuron but will not be sufficient to excite the "horizontal –line" neuron above its threshold, hence the neuron will remain silent. As experimentalist, measuring the firing of the neuron to lines of different orientations we will conclude that the neuron is sensitive (also termed "tuned") to a horizontal line at a specific location.

If we think about the significance of this "tuning" of neurons, we can say that V1 neurons are "expecting" to see lines in particular orientations and locations on the retina. It turns out that such a neuron can be active even if the stimulating line is not perfect- i.e. will be a bit crooked or will be missing a small part, as long as the template activation will be sufficient to cross the threshold of the neuron- it will fire. Thus, a nerve cell can to some extent "correct" and complete slight mismatches between its a-priory expectations (which are built into its pattern of connections) and the real optical picture that actually arrives at the retina. We see therefore, that even at the earliest stages of the visual system- some creative element, some dissociation between the external optical information and the internal neuronal image already begins to appear.

PIECING THE PICTURE TOGETHER- SEQUENTIALLY

As I said above, the visual system is not a single entity but is made of a series of visual areas. In Figure 3 (left panel) these areas are depicted as a series of climbing stairs. This picture emerges from many studies showing that a cortical area integrates and abstracts the information coming from the area upstream to it. We have already seen how sensitivity to an elongated line can be generated using integration of signals. Let's now consider how this process can be extended along the cortical hierarchy. We start with the fact that V1 contains neurons that are specialized ("tuned") to different line orientations. Consider now that two such neurons, one tuned to a vertical line and the other to a horizontal line send their output branches to a "higher level" neuron in area V2- what will this target neuron in V2 be tuned to? Following the logic we discussed with regards to the integration in V1, we can see that by the mere convergence of signals from these two neurons and a threshold, a new kind of tuning or expectation emerges. Importantly, this type of tuning is not present in any of the inputs this neuron received from the area below it. Instead of being tuned to a vertical or a horizontal line- the V2 neuron will be sensitive to the "gestalt" of these two elements- i.e. to the symbol "+" (Figure 3 right panel). The reason for this creative ability is, as before, the threshold- each line, on its own, will not be sufficient to cross the threshold- however, when placed together in the form of a "+"- the combined excitation of the two inputs is sufficient to make the neuron start firing. While this is of course a simplified model, it illustrates how the visual system, through a sophisticated choice of connections between nerve cells can start solving the problem we considered in the beginning- how an isolated patterns of dots in the retina become a holistic picture in the mind of the observer.

In this example we used a simplified model built of only three nerve cells- imagine the richness and complexity that can be achieved if you consider many thousands of such nerve cells, all interconnected and integrating information at different levels of strength and of excitation or suppression.

The hierarchical process of integration does not stop in V2 and continues through the hierarchy of visual areas all the way to extremely complex and sophisticated template tuning. Indeed, recent experimental findings from the human cortex reveal that at the top of the hierarchy the integration process is so advanced that neurons are tuned to templates that appear similar to a full visual object- e.g. a face, a tool, a body-part or a topographical scene (Figure 3 right panel). These results show, for example that merely presenting elements of a face is not sufficient to excite the neurons at the top of the hierarchy. In order to get high firing the elements must be placed in the correct template configuration- i.e. in the correct relative position in the real object. Furthermore, if some of the picture elements are obstructed, yet we are still able to perceive the entire template, the neurons' firing continues as if the entire object was present- indicating the ability of such holistic neurons at the top of the hierarchy to "complete" the missing template information.

In summary, it appears that the hierarchy principle, although simple in its basic operation- i.e. converging connections between nerve cells, actually achieves tremendous richness and sophistication in tuning neurons at the top of the cortical hierarchy. We can say that the tuning of such holistic or gestalt neurons in fact constitutes our visual memory or expectation of the visual object. These expectations are embedded in the specific structure of neuronal connections-ready to produce activation whenever the proper optical pattern appears on the retina.

THE COMBINATORIAL EXPLOSION

It appears then that the hierarchical integration of information succeeded in solving what is commonly termed the "binding" problem- the problem of combining the isolated bits and pieces of the retinal pattern into a meaningful image. Indeed we find that at the top of the cortical hierarchy single neurons appear to accomplish this binding. Could it be then that the hierarchical integration is the critical principle that underlies all subjective experiences? In other words- could it be that whenever such a "gestalt' neuron, that is tuned, say, to the template of Boticelli's Venus, is firing- this in fact constitutes our mental image of Venus? As a historical note, this notion, of a nerve cell tuned to a single, specific picture, has been proposed, as a thought experiment, by the scientists Barlow and Letvin who coined it, tongue in cheek, "The Grandmother neuron"- arguing that perhaps such neurons exist in our brains, and if lesioned, this will prevent us from ever perceiving our grand-mother again!

Apparently we have succeeded in formulating a single principle that potentially could account for all conscious visual images- and as such indeed satisfies our search for an elegant theory. So did we finally arrive at the sought for link between brain and mind? Unfortunately, while the hierarchical integration solves the binding problem, this solution contains within it the seed of a deeper problem that makes this potentially elegant model completely unfeasible. The reason is simple- there are just too many different pictures that we are capable to consciously perceive. If we were to allocate for each visual image that can potentially emerge in our mind even a single specialized neuron, we will quickly run out of neurons even if we dedicate our entire brain to the visual sense.

To comprehend the enormity of the space of all possible distinguishable visual images- just consider the case of human faces. We can see clearly and likely differentiate between most human faces belonging to our own race- i.e. many millions of exemplars. Note that we are not dealing here with recognition- but merely the ability of our brain to construct a clear and distinct visual image of such faces. Now consider that each of these faces could appear in thousands of different variations- with a hat and without a hat, having many different colors of hair and skin complexions, being young or old, smiling, sad or surprised and on and on. Each and every one of these permutations should be multiplied by all the faces and all other parameters leading to trillions of distinct face exemplars- and each of these unique exemplars will necessitate the allocation of a special neuron– and there are simply not enough brain cells for such a task. This enormous combination of different possibilities has been termed the "combinatorial explosion" problem.

It is clear therefore that we simply don't have enough neurons in our entire brain to individually represent each perceivable visual image. And every nerve cell, even at the highest point in the cortical "hierarchy" must be tuned to many different templates. Experimental findings from the human visual system actually support this inevitable conclusion. They indicate that each neuron, e.g. a "face" neuron is actually tuned to literally thousands or even millions of different face templates. You can metaphorically conceptualize such a neuron as a "totem pole"- i.e. a collection of different face templates stacked on top of each other. The concept is in principle similar to the notion of a "grand mother" neuron that we discussed earlier- but the totem pole neuron consists of millions of different "grand mothers". Thus when we look at a face the retinal pattern will match one of the pre-existing face templates and the neuron will start firing. Because every neuron is tuned in parallel to millions of different templates, this presumably can solve the problem of the combinatorial explosion.

However this "solution" unfortunately only leads us again to a major new conundrum. Imagine that such a "totem-pole" neuron, responsible, in parallel, for the perception of thousands and perhaps even millions of different faces starts firing at a fast rate. How can we know which of the thousands of different faces that this neuron is tuned to- i.e. that are in this neuron's library- is the one we are supposed to see? In other words, the activity of an individual totem-pole neuron, firing alone, is utterly ambiguous- it does not point to any specific visual image.

THE NEURAL "WORD"

Here we come back to the idea of the written word and the Alphabet that I introduced in the beginning. The beauty and power in the invention of the written word is that it was a way to unambiguously represent millions of different concepts with just a handful of basic building blocks. The idea here is that in the same way that a written word is created by the combination of different letters, so is the visual image created by neurons- i.e. it is generated by the unique combination of a specific group of active nerve cells.

To see how this could happen, let's consider again our "totem-pole" neuron, and assume it is tuned to the face of an old friend of ours- "David". However, this is a totem pole neuron, so it is also tuned to many others faces, among them are also Doris, Debby and Dov. Let's call it for convenience the "D" neuron. When this neuron fires, how can we tell which of all these persons is the one we are supposed to see? The answer is that the D neuron is not acting alone: another neuron is located nearby and it is also active when we see David. However, in contrast to the D neuron, this neighbor neuron is tuned to another set of faces- in this case Mary, Jack and Larry. For convenience we will call this neighboring neuron the A neuron. Critically, as you probably noted, neuron D and neuron A are tuned to overlapping yet different libraries of faces. This is where the resemblance to the alphabet becomes apparent- just as it is with the alphabet; every neuron participates in the representation of thousands of images. However, the ambiguity is resolved through the *joint* activity of the neuronal group. We will use the term "cell assembly" for such a group- in honor of D. O. Hebb, a leading psychologist who coined this term. Note that in our toy model, in which the cell assembly consists of just of neuron D and neuron A- it is sufficient that both these neurons will be active, for the cell assembly to resolve the ambiguity – ruling out Doris, Debby and Dov and leading to the unambiguous emergence of *David's* face in our mind.

It is possible to experimentally estimate (very roughly) how many neurons are active when a single image is perceived -and the number is likely to be no less than a million. In other words, a neuronal word in the visual system consists of about a million different letters. Imagine the vast, truly awesome number of possible "words"- i.e. visual images- that can be generated in the brain with such a vast and powerful writing system.

MIND READING

For a moment it may seem that our"alphabet" model of neuronal activity finally solved the combinatorial problem- and hence the problem of conscious vision: neuronal "words" in the form of active cell assemblies are created and form the basis of the visual images that emerge in our mind. But, again, this solution is deceptive, and only leads us into an even deeper problem, which is perhaps the most mysterious and unresolved issue in consciousness research. We should recall that the meaning of a written word is not available to the word "on its own". There is no meaning to the collection of letters "D-a-v-i-d" without a reader- some process that receives these letters and binds them into a meaningful entity,

In the brain, then, there must be some mechanism that "inspects" the activity of the individual totem-pole neurons in the assembly and recognizes this pattern of activity as representing "David". The mere isolated activity of the nerve cells, even when they are active together, remains ambiguous- only when a process "reads-out" the pattern of activity and somehow integrates it into a holistic entity the ambiguity in the neuronal activity can be resolved. In a sense our situation has not advanced a bit beyond that of the pattern of activity in the retina- we simply substituted the isolated signals of the photo-receptors with that of highly sophisticated "totem pole" neurons. While the high level neurons indeed succeeded in binding the isolated pixels into a holistic gestalt- as isolated nerve cells their activity remains as ambiguous and meaningless as that of a photoreceptor in the retina.

Identifying the "reader" that binds and assigns meaning to the neuronal representations in the visual system is thus a fundamental step in our search for a neuronal theory of visual awareness. Influential theories attribute such read-out function to the frontal and parietal lobes of the brain. It is known that there are extra-visual frontal and parietal areas which are specialized for the highest cognitive roles- such as attentional control and decision making. A particularly intriguing, although poorly understood, high order function attributed to the frontal lobes is the representation of the sense of self- our ability to be aware of who we are. Although not well-defined, we do have a strong intuition that the visual image we see does not simply hangs up out there- but is actually owned by "me", the observer. Putting it in neuronal terms- it is an attractive possibility that the same frontal brain areas that represent the sense of self- our awareness of our own being- are also the ones that are engaged in reading out and binding the fragmented pattern of neuronal activity in the visual areas into a unified conscious percept. In our concrete example- *I* am the one who sees David's face and it therefore makes sense that frontal brain regions that are associated with my awareness of myself are the ones that integrate the isolated neuronal letters into a meaningful word and in the process endow the visual experience with its strong subjective flavor and perspective.

To examine if this intuition is correct we must rely on experiments – mainly with humans who can communicate to us such self-related mental states most accurately.

An important set of experiments that examine conscious vision are threshold experiments. In such experiments pictures are presented in such a way that they are difficult to perceive. For example, in a paradigm called "backward masking" we present a picture of a target object for a very brief period of time, a fraction of a second, and then we obstruct this picture by replacing it with a high contrast meaningless pattern of randomly oriented lines.

Subjects sit in front of the computer screen and indicate, either vocally or through button presses whether they succeeded in seeing the target image. By manipulating the duration of the target exposure, the experimenter can bring the subjects to the threshold of their perception- i.e. arrive at an exposure duration in which the subjects sometimes see the target and sometimes fail to see it. If we now examine what happens in the brain of these subjects at such threshold experiments- we find enhanced neuronal activity in high order areas of the visual system when the subjects succeed in seeing the targets, but at the same time we often see increased activity in the frontal and parietal parts of the brain as well.

It will seem then that indeed, theories that posit a higher order "readout" mechanism in the frontal part of the brain are supported by these experimental findings. However, closer look at this problem reveals that the situation is not so clear cut. The problem is that, while we indeed see brain activation in the frontal and parietal lobes when subjects succeed in seeing pictures, the cognitive process that is associated with this fronto-parietal activity is not clear.

Alternative theories could be brought up to explain such extra-visual involvement. For example, it may be the case that the frontal lobes, which are known to play a role in short-term, or "working" memory, may be related to an automatic tendency to memorize, at least briefly, the visual information we just consciously perceived. Even more general aspects, such as increasing our attention, or even the sense of reward a person experiences when successfully performing a difficult task – all these may produce the fronto-parietal brain activations.

Finally, and perhaps most significantly, such activity may be associated with our ability to reflect upon our own visual experience. This mental capacity, also termed "meta-cognition" is the process by which we not only experience the direct perceptual aspects of the image, but we also have the capacity to become aware of the fact that we are experiencing the percept. Note that when subjects are asked, in the course of an experiment, whether they saw a target or not, they are in fact requested, to some extent, to engage such meta-cognitive processes as well.

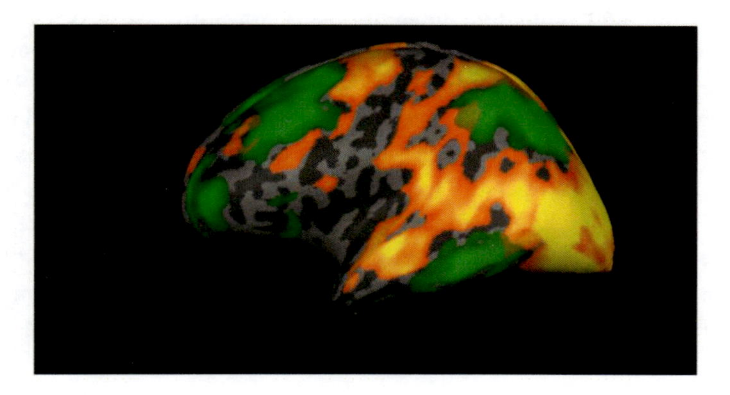

Figure 4.

Thus, we can see that the conventional visual research paradigm, where the scientists make every effort to control all sensory and behavioral parameters and obtain precise evaluation of the subjects' experience, is problematic. It leads to many additional layers of auxiliary cognitive processes that may not necessarily be an integral part of the perceptual process itself.

In a sense, this conundrum is reminiscent of what is called "the measurement problem" in quantum physics, where the experimental settings and the attempt to be obtain measurements that are as precise as possible actually interferes and disrupts the phenomena to be measured. In the cognitive domain our attempt to probe as accurately as we can the mental state of the subjects forces us to interfere with this very mental state and hence to distort the cognitive state we attempt to study.

It seems we are faced with a built-in limitation of cognitive research that is embedded, just as in physics, in the very nature of conscious awareness. Is it then possible to circumvent these methodological limitations by studying visual experience in a bit more open and spontaneous conditions? Can we create a situation in which the subject is free to focus on the visual experience itself and is not requested to constantly reflect or introspect about it?

GETTING ABSORBED

One psychological state that appears to potentially resolve this "measurement problem" is a condition that can be loosely called "absorption". This is the state you enter when you are swept away by the outside world. During absorption, we are so engaged and drawn into the immediate sensory experience that we lose all sense of self. We experience such states in moments of emergency, but often this can happen to us even during a highly engaging and captivating times such as when watching a thrilling movie. In such absorbing moments we lose all sense of time and reflection- metaphorically we "lose ourselves" in the experience. During absorbing moments the subjective experience is total and immediate, and we don't have time for introspection or meta-cognition. To study what happens in the brain during such moments, it is necessary to simulate them inside a brain scanner. While it is obviously difficult to recreate total absorption in such an enclosed and noisy environment – an absorbing state can be approximated. One particularly successful approach, we found, was to show subjects a highly engaging movie. We essentially take advantage here of the genius of great movie directors (e.g. Sergio Leone in the "Good the bad and the ugly") to take hold of our focus of attention. Other, less colorful ways are to engage the subjects in extremely rapid and difficult sensory-motor tasks, which require their full attention and do not allow any time for more introspective interferences.

When examining what happens in the brain during such absorbing moments, we find a surprising and intriguing phenomenon. First, as expected from previous research, large parts of the cortex, particularly towards the back of the brain (more technically, the occipital, parietal and temporal lobes) including the visual system, are highly activated by the sensory stimuli. However, no less interesting than these rich patterns of sensory-driven activations is the observation that within the "sea" of sensory and motor areas, there were distinct islands that consistently *failed to* respond to the external stimulation. These were cortical areas that, instead of being excited by information coming from the external world, appeared to be tuned to aspects of a more internally-oriented information (Figure 4).

While the role of each of these non-responsive islands is not fully understood yet-ongoing research is gradually revealing that indeed these regions are associated, among other functions, with self-related tasks, such as self-evaluation, voluntary decisions and autobiographical memories. To reflect their self- orientation we have termed this system of areas- the "Intrinsic" system. It is also commonly termed, for other reasons, the "Default mode" network.

THE INS AND OUTS OF THE HUMAN CORTEX

We see that using engaging movies and other highly absorbing sensory-motor tasks has, unexpectedly, uncovered a fundamental subdivision of the human cortex into two global systems. One, the "Extrinsic" system is oriented *outward*- it deals with information derived from the external world. The other, the "Intrinsic" system is oriented *inward*, and is related to information derived from the subject- the individual's self (Figure 4).

To illustrate the complementary functional specialization of the two systems lets consider two extreme examples: consider a Samurai warrior aiming an arrow, becoming one with his external target. In such a complete focus on the external world, we can presume that the *Extrinsic* system is at its highest activity. In contrast, consider attending a boring lecture, when the mind "wonders" and we find ourselves detached from the external world, and instead worry about some pressing decisions we need to make. At such moments, the *Intrinsic* system is turned on while the Extrinsic system reduces its activity.

We can actually observe such swings in activity during brain imaging experiments. When subjects' brains are scanned while they perform an demanding externally-oriented task, we see the Intrinsic system actually lowering its activity. In contrast, when we detach from the external environment and engage in thinking about ourselves-planning or recalling some personal episodes, we see the swing reversing and the Intrinsic system activated while the Extrinsic system reducing its activity.

This antagonistic relationship between the two systems occurs only under extremely focused moments. Under most daily life situations both systems are co-active at various degrees, reflecting the needs for combined intrinsic and extrinsic information.

However, there is one mental state which is particularly intriguing and which is worth considering separately. This is the condition called "introspection"- when we focus at the same time both on an external object- say a cup of coffee but at the same time we are also fully alert to the fact that we, as observing subjects, are present and are the ones who experience this cup. Interestingly, we see that under such special moments, both Extrinsic and Intrinsic systems are co-active in a fairly balanced manner. It is interesting to consider that such a particularly balanced mental state constitutes the starting point of all philosophizing about the mind. This is the reflective state in which we become simultaneously aware both of ourselves and our sensory experience – leading to the deep puzzlement about how such experiences could be generated.

Another fascinating question concerns the activity of the Intrinsic and Extrinsic systems during other unique mental states such as various meditative practices. It is interesting that within meditative traditions, some emphasize absorption and merging with the external world- i.e. purely Extrinsic functions- for example in certain Zen practices. In other types of meditation, where the aim is a more inner focus and detachment from external disturbances, we would expect the opposite -Intrinsic - emphasis.

And in-between there are the fascinating practices that aim at a perfect balance, at some merging or blur of the boundaries between the external and internal worlds. We still don't know what happens in the brain during such states- but this is experimentally tractable and constitutes a fascinating topic for future research.

How are these new insights into brain organization impact on our original problem of the "reader" and its role in conscious perception? We see that contrary to the intuitive sense we have that seeing must always involve some aspect of self-awareness- the experimental findings indicate that the opposite is true. Regions in the brain associated with accessing the self appear to be part of a specialized system that often shows *antagonistic* activity to that found in sensory areas. Thus, the self-oriented Intrinsic system appears to shut off precisely when the visual experience is particularly intense, rich and engaging. Conversely, sensory experience is diminished when we are disrupted by self-related processes. Clearly then, the Intrinsic, self-related system does not appear to serve in the role of the read-out mechanism that is essential for visual perception. The principle that appears to dominate brain organization is thus the principle of specialization- in which different brain systems specialize in different cognitive functions. This specialization dominates also the relationship between visual perception and self-awareness- these two mental states will be joined and split as the situation calls for, and one can not therefore be fully dependent on the other.

ACTIVE MIRRORS

From all the above it appears that the notion that "subjectivity"- i.e. self-representation is the mechanism that "reads out" and makes sense of the neuronal activity in the visual system is actually contrary to recent experimental observations. Thus, we are back- facing the main problem with which we began our enquiry- where is the "readout" mechanism that can combine the isolated neuronal letters into meaningful holistic entities? In other words, what is the mechanism that binds or glues together the isolated activity of the individual "totem-pole" visual neurons? As we saw above, the activity of isolated neurons remains ambiguous and meaningless. The binding mechanism therefore is the crucial process that endows visual meaning to the neuronal activity and thus is the key mechanisms that leads to the emergence of a conscious visual image in our mind.

I would like to propose that we should not look for the readout and binding outside of the visual system - by some yet higher order brain area. Rather, the read-out and the neuronal binding are accomplished *locally* by the very same high order visual neurons that form the active assembly to begin with. In other words, the same "totem pole" neurons- the neurons that constitute the individual letters of the neural "word" become a unified meaningful entity when they "read out" each other (Figure 5). To understand how such self readout can be implemented, let's begin with the very simple notion of mirroring or reflection. But we will consider a special type of such reflection – an *active* reflection. This is a sort of a dialogue in which both the mirror and the mirrored actively take part.

It is interesting that one of the most familiar mirroring in western folklore is that of the evil queen in the fairy tale "Snow-white". Note that the mirror in this archetypical story is not a simple mirror, of the kind we encounter in the bathroom. This is an *active* mirror, one that is both influenced by and influences back the mental state of the mirrored- the queen. In other words, among the various factors that define the mental state of the queen, we must include also the state of the mirror. On the other hand, the mental state of the mirror itself also reflects the queen's which in turn incorporates the mirror's and so on. We see that such a process of mutual influence (also called recursive interaction) is continuous and blurs the distinction of the mirrored from its reflection. Note that from an information point of view- that is, considering our ability to know the state of affairs, such a process of recursive interaction could be considered as binding the queen and the mirror into one informational entity. Thus, these two agents that started, before the mirroring process, as completely separate entities- in the sense that knowing the state of one had no relevance to the state of the other- now become, following the reciprocal exchange of signals between them, a joint entity- so that knowledge gained about the mirror, will also tell us something about the mirrored queen, and vice versa. Importantly, in such recursive process- we lose the ability to separate out the isolated contributions of the mirror and the mirrored- because they are entangled beyond separation through their mutual dialogue.

It is not difficult to see how this principle of mutual "mirroring" can be expanded to a group of agents. Think, for example, about a discussion group, in which each of the participants truly listens and is sensitive to the others. In such group discussion the mental state of each of the participants, can not be considered any more as isolated from the others. The state of each participant is actually a weighted sum of all the mental states of the other members in the group. This is a process of a "group mirroring" – every member in the group is not isolated any more but is both affecting and being affected reciprocally by all others- in a recursive continuous process of mutual interactions. Note that such group mirroring does not imply that the individual members in the group become identical. Each member has its own unique individuality reflected, for example, in each person's unique sensitivity to the messages delivered by specific individuals in the group. However, through the exchange of signals, a new entity is now created from the previously isolated members of the group- a holistically unified assembly.

Finally, a critical condition for group discussion to be successful is reaching a stable consensus. To see why this is important, consider a situation whereby a number of the members in the group keep vacillating, and constantly change their mind. If this happens fast enough, by the time a message from one of these vacillators is delivered to the group, they already change their mind. This leads to a mismatch between how the mental state of an individual is perceived by the group, and the true mental state of the individual. In order for all mental states to be correctly coherent with how they are perceived there must be at least a moment of stability- in which all messages have sufficient time to arrive before some members change their minds- so that all group members "agree" on what each one of them is thinking about.

With a bit of creative imagination it is hopefully possible to see the parallels between such group discussions and the exchange of signals between neurons in an assembly.

As I noted earlier, nerve cells were optimized by evolution precisely for the task of social group communication. With their numerous branches and contacts to other neurons, each cell is capable to "listen" in parallel to thousands of other group members and integrate their messages. The "mental state" of a neuron is its state of excitation which is reflected in its spiking activity. Note that similar to the group discussion metaphor; the internal state of each neuron recursively incorporates in it the states of all other neurons connected to it, including its own prior state.

Figure 5.

Group Reflection and Conscious Experience

Now that the concept of group mirroring or reflection has been described, the central hypothesis of this chapter can be formulated: subjective experience emerges when group mirroring binds neurons into a meaningful entity. Put differently- the critical condition for the emergence of subjective experience is the integration of neuronal "letters" into meaningful "words" through the reciprocal exchange of signals.

Let us examine this notion in the concrete example we discussed above – the emergence of the face of "David "in the mind of an observer. We saw earlier that such a face elicits a pattern of activity in a large number of "totem-pole" neurons- neurons that are tuned to many different faces including "David". However in order to resolve the ambiguity inherent in the multi-face sensitivity of each isolated neuron (the letters) - a group of such neurons must be bound into a single assembly or entity- in other words to become a "word".

The suggestion is that this binding process does not depend on an external readout mechanism, but it occurs locally within the visual system itself. Such assembly binding happens when a number of neurons that share the same template in their library activate each other. If such mutual activation forms a stable group, this assembly unambiguously defines a unique percept, and at this moment a visual image appears in our mind (Figure 5). Note that this process is continuously recursive and can not be broken into discrete steps- thus it, in principle, could not be replaced by a conventional digital computer that depend on temporally discrete symbols. To summarize, it is proposed that subjective experience is the product of stable recursive activation within a cell assembly.

SPLITTING THE MIND

As stressed above, any hypothesis has merit as long as it is compatible with experimental observations. Unfortunately human brain research is at an early stage of development and conclusive evidence with regards to issues of consciousness is still lacking. Thus, the notion of neuronal reflections as the basis of subjective experience, while leading to testable predictions, is only a hypothetical framework at this stage. A strong prediction is of course that preventing reciprocal interactions among neighboring neurons should abolish all subjective experience. However, we are far from being able to generate such massive disruptions, particularly in conscious humans. Nevertheless, the reflection hypothesis allows us to explain a number of known observations. I will touch here on just two of them. The first one relates to the anatomical structure of the cerebral cortex. The cortex is the main "suspect" when it comes to the generation of subjective experience, especially since, as I pointed out above, local damage to this tissue leads to immediate and often specific deficits in awareness. For example, local damage to the temporal lobes may cause a loss in the ability to see colors or to recognize individual faces.

But what is it that makes the cerebral cortex so special compared to other brain structures? In recent years it was established that one of the unique features of the cerebral cortex is a particularly dense network of local branches inter-connecting neighboring cortical neurons. This network, called intrinsic connections (since they are intrinsic to a single cortical area), provides an optimal hardware for the type of local group reflections hypothesized as critical to subjective awareness.

However, because of the way the brain is built, each cortical area is actually split in half, with each half placed in a separate brain hemisphere. Because of this anatomical quirk, at the places where a cortical area is split- the intrinsic connections need to extend long distances (from one hemisphere to the other) so as to connect neighboring cortical points. Thus, we have here a unique situation in which connections between neurons that are functionally local are physically "stretched" and now need to send long distance branches to interact with their "neighbors".

An intriguing possible link between this interhemispheric network of connections and conscious awareness has been discovered in one of the most significant observations in the history of consciousness research. It was found in patients that, for clinical purposes, had the inter-hemispheric system of connections severed (to prevent spread of epileptic seizures). Detailed research conducted in such "split brain" patients revealed the astonishing observation that this operation leads to the separation of conscious awareness into two isolated sets of subjective awareness states. In such split brain patients each hemisphere became unaware of the subjective state of the other. In other words, preventing the mutual exchange of signals between neurons located across the two hemispheres led to the split of the commonly unified conscious awareness into two separate subjective experiences co-existing within a single person's head!

How can this amazing observation be explained within the framework of neuronal reflections? In the "split brain" operation, the ability of a cell assembly to exchange signals across the two cortical hemispheres and in this manner to bind the groups of neurons into a unified entity is surgically interrupted. Because of this physical block, neuronal reflection is possible in these patients only on one or the other side of the brain and can not occur across hemispheres. According to the reflections hypothesis such separate assemblies indeed should lead to the emergence of two separate subjective states. Thus, the split brain phenomenon provides a striking demonstration of the critical role neuronal interconnections play in the formation of a unified conscious experience.

Neuronal Ignitions

A second experimental observation is the repeated findings that conscious awareness is always associated with intense and sustained bursts of spikes emitted by the relevant neurons. In threshold experiments of the kind described above, we see evidence that crossing the visibility threshold is associated with a rapid and intense "ignition" within local neuronal groups. This observation points to a tight link between perceptual awareness and intense and long lasting bursts of neuronal activity. The emergence of such "ignitions" is nicely compatible with neuronal reflections. Note that when neurons reciprocally excite each other-this can lead to an explosive process- also termed "positive feedback" - so in fact it is expected that neuronal reflections will naturally lead to such ignitions. Furthermore, the rapid firing is critically needed for the spiking activity to successfully perform its binding function. Note that the main way in which nerve cells can reveal their functional state to other neurons is through the signals (spikes) they emit. A substantial part of this information is coded in the length of the interval separating subsequent spikes – essentially the instantaneous firing rate of the neuron. If the neurons will use low firing rates for communication it will take too long for any unambiguous consensus to be reached by the neuronal assembly. It is as if, in our metaphorical group discussion, each person will emit only one word every hour – imagine the length of time it will take to then reach a consensus. Rapid firing of neurons is thus essential for effective neuronal reflections, and if the reflection hypothesis has any merit, also to the emergence of a conscious visual percept.

Summary

We certainly face a long, tortuous road in our journey towards understanding the mystery of conscious awareness. I tried to illustrate in this chapter how advancements in research methods, particularly in human brain research, opened an exciting and promising era of progress in this endeavor. The idea I try to promote is that the neuronal "group reflection"- i.e. a reciprocal information exchange in local neuronal assemblies is at the heart of conscious experience. While I emphasized here concepts and knowledge obtained from the visual system- the principle of neuronal reflections could be extended to other sensory modalities or conscious cognitive states. The difference between different subjective experiences states depends on the identity of the neurons that are united by the reflections and their relative position relative to the rest of the brain. It is interesting to note, in this context, that the word 'Consciousness' literally means "common knowledge". It seems that the intuition of ancient thinkers anticipated modern brain research- both in the realization of the power inherent in combinatorics - as in the invention of the alphabet, and in the critical role of cooperation in the emergence of conscious experience.

REFERENCES

Baars, B. J., Ramsoy, T. Z. and Laureys, S. (2003). Brain, conscious experience and the observing self. *Trends in Neurosciences, 26*(12), 671-675.

Block, N. (2005). Two neural correlates of consciousness. *Trends in Cognitive Sciences, 9,* 41–89.

Block, N. (2007). Consciousness, accessibility, and the mesh between psychology and neuroscience. *Behavioral and Brain Sciences, 30*(5-6), 481-548.

Dehaene, S. and J. P. Changeux (2011). Experimental and theoretical approaches to conscious processing. *Neuron, 70*(2), 200-227.

Haynes, J.-D. (2009). Decoding visual consciousness from human brain signals. *Trends in Cognitive Sciences, 13*(5), 194-202.

Lamme, V. A. F. (2006). Towards a true neural stance on consciousness. *Trends In_Cognitive Sciences, 10*(11), 494-501.

Rosenthal, D.M. (2005). *Consciousness and mind.* Oxford, UK: Clarendon Press

Searle, J.R., Dennett, D.C. and Chalmers, D.J. (1997). *The mystery of consciousness.* New York: New York Review of Books

Rees, G., Kreiman, G. and Koch, C. (2002). Neural correlates of consciousness in humans. *Nature Reviews Neuroscience, 3*(4), 261-270.

Tononi, G. and C. Koch (2008). The neural correlates of consciousness - An update. *Year_in Cognitive Neuroscience, 1124*, 239-261.

Zeki, S. (2001). Localization and globalization in conscious vision. *Annual Review of Neuroscience, 24*, 57-86.

The Physical Approaches

In: Consciousness: Its Nature and Functions
Editors: Shulamith Kreitler and Oded Maimon
ISBN 978-1-62081-096-5

Chapter 20

A Quantum Physical Perspective of Consciousness

Ron Kreitler*

Independent Researcher, 5 Spinoza Street, Tel-Aviv, Israel

Abstract

This chapter analyzes similarities and differences between quantum-physics and consciousness, using non-mathematical language. It will be argued that the mind-body schism, which puzzles psychologists and neurologists alike, can be resolved by applying extra-dimensions to the study of consciousness. This set of dimensions provides a unique and non-classical view of the universe at large and of the human brain in particular. An hypothesis regarding the localization of consciousness in the brain will hence be presented. Finally, an experiment designed to prove the extra dimensional nature of cognitive activity will be described.

Introduction

Consciousness is among the most contentious issues in modern science because it is one of the least coherently defined concepts. Up to date, consciousness has not even been located. Therefore, scientists are deeply divided concerning its mode of operation. Reductionists, most notably Crick (Crick and Koch 1998), assume the mind is solely a product of neural activity, while most psychologists argue the mind includes more than its physiological constituents, suggesting that cognitive processes underlie consciousness. But both the reductionist and the cognitive schools do not define the mind as such, focusing their research either on operational aspects i.e., brain waves, attitudes etc., that are accessible to observation, or on comparisons to other fields of science. In past decades it was customary to search for biological models of consciousness. Recently, several quantum-mechanical models of consciousness were proposed.

* E-mail: kreitler@netvision.net.il; Tel: +972-3-5238040; Mobile +972-544590725.

NON-LOCALITY

The connection between consciousness and quantum-mechanics has been suggested by none other than the founding father of quantum-mechanics, von Neumann, who showed that only an act of observation can collapse a superposition wave-function. For example, an electron firing device in a so-called double-slit experiment in which an electron is fired at a screen with two slits, produces a wave-function, as long as it is not directly observed. Only when an act of observation is conducted, do electrons appear as particles, passing through either one of the slits (Bohm, 1980; Sakurai and Napolitano, 2011). Therefore, the experimenter who determines whether or not an act of observation will be conducted is not an objective entity recording results, but an integral part of every quantum physical experiment (Rosenblum and Kuttner, 2006).

The consequent role of an observer in an experiment, as well as the capacity of particles to be exposed as waves – a phenomenon called particle-wave duality – when an act of measurement is not conducted, led theorists such as Young to suggest that indeterminism is a characteristic of particles. Young pointed out that Heisenberg's Uncertainty principle, which states that the momentum and location of a particle cannot be simultaneously assessed, indicates indeterminism is an integral part of quantum-mechanics. Hence, the universe is comprised of a set of probabilities. Young argued that since particles, which are the building blocks of all matter are not predetermined, they might possess intentional aspects (Young, 1975).

Orlov proposed a quantum model of doubt-states which compares indecisiveness with photon polarization. In Orlov's model of quantum thinking, the two options a person has according to Boolean logic arise from an ambiguity to which he refers as "the wave logic of consciousness". Orlov's wave logic presents a sphere of options, called an Orlov Sphere, which translates states of indecision into wave-functions. For example, if we hear a sound at the door we may be in doubt as to whether our partner just returned home or whether someone is attempting to break in. Orlov considers these two alternatives as being not exclusive but rather as complementary, thus making them wavelike. Technically this is achieved by drawing a wave-function, which outlines the percentage attributed to both possibilities, namely, finding our partner or a robber at the door. Consequently, there are two waves. Since waves have a degree of freedom called phase, that measures the extent to which the two waves are in phase with one another, they can be mapped into a sphere. The numbers N (referring to the chance we attribute to any of the two possible outcomes) and P (referring to phase) make it possible to map the entire surface of the sphere according to varying states of uncertainty. Locations on the surface of the Orlov Sphere correspond to degrees of certainty, e.g., the north-pole corresponds to our being sure our partner came home, whereas the sphere's south-pole corresponds to absolute certainty as to there being a robber. All north-south parallels represent intermediate degrees of uncertainty, and the equator designates a 50 percent state of doubt. The phase variable measures longitude, similar to Earth's longitudinal lines, with the Greenwich meridian and the opposite Date line corresponding to 0 and 180 degrees in tandem. Hence, all locations around the sphere can be attributed to specific states of doubt. Otherwise put, every possible state of doubt can be represented by a single point on the Orlov Sphere (Herbert, 1994).

Quantum aspects of consciousness were also studied by Eccles (Beck and Eccles, 1992). He suggested that conscious events occur at a vesicle of the nervous system's synapses. These synaptic emissions are small enough to fall within the realm of quantum physics, which has been found to be below the Plank constant (10^{-33}cm.). Hence, they are subject to Heisenberg's Uncertainty Principle, which applies to entities below the Plank constant. According to Eccles, what the mind does is to alter the probabilities of emission in such a way as to make the brain perform the desired physical tasks. More precisely, mental activity is being defined as a selection of vesicles which already have a probability of firing (Paster, 2006).

Much attention has lately been devoted to quantum activity in the brain. Cambridge University physicist Donald proposes the brain has several quantum switches, in particular sodium channel proteins. State University of New York physicist Mould introduced the term "common mechanism" to define biological processes that participate in the process of collapsing wave-functions. Among these he includes endorphin proteins, whose operation takes place within the size-range of quantum effects. Berkeley physicist Stapp (2007) suggests calcium ions in neurons are biological entities that participate in the quantum process of collapsing wave-functions. Stapp, like Penrose (whose work will be discussed later), compares the mind-body schism to the divide between classical and quantum mechanics, suggesting that the study of consciousness can help create a theory of everything – a term often used to describe a theory that unites quantum mechanics with classical physics (Paster, 2006).

Attempts to construct quantum computers indicate how deeply rooted the notion of a quantum mind has become in recent years. Although quantum interactions within the brain have not yet been fully understood, it is widely accepted that the mind operates according to quantum principles. This notion can be traced back to the physicist David Bohm who introduced holography into physics. According to Bohm, the tangible reality in which we live is an illusion, similar to a holographic image. Like most illusions also this one is a product of our mind. Underlying the illusion is a hidden reality that merges the entire universe, even uniting past, present and future, producing all objects we perceive. Bohm calls this level of reality the "Implicate" which means enfolded. To our own level of existence he refers as the "explicate" or unfolded level of reality. His theory regards the manifestations of all forms in the universe as resulting from countless enfoldings and unfoldings between these two orders (Bohm, 1980). Essentially, Bohm claims, the mind produces the reality we observe which in turn is governed by quantum principles.

Pribram clarified this concept by proposing the brain operates according to holographic principles. More explicitly, he suggested that in addition to the circuitry based on the large fiber tracts in the brain, processing occurs also in webs of fine fiber branches i.e., dendrites that form webs. This type of processing is described by Gabor quanta of information. These are wavelets, which are used in quantum holography, and form the basis for fMRI, PET scans and other image processing procedures. Gabor wavelets are windowed Fourier transforms that convert complex spatial (and temporal) patterns into component waves whose amplitudes become reinforced or diminished at their intersections. Fourier processes are the basis of holography. Holograms can correlate and store a huge amount of information (Pribram, 1971, 1982).

In a set of dramatic experiments, which strongly affected Pribram, Lashley (1950) demonstrated the holographic nature of brain activity. Lashley trained rats to perform tasks such as maze running, before he removed parts of their brains. His initial goal was to locate the area of the brain responsible for memory, but to his surprise he discovered that the rats' memory could not be impaired. Even after removing 90 percent of a rat's visual cortex,

Lashley found that the rats could still perform tasks requiring complicated visual skills, indicating they could see (Talbot, 1992). Further research conducted by Pribram revealed that 98 percent of a cat's optic nerves can be severed without hindering its visual capabilities (Pribram, 1971).

One of the most important aspects of holography concerns non-local influences. These were introduced into psychology by Jung. His proposition of archetypes, originating from the collective unconscious, presupposes the existence of non-local influences. According to Jung, archetypes are shared by the majority of people and are transmitted from one generation to another (Jung, 1981). Suggesting that thoughts do not necessarily originate from the experiences and personal history of an individual is akin to proposing a non-local influence. In contrast to Jung, Pribram extended the study of Gestalt to the field of perception. He proposed that we recognize alphabetical letters not according to their specific form, but according to our expectations regarding what their form should be (Pribram, 1991). Since our recognition of shapes results from gestalt, it is evident that clues which do not concern a present external situation influence our thoughts. Recently, the study of non-local influences has been further extended to the field of memory retrieval. According to Norman, environmental clues often awaken memories to which they are only associatively connected (Polyn, Natu, Cohen and Norman, 2005). Sitting at a table in front of our boss we may recall ourselves being scolded by a teacher at school; hence we may develop a defensive predisposition, which would not have resulted from anything our boss has done. The similarity between the two scenes might be reduced to our position at the end of table or to the color of the chair we were sitting on.

No unequivocal explanation has hitherto been provided as to how the connection between an external clue and a particular memory is facilitated. Studies of creativity, which attempted to localize talents, did not point to a particular brain center, but rather to a number of centers (Gardner, 1983, 1993). The so-called whole-brain theory proposes brain functions are inseparable. The theory suggests reciprocal activation within the sub-processes of attention, perception, learning and memory (Besar, 2006). Proponents of the whole-brain theory maintain that brain activity while remembering something is similar to what it is during a first experience. Also Pribram (1991) claims memories are generally recorded all over the brain, and information concerning a specific object or event is not stored in a particular cell or localized part of the brain (Bohm, 1980).

The preposition of non-local influences is not unique to mind-related studies. Also quantum physics relies in part upon so-called hidden variables (Bohm, 1980). These are variables e.g., particle-spin axes that cannot be simultaneously measured. Bohm claims particles have both a definite position and a definite velocity even though we can never measure both of them at the same time. The explanation for the existence of hidden variables is based upon non-local influences. According to Bohm the forces acting upon a particle at one location depend on conditions at a distant location (Bohm, 1980). Non-locality is a phenomenon that has bewildered physicists for over a generation. Already Einstein was appalled by the proposition of non-local contact and together with his colleagues Podolsky and Rosen devised the EPR (Einstein- Podolsky-Rosen) experiment, which was conducted by Aspect in 1982 when particle accelerators were available (Sakurai and Napolitano, 2011). Einstein et al. suggested an experiment in which photons would be emitted in opposite directions within an accelerator. After traversing a certain distance at the speed of light (photons always travel in the speed of light) within the accelerator, the route of one of the two photons would be intercepted, thus altering its angle of polarization (a term vaguely referring to a particle's route). Quantum theory predicts the other photon should change its angle of polarization in accordance with its distant counterpart. Einstein et al. (Einstein, 1935) argued any contact between the two photons traveling (as photons do) in the speed of light, would necessitate faster than light communication, which contradicts the laws of physics (Einstein, 1935). Bohr responded by proposing the two photons did not need to communicate over space because both constitute parts of a larger system: if two particles have once been part of a single system, they always maintain a connection with one another (Bohm, 1980; Bohr, 1958). This connection is called the quantum connection. Evidently, the quantum connection has non-local attributes.

Einstein and most physicists after him were puzzled by non-locality. In classical physics one object can act upon another only by means of a force field like gravity that traverses space at a finite velocity, but the quantum connection violates this principle. Unlike any other force that acts upon its immediate field, the quantum connection is unmediated, which means it jumps directly from one point to another. Moreover, it does not weaken over distance. Since it does not actually traverse space, it also cannot be intercepted. No shield can therefore prevent its passage. Unlike other forces which travel over time, the quantum connection is immediate. Last but not least, it discriminates between systems with which it has interacted since it was last measured, affecting only them, and not bearing any effect on other systems with which it has not interacted. Once an act of observation is conducted, all former connections are severed and new ones are created (Herbert, 1994).

Two generations of physicists wondered at these elusive characteristics of the quantum connection. Likewise, several generations of psychologists and neurologists were unable to localize the process of associations. Evidently, both quantum-mechanics and psychology are based upon connections, which are to date only partly explained. As mentioned above, the quantum connection though empirically verified, is not entirely understood. Similarly, the process of memory retrieval is not fully comprehended. Even though the principles of memory retrieval are generally understood, the process of association remains unexplained. Apparently, both fields rely on connections that can be empirically verified and can therefore be objectively studied, but cannot be unequivocally explained. Therefore, the main similarity between the two fields lies in their ambiguity. Essentially the problem seems to be one of interconnectedness. Both quantum-mechanics and the study of cognitive processes are obstructed by non-local influences.

Interconnectedness and Dimensional Dilation

In his theory of general relativity Einstein showed that space is the product of gravity. More explicitly, it is gravity that shapes the universe, and produces the forms we observe, including the orbits of planets (Resnick, Halliday and Krane, 2002; Green, 1999). Hence, the geometry we see everywhere around us results from connections gravity produces. Other connections, which the gravitational force does not allow, are non-existent insofar as we may be able to judge. The frame of reference (FR) in which we are situated prevents us from seeing or otherwise perceiving connections that violate the geometry gravity produces. However, the theory of general relativity also postulates that a change in FR can provide us with a different vantage point of the universe (Einstein, 1920; Resnick, Halliday and Krane,

2002). Hence, it may be speculated that variations in gravity allow connections that would otherwise be precluded. For example, within a black hole where gravity is immense, relativity fails (Randall, 2006). This situation is referred to as a black hole singularity. What happens in a singularity is that space-time is thus warped by the forces of gravity, that all dimensions of space are dilated. Assuming gravity produces space, a significant increase in gravity should curve space. The stronger gravity becomes, the more warped space is expected to be. Consequently, distances, when measured by the shortest straight line between two points, shrink. When gravity becomes extremely strong – as it does within a black hole – distances disappear, and the dimensions of space simply vanish. The term singularity originates from the assumption that everything is entangled with everything else. This is the result of dimensional dilation. It is dimensional dilation that creates connections which are forbidden outside a black hole.

So, one could be tempted to assume that gravity holds the key to explaining the quantum connection. According to this line of thought, if gravity had been significantly weaker than it is, it would have allowed connections that, from our FR, seem non-local. Indeed, gravity barely affects quantum processes and is usually ignored in particle-physical calculations (Randall, 2006). Gravity is a force that acts upon objects. At the size scales of particles, its influence is negligible. Penrose claims that above a critical size (the Plank constant) space-time curvature effects cause the system's wave function to collapse under its own weight (Herbert, 1994). Below the Plank constant, which is 10^{-33}cm, the waves that particles emit do not collapse, and can therefore be measured, which comes to show how weak gravity is in the particle FR. Its weakness can also be derived from de Broglie's equation. De Broglie showed that every object of mass emits waves. His equation predicts an inverse relation between the momentum of an object and the length of its wave (De Broglie, 1949). In simpler words, the larger an object is, the smaller are its waves. Thus, when extremely small entities such as particles are concerned, waves are relatively large and momentum is small enough to permit the emission of waves. When objects of any kind emit waves, their boundaries cannot be defined. It is reasonable to assume, a wavelike surrounding permits connections that would not seem possible for objects with clear boundaries. Since the size-scale of particles never exceeds the Plank constant, it is tempting to speculate that particles inhabit a gravity-free realm.

Even though most quantum calculations ignore the effects of gravity, an assumption according to which gravity is non-existent within the FR of particles would be entirely false. Indeed, particles do not exert gravity, and therefore its effects can safely be ignored in quantum mechanical calculations, but that does not imply particles are not influenced by gravity. The effects gravity does bear upon particles can be seen in a phenomenon called gravitational redshift. The redshift of light is created because photons decrease their frequency as they escape from a gravitational field. Since a photon always travels at the speed of light, it cannot slow down, as larger objects do, when subjected to the force of gravity. So photon-waves are simply stretched out and this phenomenon is referred to as redshift. Hence, we can see that though elementary particles do not exert gravity, they are affected by it. This should come as no surprise since Einstein predicted by what amount light should bend as a result of the sun's gravitational field (Randall, 2006). Contemporary research is aimed at verifying the speculated existence of the Higgs particle which is supposed to endow other particles with gravity.

Nevertheless, scientific attention is focused on the relative weakness of gravity in the particle FR. Sometimes it is referred to as the "hierarchy problem of forces" and in other contexts it is termed "background independence". The latter term refers to the reliance of theories upon the background of space-time. When string theorists try to unite quantum-mechanics with the theory of relativity, they face the following theoretical problem: Most versions of string theory take space-time for granted, presupposing its existence and drawing their conclusions on the basis of this assumption. But according to general relativity the force of gravity shapes space-time, and so it follows that gravitational influences should be significant in the particle FR, which they are not. The evident weakness of gravity in relation to the other three forces is considered as the hierarchy problem (Smolin, 2007). Attempts are made to formulate a theory which will not be reliant upon space-time.

One reason for the evident and undisputed weakness of gravity in the particle FR might be related to their relative speed. To explain the connection between speed and gravity it is advisable to envisage an airplane. Airplanes ascend due to their speed. The same is true for rockets and spaceships, which stay in orbit thanks to their acceleration. Since the relative speed of most particles is significantly higher than that of any larger object we can see, it is logical to assume that the effect gravity exerts on them is reduced. This is evident from a phenomenon referred to as *time dilation*. In his Special Theory of Relativity, Einstein showed that in high velocities, time slows down. This phenomenon is called *time dilation* (Einstein, 1920). If we place an atomic clock on a spaceship, and then compare it to an atomic clock that remained on Earth, the latter would have advanced further. Verification for time dilations also comes from high energy particle physics: when Muons move at 99.5 percent of the speed of light, they decay (turn into another lighter particle) more slowly than do slower moving Muons. The lifetime of the fast-moving Muons increases by a factor of approximately ten. According to the Special Theory of Relativity, the actual lifetime of these Muons and other fast-moving particles is not longer. It is their relative time-flow that slows down (Sakurai and Napolitano, 2011). Einstein's famous twin paradox illustrates this strange phenomenon. If one twin left Earth and traveled at half the speed of light, he would return younger than his twin brother who stayed on Earth. From the traveler's point of view, less time has elapsed than has been measured by his Earth-bound twin. It can therefore be concluded that relative speed of motion determines not only our acceleration toward Earth, hence the pull of gravity upon us, but also our exposure to the motion of time. So, the high speed of particles could likewise reduce the influence of gravity upon them.

According to Einstein's principle of equivalence, the forces a person or any other object experiences due to gravity and to acceleration are the same. These two forces can therefore be regarded as being equivalent. Hence, both gravity and acceleration, which is a form of motion, bend time (Resnick, Halliday and Krane 2002). The bending effect is evident from time dilation: the stronger gravity is, the slower time moves. Within a black hole, time is speculated to be in a standstill. Likewise, at the speed of light, time is motionless (Greene, 1999). Therefore the photon, which constantly moves at the speed of light, is situated in a timeless FR. Consequently, instantaneous contact between two photons such as occurred in the EPR experiment is possible. This might serve as further demonstration for the principle of equivalence, because both speed and gravity have an identical influence on the motion of time[1] (Resnick, Halliday and Krane, 2002).

A most dramatic time-bending effect was identified by Wheeler (1990). Various attempts to obtain unambiguous results in the double-slit experiment led to abundant repetitions of the experiment in which the distance between the slits and the monitoring device was altered, but to no avail. As much as the distance between the monitoring device and the slits was increased, thus creating a time gap between emission and measurement, particles always responded in the same manner: when observed as particles and when unobserved as waves. From these experiments Wheeler concluded that the photons acted according to what might seem like prior knowledge. When they started behaving as particles, the experimenter had not yet turned his monitoring device on. Therefore, it appears that photons can be affected not only by influences deriving from the past but possibly also by future causes (Wheeler, 1990). Thus, the arrow of time can be considered meaningless from the viewpoint of a photon.

The quantum connection involves particles that move in relatively high speeds. The speed of most elementary particles is significantly lower than that of the photon but even slower moving particles move significantly faster than objects exceeding Plank size. Therefore, the FR in which they are situated is relatively timeless. And in a relatively timeless FR it is not surprising to find effects like those that are detected in the double-slit and the EPR experiments. What seems to be simultaneous coexistence is actually the result of an attempt to superimpose the time dimension upon a FR where time is non-existent or at least meaningless. When time ceases flowing as it does at the speed of light, coexistence can no longer be defined as being simultaneous. Simultaneity in general is a meaningful term only in a temporal FR. If time were excluded from our calculations, there would be no standard by which simultaneity could be defined (R. Kreitler, 2012). Therefore, interconnectedness of the kind that was discovered in the EPR experiment should be considered a feature of particles, especially of high speed particles like the photon.

Dimensional dilation is not confined to time. When time is dilated, all other spatial dimensions are dilated as well, and the entire spectrum of space-time is transformed. To understand this phenomenon it is enough to recall that distances are assessed according to the time it takes to bridge them, e.g. the speed of light is approximately 365 thousand km per second. All reference to speed, infer time. So, if there had been no meaning to the unit of 'second', we could not have measured any unit of length.

Evidently, dimensional dilation is a phenomenon related to motion in high speeds, but according to the equivalence principle it can also be produced by variations in gravity. Experiments carried out on atomic clocks indicated that when two clocks were situated at different altitudes, e.g. one at a mountain peak and one on the plain, the former moved slightly faster (Pound and Rebka, 1959). This phenomenon is explained by the varying distance between the two clocks, and the core of the earth. The clock that was further away was slightly less affected by gravity.

Since the dilation of one dimension entails the dilation of all other dimensions in a specific FR (this principle will be discussed in the section 'Characteristics of a Non Temporal Frame of Reference'), it may be assumed that in a hypothetical FR where gravity were non-

[1] Not only time slows down when the speed of light is approached. Also, the mass of particles and all other objects increase with their speed. Therefore, physicists specifically refer to the rest versus the relativistic mass of particles. At 99 percent the speed of light, the relativistic mass of a particle is approximately seven times larger than when the same particle is at rest (Sakurai and Napolitano, 2011).

existent, all dimensions would be dilated. Such a FR is a purely theoretical construct. As explained above, gravity and space-time are essentially one and the same thing. So without gravity there would be neither space nor time. Hence, one may wonder what that FR would be made of. How could it be defined, if it lacked the dimensions of space? Regardless of whether it could exist, it is obvious such a FR would be dimensionless, meaning it would resemble a singularity in that it would lack the classical dimensions of space. Fortunately, particles are not immune to gravity. Therefore, their FR can still be defined within the confines of space-time. But, as we have seen, the effects of gravity upon them are diminished, partly due to their sub-Plank size, and partly due to their relative speed of motion. Consequently, various degrees of dimensional dilation are apparent in quantum experiments.

It is noteworthy that dimensional dilation does not occur in everyday life. Although atomic clocks that were placed on jet-planes were slightly retarded in comparison to atomic clocks that remained stationary on Earth, dilation effects people encounter, are negligible. This is not the case for particles. They are much more susceptible to dimensional dilation effects. Therefore, various weird effects such as simultaneity, particle-wave duality and hidden variables have been associated with them.

TWO CONVERGENT FRAMES OF REFERENCE

From our human vantage point the above-mentioned phenomena seem incomprehensible. It is immensely difficult to imagine a FR where the dimensions according to which we are used to think are dilated. Our perception of the universe is thus based on the three spatial dimensions, and on time, that most of us are incapable of imagining a universe in which these dimensions are not to be found. This comes to show how detached two FR can be, even though both are situated in physical proximity. Apparently, our universe is comprised of two realms. One with discrete entities, which have clear boundaries and one in which all boundaries are wavelike and fuzzy; one that is shaped by gravity and one where gravity is barely noticeable; one in which time is the organizing dimension and one in which time is non-existent; one in which Newtonian physics is valid and one in which the theory of relativity fails and forces seem non-local. The differences between the realms could have rendered them parallel universes were it not for the fact that particles are the building blocks of the universe. So they cannot be confined to a separate universe. They are the universe.

The incredible degree of entanglement between the two realms can be derived from the failure of reductionists to explain quantum physical processes. According to the reductionist school, all psychological processes result from biological phenomena which in turn can be explained by chemical reactions. These result from interactions between atoms which are comprised of subatomic particles obeying quantum-mechanical rules. According to these rules it is an act of observation that collapses wave-functions. Hence, the so-called 'reductionist pyramid' is challenged by the evident dependence of the most elementary particles in nature on a much more complex and non-elementary system: the human mind. This reliance produces a kind of vicious circle that prevents any further reduction (Rosenblum and Kuttner, 2006).

A similar problem confronts physicists while conducting quantum experiments. At the Plank scale length (10^{-33}cm), which constitutes the threshold between the two realms, gravity is a significant force that cannot be neglected. Therefore, the energy required to probe the

Planck scale length disrupts quantum processes that occur below that crucial size. Thus, at the Planck scale length gravity erects barriers that prevent scientists from probing the realm below. This is surprising because one could have expected relativity to concern larger objects like planets, and quantum mechanics to be confined to subatomic entities, but the boundary between the two realms seems to be fuzzy (Randall, 2006).

Similar convergence looms over the mind-body divide. By monitoring either side of the divide, predictions concerning the other can be made. Hormones, which are bodily substances, are used to assess our state of mind e.g., hormone levels indicate whether we are depressed (Kirkegaard, 1998). And conversely, by assessing our level of mental stress, the degree to which we are prone to specific bodily diseases can be predicted. A common example is the well-established correlation between personality and coronal disease. Type A personality is widely accepted as a major risk factor for the development of heart disease (Friedman, 1996). Likewise, post traumatic stress disorder was found to be associated with the occurrence of inflammations (Gill, Saligan and Woods, 2009). The correlation between stress and decreased immune activity is not unique to human beings; it has been found also among Cod fish (Marlowe, 2009). So it is evident that mind and body influence each other.

Penrose (1994) suggested the mind-body schism is similar to the divide between classical and quantum physics in that both share a common threshold: the Planck constant. As mentioned above, when objects exceed that particular size (10^{-33}cm), gravity causes their wave-functions to collapse and one of their various potentialities to actualize. Since the size of nerve cells does not exceed Planck size, Penrose claims the activity of synapses and nerve cells is carried out in the realm of potentiality, which essentially means our mind operates according to quantum rules (Herbert, 1994).

Quantum Weirdness in the Psychological Realm

Some of the weirdest features of particles have parallels in human behavior. For instance, people demonstrate a quality that is highly reminiscent of the previously mentioned particle-wave duality. This phenomenon concerns the tendency of particles to change their appearance while observed. Apparently, most people do the same. When humans are concerned, this phenomenon is called Personae. In different situations people adopt different modes of behavior, which are sometimes referred to as masks. While at work most of us are more assertive than at home. When we are being photographed, we try to smile and when we are on a date, we do our best to be good humored even if we happen to be sad. There is nothing astonishing about this all-too-natural behavior of human beings. It is generally accepted that different modes of behavior are being adopted in different social situations.

Human behavior also provides a parallel to Heisenberg's Uncertainty Principle, which states we cannot simultaneously assess both the location and the momentum of a particle. For clarification let us imagine a couple, Paul and Ann, who share an apartment. They have had a major fight. Following the fight Ann calls her best friend, and tells her all about the fight, but says she is intent on giving the relationship another chance, no more than three months. If by that time the situation does not improve, she will leave Paul and that will be it! The two friends devise a plan of action designed to improve relations. Later on, when her mother calls and asks why she sounds so sad, Ann responds by telling her mother that although she and Paul fought, she must stay with him until the end of the rent period, which is due in three months. Adopting a psychological perspective, we can say Ann's decision to stay with Paul is multi-determined. When two different causes lead up to the same end-result, namely, that Paul and Ann stay together, we can say the decision was multi-determined. Evidently Ann, like most people, has multiple causes for her decisions. Furthermore, she did not lie because in both conversations she specified the same time duration, but one may wonder whether she and Paul can still be considered a couple. To her friend she said they are, and to her mother she said more or less the opposite. So, different people can place Ann in different places insofar as the relationship is concerned. Her mother may truthfully say she is no longer in love with Paul, while her friend may just as truthfully vouch for that she is so deeply in love with him that she is ready to give him another chance (R. Kreitler, 2012).

To complicate things further, let us imagine that Ann's sister Linda, overheard her mother talking on the phone. Toward the end of the conversation, she picked up the receiver and asked her sister bluntly whether she intends to give "that scoundrel" another chance. Ann's response to such a question is easy to predict. She would probably say: Never ever, after what he has done! She would never admit to her mother and sister that she intended to do the opposite. Such a turn of events would interfere with the plan of action Ann and her friend have previously devised. Very often a question we ask influences the response we get. If we ask a custom's officer at the airport whether we should open our handbag the most likely response would be positive, although prior to our question the officer may have barely glanced in our direction. Apparently, questions, which resemble acts of observation, have the capacity to alter the observed situation, much like in a quantum experiment. The process by which this occurs can be derived from the above example. When Linda asked her sister a blunt question, Ann recalled the details of the fight, thus reawakening her anger which has begun ebbing.

Critically, it can be argued that all references to personality are biased by the presence of an introspective observer. When we make an observation that concerns traits or feelings we ourselves are likely to possess, we risk being biased. Therefore, it is worthwhile to extend the above comparison beyond the scope of human relationships which are naturally subjective. Economic evaluations provide an objective terrain. Also here quantum principles are often used in a variety of models, for instance, in the Quantum Bohemian model for financial market (Choustova, 2006). Value evaluations are likewise a subject where quantum weirdness is particularly evident. Suppose we want to assess the value of a piece of art, i.e. a painting in our possession. We may ask for an evaluation and thus obtain a document claiming the painting is worth the amount X. Based on this evaluation we may ask our bank manager for a loan for which the painting would serve as collateral. We may also auction the painting obtaining much more money for it. Alternatively, we may cheat someone into believing it is an original piece of great value and tempt the "idiot" to pay a fortune for the same painting. Thus, the final value of a painting cannot be assessed until we actually sell it, that is, perform an act equivalent to the quantum researcher's act of measurement. Even shares that can be evaluated at every given moment include a potential unrealized value which is referred to as their *option* or *future*. Stocks may one day be evaluated at amount X and the next day at amount X+3 and a week later at a 100 X. So their potential value is speculated and traded. But the prescribed value is irrelevant to a holder who does not intend to sell the share. Only on observation, that is on selling, does the value become real and meaningful. Till then, it is a potentiality. Therefore, it is a deliberate act, which a human being performs, that turns potentiality into reality.

The explanation for the tendency of things and people to change while observed is less significant than the realization that nature as a whole is subject to the Uncertainty Principle. The reality in which we live is evidently being created by acts of observation. Until an electron is observed, it has many possible routes. While observed it performs what is referred to as a quantum leap from a random potentiality-wave consisting of all the possible routes it might take into the one single course it has actually taken. Likewise, until a painting or share is sold, we have no way of assessing its final value. All prior references to value are no more than estimates and as such they resemble our expectations regarding the possible distribution of particles on the screen in a double-slit experiment.

Apparently, phenomena such as uncertainty and particle-wave duality, which are often labeled 'quantum weirdness', are neither confined to the realm of subatomic particles nor to human behavior. These are general phenomena that concern a much wider variety of entities. Regardless of whether we are observing friendship, or an economic evaluation, Newton's laws are neither relevant nor valid. This is the case because the laws of physics were made up for 'things' – these are objects with mass that exceeds the Planck constant i.e., apples falling from a tree as in Newton's famous experiment. Ideas, aspirations, feelings and also economic evaluations are mass-less. Therefore, the rules that describe objects with mass do not apply to them. The possession of mass seems to distinguish objects from theoretical entities. Thoughts and all other theoretical entities have no mass. So they cannot be expected either to exert or to be susceptible to any kind of gravitational forces. Unlike particles which are affected by gravitational forces, theoretical concepts are entirely immune to all such influences. Therefore, the FR, in which theoretical entities should be examined, is entirely gravity-free hence, non-temporal. As we have seen, time is a product of space, that is, of gravity. So in a FR where gravity is entirely non-existent, time is bound to be meaningless as well.

The described similarity between elementary particles and theoretical entities indicates that both occupy the same FR. In itself, this does not imply they are identical. Just as biological principles cannot be applied to non-living objects of mass such as automobiles or tables, although they share the same FR, not every rule of quantum-mechanics can be applied to theoretical entities and vice versa. For instance, gravity affects particles, but not theoretical entities.

In the quantum-physical literature the term 'non-things' was figuratively coined to describe the weirdness of particles. The same term can be used in reference to theoretical entities because due to their immunity to gravitational influences they disobey the rules that were made up for objects. Theoretical entities occupy a realm where none of the classical dimensions is meaningful. Apparently non-things turn into things as a result of human that is, conscious acts of observation.

Characteristics of a Non-temporal Frame of Reference

One may wonder what a non-temporal FR might look like. We are all accustomed to thinking in terms of temporal causality. The future-oriented arrow of time constitutes a cornerstone in our perception of the universe. Nonetheless, in our everyday life we often use a technique that distorts time. It is called statistics. We are all used to reading that millions of people used a particular railroad line. No one wonders how come so many people could have embarked on a single train. It is evident to us that the numbers refer to an unspecified period.

Further, they refer to large numbers of people disregarding the fact that many of these people embarked more than once. Apparently, when time is disregarded – as it is in most statistical analyses – reality spreads out sideways. The temporal reality, in which individual people board the train once a day, is reduced to a scheme in which neither individuals nor specific days are meaningful. Only the total number of passengers counts. Thus, a person who uses the train over and over again is transformed into a large number of people each of whom uses the train only once. Otherwise phrased, a person who leads a continuous life along a continuous timeline is transformed into a large number of doubles with an instantaneous existence. Evidently, when a statistical point of view is adopted, time is distorted and people lose their individual identity. In statistical analyses, the consistency of the universe is sacrificed for an accurate representation of particular characteristics e.g., the number of people using a train. It appears to be a matter of choice, whether we highlight the timeline, opting for a consistent universe, or whether we prefer a statistical viewpoint. This comes to show that a cognitive process can determine which dimensions will shape our picture of reality. Just like in the double-slit experiment, it is a conscious choice of measurement procedures that determines what result will be obtained, and what angle of reality will be exposed.

The above example suggests that in a non-temporal realm, where the time dimension is dilated, it is replaced by another dimension, namely *potentiality* that can keep track of quantum jumps in-between unrelated states of being (R. Kreitler, 2012). In quantum physics the term Potentiality refers to optional or transitional states. String Theory suggests all particles arise from different vibrational modes of an oscillating string. A string can vibrate in many ways, producing different sounds. Likewise a single string can produce different particles (Sakurai and Napolitano, 2011). This implies that particles can be reduced to energy patterns. A slightly different vibration will create a different particle. Since energy is liable to transformation, the nature of things is transitional, meaning one thing can be transformed into another. This is an elaboration of an older concept: probability waves. In the context of these equations each potential route a particle might take is considered a potentiality. Born proposed that probability waves indicate the potentiality of finding a specific particle in a definite location along the wave. For instance, the wave specifies a 30% chance of finding it in location X and a 50% chance of finding it somewhere else. Hence, the shape of the wave outlines the probability of localizing the particle, which gave rise to the wave. In other words, the wave illustrates different potential locations of the particle (Sakurai and Napolitano, 2011).

The relation between time and potentiality can be explained by the Uncertainty Principle: the more we know about time, the less we know about potentiality and vice versa. Therefore, time and potentiality can be considered *complementary dimensions*. Complementation requires that motion in either dimension should infer and be equal to motion in the other. Either we move along a continuous timeline, or we contrive quantum jumps in-between separate potentialities, but we cannot do both at once. One mode of motion excludes the other (R. Kreitler, 2012).

In a non-temporal FR we can expect quantum leaps in-between parallel potentialities to occur. Lacking a temporal sequence, events unfold sideways, providing parallel and in-continuous images of the universe, e.g., the doubles of the same person who uses the train on a daily basis. A person using the train twice a day going to and from work would seem to be two different people. No connection between the morning rider and the evening rider would be evident. Interestingly, the human mind also demonstrates a tendency for thinking about parallel possibilities. This is primarily evident from the human capacity for multi-tasking. People are able to divide their attention between different tasks (Ayres and Sweller, 2005). For example, we are all used to conduct a conversation while washing the dishes. In trying to solve a problem we often examine several alternatives. Studies indicate that only after examining several scenarios, which can be viewed as potential courses of action, we make a decision (Dörner and Wearing 1995). We are also able to put ourselves in the place of other people, imagining their feelings or motivations. All these capacities are made possible because our thoughts occupy a non-temporal FR, where the predominant dimension is potentiality, and not time. If our thoughts had been solely organized in a temporal sequence, we could not have perceived the notion of multiple possibilities. They would have seemed just as bizarre as the hypothetical perception of color or sound being spatial dimensions. Our ability to comprehend potentialities shows that our thoughts are non-temporal. As previously explained, the main feature of non-temporality is simultaneity, and the above discussion indicates that our mind enables our thoughts to instantaneously jump from one location to another. Our body stays put, but our mind wanders from one place to another. In our thoughts we can simultaneously be in two different places (Freud, 1900).

It is noteworthy that dimensions are not random entities. Einstein defined dimensions as vectors of motion (Greene, 1999) thus distinguishing them from other axes such as temperature, color etc. Since, according to Einstein, motion cannot be restricted to one vector but must combine several vectors, the sum of which adds up to the speed of light, no dimension can be separately defined. The dilation of one dimension necessarily leads to the dilation of all other dimensions in the same FR. For example, width can only be defined according to length. By convention it is shorter. So if we do not know what the length of an object is, we cannot refer to its width. Likewise, height is a dimension that can only be defined according to both length and width. If we do not know both, we cannot determine what axis describes height. However, by changing an object's orientation in space, we can redefine its dimensions. For instance, if we rotate an object onto its width, we can turn its previous length into its height. Even though the function of dimensions (as length, width etc.) can be arbitrarily determined, they can only be assessed according to one another. The same is true for time. Time is dependent upon the existence of three dimensional objects that make it possible to assess physical processes, such as particle decay. If a two-dimensional object could have been created (and it cannot), its age would not have been measurable. Even if it existed we would not have been aware of it. So it is evident that the dilation of any dimension entails the disappearance of all other dimensions.

However, the dilation of every dimension is also compensated for by the appearance of another *complementary* dimension. This process includes a transformation of the entire FR. In the case of time, it is potentiality that fills the void. This process causes all spatial dimensions to disappear. When length and width vanish, space is described by *phase*. This wavelike dimension describes a terrain where distinct boundaries are lacking. We normally imagine waves within a confined three-dimensional space but when the spatial dimensions are dilated, phase is transformed from a parameter into a spatial dimension enabling motion. Finally, the replacement of time by potentiality does not leave the other FR without an organizational dimension. In our FR time fulfills this role. In the other timeless FR, it is spin[2] that

[2] Angular momentum includes spin and depends upon the mass, size, and rotation of a spinning object. Elementary particles only spin around one axis – the one we choose to measure at a given time. Therefore, they cannot be

discriminates one particle from another, thus keeping track of events, namely, of potentiality changes (R. Kreitler, 2012).

The above described dimensional transformations characterize a non-temporal FR. When time is replaced by potentiality, all other dimensions of space undergo various transformations. Instead of space being characterized by distinct boundaries, a wave-like terrain appears. It is noteworthy that many brain functions, such as sleep, rest, or thinking are characterized by different kinds of brainwaves. For instance, sleep is associated with theta waves and delta waves (Pinel, 1992). The study of elementary particles tells us that waves are associated with potentiality variations. So a relation between brainwaves and the occurrence of non-local cognitive processes can be speculated.

Over past decades researchers such as Sirag (1983), Samal (2000) and Pitkänen (2006) attempted to use various extra dimensions to describe inner space. Unlike the models they developed, the above-proposed model does not define consciousness as being separate from physical space. No distinct inner space is hereby proposed. On the contrary, an attempt is made to translate processes involving extra-dimensions into a classical four-dimensional context. While it may not be possible to study processes that involve extra dimensions, the process of complementation provides us with tools designed to translate extra-dimensions into a classical four-dimensional FR. For example, by assessing the connection between Phase and the two spatial dimensions, we learn that the analysis of brainwaves should focus on spatial attributes.

Dimensional complementation can loosely be compared to Einstein's principle of equivalence that was described above. Motion in extra-dimensions can be considered equivalent to motion in the classical four dimensions, except for the fact that the described dimensions refer to entities with different characteristics. A shape with contours cannot be described by waves, and conversely the contours of a wavelike fuzzy shape cannot be assessed. However, as De Broglie (1949) discovered, also objects of mass emit waves. Evidently, the two FRs describe two complementary viewpoints of the same physical environment. All the objects around us including ourselves emit waves which gravity collapses, and all wavelike entities, such as evaluations have a coherent perspective that can be actualized. Hence, the extra-dimensions that have been proposed to describe theoretical and quantum entities are crucial for describing a wide array of phenomena.

Our FR includes three spatial and one temporal dimension which can be described either as a separate dimension or as a half-dimension, because when we proceed in time we also change our state of being, hence our potentiality. The other FR includes two spatial dimensions (Phase and Spin) and the intangible dimension we have termed potentiality, which like time can also be denoted as a half-dimension (R. Kreitler, 2012). Einstein proved that motion in any dimension entails motion in all other dimensions. Overall motion always adds up to the speed of light but that overall speed is divided between the various dimensions

said to actually rotate. Nevertheless, the interaction between an electron and a magnetic field depends upon the electron's intrinsic spin which is liable to measurement, and considered pivotal for defining particles. Any change in spin causes a particle to be transformed into another particle. Spin is quantized like energy and charge, making it discontinuous, meaning it performs quantum jumps. A particle slows down and speeds up in small discontinuous steps. Not every rate of spin is possible. So-called families of particles share similar spin rates, i.e., leptons have an angular momentum half that of a photon (it is common to compare rates of spin to that of a photon which is defined as having spin 1). For particles, Spin is an independent direction of motion. And just like a person cannot move both forward and backward at once, so a particle cannot simultaneously spin around more than one axis (Bohm, 1980; Sakurai and Napolitano, 2011).

(Einstein, 1920). Assuming extra-dimensions constitute an integral part of our universe (where the speed of light constitutes the speed-limit), the same principle should apply to extra dimensions as well. Therefore, we may assume that motion in one FR should be reflected by parallel motion in the other complementary FR. This is the case because both FRs are coupled, that is, connected by a single common dimension, which both share. Only in the context of that common shared dimension, motion is actually combined. Otherwise, in each FR motion adds up separately to the speed of light. For clarification, motion in length does not directly imply motion in phase, because phase belongs to a different FR. However, motion in time (which necessitates motion in length because both occupy the same FR), does imply motion in potentiality, which, in its turn, necessitates motion in phase because both occupy the same FR. In this way, time and potentiality combine otherwise separate FRs.

A comprehensive view of the universe can be obtained by adding the two perspectives of reality. Hence, the combination of 3 1/2 dimensions (axes of motion) plus 2 1/2 dimensions describes the two FRs which have hitherto been discovered. The correlation between the two FRs can be obtained by dividing the two quantities:

2 1/2 : 3 1/2 = 1:1.4

This simple numerical phrase describes the correlation between tangible and theoretical reality or between below-Plank size and above-Plank size entities. It is interesting to note that this correlation is identical to the correlation between forces which was proposed by the SU2 and SU3 theories. According to these theories the correlation between electromagnetism and the other three forces is 1:137. When high energies (10^{15}GeV) are applied, the discrepancy disappears and a correlation of precisely 1:1.4 is obtained (Close, 2007). It is noteworthy that all criticism of the SU theories relate to the SU5 theory and focuses on predictions it made. However, the unification of three forces, which was derived from the SU2 and SU3 theories and the consequent mathematical correlations between all three and the fourth force, remains unchallenged and is included in the standard model (Smolin, 2007). Since inability to merge all four forces is considered symptomatic of the gravity-quantum divide, identity in correlations between the forces and between the dimensions supports the proposed complementary model of dimensional complementation.

Cohabitation between the Two Realms

Contemporary physicists are perplexed by their inability to merge the two most important theories of the past century, the theory of relativity and quantum mechanics. Both provide contradictory viewpoints of the universe. The relativistic perspective describes a universe with clearly defined objects which are either here or there; quantum-mechanics describes a universe with fuzzy entities that are both here and there at one and the same time. In the above section, a correlation of 1:1.4 between the two perspectives was proposed. According to the suggested approach the two theories simply address complementary viewpoints of the universe. Adopting one viewpoint, the universe has distinct shapes; adopting another viewpoint, it is wavelike and non-local.

Following this line of thought, the particle FR is entirely interconnected. However, interconnectedness is also a feature of the previously mentioned singularity. So one may critically ask how come the particle realm, where gravity is barely noticeable, is so similar to the black-hole singularity, where gravity is so strong as to demolish all dimensions of space. Likewise, one can question the feasibility of interconnectedness especially between a non-local mind and an all-too-local body. Models suggesting the mind is independent from the body sound farfetched and lack empirical proof.

A geological inspection of our planet provides a clue as to how cohabitation between the realms occurs. It is common knowledge that gravity diminishes with distance from the core of the Earth. Consequently, when we dig a hole in the ground, gravitational forces increase and obstruct our work. But it is equally indisputable that if we succeeded in penetrating the crust and mantel, we would notice a gradual reduction in gravitational resistance. Since the Earth somewhat resembles a sphere, gravitational forces from all sides would cancel each other out as we approached the core. If the Earth had been a perfect sphere, the point at the core would have been gravity-free. But since the Earth is not a perfect sphere, there are several points in which gravitational influences from two directions cancel each other out, though from other directions gravity can still be sensed. Since these points are too small for a human body to enter, they are not considered worthy of investigation although NASA experts officially recognized their existence. By studying these gravity-free points, deductions regarding the principle of cohabitation can be made.

It is important to stress that the reason for our inability to move in the speed of light is our body. Our body's mass which is evident from our multi-dimensional appearance forces us to divide our motion in-between the spatial dimensions and time. Einstein proved that motion in one dimension implies motion in all other dimensions (Resnick, Halliday and Krane, 2002). But if gravity had been non-existent, our mass would have disappeared, and so we might have been able to move in the speed of light. Hence, in a place where there is no gravity, we may be tempted to assume time would move as fast as it possibly can – i.e., in the speed of light – but that would have been an illusion. As noted above, at the speed of light, time, like all other dimensions, is dilated and hence, replaced by its counterpart, potentiality. So the other set of dimensions – or rather extra-dimensions – takes over, and motion is shared between them. Consequently, from our point of view it seems as though motion occurs at the speed of light while actually, our dimensions are being dilated and motion is divided between the above proposed set of extra dimensions: potentiality, phase and spin. Nonetheless, the disappearance of our classical dimensions gives the false impression of instantaneous communication over space.

Measurements really indicate simultaneity, hence instantaneous communication. This illusion results from a lack of gravity. In the absence of gravity, we encounter a non-temporal environment which seems non-local, since all connections it sustains are instantaneous. The temporal dimension in that environment is potentiality and so motion does not occur along a continuous timeline but rather involves quantum jumps in-between alternate states of being. Thus, as external observers we are prone to conclude motion occurs at the speed of light, while motion is actually divided among a set of extra dimensions more suited for describing an interconnected FR.

Though interconnected, the gravity-free realm at the core does not resemble a singularity. Indeed, in the case of both, the classical dimensions are undetectable, but unlike a singularity, the gravity-free realm does not exert significant gravitational influences on its surrounding. The similarity between the two forms of interconnectedness can be explained by the principle of equivalence: interconnectedness in the gravity-free realm is sustained by motion, not gravity. Therefore, the gravity-free realm may be referred to as a *quantum singularity* more similar to the Implicate Order. From our perspective, which is confined to the classical dimensions of space-time, the Implicate Order is entirely interconnected. However, there is a basic difference between a black hole singularity and a quantum singularity. The difference is that in a black-hole immense gravitational forces crushed all matter. So the dimensions of space cannot be defined. This is not the case in a quantum singularity, which can be described by its own set of dimensions. Hence, a quantum singularity constitutes a separate FR within our own universe where speed and size reduce the effect of the spatial dimensions.

The quasi-spherical shape of our planet makes the existence of a gravity-free point at the core a necessary conclusion. Interestingly, the point where gravity is non-existent is created by gravitational forces. Were it not for surrounding gravity the gravity-free point would neither be created nor could it endure. Surrounding gravitational influences would have immediately led to its collapse. But the shape of the planet, as well as its mass, maintain a gravity-free realm in a tiny section of confined space-time, thus demonstrating cohabitation.

Randall (2006) proposes the relative weakness of gravity in relation to the other forces results from the shape of the brain on which we ourselves are located. According to the above suggested model of cohabitation, the shape of the planet and of other stellar bodies determines the extent of gravitational influences. If the core of a body of mass can accommodate a gravity-free realm, then physical shape can indeed be considered a predominant factor in the exertion of gravitational influences.

THE PHYSICAL IMPLICATIONS OF INTERCONNECTEDNESS

It needs to be noted that the relatively small size of our planet indicates that the gravity-free point at its core is not entirely gravity-free. Even at the core, gravitational influences originating from larger stellar bodies i.e., the Sun should be detectable. If the point had been entirely gravity-free it would have been characterized by instant quantum communication with the cores of other stellar bodies. Due to the gravitational effects of the Sun we must assume that interconnectedness is not instant. That is also the case when particles other than the photon are investigated. Although these particles are slower and are therefore more exposed to gravitational influences, they maintain the quantum connection (Greene, 1999). Therefore, it can be assumed that regardless of the presence of reduced gravitational influences, a quantum connection is maintained between the cores of stellar bodies, e.g., planets, suns, etc.

Communication of any kind necessitates the existence of a mode of transmission. At present physicists have no idea how the quantum connection is mediated, but it is widely assumed this connection is basically holographic. For instance, the ability of a single particle to appear as a wave indicates the existence of waves that spread out infinitely (Bohm, 1980; Greene, 1999). The essence of the discussed holographic connection between planet-cores can be derived from Einstein's conjecture according to which time ceases flowing at the speed of light. This conjecture indicates that a gravity-free realm should be characterized by an interconnectedness of all matter in the universe.

According to the big bang theory of universal formation, all matter had been interconnected at the time of the big bang (Greene, 1999). Adopting a classical perspective, we may assume matter drew apart as the universe expanded. For example, the Earth split away from the Sun which in turn developed from another sun and so on and so forth. According to the quantum connection, systems that have communicated with one another maintain contact. Therefore, if we adopt a non-temporal perspective, we must conclude that interconnectedness between all of matter has been maintained. The Earth should therefore be linked to the Central Sun (the sun at the center of the galaxy) which should maintain a quantum connection with other suns from which the galaxy developed. Hence, the existence of a net of communication connecting the cores of all stellar bodies is both a necessary and a logical conclusion, although we do not know how energy is transmitted.

For the sake of clarification, let us imagine a hypothetical time-traveler. In the beginning of her journey she would visit past epochs like the Bronze or the Stone ages which are associated by geologists with particular Earth-layers. Later on, her journey would take her to the early days of our planet when the Earth was molten, just like the Sun. Assuming our time-traveler did not perish, she would end up on the Sun itself, because the Earth split away from it. From our temporal perspective the connection between the Earth and the Sun is non-detectable, but from the viewpoint of elementary particles this connection should permanently exist.

Following this line of thought it appears that the same region of space attains entirely different characteristics, when it is described by a different set of dimensions. Observing ordinary space, we can measure the light-year distances separating two suns. But if we adopt a quantum perspective, and study the net of quantum communication that apparently binds the cores of all stellar bodies, these distances shrink dramatically, thus enabling particles in two parts of the universe to almost instantaneously communicate without breaching the speed of light. This perspective provides us with an entirely different map of the universe - one which allows connections that are forbidden by the geometry which gravity creates. Without gravity all dimensions of space-time change, creating shapes we are prone to overlook.

The image of the universe which can be obtained by adopting an extra-dimensional perspective is essentially holographic. Bohm suggested that were we to view the cosmos without the lenses that outfit our telescopes, the universe would appear to us as a hologram. Pribram (1971) extended this insight by noting that were we deprived of the lenses of our eyes and the lens-like processes of our other sensory receptors, we would be immersed in holographic experiences. These lens-like processes to which he refers may include the timedimension. As long as our perception of space includes time, quantum-connections between planets remain invisible and may even seem absurd. But ignoring time, Earth should be connected with the grain of matter that preceded the big bang.

Nevertheless, cords or wires connecting the planets are nowhere to be found in our FR. Just as no connection has been detected between the two photons in the EPR experiment, no visible ordinary-space connection between the cores of planets can be predicted. The speculated connections between the cores of stellar bodies involve shapes that cannot be described by the classical dimensions of space (length, width, etc.). These extra-dimensional shapes may be referred to as wormholes.

There is at present no unequivocal proof wormholes actually exist, but their existence is allowed by general relativity. Wheeler (and most physicists after him) was convinced of their existence, and studied their mathematical properties. A wormhole is expected to connect two otherwise separate and distant regions of space, two universes or – according to Thorne (2002) – two epochs within the same universe. Generally, a wormhole is believed to be a kind of shortcut. The main difference between a wormhole and any other shortcut is that a shortcut traverses space whereas a wormhole provides passage through a region of space that would not have existed had the wormhole not penetrated it (Greene, 1999). This is an elaborate way of saying that wormholes take a route that utilizes extra-dimensions or that the space wormholes traverse can only be described by extra-dimensions. The study of wormholes usually focuses on black holes. It is widely assumed that if one were to fall into a rotating black hole, she could miss its ring-like singularity and be sucked into another part of the universe or even into another parallel universe. In this case the passage would have been contrived by means of a wormhole (Zukav, 1980). The presently proposed kind of wormhole does not involve black holes. On the contrary, it is open for inter-particle communication, hence the transfer of energy.

Many propositions regarding the abundance of wormholes in the universe were made in recent years. But it is doubtful that in normal circumstances people would be able to walk through the above-speculated wormholes, even if they could reach the core, which is presently not the case. The reason people may not be able to walk through these wormholes concerns their size. The effect of mass is that it confines our body to a Newtonian FR which is comprised of classical dimensions. Mass which endows our body with its spatial attributes prevents us from probing the Implicate realm.

This inability of ours is fundamental. The two realms constitute two different and irreconcilable viewpoints of the universe. According to the Uncertainty Principle various quantities cannot be simultaneously assessed. Among these are position and momentum, time and energy and multiple axes of spin (Sakurai and Napolitano, 2011). Also complementary dimensions adhere to the same principle, i.e., time and potentiality, as well as length and width versus phase, cannot be simultaneously measured. The mutual exclusivity of these pairs of parameters maintains the in-predictability of the universe. If one could have simultaneously observed two mutually exclusive quantities, the universe would have become predictable as it would, if we could unite the various forces of nature or simultaneously measure all directions of motion (R. Kreitler 2012). When we put two different pictures one on top of the other, both seem blurred. This is precisely what happens when we try to merge different viewpoints of the universe by simultaneously observing any of the exclusive parameters. Inexplicable ambiguities emerge whenever we try to unite the classical and the quantum domains. A good example can be seen in the previously discussed double-slit experiment. This may be the case because a theory of everything contradicts the Uncertainty principle. The divide between relativity and quantum mechanics as well as between body and mind and between entities larger than Plank and smaller than Plank, seems to be a fundamental feature of the universe.

BRIDGING THE GAP

In spite of the irreconcilable nature of the two realms, the effects of gravity can in some cases be counterbalanced. As previously explained, Einstein's equivalence principle postulates that the force of gravity and the force of acceleration are basically the same (Resnick, Halliday and Krane, 2002). This means acceleration can counterbalance gravity. NASA uses this principle while training astronauts for spaceflight. However, as long as we possess a body, our rest mass can never be fully dilated. If our acceleration were increased to the speed of light, our body would disappear. This would have happened because, at the speed of light, all matter is transformed into energy. But as long as we do not approach the speed of light, we cannot probe the gravity-free realm. Nonetheless, while accelerating, we can experience weightlessness. Many people are unaware that the reason why satellites do not fall down is not lack of gravity in space but their relative speed of motion. Weightlessness should be distinguished from lack of gravity. For instance, weightlessness can be experienced in a falling elevator.

Gravity can be counterbalanced even without acceleration. In a most dramatic experiment Geim levitated frogs, thus counterbalancing gravity, without altering their speed of motion. He did so by applying a magnetic field (Berry and Geim, 1997). An electromagnetic coil of wire in which a current flowed provided the force which held the frogs suspended in midair. The frog or actually fluid within the frog's body was magnetized by the field of the magnet. This phenomenon is referred to as 'induced diamagnetism'. Following Berry and Geim's experiment, mice have been levitated for hours at NASA's J.P.L (Jet Propulsion Laboratory) laboratory in California. It needs to be stressed that levitation is by no means a result of anti-gravity. And gravity is by no means eliminated during levitation. The magnetic force simply counterbalances gravitational effects. According to Berry and Geim, a similar phenomenon can be witnessed in everyday life. When we are standing on Earth the force of gravity which the Earth exerts on us is constantly balanced by an upward force on the soles of our feet, which is exerted by the atoms in the ground on which we are standing. This force stops us from accelerating to the core of the Earth. This is a short-range force, acting over nanometers. But magnetism is a long range force like gravity. The magnetic field which levitates the frog is only a few times stronger than the fields used in MRI tests. In principle, Berry and Geim claim a person could be levitated like a frog (Berry and Geim, 1997).

It can therefore be surmised that a magnetic field can counter-balance gravitational influences. So it might be interesting to speculate on possible magnetic influences within a gravity-free realm. Indeed, the magnetic properties of the Earth are indisputable. Our planet exerts a magnetic field that originates from molten iron in the outer core. Therefore, speculation concerning a connection between the gravity-free point in the core and possible magnetism would be logical.

In recent years evidence has accumulated concerning magnetic ferrous materials, namely, magnetite (a mineral) in the brain of different kinds of species including humans. Magnetite was first detected in the brain of birds (Edwards, 1992; Semm, 1983). More evidence was presented by Beason and Semm (1996) who found that one of the three magnetic field detectors in a bird's brain is associated with the pineal gland (Edmonds, 1992). Magnetism was not only detected in the brain of birds which are assumed to use magnetic sensors for navigation. Evidence also points to the existence of magnetism in hyppocampal neurons of mice. Goto et al. (2006) identified Ntan1 as a magnetism response gene, which is abundantly expressed in the mouse brain and was doubled, twelve hours after a brief exposure to a magnetic field at 100 mT.

Kirschvink et al. (1992) found the mineral magnetite (Fe304) in human brain tissue. Using an ultra-sensitive superconducting magnetometer in a clean laboratory environment, Kirschvink et al. detected ferromagnetic material in several human brain tissues. When brain tissue was examined with high-resolution transmission electron microscopy, electron diffraction and elemental analyses, minerals in the magnetite-maghemite family were identified. The findings imply a minimum of five million single-domain crystals per gram for most brain-tissue (Kirschvink et al., 1992). Following this breakthrough, magnetite was identified in the human pineal gland as well. Baconnier et al. (2009) found calcite microcrystals in human pineal glands. The researchers suggest a resemblance between these crystals and the octoconia of the inner ear which has been shown to exhibit piezoelectricity. In the course of their study the researchers developed a special procedure for the isolation of the crystals from the organic matter within the pineal gland. Selected area electron diffraction and near infrared Raman spectroscopy showed the crystals were calcite, which means they may exhibit piezoelectricity (Baconnier et al., 2009). Dubson and Grassi (1996) conducted a study which was designed to ascertain that postmortem contamination was not the cause for recent results indicating the presence of magnetite in human brain tissue. Low temperature magnetic properties of tissue were measured to determine the presence of ferromagnetic material in the tissue. Results showed magnetite is indeed present in living tissue.

Evidently, not only the Earth but also our brain includes a magnetized section. On the basis of the previously discussed gravity counterbalancing effects of magnetism, it may be assumed that a magnetized area in the brain would exhibit gravity-free characteristics, such as non-locality. The existence of such a brain section, where magnetism shields parts of the brain from gravitational forces, may provide the long sought after explanation for our cognitive abilities.

A gravity-free section in the brain would also provide the mind with means for communication with the universe, thus supporting an old claim made by Descartes a few hundred years ago. In his 'passions of the soul' Descartes (1989/1649) laid forth a dualistic theory of body and mind in which he suggested the pineal gland functions as an inner third eye that unites the mind with the universe. For reasons which in later centuries have been entirely refuted, Descartes speculated the pineal gland is the seat of the soul. However, Beason and Semm's (1996) findings as well as those of Edmonds et al. (1992) and Baconnier et al. (2009) suggest a connection between the pineal gland and a section of reduced susceptibility to gravity. Thus, the role of the pineal gland as the location of the mind may still be verified for reasons Descartes could not have known. It should however be stressed that no connection is hereby suggested between a gravity-free brain section and the soul. The gravity-free section does not have any metaphysical characteristics. It simply endows the brain with a capacity to master tasks associated with non-local influences. These include most kinds of cognitive activity.

Like the gravity-free point at the core of the planet, the magnetized section of the brain, which is evidently the pineal gland, can be expected to operate like a mini-wormhole enabling quantum connections. Cords may not be detected exiting our heads. The kind of communication enabled by the proposed wormhole is identical to the communication between particles in the EPR or the double-slit experiments. This communication can at present not be deciphered, but it may be studied by applying a magnetic field to the pineal gland.

Up to date no unequivocal empirical proof has been obtained for the existence of wormholes neither in space nor in our mind. However, a simple experiment designed to prove their existence has been devised and planned by the author of this chapter. In the course of this experiment an SI EMU conductor would trace the course of the speculated wormhole.

The appliance of a 0.0007^{-9} miliampere magnetic field for the duration of 0.008 seconds to the pineal gland of mammals should trigger the appearance of tiny cavities in the gland. The size of the apertures can be calculated as being 0.0009^{-8} millimeters. By inserting a microscopic SI EMU conductor into any of these cavities, the extra-dimensional route of the wormhole can be exposed. It is hereby predicted that the SI EMU conductor will exit the head into which it was inserted. Its physical location will indicate the path of the extra-dimensional wormhole. The reception of transmissions originating outside the brain into which the conductor had been inserted will unequivocally prove the existence of a wormhole. Moreover, they will prove the physical feasibility of motion along an extra-dimensional route.

The proposed experiment includes two stages. In the first stage the pineal gland of a mammal should be exposed to a 0.0007^{-9} miliampere magnetic field for 0.008 seconds. According to the hereby presented research hypothesis, this procedure should result in the appearance of tiny 0.0009^{-8} millimeter apertures in the pineal gland. The second stage of the experiment is designed to check whether the exposed apertures constitute wormholes. For this purpose an SI EMU conductor should be inserted into one of the apertures. If these are actually wormholes, the conductor should exit the brain and head into which it had been inserted. If not, transmission will only be received from within the brain. The reception of transmissions originating outside the head into which the minuscule conductor had been inserted will prove both the existence of wormholes and the extra-dimensionality of these wormholes. In case no transmissions whatsoever are received – for example, because the wormhole led the conductor into a different region of space-time – the brain should be opened and the conductor sought after. If the conductor cannot be located within the aperture into which it was inserted, that would also prove that it had been sucked into a wormhole. Planetary wormholes are presently inaccessible to us, but the mini-wormholes which are hereby predicted within the mammalian brain, make it possible to study the extra-dimensional domain.

THE PERSONAL FUNCTION

Most studies focusing on the mind have hitherto been obstructed by the elusiveness of consciousness. Inability to localize consciousness led Crick and Koch (1998) and many others to adopt a reductionist approach (Rosenblum and Kuttner, 2006), but it too failed to explain mind-related phenomena. The reason why the mind is so difficult to locate may concern its holographic nature; more explicitly, the extra-dimensional route taken by the wormholes which may form the mind. If the mind constitutes a gravity-free brain section, it follows that non-local influences, which may be mediated by wormholes, determine cognitive processes. Some of these influences may originate within the brain; others – outside. Presently even inner brain influences are not fully understood. The previously discussed whole-brain theory suggests several brain centers – and not just one – are involved in mental processes. This is an attempt to merge existing theories with the evolving holographic brain-model. Adopting an extra-dimensional approach, the sought-after connections between seemingly unrelated brain-centers can be explained. Chemical experiments which were designed to study the routes of communication between brain-centers provided considerable information concerning these routes, but mind-related processes remained unexplained. It is these processes that are most exposed to non-local influences which are likely to involve extra-dimensional routes. For example, an extra-dimensional observation may reveal chemical connections between two brain-centers that do not seem to be connected when a classical three-dimensional observation is conducted.

For the sake of clarification it is worthwhile to imagine an ocean. A classical observation would outline its boundaries. An extra-dimensional perspective would reveal energetic connections between waves. Assuming the ocean is a part of conventional space, connections between waves would seem irrelevant, but if we assume the ocean is infinite – that it is space, and there is nothing apart from it – we will be prone to describe structures within it. This viewpoint is likely to expose connections between waves, whirlpools, etc. that might otherwise be overlooked. A similar phenomenon was formerly discussed in the context of statistical analyses. When we adopt a statistical viewpoint connections between people are overlooked. These can only be perceived when a temporal viewpoint is adopted. Similarly, when we observe the brain from an extra-dimensional perspective, incorporating the dimensions of Phase and Spin, connections that have hitherto not been detected, i.e., between brain-waves are likely to emerge.

According to this line of thought, the mind has hitherto not been localized because of its extra-dimensional structure. This structure results from the fact that the mind occupies a gravity-free brain-section and is therefore prone to non-local influences. These pass through routes which can only be described by the dimension of phase. If indeed, the mind occupies extra-dimensional space its modes of operation cannot be described in a three-dimensional context.

In the book Quantum Foundations of Consciousness the gravity-free section of the brain, namely, the Pineal gland, is referred to as the Personal Function (R. Kreitler, 2004). The Personal Function is the inner brain representation of a Universal Function. A Universal Function is a bi-polar and recurrent, multi-parameter quantum structure which is liable to transformation, and is therefore holographic. Every function can be defined by a particular shape that allows for a set of specific so-called extra dimensions to be prevalent. In other words, it constitutes an extra-dimensional wormhole. According to the proposed theory, the universe is comprised of numerical correlations. These correlations can be expressed as Functions which give rise to specific Frames of Reference. Unlike the Universal Function which describes portions of the cosmos, the Personal Function is confined to the brains of living mammals. Its extra-dimensional shape provides its bearer with a mind, thus endowing it with the capacity to contrive quantum communication. As explained above, quantum communication is essentially holographic. In simpler words, it involves interconnectedness, which is enabled by extra-dimensions, i.e., phase.

Since the Personal (like the Universal) Function is not confined by our four classical dimensions, it gives an outside observer the impression of a wormhole. Actually, it is a physical structure that can only be defined by extra dimensions. Not seeing these dimensions we get the false impression of mysterious, non-local connections. All physical forces and connections are field-mediated, except for the fact that the field is defined by dimensions other than the classical ones. Therefore, the Function does not constitute an integral component of our body which is a part of ordinary space. In other words, the Function is not flesh and blood. Its connection with the body can be compared to the connection between the planet and the gravity-free point at its core. Clearly, without the brain, there would have been no Function, but the Function as such is gravity-free, non-local and especially non-temporal. Hence, it is not a part of the physical brain, and can therefore, maintain a holographic quantum-connection with all life events, both past and future.

Such trans-temporal connections result from the Function's contact with itself. According to Bohr's model of quantum-mechanics, every quantum system maintains contact with every other quantum system with which it has formerly communicated (Herbert, 1994). Assuming the Personal Function is a quantum system, it follows that the Function must maintain contact with itself at every instant in our lives. For example, the quantum system that constitutes our Function at any given age must be connected to the quantum system that formed our Function when we were born. This means that from the viewpoint of the 'younger Function' also a future-connection exists. Consequently, the Function can access the brain at times other than the present. This kind of contact is by no means different from the contact between the two electrons in the double-slit experiment. As formerly explained, particles occupy a relatively timeless FR. From the viewpoint of the particles (comprising the Function), there is no difference between present and past representation of the Function. Being holographic the Function is not restricted by the classical characteristics of the body, e.g., the position of its head. For us, the past position of our head is inaccessible, but for the Function it is constantly accessible. From its vantage point, the present position of the head is adjacent to the position of the head five minutes ago and twenty years hence.

Sustaining connections with the future, the Personal Function accommodates the dimension of Potentiality. This is the case because there is no certainty the brain and hence the Function, will actualize one of many potential locations. For instance, a person can choose to go to one place or another. Thus, all locations a person may reach should be considered potentialities. This shows the Personal Function occupies a realm in which the dimension of potentiality replaces our classical dimension of time.

Norman's model of memory reconstruction (Norman, 2005) suggests memories are not simply retrieved. They are reconstructed. Norman uses the metaphor "time-tunnel" to illustrate the process of reconstruction. According to the above-suggested model, the Personal Function has access to every instant in the history of the brain. This contact does not involve any kind of actual time-tunnel. Rather it results from the non-temporal quality of quantum-communication. Being non-temporal, the Personal Function enables the brain to maintain a huge bulk of memory without having to actively store it. The Personal Function provides the mind with the capacity to communicate with itself at different times. Hence we are endowed with the cognitive capacity to re-live instances in the past, as Norman predicts. Though our body cannot travel in time, our mind seems to be able to do so. By studying the characteristics of the personal Function, a comprehensive model of memory-retrieval and association can be developed.

THE UNIVERSAL ROLE OF CONSCIOUSNESS

The ability of a quantum system – the Function – to be maintained within a physical brain indicates how deeply interwoven the two FRs are. Though both share few if any characteristics, cooperation between them seems to be necessary for the survival and well-being of humans and other species.

Apparently, consciousness provides a context in which the two FRs not only merge but also cooperate. Merger between the two is also evident in the core of our planet, but only in the context of consciousness is it presently possible to study the means by which the cooperation occurs. Every one of us is of course free to speculate on cooperation between the two in the physical framework, but at present no empirical proof for the consciousness of planets or particles, for that matter, has been obtained. As previously mentioned, speculation has been made regarding conscious attributes of elementary particles. These speculations originate from the indeterminism which particles demonstrate (Young, 1975). For example, particle-wave duality can be used as an argument in favor of particle indeterminism. According to that line of thought, the entire universe can be regarded as a physical expression of infinite states of doubt. It would follow that stellar bodies are quasi-Orlov spheres of quantum uncertainty.

But regardless of whether particles possess a degree of free will or not, the extent of conscious activity humans demonstrate exceeds the scope of indeterminism. Consciousness does not only enable us to decide in-between two courses of action. It also permits us to analyze complex situations. It provides us with a capacity to master a wide variety of complicated cognitive tasks, such as learning, planning, and daydreaming. Perhaps most importantly, it endows us with processing capacities. Our body might be able to see, hear, taste or feel, but had it not been for our cognitive abilities, these perceptions could not have been processed. We might have perceived a taste but we would not have been able to assign it to a specific food. Thus, poisonous substances could not have been distinguished and that would have been counter-evolutionary. Our ears might have enabled us to hear sounds but were it not for our conscious mind, we could not have assigned meaning to the sounds we heard. In fact, all of the body's perceptual capacities rely upon cognitive abilities. Similarly, most of our cognitive abilities rely on physical capabilities. Learning usually involves motor capacities. For instance, a person cannot learn to run and hence escape from danger without operating his or her physical body. Cognitive determination to flee is futile if one has no legs. Although various techniques are presently being developed for utilizing brainwaves for the operation of machinery, i.e., moving a wheelchair, every kind of physical operation takes place in the FR our body occupies. Even when our hands are not directly involved in the act of moving a wheelchair, the act involves a physical object, namely the chair, which has mass and can therefore be influenced only by field-forces. Hence, the act of moving it requires a degree of cooperation between the theoretical and the actual or between mind and body, hence between sub and above Plank entities.

The significance of the Personal Function does not reside in its operational capacities but rather in its extra-dimensional nature. By introducing extra dimensions into a classical body, the Function achieves that which scientists are desperately trying to achieve for decades: it merges the Newtonian and the quantum dimensional realms. The act of merging does not introduce any new rule that replaces existing laws of physics. On the contrary, unification results from providing a context for the cooperation between two otherwise separate and even contradictory FRs. The proposed model accepts the dual nature of the universe and, as noted above, proposes a specific correlation between the two realms: 1:1.4.

The division between classical and quantum physics, hence between entities that are larger and smaller than the Plank constant, seems essential for the endurance of the universe. If this separation could have been circumvented, or if one could have observed simultaneously two mutually exclusive quantities, the cosmos would have entered a loop, and would hence become pre-determined. Thus, a single unfortunate collision could set off a process that would eventually lead to the destruction of the entire universe. Uncertainty and a consequent degree of chaos, probably allow a complex system like the cosmos to correct itself. The existence of multiple potentialities presumably provides the universe with a sufficient diversity to ensure its infinite endurance.

Therefore, the significance of the Uncertainty Principle is that it makes the universe indeterminate. That could not have been achieved without the existence of conscious creatures. Were it not for the existence of consciousness, the two realms would have remained totally separate. One would involve objects; the other would include minuscule entities. If that had been the case, no uncertainty could have been monitored. The double-slit and the EPR experiments indicate uncertainty only because of the involvement of the time dimension in a context where time is otherwise dilated. If the two realms had not been joined in any way, no such process of so-called de-coherence, in which particles are exposed to the dimensions of the universe, would have taken place. And in this case, particles would have been entirely immune to gravity, and so, their momentum could not have been measured. Hence, it is a conscious act that merges the two physical FRs. But it is noteworthy that if the experimenters' mind lacked a body, it could not have carried out the act of measurement. So neither the mind nor the body can separately ensure indeterminism.

Cooperation between the two FRs is apparently facilitated by consciousness. A conscious creature, with a quantum mind and a physical body, provides the universe with a framework for cooperation between the two FRs. Physical cooperation means that motion in either FR entails motion in the other. If there had been no creatures which used their cognitive abilities to tame their physical bodies, the two realms could not have been brought together. The division between relativity and quantum-mechanics would have been complete and the universe would have become predictable. Therefore, the universal role of consciousness may be described as merging the two FRs thus making the universe indeterminate. Hence, consciousness may be regarded as the psychological counterpart of the Uncertainty Principle.

This does not mean the universe is alive. Living creatures often regard consciousness as one of the characteristics of being alive, but this conception might result from personal experience. As Orlov's model shows, every uncertainty e.g. every super-position of an electron can be expressed as a state of doubt. This in itself does not mean particles are living creatures. It only means that one important characteristic is shared between living creatures and the universe at large.

In conclusion, consciousness may be defined as the tool by which nature binds two otherwise separate FRs, making the universe in-determinate and infinite. Consciousness imposes the concept of time upon a realm where time is dilated. Likewise, consciousness introduces the concept of potentiality into our temporal realm, where it would otherwise have been non-existent. In this way, consciousness makes both realms indeterminate. In other words, consciousness subjects the entire universe to the Uncertainty Principle. Due to the effects of consciousness the entire universe – including both larger than Plank and smaller than Plank entities – exhibits indeterminate features. In-determinism ranges from suns that may continue along their orbit, explode in a supernova, or hit another sun, to particles that may be observed as waves.

Universal indeterminism is most easily observable in the study of astronomy, i.e., we can predict astronomical alignment between two planets. These predictions may falsely be interpreted as predictions of planetary location. Indeed, they enable us to extrapolate the orbital progression of planets years in advance, but a phenomenon known as the Chandler wobble[3], which is only liable to prediction several months in advance, interferes with these predictions and prevents us from assessing the exact location of a planet at a given date. So we may be able to predict the general location of a planet but its precise position and likewise the angle between it and other planets is subject to uncertainty. Small variations the Chandler

[3] The Chandler Wobble is the change in the spin of the Earth on its axis.

Wobble creates may potentially determine whether a collision with an asteroid or comet will occur.

Indeterminism, which seems to be a fundamental feature of the universe, is enabled by the combination of two separate and even contradictory FRs. It is the irreconcilability of the two that provides the universe with a degree of uncertainty. And this unique combination of irreconcilable FRs is made possible due to the existence of conscious creatures that can collapse quantum wave-functions at will, thus actualizing potentialities and setting off processes of de-coherence. Thus, the two separate realms are merged, without being joined, making the universe indeterminate. In sum, the proposed model suggests that the existence of consciousness in general and of conscious creatures in particular, provided the universe with a unique context in which unification occurs.

CONCLUSION

Quantum mechanics and psychology are two fields that are as far apart as one could possibly imagine. One is a natural science – the other a social science. One concerns the fundamental building blocks of the universe – the other the most complicated and diverse entity: human beings. One focuses on precise measurements of minuscule processes – the other on the emotions and ambiguous multi-determined attitudes of human beings. Nonetheless, a striking similarity appears between the two sciences. First and foremost, both rely upon holographic principles. Both the quantum connection and the basic psychological assumption concerning the existence of a conscious 'mind' which cannot be reduced to its neurological constituents are essentially holographic. Being holographic the two fields focus on interconnectedness. Particle physics studies quantum interactions within the Implicate Order. Psychology focuses on a variety of connections, i.e., between associations, external clues and memory-retrieval or between attitudes and behavior.

A large body of research, which was discussed above, points to a significant degree of similarity between the two fields. But such similarity does not suffice to make the case that these are two facets of one science. Nevertheless, the discussed similarity between the fields is not limited to holography and interconnectedness. The two fields focus on a FR that is essentially non-classical and therefore separate from the classical four-dimensional space, in which we live. The FR that is studied by both quantum physicists and psychologists involves extra-dimensions. But also this similarity is not unique because economics also probes the extra-dimensional realm. Since the extra-dimensional realm includes all theoretical entities which lack mass and are therefore gravity-free, it involves all theoretical sciences, not only psychology.

However, indeterminism provides quantum-mechanics and psychology with another common denominator that is not shared by any other field of science. The above comparison between universal and personal indeterminism sets the two fields apart from all other fields of science. No other field is thus plagued by inherent ambiguity. Likewise, no other science explicitly studies indeterminism. This is not the case regarding multi-determinism. A variety of sciences ranging from medicine to meteorology analyze multi-determinant processes, but no sciences other than quantum-mechanics and psychology explicitly probe the realm of uncertainty and indecisiveness. Therefore, it may be speculated that the study of quantum interconnectedness is in fact the study of the physical attributes of cognitive processes. And conversely, that the study of human attitudes, beliefs, emotions and conflicts, concerns the macroscopic representation of quantum-mechanical phenomena.

When we observe the seemingly random motion of particles we do not know to what non-local system they belong. Larger physical processes bear no resemblance to quantum processes and therefore no real-world representation of inter-particle interactions is currently available. Similarly, when we observe human beings we only see the outward expression of cognitive processes (what a person says or does), but we know relatively little about the physical or rather physiological interactions governing these processes. Therefore, the above-suggested model opens a wide array of experimental possibilities for both sciences. Psychologists will be able to formulate research hypotheses regarding the processes underlying emotions, attitudes, etc. Physicists will be able to observe the processes to which quantum interactions give rise, and thus formulate an extra-dimensional geometry enabling quantum connections.

The feasibility of extra-dimensional connections will become evident from the previously proposed experiment that is designed to unequivocally prove the existence of mini-wormholes in the brain. A successful conclusion of this experiment will verify the extra-dimensional nature of mental activity.

REFERENCES

Ayres, P. and Sweller, J. (2005). The split-attention principle in multimedia learning. In R.E. Mayer (Ed.), *The Cambridge handbook of multimedia learning*. New York: Cambridge University Press, pp. 135-146.

Baconnier, S., Lang, S. B., Polomska, M., Hilczer, B., Berkovic, G., and Meshulam, G. (2009). Calcite microcrystals in the pineal gland of the human brain: First physical and chemical studies. *Bioelectromagnetics, 23*, 488-495. [published online Sep. 10, 2002] [DOI 10.1002/bem.10053].

Basar, E. (2005). Memory as a "whole-brain-work": a large-scale model based on "oscillations in super-synergy". *International Journal of Psychology, 58*, 199-226. [DOI: 10.1016/j.ijpsycho.2005.04.008].

Basar, E. (2006). The theory of whole-brain-work. *International Journal of Psychology, 60*, 133-138. [Doi: 10.1016/j.ijpsycho.2005.12.007].

Beason, R. C. and Semm, P. (1996). Does the avian ophthalmic nerve carry magnetic navigational information? *Journal of Experimental Biology, 199*, 1241-1244.

Beck, F. and Eccles, J. (1992). Quantum aspects of brain activity and the role of consciousness. *Proceedings of the National Academy of Sciences of the USA, 89*, 11357–11361.

Berry, M. V. and Geim, A. K. (1997). Of flying frogs and levitrons. *European Journal of Physics, 18*, 307-313. [doi: 10.1088/0143-0807/18/4/012]

Bohm, D. (1980). *Wholeness and the implicate order*. London: Routledge.

Bohr, N. (1958). *Atomic theory and human knowledge*. New York: Wiley.

Carlson, N. R. (1986). *Psychology of behavior*. Newton, MA: Allyn and Bacon.

Caipang, C. M., Berg, I., Brinchmann, M. F., and Kiron, V. (2009). Short-term crowding stress in Atlantic cod, Gadus morhua L. modulates the humoral immune response. *Aquaculture, 295*, 110-115.

Choustova, O.A. (2006). *Quantum Bohmian model for financial market*. New York: Elsevier.

Close, F. (2007). *The new cosmic onion: Quarks and the nature of the universe*. Boca Raton, FL: CRC Press, Taylor & Francis Group.

Crick, F. and Koch, C. (1998). Consciousness and neuroscience. *Cerebral Cortex, 8,* 97-107.

De Broglie, L. (1949). A general survey of the scientific work of Albert Einstein. In P. Schilpp (Ed.), Albert Einstein: Philosopher-scientist, Vol. 1. New York: Harper & Row, pp. 107-127.

Descartes, R. (1989/1649). *Passions of the soul* [Les passions de l'ame]. Indianapolis, IN : Hackett Publishing Co. [English edition, trans. Stephen H. Voss].

Dobson, J. and Grassi, P. (1996). Magnetic properties of human hippocampal tissue – evaluation of artifact and contamination sources. *Brain Research Bulletin, 38*, 255- 259. [Doi 10.1016/0361-9230(95)02132-9]

Dörner, D. and Wearing, A. (1995). Complex problem solving: Toward a (computer-simulated) theory. In P. A. Frensch and J. Funke (Eds.), *Complex problem solving: The European perspective*. Hillsdale, NJ: Lawrence Erlbaum Associates., pp. 65- 99.

Edmonds, D. T. (1992). A magnetite null detector as the migrating bird's compass. *Proceedings: Biological Sciences, 249*, 27-31.

Edwards, H. H., Schnell, G. D., DuBois, R. L., and Hutchison, V.H. (1992). Natural and induced remanent magnetism in birds. *The Auk, 109*, 43-56 [published by University of California Press on behalf of the American Ornithologists' Union].

Einstein, A. (1920). *Relativity: The special and the general theory: A popular exposition.* London: Methuen.

Einstein, A., Podolksy, B., & Rosen, N. (1935). Can quantum-mechanical description of physical reality be considered complete? *Physical Review, 47*, 777-780.

Freud, S. (1900) *The interpretation of dreams.* New York: Macmillan.

Friedman, M. (1996). *Type A behavior: Its diagnosis and treatment*. New York, Plenum Press (Kluwer Academic Press).

Gardner, H. (1993). *Frames of mind: The theory of multiple intelligences.* (10nd Ed.). New York: Basic Books.

Gill, J.M., Saligan, L., Woods, S., and Page, G. (2009). Post-traumatic stress disorder is associated with an excess of inflammatory immune activities. *Perspectives in Psychiatric Care, 45*, 262-277.

Goto, Y., Taniura, H., Yamada, K., Hirai, T., Sanada, N., Nakamichi, N., and Yoneda, Y. (2006). The magnetism responsive gene Ntan1 in mouse brain. *Neurochemistry Intenational, 49*, 334-341.

Greene, B. (1999). The elegant universe: Superstings, hidden dimensions, and the quest for the ultimate theory. New York: Vintage Books.

Herbert, N. (1994). *Elemental mind.* New York: Plume/Penguin.

Jung, C. G. (1981). *The archetypes and the collective unconscious* (Collected Works of… Vol 9, Part 1; 2nd ed.). Princeton, NJ: Bollingen. [Translator R.F.C. Hull]

Karch, A. and Randall, L. (2001). Locally localized gravity. *Journal of High Energy Physics, 105*, 1-22.

Kirkegaard, C. and Faber, J. (1998). The role of thyroid hormones in depression. *European Journal of Endocrinology, 138*, 1–9.

Kirschvink, J. L., Kobayashi-Kirschvink, A., and Woodford, B. J. (1992). Magnetite biomineralization in the human brain. *Proceedings of the National Academy of Sciences, 89*, 7683-7687.

Kreitler, R. (2004). *Quantum foundations of consciousness.* Tel-Aviv: Author's Edition, [www. Ronkreitler.com ISBN 965-90930-0-4]

Kreitler, R. (2012). Quantum logic and the principle of extra dimensions. In S. Kreitler, L. Ropolyi, G. Fleck and D. Eigner (Eds.), *Systems of logic and the_construction of order*. Bern, New York, Vienna: Peter Lang Publishing Group, pp. 95-129.

Lashley, K. S. (1950). In search of the engram. *Society of Experimental Biology, Symposium 4*, 454-482.

Paster, R. (2006). *New physics and the mind*. New York: Author's Edition [www.booksurge.com].

Penrose, R. (1994). *Shadows of the mind:Aa search for the missing science of consciousness*. New York: Oxford University Press.

Pinel, J.P.J. (1992). *Biopsychology*. Needham Heights, MA: Allyn and Bacon.

Pitkänen, M. (2006). Topological geometrodynamics. Beckington, Frome, UK: Luniver Press.

Polyn, S. M., Natu, V. S., Cohen, J. D., and Norman, K. A. (2005). Category-specific cortical activity precedes recall during memory search. *Science,* 310, 1963-1966.

Pound, R. V. and Rebka, Jr. G. A. (1959). Gravitational red-shift in nuclear resonance. *Physical Review Letters, 3,* 439–441. [doi:10.1103/PhysRevLett.3.439].

Pribram, K. (1971). *Languages of the brain: Experimental paradoxes and principles in neuropsychology*. Englewood Cliffs, N. J.: Prentice-Hall.

Pribram, K. (1982). What the fuss is all about. In K. Wilbur (Ed.), *The holographic paradigm and other paradoxes*. Boulder and London: Shambhala, pp. 27-34.

Pribram, Karl (1991). *Brain and perception: Holonomy and structure in figural processing*. Hillsdale, NJ: Lawrence Erlbaum and Associates.

Randall, L. (2006). *Warped passages: Unraveling the mysteries of the universe's hidden dimensions*. New York: Harper Perennial/Harper Collins Publishers.

Rath J. F., Langenbahn, D. M., Simon, D., Sherr, R. L., Fletcher, J., and Diller, L. (2004). The construct of problem solving in higher level neuropsychological assessment and rehabilitation. *Archives of Clinical Neuropsychology, 19*, 613-635.

Resnick, R., Halliday, D. and Krane, K. S. [with assistance of P. Stanley]. (2002). *Physics* (Vol. 1, 5th Ed.). New York: John Wiley and sons.

Rollero, A., Murialdo, G., Fonzi, S., Garrone, S., Gianelli, M. V., Gazzerro, E., Barreca, A., and Polleri, A. (1998). Relationship between cognitive function, growth hormone and insulin-like growth factor I plasma levels in aged subjects. *Neuropsychobiology*, 38, 73-79.

Rosenblum, B. and Kuttner, F. (2006). *The quantum enigma*. New York: Oxford University Press.

Sakurai, J. J. and Napolitano, J. (2011). Modern quantum mechanics (2nd Ed.). Boston, MA: Addison-Wesley [Pearson Education].

Samal, M. K. (2000). Can science 'explain' consciousness? In B. V. Sreekantan et al., (Eds.), Proceedings of national conference Scientific and Philosophical Studies on Consciousness. NIAS, Bangalore, India: NIAS, pp. 1-7.

Semm, P. (1983). Neurobiological investigations of the magnetic sensitivity of the pineal gland in rodents and pigeons. *Journal of Comparative Biochemistry and Physiology, 76,* 683–689.

Sirag, S.-P. (1983). Physical constants as cosmological constraints. *International Journal of Theoretical Physics*, *22,* 1067-1089.

Smolin, L. (2007). *The trouble with physics: The rise of string theory, the fall of a science, and what comes next.* Boston, MA: Houghton Mifflin Harcourt (Copyright 2006 by spin Networks Ltd).

Stapp H.P. *Mindful universe: Quantum mechanics and the participating observer*. New York: Springer, 2007.

Talbot, M. (1992). *The holographic universe*. New York: HarperPerenial.
Thorne, K. S. (2002). Spacetime warps and the quantum world: Speculations about the future. In R. H. Price (Ed.), *The future of spacetime*. New York: W.W. Norton, pp. 109-152.
Wheeler, J. A. (1990). *A journey into gravity and spacetime*. San Francisco, CA: W. H. Freeman, Scientific American Library.
Wiltschko, W., Munro, U., Wiltschko, R., and Kirschvink, J.L. (2002). Magnetite- based magnetoreception in birds: the effect of a biasing field and a pulse on migratory behavior. *The Journal of Experimental Biology, 205*, 3031-3037.
Young, A. (1975). *Geometry of meaning*. New York: Delacorte press.
Zukav, G. (1979). *The dancing Wu Li masters: An overview of the new physics*. New York: Bantam.

In: Consciousness: Its Nature and Functions
Editors: Shulamith Kreitler and Oded Maimon

ISBN 978-1-62081-096-5

Chapter 21

SUPER TURING AS A COGNITIVE REALITY

Hava T. Siegelmann
Dept. of Computer Science, Program of Behavior and Brain Science
UMass Amherst, MA, US

The 1950s witnessed an explosion, which has propagated down the succeeding decades resulting in one of the most significant changes in human history: the era of the computer. Virtually, all current computational systems are based on Alan Turing's 1936-8 Machine model, which described the human clerks of his time, who perfomed rote calculations following a fixed set of instructions – and was designed ultimately replace them (Turing, 1936); Turing's second model, the Universal Turing Machine is a programmable version of the first model, and it became the basis for realization it. Programmability was a great innovation allowing the hardware to remain constant while producing, in effect, different machines (i.e. different programs) to apply to differing tasks. While pundits promised the full integration of computers and robots in our lives as intelligent collaborators, the real capabilities have not lived up to this promise. Often, an attractive interface or friendly looking robot that implies innovation, masks software that could have been written years ago and the appearance of machine intelligence is substituted for intelligence itself.

Super-Turing Computation (STC) is a new brain inspired computational method. Unlike the ubiquitous Turing Machine method, which is fully programmed, the Super-Turing paradigm is a complex dynamical system that continuously tunes its instruction set to the environment with autonomy.

The Artificial Intelligence and Machine Learning world was captivated recently, when IBM's Watson computer won a series of Jeopardy games against impressive human Jeopardy champions - seemingly a feather in the cap of machine intelligence. Indeed, Watson pushes Turing computation toward its limits using expensive computer clusters, increasingly large amounts of memory, enormous databases and clever programming with myriad "if-then-else" statements. But, Watson, like other similar Artificial Intelligence (AI) demonstrations, and in fact, all Turing-type computational systems, needs to operate in very orchestrated, specific environments, for which it was fully designed in advance. No matter how resources are increased, a Turing computer's program is incapable of updating itself to accomplish other tasks or work in an environment other than what it was programmed for. Security systems can only consider previous security breaches rather than take hints and predict new ones. Robots designed to escort patients or soldiers have fixed memory and do not learn from the experience as the human near them does.

Turing himself, as witnessed for example by his 1948 notes, saw his "Turing Machine" only as a beginning and, in fact, searched continuallly for a superior system paralleling the cognitive mechanisms of the human brain, capable of dramatically increased flexibility, the ability to learn and infer. While Turing strongly believed that such adaptive machines can be found, he died before he was able to develop his concept, which remained dormant until the Super-Turing Theory of Computation was proposed in my 1993 Ph.D theses (Eduardo Sontag – advisor), a series of papers, e.g., (Siegelmann and Sontag, 1994, Siegelmann, 1995) and the detailed book, "Neural Networks and Analog Computation: Beyond the Turing Limit" (Siegelmann, 1998).

Super-Turing computation is a flexible version of Turing computation; it can be thought of as a machine capable of reprogramming itself (creating a new Turing Machine) at each computational step; The ability to create a new Turing machine provides a profoundly stronger version of computation. Super-Turing machines compute all binary input-output functions of Turing machines and a vast number of functions which Turing machines cannot compute even given infinite amount of resources. A Super-Turing machine can compute all binary input-output functions in exponential time, it computes a very well defined and bounded class of binary functions in polynomial efficient time which we refer to as Analog-P or P/poly. Furthermore, it can also compute functions, which are not bounded to binary.

Super-Turing's flexibility may stem from the fact that memory and program exist in an interlaced fashion, adapting and changing together dynamically: reminiscent of neural networks, adapting to the situation. One of the primary questions computer science faces today in attempting to move beyond Turing-type computers is the question of how a computational system with finite resources can deal with the infinite information and precision inherent in the real-world. Super-Turing computation can be achieved in numerous ways; each has the quality of reducing infinite real world data to a manageable level while maintaining sufficient information:

(a) Analog recurrent neural networks with real number synaptic weights are Super-Turing. The neural activity can be constrained to rational numbers; real activations do not increase the computational capabilities of the analog networks. To contrast, recurrent Neural Networks with Rational Weights and Rational Activations are equivalent to Turing Machines. The complexity class of problems computable by these machines in polynomial time is the classical P.
Importantly, an analog Super-Turing device (e.g., a neural network) adheres to the rule, "linear precision suffices," meaning that the machine requires only t bits of precision for t computational time steps, both in the machine's description and in its calculation; this rule is equivalent to known exponential divergence in chaotic systems, supporting the physical plausibility of Super-Turing computation, as detailed in my 1995 Science article (Siegelmann, 1995). This means that the network can start with rational numbers (both in the description of the machine and in the calculation) and adds to the rational number precision needed. Even though precision

will never be infinite and the numbers will in a practical sense remain rational, the mere ability to not bound them in advance provides Super-Turing power.

(b) A Turing machine can achieve Super-Turing computation with an extra advice line, which is of non-uniform polynomial length; this is equivalent to a Turing machine that receives one bit from the environment at each computational step via open sensors (as long as polynomial non-uniformity applies). Turing noted the quality of non-uniformity for increasing the power of the machine toward intelligence.

(c) Stochastic Turing machines (or digital computers) incorporating a binary coin, which is itself biased by a real number value are stronger than the Turing Machines. As the real values are not seen directly by the network, which sees only the binary value of the coin, the associated computational class is along the Super-Turing hierarchy (mentioned below), falling exactly on BPP/log (Siegelmann, 1998). If the bits of the bias in the coin are given instead, the machine reached the top oh the hierarchy, Analog-P.

(d) Interestingly, most practical methods to attain Super-Turing computation correlate with adaptive biological brain processes: as we recently proved, evolving neural networks where their weights can change are Super-Turing (Cabessa and Siegelmann, 2011). The evolving weights do not need to be real, but can be rational numbers. Evolving neural networks like the reconsolidation process, updates memories, enables perception of the presence as well as prediction to future based on experiences. Evolving weights also have the property of practicality, the machine update, never being infinite.. Since memory and computation are strongly integrated in neural networks, evolving networks correspond to machines that change their program continuously.

The crucial source of information is not the precision, but rather the information contained in the machine description or equivalently in the external bits from the environment. The 'linear precision suffices' theorem means that the real world (real numbers) can be approximated by small, but growing precision, which is sufficient for all practical purposes for decision-making. I proved with Jose Balcazar, Ricard Gavalda and Eduardo Sontag that an infinite hierarchy of efficient computation, bound by Turing at one end and Super-Turing at the other, exists corresponding to the amount of information contained in the environmental bits. The complexity of the machine is thus dependent on the quality of the information relative to the machine: If the bits can be efficiently computed by a TM, they do not add to the power of the machine; if they can be computed but require significant time, their inclusion may add efficiency. If the bits contain randomness or otherwise richer sources, the resulting machine can be even stronger.

It is possible that the same qualities of efficiency, learning, increasing precision and continuous dynamics that describe Super-Turing computation – may underlie the natural world, from cellular biology to gravitational physics. Natural Selection tends toward efficiency; the world, through billions of years of evolution, may have arrived at the same efficient Super-Turing method originally uncoverd as an efficient and more capable computational method.

a) Super Turing in Cognition

To begin to grasp the significance of Super-Turing processes, it helps to keep in mind that the Universe is, in a practical sense, infinite; any bit of it can be described by an infinite amount of information. Organic life however, is finite; the animal brain cannot process the infinite, dynamic universe. Instead the brain creates a highly realistic proxy of the real world. The model we create in our minds is highly efficient; in true Super-Turing fashion, it is detailed only in those parts needed for a particular purpose - devoting only the resources necessary to the task at hand and adding more information as needed, ; it learns and increases precision as time goes on. It is this parallel between the cognitive processes and Super-Turing computational systems that suggests Super-Turing systems as a feasible basis of computation in human-like, artificial intelligence.

Intelligence depends on adaptivity of memory, enabling the individual to better fit and adapt to changes in his environment. These Super-Turing-like qualities charaterize Reconsolidation (ReC) first defined in the seminal work of Sara (2000). ReC enables adaptivity in memory; Sara asserted that learning and memory adaptation require the comprehension and use of previous memories, just as occurs in perception and other intelligent processes. She argued further, that when a memory trace is activated it becomes labile and another process, which she named reconsolidation, must be activated to preserve the recalled memory. During reconsolidation the trace can change – causing adaptivity in memory and consequently in behavior. Most research over the past ten years focused on the outcome occurring when the reconsolidation process is externally being interfered with (e.g., by inserting intra-hippocampal anisomycin treatment) and active memory traces are lost; this interference in ReC was proposed to have application in the treatment of phobias, PTSD, and addiction (Nader, Schafe and LeDoux, 2000; Stollhoff, Menzel and Eisenhardt, 2005; Taubenfeld, Muravieva, Garcia-Osta and Alberini, 2010).

But these results do not explain the day-to-day role of reconsolidation when there are no severe chemical disturbances. We suggest that it is likely that ReC's main role is to lead to intelligent adaption of memories; repetitive occurrences of ReC have been demonstrated causing monotonically ordered modifications correlating to changes in the subject's environment, both in electrophysical measures in hippocampus place cells CA3 and CA1, as well as in psychophysical experiments based on morphing faces (Leutgeb et al., 2005; Preminger, Blumenfeld, Sagi and Tsodyks, 2009).

Another process typifying Super-Turing is active attention, or what is called in cognitive psychology "cognitive misers" (Fiske, 1992). These reveal the principle, in which the brain processes only those details mandatory for perception. Elements less necessary to the present task are processed in a less detailed manner fitting the Super-Turing quality of of minimalistic efficiency when processing data.

B) Super-turing Theory via Neural Networks

Computational neural networks (NN) have their origin in the 1940s; their design attempts to mimic the brain's processing abilities. Neural networks provide an alternative to the older psychological approach that thoughts and understanding follow the manipulation of symbols: They are a semi-parametric, non-symbolic form of computation, where memory is distributed and there are no separate areas for instructions; this more closely parallels current thought on brain structure.

With all Turing's matchless contributions to computer science, it is less well known that in 1948 he also originated random neural networks, which he called "Unorganized Machines" and associated them with developing machine intelligence. The 1980s saw the advent of Analog Neural Networks (ANN), which added significant learning capabilities to the previous models of the yes/no binary firing neurons. The neural net has particular advantages in achieving Super-Turing computation: unlike the UTM, which cannot update its instruction set, neural nets update their instruction set with each memory update since ptocessing and memory are fully interleaved.

My journey into the world of Super-Turing computation started as an effort to understand NN processing abilities or lack thereof. To my surprise, I found that rather than being significantly weaker in terms of information processing than Turing-type computers, the accepted notion at that time, they were stronger; and that while NN could simulate Turing computation, they had abilities Turing computers couldn't duplicate. My prototype network, specified in equation (1) was subsequently proposed to serve as the standard in analog computation systems, paralleling the role of the Turing machine in digital computation.

The Analog Recurrent Neural Network (ARNN) prototype consists of an interconnection of *N* neurons, where each updates by the following equation:

$$x_i(t+1) = \sigma\left(\sum_{j=1}^{N} a_{ij} \cdot x_j(t) + \sum_{j=1}^{M} b_{ij} \cdot u_j(t) + c_i\right), \quad i = 1, \ldots, N \tag{1}$$

All a_{ij}, b_{ij}, and c_i are numbers describing the weighted synaptic connections and weighted bias of the network, u are inputs arriving on M input lines in a stream, and P, $(P<N)$ of the neurons will serve as output neurons, whose values will propagate on P output streams. The activation function *sigma* applied to each neuron is the classical saturated-linear function defined by:

$$\sigma(x) = \begin{cases} 0 & \text{if } x < 0, \\ x & \text{if } 0 \le x \le 1, \\ 1 & \text{if } x > 1. \end{cases}$$

Prior to my work, ARNN were assumed to be a weaker form of computation than the Turing machine in that they cannot have separate memory, cannot compute "if-then-else" statements and due to continuity, cannot test internal values for exact comparisons. It was indeed surprising when I proved that the ARNN with a few dozens of neurons and when all weights were taking simple rational values, was the computational equivalent to the universal Turing machine (Siegelmann, 1995; Siegelmann and Sontag, 1995). This veneer of equivalence is stripped away with the introduction of either one real value or of dynamic changes in synapses. In this case, the ARNN can compute such complex functions that no

Turing system can, e.g., the halting problem (Siegelmann, 1995; Siegelmann and Sontag, 1994).

We evaluated the STC for practical reliability in the mathematical lemma, "Linear (bit) Precision Suffices" (Siegelmann, 1998; Siegelmann and Sontag, 1994). Neither real weights nor highly precise rational weights are necessary for Super-Turingness. Super-Turing method relies on only k precision for the calculation of k computational steps. Many physical systems, including chaotic and quantum, have the same exact feature, which is there considered (only) "exponential (value) sensitivity to initial conditions." This is in fact the basis for the proof in recent paper on evolving networks (Cabessa and Siegelmann, 2011).

In the intervening years, I became interested in neural and cognitive functions. During these studies, it became evident that many biological functions shared properties with Super-Turing. The computational principles underlying STC can be explained using the ARNN example:

1) Super-Turing processes are value based (as opposed to bit-based). Yet they are efficient, starting with a minimal amount of precision and increasing precision only as needed.
2) Flexibility is a key Super-Turing trait and is present in memory (c_i from equation (1)), and memory associations (a_{ij} from equation (1)). When Super-Turing systems lose flexibility they cease to be Super-Turing and become Turing; in organic systems, this leads to illness and eventually death. Similarly, a common trait of many forms of mental illness is a lack of flexibility in perceiving, interpreting, and/or remembering information.
3) Super-Turing processes may be present in brain where memory and processing do not exist separately as in Turing machines, but are interleaved, while dynamically updating and tuning. Reconsoliation updates memory and processing.
4) Dynamical weighting of different inputs (b_{ij} from equation (1)) such as in active attention leads to Super-Turing power. And lastly,
5) Collective super-Turing computation arises from the interaction of nodes: Even a single neuron, with richer synaptic and memory structures can transform otherwise simple neurons in its network to form a collective super-Turing computation network.

C) Symbolic and Subsymbolic Processing

Turing computation is an entirely logical, "amodal" symbolic process, well founded within the field of mathematical logic and in philosophical and computational theories of language. But, symbols in organic intelligence, are based on complex, motivated, correlated sub-symbolic associations; without the sub-symbolic, symbols are empty containers, lacking meaning, the ability to associate or to adapt. Organic symbols change via their sub-symbolic relationships based on circumstance, e.g. stress or ease, danger or safety, trust or distrust, etc. Super-Turing computation combines the analog sub-symbolic representation with symbolic logically inferable parts.

Symbols seem to reside in different contexts in brain. The labs of Fried and Koch showed that the identity of visual stimuli seen by a patient can be predicted from the firing rate of four medial temporal lobe (MTL) cells triggered between 300 and 600 msec after image onset. These cells were named concept neurons (QuianQuiroga, Reddy, Koch and Fried, 2007). The same labs recently showed that the patients can control a display by modulating the firing rate of four MTL neurons using thoughts regarding particular concepts (Cerf et al., 2010). While questions were raised regarding the interpretations of these experiments, concept cells appear to be high-level abstractions – some type of symbols.

A functional magnetic resonance imaging (fMRI) study revealed six small highly face-selective regions in the cortex of the macaque temporal lobe. Single-unit recordings within two of these areas showed that almost all visually responsive cells are face-selective (Freiwald, Tsao and Livingstone, 2009).

These two studies consider symbols as point neurons, but other studies consider the strong interconnection to provide a concept of meaning (Tononi, 2004).

Concepts are also related to categorizing, and the ability of humans to categorize data and the relations between concepts and categories has been studied in the field of psychology (e.g., Gold, Cohen and Shiffrin, 2006). Abstraction is another way to understand the processes creating concepts out of general data (Barsalou, 2003). Recent theoretical research based on category theory relates synaptic assemblies and concepts (Healy and Caudell, 2006).

Symbolic AI operated on a different notion of symbols, also referred to as “logical rules" ("if A then B") or "amodal symbols" (Barsalou, 1999). "Amodal" refers to "disembodied" or "non-motivated," not related to emotional or perceptual processes. This is a very different notion from the one used in neuroscience and psychology where "grounded symbols" are connected to emotional (motivational) processes, "grounded" in bodily processes (it has to be a complex process, not just a notation).

These different meanings of “symbols,” caused disconnects between the AI philosophy and brain research. A highly important subject that needs explanation concerns how neural mechanisms transform body-related signals into amodal logic-like signs, or even more challenging: how higher mental concepts, such as beautiful and sublime, meaning of life... concepts, which unify amodal and emotionally grounded processes arise. This duality of representation and inference, naturally combining analog with digital, static with learning – is expected to appear in super-Turing.

D) Thesis of Computational Limits

Following the Super-Turng theory of 1993, the term, ‘Hypercomputation’ was coined by philosophers Jack Copeland and Diane Proudfoot in (Copeland and Proudfoot 1999). It has become a catchall encompassing diverse concepts that are either Turing as in the case of Quantum Computation, but differ on a mechanical level; or are not Turing, and in this case may be more thought experiments, than plausible system. Super-Turing, on the other hand, is modeled after nature and is potentially realizable: A good mathematical model of a natural system is not expected to fully describe the system, but to capture sufficient system attributes to be useful in explaining it; when building a realization from a mathematical model, it is the reverse engineering of the same process; there is not necessarily a one to one correspondence between the model and the realization. This is the situation with the Universal Turing Machine model and the many digital computers that are realizations of it. This may potentially occur with the Super-Turing model.

It is the natural limit on the computational power of super-Turing computers that makes them a feasible machine model. The Thesis of Analog Computation (Siegelmann and Sontag, 1994; Siegelmann, 1998) suggests that "No Reasonable Abstract Analog Device Can Have More Computational Capabilities (Up To Computational Time) than the Analog Recurrent Neural Networks." It is hypothesized that feasible analog computers and chaotic physical systems will compute no more and no much faster than the ST model, exemplified by the ARNN. This means that the ST theory is relevant to describe physical systems.

E) The Future of Artificial Intelligence

Computer Science and the industry surrounding it are investing ever-increasing amounts of money, time and research at an effort to push Turing-type computation ahead. Yet, a fixed Turing program has not yet managed to pass the Turing test, which necessitates a computer to act so intelligently that it can be mistaken for a human. It may be that a Super-Turing computer with its evolving properties may have the capabilities to bring true intelligence to artificial systems; that a ST computer will allow the creation of inferential, adaptive machine intelligences that will work robustly and autonomously in real-world, dynamic environments; that it will learn and communicate effectively with humans and make truly autonomous robots and intelligent systems a reality.

REFERENCES

Barsalou, L.W. (1999). Perceptual symbol systems. *Behavioral and Brain Sciences, 22*, 577-609.

Barsalou, L.W. (2003). Abstraction in perceptual symbol systems. *Philosophical Transactions of the Royal Society of London: Biological Sciences, 358*, 1177-1187.

Cabessa, J. and Siegelmann, H. T. (2011). Evolving recurrent neural networks are super-Turing. *International Joint Conference on Neural Networks,* July, San Jose, USA.

Cerf, M., Thiruvengadam, N., Mormann, F., Kraskov, A., QuianQuiroga, R.., Koch, C., and Fried, I. (2010). Online, voluntary control of human temporal lobe neurons. *Nature, 467*(7319), 1104-1108.

Copeland, J. and Proudfoot, D. (1999). Alan Turing's forgotten ideas in computer. *Scientific American*, 99-103.

Fiske, S.T. (1992). Thinking is for doing: Portraits of social cognition from Daguerrotypes to Laserphoto. *Journal of Personality and Social Psychology, 63*, 877-839.

Freiwald, W. A., Tsao, D.Y., and Livingstone, M.S. (2009). A face feature space in the macaque temporal lobe. *Nature Neuroscience, 12,* 1187 – 1196.

Gold, J., Cohen, A., and Shiffrin, R. (2006). Visual noise reveals category representations. *Psychonomic Bulletin& Review, 13*, 649-655.

Healy, M. J. and Caudell, T. P. (2006). Ontologies and worlds in category theory: Implications for neural systems. *Axiomathes, 16*, 165-214.

Leutgeb, J. K., Leutgeb, S., Treves, A., Meyer, R., Barnes, C.A., McNaughton, B.L., Moser, M.B., and Moser, E.I.(2005). Progressive transformation of hippocampal neuronal representations in morphed environments. *Neuron, 48*, 345–358.

Nader, K., Schafe, G. E., and LeDoux, J.E. (2000). Fear memories require protein synthesis in the amygdala for reconsolidation after retrieval. *Nature, 406*, 722-726.

Preminger, S., Blumenfeld, B., Sagi, D., and Tsodyks, M. (2009). Mapping dynamic memories of gradually changing objects. *Proceedings of the National Academy of Sciences, 106(13)*, 5371-5376.

QuianQuiroga, R., Reddy, L., Koch, C., and Fried, I. (2007). Decoding visual inputs from multiple neurons in the human temporal lobe. *Journal of Neurophysiology, 98*, 1997–2007.

Sara, S. J. (2000). Retrieval and reconsolidation: Toward a neurobiology of remembering. *Learning and Memory, 7*, 73–84.

Siegelmann, H. T. (1995). Computation beyond the Turing limit. *Science, 238 (28),* 632- 637.

Siegelmann, H. T. (1998). *Neural networks and analog computation: Beyond the Turing limit.* Boston, MA: Birkhauser.

Siegelmann, H. T. and Sontag, E.D. (1994). Analog computation via neural networks. *Journal of Theoretical Computer Science, 131*, 331–360.

Siegelmann, H. T. and Sontag, E.D. (1995). On the computational power of neural nets. *Journal of Computer and System Sciences, 50*, 132–150.

Stollhoff, N., Menzel, R., and Eisenhardt, D. (2005). Spontaneous recovery from extinction depends on the reconsolidation of the acquisition memory in an appetitive learning paradigm in the honeybee (Apismellifera). *Journal of Neuroscience, 25*, 4485-4492.

Taubenfeld, S. M., Muravieva, E.V., Garcia-Osta, A., and Alberini, C.M. (2010). Disrupting the memory of places induced by drugs of abuse weakens motivational withdrawal in a context-dependent manner. *Proceedings of the National Academy of Sciences, 107*, 12345-12350.

Tononi, G. (2004). An information integration theory of consciousness. *BioMed Central Neuroscience, 5*, 42.

Turing, A.M. (1936). On computable numbers with an application to the *Entscheidungsproblem. Proceedings of the London Mathematical Society, 2 (42)*, pp. 173-198.

In: Consciousness: Its Nature and Functions
Editors: Shulamith Kreitler and Oded Maimon
ISBN 978-1-62081-096-5

Chapter 22

A Novel Theory of Consciousness Based on the Irreducible Field Principle: The Concept of "Geometrical Feeling" Leading to Consciousness Definition

Michael Lipkind

Unit of Molecular Virology, Kimron Veterinary Institute, Beit Dagan, Israel
International Institute of Virology, Neuss-Hombroich, Germany

Abstract

Irreducibility of consciousness to physical fundamentals, that is displayed by the conscious free will and mental causation, looks as violation of the physical laws (psycho-physical gap) that led to proclaiming the "Hard Problem" of consciousness (Chalmers, 1994). Consequently, the consciousness' analysis beyond the strict physical basis became a challenge for induction of fresh ideas. The field concept characterized by the physical action-at-a-distance as opposed to the chain-like diffusion-based chemical reactivity is a novel approach for consciousness theorizing. However, the existing field-grounded theories of consciousness based either on electromagnetic field (EM), or on hypothetical abstract fields irreducible to the physical fundamentals do not save the situation: the EM field-based theories cannot solve the Hard Problem, while the suggested concepts of "irreducible" field are vaguely characterized, loosing a strict formal meaning of the field notion.

The suggested theory of consciousness is based on the concept of "extra ingredient" as a new basic fundamental, additional to the existing physical fundamentals (mass, space-time, and charge), that is expressed by the postulated field characterized as irreducible, vectorial, repulsive, anisotropic, and species-specific. The concept of "geometrical feeling" proposed as a novel "protophenomenal fundamental" is defined as a feeling of non-congruence between the dynamic anisotropic configuration of the postulated abstract field and its "physical realization" expressed by actual distribution of the respective molecular substrate of the living system. The theory provides definitions of protoconsciousness, primordial consciousness, and consciousness *per se*. The conception of the vectorial field by A.G. Gurwitsch (1944) was a principal theoretical basis for formulating axiomatics of the novel fundamental.

PREFACE

On the first view, a *"geometrical"* representation of Consciousness[1] may look like an extravagant, if not an artificial, idea, which perhaps might be employed as an illustration or ornamentation of the Consciousness problem but not as its real essence. Although both the geometry and Consciousness are abstract notions, the former is characterized by the absolutely determined mathematical strictness, while the latter, in sharp contrast, is a dim, tender, uncertain, non-delineated, diversified, multiform, half-spontaneous, unpredictable, hardly perceptible, and *undefined* notion, which cannot be confronted to the geometry even for the sake of the illustration or ornamentation. Therefore, the suggested geometrical representation of Consciousness may cause a reasonable perplexity. Certain alleviation is that the proposed conception of the "geometrical feeling" is associated with the concept of *field*, which has a strict physical meaning and which has already been employed in some theories of Consciousness (Libet, 1996a,b,c; Searle, 2000a,b,c; Pockett, 2000, 2002; John, 2001; McFadden, 2002a,b; Romijn, 2002; reviewed by Lipkind, 2005).

The present article advances a novel approach to the Consciousness problem based on the irreducible field principle. The central point of the proposed theory is the concept of "Geometrical Feeling", which is considered as an intrinsic essence of any living entity, the human Consciousness being evinced as its paramount expression.

I. INTRODUCTION

1. Enigmatic Nature of Consciousness

> "How the conscious experience comes from the irritational processes in the brain is of the same mechanism as that of the appearance of Jiin when Aladdin has rubbed the lamp". Thomas H. Huxley (1866)

> "It is in no way intelligible how consciousness might emerge from the coexistence of carbon, hydrogen, nitrogen, oxygen etc. atoms in my brain. It is entirely and forever incomprehensible. *Ignoramus et ignorabimus*". Emil Du Bois-Reymond (1872)

> "The passage from the physics of the brain to the corresponding facts of consciousness is unthinkable. Granted that the thought and a definite molecular action occur simultaneously, we do not possess an intellectual organ, or apparently any rudimentary organ, which would enable us to pass, by a process of reasoning, from one to the other". John Tyndall (1879)

The above epigraphs demonstrate a discouraging agnostic perspective for scientific solution of the Consciousness problem. Although those pessimistic judgments were uttered in the 19th century, while since then the tremendous development of all the branches of the natural science was achieved, the situation in the Consciousness problem did not changed much: the highest expression of the Consciousness' "state of art" is the notoriously famous

[1] Some main notions related to the topic and acquired terminological nuances are used in capitals, e.g. Consciousness, Life, Hard Problem, Binding Problem, Extra Ingredient, Psycho-Physical Gap, Protophenomenal Fundamental, Geometrical Feeling, Whole, etc.

"Hard Problem" of Consciousness (Chalmers, 1994, 1995) emphasizing the same helpless situation, which may be illustrated by the following utterances: “There can be no solution to the Hard Problem within our current conceptual framework” (Robinson, 1996); "[T]he emergence of experience goes beyond what can be derived from the physical theory" (Chalmers, 1995); “[T]here is in the world a category of relations forever unknown to us” (Bieri, 1995); "It seems we do not at present have any intuitively compelling explanations of how to bridge the psychophysical gap" (Van Gulick, 1995); “[T]he explananda still remain undefined: it is not at all clear what it is to be explained” (Metzinger, 1995); "[H]ow the water of the brain is turned into the wine of consciousness?” (McGinn, 1991).

All the above utterances manifesting a gloomy situation in the problem of the Consciousness nature may be concluded by another one made in the form of the illative question: "Is it really the case that feelings of joy, sorrow, or boredom are nothing but chemical reactions in the brain – reactions between molecules and atoms that, even more microscopically, are reactions between some of the fundamental particles, which are really just vibrating strings?" (Greene, 1999).

2. Appeal for Novel Approaches Based on Counter-Intuitive Ideas

The recently proclaimed massive offensive on the Consciousness problem under the aegis "Toward a Science of Consciousness" (Tucson, 1994, 1996, 1998, 2000, 2002, 2004, 2006, 2008, 2010) involving representatives of all the branches of the natural science and philosophy has resulted in no decisive progress: attempts to explain the ‘puzzle’ of Consciousness have come to the same dead end – the *psycho-somatic gap*. (McGinn, 1991; Chalmers, 1995; Metzinger, 1995; Biery, 1995; Van Gulick, 1995). A responsive reaction to such a dreary agnosticism is a desperate appeal for creating a totally new paradigm based on counter-intuitive ideas (Shear, 1995). Meanwhile, however, the expectation for an essential solution of the Consciousness problem resembles a scene from the medieval theatre when an entangled inextricable plot's situation is solved by a *"Deus ex machina"* descending from the stage ceiling and turning the performance's pell-mell situation into a happy end. In the frame of such an allegory, the research goal is a *scientific embodiment* of such *"Deus ex machina"* using the *counter-intuitive* approach that would lead to a substantial revision of the established scientific axiomatics with a visionary hope to formulate an adequate definition of the Consciousness (Sutherland, 1989; Lipkind, 1998, 2003).

I have to confess that the suggested here a novel approach to the Consciousness problem is associated with a revision concerning essential *biological* grounds of Consciousness. An undertaken criterion of the validity and fruitfulness of the suggested theory is based on a possibility to formulate that slipping away a non-tautological *definition* of Consciousness.

II. THE MAIN PHILOSOPHICAL AND BIOLOGICAL ASPECTS OF CONSCIOUSNESS

1. The Psycho-Physical Gap and Gödel's Theorem

Any interpretation of the Consciousness problem inevitably sets against the eternal conundrum designated as the Psycho-Physical Gap. The various designations of the Gap reveal certain universality together with some differences, that needing an appropriate classification.

The Explanatory Gap (Levine, 1983) has a general epistemological significance expressing limitation of human cognitive capacities in view of the existing state of knowledge and scientific axiomatics, i.e. expressing the agnostic worldview toward such a particular topic.

The Psycho-Physical Gap emphasizes the in principle difference between the material (physical) world and the immaterial (by definition) phenomena demonstrated by human experiential and spiritual manifestations.

The Somato-Mental Gap concentrates on the difference between the brain's mental activities and its somatic functioning.

The latter (chronologically the first) version of the Gap was introduced by A. Gurwitsch (1954/1991[2]) who considered and analyzed the process of perceptions' formation as a chain of events occurring along the whole pathway starting from an initial receptor irritation by a physical stimulus and ending by the respective conscious experience at the final destination associated with the brain cortex. The specific analysis was applied to the pattern of formation of visual perception that starts at the retina and via thalamus reaches the respective cortex area. Such analysis harbors the potential possibility of touching that sacred bridge at which the mystical physical-to-mental switch occurs. This time/locus moment – *sanctum sanctorum* of the Consciousness problem – was designated by Gurwitsch as "the *break of continuity*" or "the *abrupt of entirety*" – an immanently imperfect translation of Gurwitsch's original Russian expression (1954/91). Thus, Gurwitsch's Gap relates to a discrete somatic-psychic act that starts from the excitation of a receptor by an external (physical) stimulus, reaches the "physiological end", "jumps" through the Gap ("the abrupt of entirety"), and continues its further development on another shore (i.e. on the psychic level), this including its binding to the other such discrete psychic acts, simultaneously occurring or previously occurred and memorized, as well as to the whole memory store and the general personal background however identified ("I", "self", "psyche", etc.).

In classical physiology, the above pathway is explained using the concept of neuronal firing, which is one of the main concepts of the neuron doctrine (Ramon y Cajal, 1899-1904) – the neuroscience's stronghold. In particular, the theorizing is based on combinatorics, applied to the incredibly complicated networks of the enormous (practically endless) number of possible neuronal interconnections that represent the brain as "the most complicated of all material objects" (Edelman, 1993, p.7). In fact, the physiological expression of the neuronal

[2] The cited work was written by A. Gurwitsch in 1954, but could be published only in 1991 (after Gorbachyov's "perestroika"); however, the manuscript was circulating among the scientific circles in Russia, being cited and commented, so that the date 1954/1991 seems to be reasonable for indicating Gurwitsch's priority.

firing is reduced to the process of *conduction* of an impulse along a nerve fiber bringing the receptors'-caused physical stimulation up to the psychic sphere. In the framework of the triad – *external (physical) stimulus - physiological conduction - mental experience* – there is no logical "jump" between the 1st and the 2nd components: the external stimulus is fully characterized by the physical language, while the physiological conduction is described by means of the whole glossary of physical-chemical, biochemical, biophysical, and physiological notions which, in principle[3], are reducible to the same physical fundamentals. Thus, the Gap inevitably occurs between the 2nd and the 3d components of the triad. After the "break of continuity" (using Gurwitsch's designation), the description can be done only in psychological language with no Rosetta stone for its translation into the biological glossary.

However, the 3rd component of the above triad may be expanded by including the *psychic (re)action* as an additional act to the *mental experience*, which may occur when the experience expressed, for example, by self-analysis, imagination, thinking, intention, etc., may lead to a *volitional* act realized through the body's action described again by the physical language. Then, such a physical action stimulated in response to the respective experiential process means another jump over the gap – this time in the opposite, *mental-to-physical* direction. This belongs to another aeonian enigma which may be called, after A.J. Rudd, "phenomenal judgment and mental causation" (Rudd, 2000) and can be expressed by the question: How can immaterial intention originated from (within) the physical stuff of the brain make further influence upon the physical stuff of an efferent nerve conducting an immaterial intention to the somatic organ(s) that realizes it?

Anyhow, the main approach to the Somatic-Mental Gap problem in the light of the classic neuron doctrine of the brain functioning can be described by two different ways (Gurwitsch, 1954/1991), one – more radical (formal, definite) and the other – more moderate (rather intuitive), namely:

A. *Radical conclusion.* All the events within the somatic part of the chain, starting from the initial one just after the excitation of a receptor and including the last one which is that just *before* the "break of continuity" do *not* differ in principle from each other. In accordance with this conclusion, the task is to establish an unequivocal relationship between the content (expression) of the *last* event (process) in the somatic part of the chain and the essence (expression) of the corresponding experience at the psychic end of the chain. Then, the somatic part of the chain can be reduced to the *conduction* of the stimulus from the receptor to the "end" of the somatic part (i.e., "before" the "break of continuity"). In such a case, the "break of continuity" is so drastic that there is no much hope for establishing any unequivocal connections between the two spheres.
B. *Moderate conclusion.* The processes in the somatic part of the chain become more and more changed, so that the last one before the "break of continuity" must be *fundamentally* different from the earlier events and maximally close to the first event at the psychic part. In this case, the somatic part of the chain is *not* a mere *conduction* but includes a certain *qualitative* processing of the moving stimulus with the result

[3] "In principle" in this context is valid only within the framework of the DNA-triplet-code-based molecular genetics, which is the dominating trend in contemporary biology. This is an important comment for the further analysis, which is based on a "non-dominating" trend in biology.

> that the latter gains an *extraordinary quality* of being able to "jump over" (to cross through) the Brain-Mind Gap[4]. However, neither in the arsenal of physicochemical notions, nor in the neural-physiological data there are means for an adequate description of the "pre-Gap" development of such *extraordinary quality*.

In the present article, Gurwitsch's approach is taken as a starting point for the *further* analysis, which includes an attempt to determine the "pre-Gap" and "post-Gap" provinces within the whole receptor-to-cortex "silk way" of the perception formation. The peculiarity of the visual perception is that all the five neurons of the visual analyzer – not only the proximal neurons (thalamic and cortical) but also the distal ones – the 1st, 2nd, and 3rd, which anatomically are located within the eye's retina – in fact, belong to the brain. Hence, the first neuron, which during the ontogenetic development differentiates (changes) into a retinal rod or cone, is a part of the brain, in contrast to the other analyzers – acoustic, tactile, gustatory, olfactory, and equilibrium – whose receptors are connected with the brain via peripheral nerves. Such peculiarity sharpens the unimaginable task of the pre-Gap/post-Gap territory division by the mysterious *physical stimulus – mental perception* transformation, which must occur *within* the brain, as opposed to the other analyzers where such division could be suggested to occur between the peripheral neural ("pre-brain") part and the adjacent ("within-brain") part of the perception formation. Hence, there are two entangled questions concerning the Gap: *where* and *when* within the brain, a magic conversion of the neural-physiological excitation into a conscious percept (crossing the Gap) occurs. The response is to indicate this space-time moment, i.e., *anatomical "localization"* and *temporal occurrence* of the physical-*versus*-conscious transformation within/during the receptor-to-perception pathway. The possibilities of such occurrence may be imagined to include a wide range of the space-time extent: from an instantaneous event in a highly specified spatial point (according to the above-described *radical* conclusion) to a gradual metamorphosis amplifying throughout the whole time-space extent of the physical-mental transformation (according to the above-described "*moderate* conclusion").

If the Consciousness problem is considered from the point of view of mathematical formalism, its "hardness" falls into limitations to the decidability or verifiability pointed out by Gödel's theorem (Gödel, 1931; Nagel and Newman, 1958). According to Gödel's theorem, no formal system (within itself) can fulfill the following three conditions altogether, namely, it cannot be (a) finitely describable, (b) consistent (free of contradictions), and (c) complete (be proven to be true or false). The Chalmers' "Hard Problem" falls into the trap of the "undecidable Gödel sentences", thus, bringing the problem to the "Gödel's Gate" (Antoniou, 1992), from which there are several paths depending on which from the above three conditions is given up. The path based on giving up the second condition (consistency, absence of contradiction) should be beyond consideration as incompatible with the scientific method in general. The path based on giving up the third condition (completeness) means choosing the consistency and finiteness and, thus, dealing only with constructive proofs (Beeson, 1985). The tasks decidable by this path belong, using Chalmers' terminology, to the sort of "easy" problems. The path based on giving up the first condition (finite description) means choosing the consistency and completeness and, hence, losing the finite description, so that the one choosing this path gets no constructive proofs and, thus, obtains the notorious

4 This is similar to the views by F. Crick about the "transfer from implicit to explicit" (Crick 1994b, p.11).

Hard Problem, which, according to the Gödel's theorem, is non-solvable within the frame of the considered formal system. The golden middle path between the finite but incomplete and infinite but complete is to try to go beyond the theory (beyond the given formal system) by constructing a larger one. "Any apparent disharmony can be removed only by appropriately widening of the conceptual framework" (Bohr, 1958).

2. Analysis of the Notion of Physicalism: Hempel's Dilemma

The hope for "future discoveries" that would settle any inconsistencies of today's physics becomes rather doubtful in the light of a recently raised problem of the meaning of the notion of "physicalism" itself (Montero, 2003). This is symbolized by what has come to be known as Hempel's dilemma (Hempel, 1980), which concentrates on strictly formal investigation of what *"physical"* means[5]. Namely, if the term "physical" is to be defined via contemporary physics, then it would appear that physicalism is straightforwardly false, since today's physics is *a priori* incomplete and does not express the whole truth. However, if, instead, it is attempted to define physicalism by reference to what physics may become in some future, any conclusion is incompetent, because it is not known what form physics might take in the future. On the other hand, in principle, any new scientific regularities yet to be discovered must be based on such new laws, which, although being natural, are in a way "supra-physical" if the notion "physical" relates to the presently established physical laws (Montero, 2003). Such a "supra-physical" meaning would refer to any suggestion regarding a new fundamental, which would be additional to the established physical fundamentals (mass, charge, and space/time).

3. Holistic Principle *versus* Deterministic Canon

The theoretical basis of the modern biology is based on the reductionist philosophy, which is expressively "anti-holistic". Such inclination seems to be originated from the innate human impulse towards the Whole that can be expressed by the childish "break-and-see-what's-inside" as compared to the scientific "uncover-and-illuminate-the-black-box". Respectively, any yet unknown Whole, especially that having dynamic expressions, inevitably induces irresistible urge to open, to break, to destroy, to dismember, to decompose it, or to get inside, to penetrate into it – all that for discovering how it is built, composed, i.e. what its structure is, or understanding how it "works" or "acts", i.e. how it functions. This is an intrinsic human instinct, which becomes evident since the early childhood when a curious infant breaks toys in order to see what is inside. This instinct is not only a powerful factor of exploration of the environmental reality, but is a deep internal inclination of *Homo sapiens* as

[5] Lord Kelvin had said: "I never satisfy myself until I can make a mechanical model of a thing. If I can make a mechanical model I can understand it. As long as I cannot make a mechanical model all the way through, I cannot understand" (cited from Barber 1961, p. 598). Analogously, Noam Chomsky noticed (in private conversation with John Searle [1994, p.25]) that as soon as we come to understand anything, we call it "physical". Consequently, anything in the world is either physical or unintelligible and, hence, the "understanding" of any natural phenomenon, including both Life and Consciousness, means ultimate reduction of the observed phenomenon to the physical fundamentals.

a species. In science such a "break-and-see-what's-inside" instinct has been designated as reductionist analysis, which is considered as the most direct way towards *"De rerum natura"* according to the ancient classics by Titus Lucretius Carus.

The role of the above "analytical instinct" is to gain understanding via decomposition of the Whole into its parts, portions, components, constituents, fractions, ingredients, elements, etc. In physics, this is the most powerful if not the only way of scientific exploration, the epistemic principle being remarkably straightforward: As comprehensive as possible knowledge about the Whole's fragments (*micro*-level) determines the respectively comprehensive description of the Whole's properties to be manifested on a higher *macro*-level, but not *vice versa:* If at the higher level anything still remains unclear, this means that the knowledge obtained on the lower level was not enough comprehensive. Such reductionist principle is a corner stone of the modern science, physics in particular, since the time of Galileo and Newton.

In accordance with the reductionist analysis, the dominating trend in understanding the Whole is directed *not* to the relationships *between* the *Whole* and *its parts* (the holistic principle) but *between* the *parts,* themselves, *within* the *Whole*, the latter being only a kind of a receptacle, or just a zone of the space occupied by these parts. Consequently, the Whole (its properties) is considered as an immediate result of the assembly of the parts, i.e. the properties of the Whole are totally reduced to the properties of its parts (the Lego toy code).

4. Inevitable Choice: Radical Emergence *versus* Panexperientialism

The reductionist approach to the Consciousness problem consists in finding such elementary factor (component), which would be basic for the whole conscious phenomenology. It is natural to hope that a suitable way of formulating an elementary Protophenomenal Fundamental is an attempt to search for the most rudimentary forms of Consciousness, which may be found by investigating human ontogenetic development. Yet, considering ontogenetic appearance of Consciousness, one cannot escape the choice between two alternative explanations (*tertium non datur!*), namely, radical emergence (Kim, 1993), or generation problem (Seager, 1995), on one side, as opposed to panpsychism (Clifford, 1874), or panexperientialism (Griffin, 1997), or pan-proto-psychism (Van Gulick, 2001), on the other side.

Putting apart all the historical premises, numerous varieties of reduction and emergence (Van Gulick, 2000, 2001), and recent discussions on different variants of the radical emergence (Silberstein, 2001; Hagan and Hirafuji, 2001; Hunt, 2001), the problem could be shortly outlined as follows.

The radical emergence means that particular material configurations and processes in the brain produce conscious experience, i.e. the mentality ultimately emerges out of the *wholly insentient* matter, while the panexperientialism means that *every* element of physical reality is associated with mental (protomental) capacity (e.g. "*units-events*", Griffin, 1997), i.e. the mentality is 'mental' all the way down. The former inevitably leads to the Hard Problem and Explanatory Gap, while the latter bears the 'combination problem' (James, 1890; Seager, 1995), which means merging of units of 'atomic Consciousness' into higher forms of Consciousness up to the Consciousness *per se.* A fantasy based on reality can ask an enormous lot of questions[6] that make the whole doctrine of panpsychism "logically unintelligible" (W. James, 1890), so that instead of "Unsnarling the World Knot" (Griffin, 1997), the panpsychism brings back the generation problem, leading to the same Explanatory Gap with the same lot of questions, i.e. again "the inconceivable remains inconceivable" (De Quincey, 2000, p.81).

5. Supervenience and Isomorphism

The idea of supervenience initially mentioned by Moore (1922) and later employed by Hare (1952) in philosophical debates was applied to the mind-body problem by Davidson (1970). The modern concept of supervenience (Kim, 1978, 1984, 1987, 1990, 1993; Horgan, 1978, 1982, 1984, 1993; Hare, 1984) is outlined by Chalmers (1996) by the following way: "The notion of supervenience formalizes the intuitive idea that one set of facts can fully determine another set of facts" (Chalmers, 1996, p.32). However, further consideration reveals a certain contradiction. On the one hand, "[i]t is widely believed that the most fundamental facts about our universe are physical facts, and all other facts are dependent on these", whereas, on the other hand, "the kind of dependence relation that holds in one domain, such as biology, may not hold in another, such as that of conscious experience" (Chalmers, 1996, p.32). Accordingly, "[t]he physical facts about the world seem to determine the biological facts, for instance, in that once all the physical facts about the world are fixed, there is no room for the biological facts to vary. This provides a rough characterization of the sense in which biological properties supervene on physical properties" (Chalmers, 1996, p.32-33).

This means that in the framework of the supervenience concept, the morphological (organismic) high-level properties (B-properties) can be considered as supervenient on the cellular low-level properties (A-properties). Together with this, by Chalmers' consideration, Life is supervenient on the physical world, whereas Consciousness may not be supervenient on Life that is the ground of Chalmers' "innocent dualism" (Chalmers, 1995, p.210)[7]. Meanwhile, it is important for the further analysis to dwell on a problematic connection between the concepts of supervenience and isomorphism.

Formally, isomorphism means likeness or similarity and, according to a certain intuitive feeling, it may be associated, somehow, with either reduction, or even definition. According to the statement by Hofstadter, the "*isomorphism* [is] another name for coding" (Hofstadter, 1985, p.445), which sounds intriguing while needing clarification. For example, as far as the Life is concerned, the genetic triplet code could be considered as the "isomorphic prototype" of the Life but that is not true since the Life as a domain is much richer and not confined to

[6] Is there an elementary mental unit? Are the elementary "physical-mental" units homogeneous or heterogeneous? For example, is "atomic consciousness" of hydrogen atom the same in any compound in which it could occur to be, let it be water, nitric acid, carbohydrate, protein, DNA, or a mineral crystal deep within a mountain or in an ammonite fossil, or a nucleoprotein in a neuron of a dead brain, or alive brain belonging to an imbecile or an Alzheimer patient, or a "normal" person who is dreaming, or meditating, or experiencing Bach's Chaconne, or being under anesthesia, or LSD, or typhoid fever, or being in a number (close to infinity) of other possible states. The same could be asked in the case of any element, any molecule, any polymer, etc., etc. And does this "atomic consciousness" depend on the "activity" of the respective atom whose fate might be deadly "passive" for millennia within a geological layer contrary to that involved into metabolic machinery? Etc., etc.

[7] This assertion, inconsistent with the spirit of the present article, will be discussed in detail later.

the genetic code. As to the Consciousness, the question is concentrated on what its "isomorphic prototype" is, i.e., what is that which "codes" the Consciousness.

This question resulting from the combination of the above two notions – supervenience and isomorphism – becomes the culminating point: the above *isomorphic prototype* of the Consciousness seems to become a knot fixing the source of the whole enigma. "The problem of isomorphism is that if consciousness literally resides in the brain, then there must be *something* in the brain that literally resembles or is similar to consciousness – that is to say consciousness itself. But we have no idea how that could be the case" (Revonsuo, 2000, p.67).

In this respect, the experiments by Driesch (1891) are of crucial significance: they have clearly demonstrated the absence of unequivocal connection between the cellular and morphological levels in the course of the ontogenetic development of morphological traits, which is realized neither by the Lego toy mosaic principle, nor by the stimulus-response reactivity chain. This disproves any claim that the morphological level is supervenient on the movement trajectories of individual cells which means that the morphogenesis is determined not by the internal properties of the cells but by the regularities associated with the developing "whole" related to the morphological level, itself, that is, essentially, the expression of the vitalistic principle (Driesch, 1891, 1908, 1915).

However, apart from the philosophical and biological grounds, I want to indicate the robust evidence demonstrating the *absence of isomorphic identity* between the living manifestations expressed on the different levels of biological organization: organismic (morphological), cellular, and molecular. Namely, any macro-expression of a developing morphological organ, e.g., its evolving species-specific morphological contours, may be realized via a number of different structural patterns observed on the cellular (microscopic) level. Similarly, a macro-expression of any living state observed on both the morphological and the cellular levels demonstrates the same absence of isomorphic identity with the respective processes occurring on the molecular level, e.g., current chemical (enzymatic) reactions. The claim is that the same living manifestation that is observed (and described by strict criteria) on a higher level can be expressed via a (rather high) number of quite *different* possible microstates expressed on the lower level. Such a higher level may be the morphological (organismic) level *versus* the cellular one as well as both the cellular and morphological levels *versus* the molecular one. It is important to emphasize that the absence of isomorphic identity is not due to the "ordinary" statistical fluctuations, which would be assumed within the framework of cellular-*versus*-molecular levels. Indeed, it is possible to argue that since the whole mass of the molecules constituting any cell are under chaotic movements, which are superimposed upon any observable "organized" molecular movements causing a structurally specific cell morphology, the absence of the absolutely strict isomorphic identity may always be considered as being due to such statistical fluctuations. In this respect, the framework of the cellular-*versus*-morphological levels is the most unequivocal relation, since within this framework the acting units are cells, which, in contrast to molecules, are not subjects to chaotic movements. Therefore, the fact of the *absence* of isomorphic identity during formation of geometrically *specific* morphological contours from *equipotential* elements – cells (this fact usually is neglected in most theoretical considerations of molecular biology), is a strong evidence that the Life phenomenology cannot be explained "on the cheap" (Chalmers, 1995, p.208), i.e. by a mere reduction to the physical fundamentals.

Thus, the robust biological reality demonstrates the absence of isomorphic identity between the morphological (organismic) macro-expression and the respective processes realized on the cellular micro-expression. Similarly, there is no isomorphic identity between the cellular and the molecular levels. Such influence of a higher level of a living entity on its lower level in both the morphological-cellular and cellular-molecular contrapositions is an expression of the "*up-down causation*" (Hofstadter, 1982, p.197). On the molecular level, the supervenience is realized through restraining chaotic movements of the excited molecules by influencing on the molecules' trajectories, which become to be under certain preferable orientation. Theoretically, such "vectorization", according to Gurwitsch's expression (Gurwitsch, 1944), is associated with the postulated *vectorial field* principle.

6. The Concept of the "Extra Ingredient"

The irreducibility of Consciousness to the established physical laws is considered in the light of the recent version of the dualistic approach to the Consciousness problem that is associated with the concept of the "*Extra Ingredient*" (Chalmers, 1995) as an *irreducible* basic fundamental, additional to the established physical fundamentals. The effort is to concretize the concept of the Extra Ingredient (Lipkind, 2003, 2006, 2007, 2008, 2009) in such a way that all the Consciousness manifestations could be reduced to this basic fundamental, which is designated, accordingly, as the Protophenomenal Fundamental (Lipkind, 2006).

A tentative candidate for the role of the postulated Extra Ingredient was proposed by Chalmers, himself: That is a general notion of information, which is considered by Chalmers as having two basic aspects – physical and phenomenal (Chalmers, 1995).

According to the suggested theory, the Consciousness *as it is*, being a "heterogeneous hodgepodge" (Flanagan 1992, p.213-14), i.e., highly variable and dynamically changing in its attributes and manifestations, *cannot* in principle be in the *role* of a basic *fundamental,* alongside the established physical fundamentals – mass, charge, and time/space. Then, there is a possibility that a certain *attribute* of the hodgepodge Consciousness may fill the role of such a fundamental.

However, in accordance with the declared task, the Extra Ingredient designated as the Protophenomenal Fundamental has been postulated as a *common basis* for *any* manifestation of the Consciousness hodgepodge (Lipkind, 2006). Since the difficulties of such endeavor look insuperable within the existing paradigm, any success on this way is possible only in case of the development of a novel approach for attacking the Consciousness enigma. Such an approach has been suggested and further elaborated (Lipkind, 1998, 2003, 2006, 2008, 2009), and the present exposition is a necessary step aiming the axiomatic analysis of the newness of the suggested approach. The latter, used as a "working" tool, is represented by the proposed *morphic* principle (Lipkind, 1998, 2003), according to which the species-specific *Form (Morpha)* of a living entity is taken as a basis for an abstract definition of Life as the first step for the further formulation of the Protophenomenal Fundamental. The reason for such a choice stands in accordance with my conviction, that namely morphology can be accounted as an intrinsic feature of any living system since it cannot be derived from whatsoever property of its physical constituents based on canonic properties of the involved molecules[8].

However, if the concept of the *Extra Ingredient* considered as an *irreducible* fundamental, additional to the established physical fundamentals, is specifically associated with *Life* as a realm, the whole idea belongs – *just formally* – to the framework of the vitalistic trend, which with fatal inevitability induces traditional dislike *before* any specific consideration.

7. The Binding Problem of Consciousness

From historical perspective, the Binding Problem was connected first with psychology being dependent on general philosophical views on spatial-temporal contiguity of mental representation of the external world (Hume, 1735/1958, 1777/1975). The modern version of the Binding Problem is expressed on neurological level being based on the well established evidence on the disjunctive way of processing of visible percepts realized within different brain cortex areas. In spite of such spatially segregated processing of particular features of the object, it is perceived as unitary. Therefore, the synthesis of all the disjointed, dispersed, and separately processed elements of the complex signals from the continually changing picture of the external world must be realized by *binding* together neurological states occurring in different brain areas. In particular, the visual data are processed separately within about fifty functionally segregated specialized cortical areas (Hubel and Livingstone, 1987; Livingstone and Hubel, 1987), each one being responsible for a specific feature, like movement, colour, texture, size, curvature, some topological properties like height/width ratio, stereoscopic depth, orientation of lines and edges, and so on (Treisman, 1986; Ramachandran and Anstis, 1986; Hubel and Livingstone, 1987; Livingstone and Hubel, 1987; Ramachandran, 1990; Zeki, 1992, 2000). At the same time, multi-modal association areas in the cortex in which single perceptual features could be unified into a final perceptual image have not been found, so there is no explanation how the disjointed features of any perceived object are linked together. This binding problem looks more keen since there is *no* locus in the cortex, which could be called either *"master map"* (Treisman, 1986), or *multi-modal association areas* (Damasio, 1989), or *central cortical "information exchange"* (Hardcastle, 1994).

The binding problem is particularly complicated in the case of the visual system. The histological fact is that about 300 retinal rods (the 1st neuron of the visual network way) are structurally (histologically) connected via bipolar cells (the 2nd neuron) with one ganglion cell (the 3d neuron). Consequently, these 300 adjacent rods form a microarea within retina which during the vision process may contain heterogeneous picture as projected from a (micro)part of the visible object. Thus, the axon of the ganglion cell must conduct forward an impulse carrying such complex (integrated) visual information. This already is incompatible with the classic neuronal theory according to which the neuron firing means only the *conduction* of a signal that is realized according to the 'all-or-none' principle.

---

[8] Such a view can be disputed: it is possible to claim that the morphological shape can, in principle, be inferred from the properties of the whole totality of all the molecules involved into a living system having that shape. However, such a case must be compatible with *isomorphic identity* between the three levels of biological organization (molecular, cellular, and morphological) that is not the case (Lipkind, 2009) as it will be considered in detail later.

Thus, two opposite processes associated with the same current visual signal go on simultaneously through the whole way from the receptors to the brain. One of them consists of the anatomically determined *confluence* of distinct signals related to different spatial parts of the perceived object into a *single but complex* signal, its complexity being based on the additivity (puzzle-like summation) principle – integration of different parts into a certain whole. The other one consists of the above-described *splitting* of the perceived object's image into dozens of quite different object's features, like shape, movement, color etc. which are processed separately in distinct cortical areas. The possibility of the coexistence and simultaneous realization of these two antidromic processes within the same anatomic unit (system) is totally incomprehensible. On one hand, confluence (merging, junction, maybe synthesis) of a number (hundreds!) of axonal impulses reflecting *parts* (portions, pieces) of the object, and, on the other hand, disjunction (splitting, breaking, maybe analysis) of the object as a whole – *not into parts* (portions, pieces) – but into so drastically different (somewhat category-like) and causally disconnected features, like color, form, texture, movement, spatial relationships, etc. And after (or in parallel to) that, there is binding of such disjoined features into an integral coherent image. Such antidromic way of stimuli processing within the neural networks displays the bias/variance dilemma that is incompatible with the neuronal theory. According to the modern computational language, the problem is formulated as claiming that the objects or their different aspects have to be represented (to co-exist) within the same physical "hardware" (brain), that resulting in the *"superposition catastrophe"* (Von der Malsburg, 1987). Therefore, if to define the main postulate of the neuronal theory in the way that each mental representation ("symbol") of the external objects is represented by the corresponding subsets of coactive neurons within the same brain structure, then if more than one of such "symbols" become active at a moment, they become superimposed by coactivation. In such a case, any information carried by the "overlapping" subsets must be lost, that being the mortal verdict for the classic neuronal theory.

Consequently, the Binding Problem, being not a purely theoretical construction but arisen from the very heart of the neurobiological and psychological reality, looks incomprehensible within the framework of the anatomic structure and physiological regularities associated with perception. An especial question concerns the relation of the Binding Problem *per se* to the somato-mental gap, which is so fathomless in the case of the Hard Problem. In this respect, the main question concerns the "localization" of the events associated with splitting the signals into different components (constituents, features) and their consequent binding, i.e. timing of this signal processing in relation to the somato-mental gap. A reasonable supposition is that, the integration of the initial signals within the retina level precedes their further splitting which, hence, occurs still within somatic level (thalamo-cortical framework), i.e. *before* the Gap, while the binding is realized within the mental level, i.e. *after* jumping over the Gap. The in principle possibility that *both* the splitting and binding take place within the mental sphere (*after* the Gap) is hardly probable since the anatomical areas in the brain cortex correspond to the already split components/features (Hubel and Livingstone, 1987, Livingstone and Hubel, 1987; Ramachandran and Anstis, 1986; Treisman, 1986; Ramachandran, 1990; Zeki, 1992).

Thus, all the above considerations connected with both the Hard problem and the Binding problem are closely associated with some particular actual questions formulated in the sphere of the experimental neuroscience (John, 2001), e.g. the synchronization of neural units activities within a brain area, coherent activity between the brain areas, the role of coherent oscillations in binding, the functional significance of distributed coherence, spatiotemporal patterns of coherence, and the role of coherence in brain encoding of information.

8. The Hard Problem of Consciousness

The Hard Problem of Consciousness was formulated (Chalmers, 1994, 1995, 1996) by exposing a simple question to which there was (and has yet been) no answer in the contemporary state of art: Why does neural activity give rise to subjective experience? This question can be paraphrased in a more general form: How at all can consciousness emerge from a physical system of whatever high degree of complexity? According to the modern anti-vitalistic natural science, any biological system is considered as a physical system of an especial sort. Hence, the above question calls into doubt the very possibility of the emergence of a conscious flash from the entrails of a physical system. By critical consideration of a number of works on Consciousness, which are of theoretical significance (Jackendoff, 1987; Allport, 1988; Baars, 1988; Wilkes, 1988; Edelman, 1989; Crick and Koch, 1990; Crick, 1994a; Dennett, 1991; Flohr, 1992; Clark, 1992; Humphrey, 1992; Hardin, 1992), Chalmers demonstrates their impotency to solve the Hard Problem (Chalmers, 1995, p.207).

However, an "opposite" phenomenon to be considered is expressed in the next question: How can the emerged Consciousness exert any effect upon the physical stuff ? The initial "recipient" of such a sprung-up conscious "impulse" is a molecular substrate in the physical vicinity (though it were a neuron or neuronal associations) in which an arisen conscious impulse is to be further transformed into a neural (efferent) signal. Since the concept of the Extra Ingredient meaning that the Consciousness is irreducible to the established physical fundamentals, other possible candidates for such an Extra Ingredient, namely, non-algorithmic processing, nonlinear and chaotic dynamics (Penrose, 1989, 1994), quantum mechanics (Hameroff, 1994), and even "future discoveries" in neurophysiology may be considered. Chalmers convincingly proves that there is "nothing extra" in those ingredients as far as their explaining power is concerned.

III. The Origin of Consciousness

1. Hippocratian Preformism and Aristotelian Epigenesis

The problem of the Consciousness' origin is closely connected with the problems of the developmental biology, which from the very beginning was characterized by antagonistic confrontation of two basic theoretical concepts – preformism and epigenesis. Preformistic roots can be retraced to Hippocrates, while the epigenetic idea originates from Aristotle.

According to the logical structure of the preformism, the zygote (the fertilized egg cell) contains all the potential pre-requisites for the development from it of a future organism with its entire species-specific and individual patterns, features, and properties. As to the classical (embryological) preformism, the zygote presents a puzzle-like (mosaic) spatial distribution of the entities inside it, each entity presenting (being responsible for) a certain morpho-anatomical part of the future organism. This original, "embryological", definition of the preformistic principle was transformed into "genetic" definition, according to which all the observed specific complexity of the organism can be reduced to independent separate features, which can be projected onto strictly determined separate entities contained in the zygote and defined as genes. Consequently, the classic Mendelian genetics can be considered as a pure form of this kind of preformism.

According to the logical structure of the epigenesis, any momentary stage of the embryo development can be derived from an immediately previous stage but not from a stage remote by several steps earlier. Hence, any stage of the whole chain of the development contains only an actual pre-requisite for the immediate next one, such epigenetic *actual* pre-requisites being opposed to the preformistic *potential* pre-requisites. Thus, the epigenetic principle gives much more freedom to the elements (cells) for their "behavior" during the formation of any structure from these elements as opposed to the predetermined interconnections between the elements according to the preformistic principle. This means that the same macro-form (macrostructure) may be formed by somewhat different ways as opposed to the rigid preformistic rules. Hence, the epigenetic principle in connection with the individual embryo development is in full connection with the holistic principle.

2. A Riddle of Consciousness' Ontogenesis

The basic question related to the Consciousness ontogenesis is at what stage of human individual development a however simple primitive *Protoconsciousness* emerges. The present analysis is based on the assertion that the ontogenetic developments of the somatic and mental spheres go in inalienable association, and the question is how this inalienability is realized.

According to the initial basis, a mature individual's Consciousness associated with the functioning brain and expressed as a continual stream of variegated experiences (only partly depending on the environments), lays upon a certain personal "*psychic background*", which is slowly changing and advancing along with the age and which can be identified with the individual's *"I"*. The psychic ontogenesis relates to this background.

The analytical approach is based on imaginable backward evolution of Consciousness meaning to find out a stage of its emergence. However, the consistent analysis exhibits impossibility to indicate a moment of the first appearance (emergence) of Consciousness *per se* during the brain development including the antenatal period. Moreover, such retrospective analysis cannot be limited to the developing brain: consistently performed retrospective examination inevitably brings to the individual's *zygote* – the fertilized egg cell – as the *initial* point of the individual psychic development.

Evidently, such "back evolution"-based analysis must include the respective successive "simplification" of the psychic phenomenology. It is not *a priori* clear till what level such conceivable involution is realizable, especially during the early – "pre-cerebral" (before the cell differentiation) period of the development, i.e. till what level the corresponding "simplified" ("rudimentary") psychic phenomenology would still have its specific attributes ("psychic gleams"). The success in this way depends on formulation of such a *rudimentary psychic act*, which would be taken for the further analysis of the notion of *Protoconsciousness* to be postulated and defined. Such rudimentary psychic phenomenon has been suggested (Gurwitsch, 1954/91) to be defined as *knowledge*. The choice of this notion was conditioned by two advantages: (a) the knowledge can be considered as the ground of any expression of the psychic phenomenology; (b) the notion of the knowledge permits quantitative estimation like “less – more” throughout the evolution process without any “break of continuity” (Gurwitsch, 1954/91).

Thus, at this point of the analysis, two conclusions are taken together: in the human individual *psychic* development, the initial point is the zygote where the psychic activity is reduced to the knowledge[9]. But *what* is this knowledge about? This is a crucial point determining whether the proposed formulation of the Extra Ingredient is indeed non-tautological. In the Drieschian style, it would sound as follows: “An embryo (in our case, a zygote) *knows its own state*” (Driesch, 1908). The tautology here is clearly evident, being connected with the word “state” which is too general. Therefore, if the “knowledge” is accepted as a rudiment of Consciousness, the main task, in order to escape the tautology, is to indicate such specific quality of the notion of *“state”* to which the *“knowledge”* must be related.

3. A Puzzle of Consciousness' Phylogenesis: Discrepancy between the Hominid Brain's Anatomical Evolution and the Respective Mental Development

The puzzle of Consciousness' phylogenesis faces the mysterious problem of the evolutionary *trend*, which stands in contradiction to the classic Darwinian paradigm, including the neo-Darwinist version, based on the triplet code – the putting-stone of the modern molecular genetics. The evolutionary “trend” as a concept is in divergence with the central principles of genetics – the gene notion and the blind mutations' fortuity, i.e. the "trend" as a notion is opposite to the "fortuity" as a notion.

Among such evolutionary “trends”, the most striking (and the most compatible to the present study) is the so-called “Great Encephalization”, which is a drastic extension of the brain size, which occurred during a very short (in evolutionary scale) period of the hominid phylogenetic evolution. The most remarkable phenomenon associated with the “Great Encephalization” is that such extraordinary brain size's rapid extension was *not* accompanied by the respective adequate enrichment of the brain functional (behavioral) activity.

Consequently, the “Great Encephalization” reveals two paradoxical phenomena (Lipkind, 2008): (1) There is *no* accord between the *monotonous* course_of the “ordinary”, genetically determined, “trend-free” biological evolution and the *accelerating* pace of the evolutionary “trend-determined” hominid brain development up to the *Homo sapiens'* size, and (2) There is *no* accord between such *accelerating* pace of the hominid brain development and the *absence* of any immediate adequate expression of the psychological potential of the developing "big" human brain.

The fact is that anatomical completion of the *Homo sapiens* is accounted to occur about 130,000 (Africa) – 90,000 (Near East) years ago, i.e. before the development of language, of cooking, of agriculture, while the vertiginous burst of the human mental development happened during historical (civilization) period (about 10,000 years ago). This means that the

[9] This is to emphasize the difference between the notion of knowledge in this context and the notion of information suggested by Chalmers (1995). The latter has a certain abstract mathematical cybernetic meaning and can be applied to the "dead" world as well. In the present context, the knowledge means (is inalienable of) its immediate experience, i.e. the knowledge of a momentary “state" means a kind of its “possession” and the possession is already a feeling of belonging to a certain possessor which is a kind of the “Self”.

evolutionary augmentation of the brain size, meaning the increase of the overall number of neurons and the respective increase of the inter-neuronal connections, proceeded with *no* functional load, i.e. throughout a long period (about 100,000 years), the anatomically perfect human brain was out of full use, while ready for the maximal use up to the full potential realization of the mental capacity. Such a situation leads to the assumption about the evolutionary *trend* consisting in a *preliminary* development of a highly specific anatomical basis – brain, long *before* (as if *in advance*!) its psycho-physiological functioning as a sufficient basis for the *subsequent* origination and development of the Consciousness *per se* expressing a full spectrum of the potential human mental activities. The puzzling conclusion that during the hominid evolution the "Great Encephalization" was *not* accompanied by the immediate adequate enrichment of the brain functional activity leads to the awkward situation of anatomy *without* physiology – like appendix, but the brain is not appendix.

The above conclusion is *not* in accordance with the classic Darwinist principle of traits' selection providing *immediate* advantage for the fittest. Such trend – the rapid phylogenetic development of the anatomical brain – is *not* in accordance with the principle of survival of the fittest, since, due to the absence of any functional advantage, the large brain did *not* make its host the fittest[10].

When the anatomically developed but "sleeping" brain was suddenly awaken – about 10,000 years ago, – the *Homo sapiens* went out of the Darwinian evolution, since the humanist principle of equal rights for any individuals – for the fittest as well as for the weakest – is against the principle of the struggle for existence.

Accepting a view, that scientific progress grounded on human mental potential will *never* reach the situation when "everything has been discovered", one comes to the conclusion that the process of the scientific exploration has no end, and, hence, correspondingly, the brain capacity for memorizing may be practically also endless.

IV. The Field Concept Employed for Consciousness Theorizing

The conception of the abstract field principle in physics can be considered as the highest flight of the human intellectual capacity. The Newtonian dynamical laws, based on the *direct* mechanical collisions of moving bodies, were amended (by the same author!) with quite a different kind of a physical force – gravitation, based on an entirely new canon – *action-at-a-distance*, i.e. instantaneous attraction between the masses of interacting bodies as opposed to their direct *physical* collisions as well as the chain-like *diffusion-based* chemical interactions. However, such at-a-distance attraction had not yet included the notion of field as a particular property of the *space between* the interacting bodies. The notions of the 'field' and 'wave' first uttered by M. Faraday for the explanation of electric and magnetic phenomena were further developed and formalized by J.C. Maxwell in the form of the electromagnetic *field*. Further development of the field principle has resulted in profound generalization of the physical reality based on the strictest formalism (equations by Maxwell, Lorentz, Planck,

[10] The modern development of the *non-genetic* inheritance systems, which are based on a new principle that a "gene" has meaning only within the system as a whole, that including epigenetic inheritance, behavioral inheritance, and symbolic communication (cultural) inheritance including the concept of *meme* (R. Dawkins, 1982), do not exclude the logic of the present analysis.

Einstein, Lagrange, Schrödinger, and others). All that concerned the 'classic' period of the field concept development, which was symbolized by Einstein's life-time efforts toward a unitary field theory incorporating electromagnetic and gravitational fields (Bergmann, 1979). Moreover, the field principle penetrated the modern theoretical summits based on hyper-dimensional spaces (Pagels, 1985; Kaku, 1994) including 'reflection space' and 'catastrophe structures' (Sirag, 1993, 1996) that was crowned by the Ultimate Theory of Everything based on the 'superstring revolution' (Callender and Huggett, 2001) aiming "a quantum-mechanically consistent description of all forces and all matter" (Greene, 1999, p.368).

Thus, the field concept is the most universal notion transfixing scientific description of the whole physical world – from elementary particles to the cosmic level. As to the Theory of Everything, it has an attractive taste of the Promised Land – quite a reasonable belief that such "Everything" must include Consciousness as well. Here, however, there is a snag. The development of the Ultimate Theory unconditionally belongs to the Kingdom of Physics-Mathematics where the Mathematics is leading: both the Philosophy and Biology are not only beyond the vanguard – they have no say in the race toward the Ultimate Theory but have to wait passively and simply follow the race progress. Moreover, even in the positive case, the only, rather questionable, contribution, which such "Everything" could provide for Consciousness understanding, is a highly general 'anthropic principle'[11] (Greene, 1999, 2003). Since the Ultimate Theory *a priori* is based on the physical fundamentals, any however complicated formalization based on whatever 'super-version' of the superstring notion (Barrow, 1992; Greene, 1999, 2003; Weingard, 2001) has no potentiality for deduction of such non-physical entities, like experience, awareness, intention, volition, imagination, and many other Consciousness' hodgepodge manifestations. That lies in the fundamental impossibility to ascribe to a system characterized by purely physical parameters the property of conscious experience. Accordingly, any comprehensive analysis of the 'puzzle of consciousness' inevitably leads to the dead end, which is the above-described notorious Psycho-Physical Gap confronting mental-physical, or, as more recently uttered, mental-nonmental (Montero, 2003) realms, which means "finding a place for the mind in a world that is fundamentally physical" (Kim, 1998).

1. Analysis of the Existing Field-Grounded Theories of Consciousness

The modern theories of Consciousness using the field principle (for the detailed description, see Lipkind, 2005) fall into two groups: (1) those based on the established physical fields, and (2) those grounded on the postulated autonomous fields irreducible to the physical fundamentals. The theories of the 1st group are based on the reductionist approach explaining Consciousness from the physicalist point of view with no need for any Extra Ingredient, while in the theories of the 2nd group the field concept has a capacity to express the very meaning of the Extra Ingredient. Hence, the theories based on the physical fields may, in principle, elucidate only *neural correlates* of Consciousness (NCC) [Crick and Koch, 1998; Frith, et al., 1999; Metzinger, 2000], leaving the Hard Problem unsolved, while the

[11] "[O]ne explanation for why the universe has the properties we observe is that, were the properties different, it is likely that life would not form and therefore we would not be here to observe the changes" (Greene, 1999, p.413). This view, however, later was considered as doubtful (Greene, 2003, p.53).

concept of the irreducible field looks as a potential tool for solving the Hard Problem. However, some of the physical field-based theories have approached to the Binding Problem which is entangled with the Hard Problem ("localization"/timing in relation to the Somato-Mental Gap), and, hence, both kinds of fields are to be considered. The former kind of the field includes the electro-magnetic (Pockett, 1999, 2000, 2002; McFadden, 2000, 2002a,b; John, 2001; Romijn, 2002) and quantum field-based theories (Hameroff and Penrose, 1996), while the autonomous (irreducible) field-based theories are those suggested by Libet (1993, 1994, 1996a,b,c, 1999, 2001, 2003), Searle (2000a,b,c), and Sheldrake (2005).

1.1. Electromagnetic Field-Based Theories of Consciousness

The theories based on the EM field and quantum mechanical (QM) approach implicating the field principle started from the re-comprehension of the classic neuronal conception based on a discrete connectionist network of synaptic transactions – "the doctrine of the neuron" established by Ramon y Cajal (1899-1904). It became evident that the classic neuronal theory could not explain regularities of fractionation, selection, and re-assembly of the signals moving from different receptor zones toward the respective brain cortex areas, that is signified by the Binding Problem.

According to the EM theories of Consciousness, the whole incredibly complicated network of the inter-neuronal connections within the brain is considered as a source of the respectively complicated continual electromagnetic field, which is considered as the physical basis of the subjective experience. The electromagnetic manifestations of the brain cortex have been recently analyzed using modern techniques, e.g. electroencephalography and magnetoencephalography as well as newly established highly sophisticated scanning (imaging) methods like positron emission tomography and functional magnetic resonance imaging. By means of those techniques, a certain correlation between the functioning of neurons in the cerebral cortex, on one hand, and the corresponding subjective experiences, on the other hand, has been demonstrated (Näätänen et al. 1994; Raichle, 1998; Schacter et al., 1998; Frith et al., 1999; Mölle, 1999; Hughes and John, 1999).

It was found that evoked potential wave-shapes in different brain regions became closely similar showing remarkable correlation with the stimuli connected with perception (John, 1968b, 1972; John and Schwarz, 1978). The found regularities were explained by the experimentally demonstrated *synchronization* of neuronal activity *within* and coherence *among* spatially separated brain regions that was supposed to be a factor binding distinct attributes into an integral percept (Engel et al., 1997). Such binding was considered as a prerequisite for the subjective awareness (Engel, 2000; John, 2001). In this respect, the facts of importance are those connected with oscillations synchronized across brain regions that have been found in human intracranial recordings enhanced during cognitive tasks and abolished by anesthesia (Varela, 1996; Tallon-Baudry, 2000). The essential finding is that these phase-locked oscillations occur with *zero time delay* between the involved regions (Desmedt and Tomberg, 1994).

Such a synchrony between *distant* brain regions with zero time delay is unexplainable by discrete synaptic transactions, which require appreciable time. This fact is a crucial theoretical obstacle for any comprehensive theory of Consciousness: "[I]ntegration of this dispersed information into global consciousness is not compatible with the capacities of any single cell" (John, 2001, p.185). Together with this, a proposition about the existence of a remote common source from which "parallel influences" may arrive simultaneously to all the areas involved in analysis of the attributes of the corresponding signal is improbable: Such "hypothetical generator must be credited with *a priori* knowledge of what features are to be bound, requiring multimodal sensitivity to those attributes which are fractionated. This paradox rules out a common source as a plausible explanation..." (John, 2001, p.185). Such logical conclusion was supported by the experiments using multiple moving microelectrodes chronically implanted into conditioned cats (John, 1972). It has been shown that the observed covariance was due to local neuronal activity rather than to volume conduction from some common source in a distant region.

The experimental basis of the EM-based theories was first grounded on the 40 Hz oscillation-based synchronization of neuronal firing (Galambos et al., 1981; Gray and Singer, 1989; Gray et al., 1989; Crick and Koch, 1990; Engel et al., 1991a,b; Ribary et al., 1991; Desmedt and Tomberg, 1994), while a certain field theory version included "steady-state neuromagnetic field" as the basis of synchronization (Tononi et al., 1998). It has been proposed that 40 Hz oscillations may underlie short-term memory by binding together the fragmented attributes of a complex stimulus (Lisman, 1998)[12].

1.2. The QM Field-Based Theories of Consciousness

The connection between QM and Consciousness is of an especial origin. The initial work by Ricciardi and Umezawa (1967) was motivated by the question how the excitation of myriads of individual neurons results in coherence and integrity of conscious activity. During the further development of these studies (Stuart et al., 1978, 1979), the quantum model of brain was based on the combined formalism of the many-body physics problem and spontaneous breakdown of symmetry. Such comprehensive prevalence of the physical aspects in the first QM theory of brain functioning has remained in the next generation of the QM theories of consciousness (Stapp, 1993, 2001; Jibu and Yasue, 1995; Jibu et al., 1996; Penrose, 1989, 1994, 1996, 1997; Vitiello, 1995). The theoretical considerations included quantum measurement problem, e.g. collapse of the wave function (Penrose, 1994, 1996, 2000; Bierman, 2003), Bose-Einstein condensates (Marshall, 1989; Zohar, 1996), spontaneous breakdown of symmetries (Pessa and Vitiello, 2003), the extension of the quantum field to dissipative dynamics (Vitiello, 1995; Pessa and Vitiello, 2003; Alfinito and Vitiello, 2000), time entanglement between mind and matter (Primas, 2003; Mahler, 2004), quantum noise and chaos (Pessa and Vitiello, 2003), synchronization of homoclinic chaos (Arecchi, 2003).

Thus, in the most of the QM theories of Consciousness, the burning fundamental problems of quantum physics itself were considered, the Consciousness serving either as an inspiring enigma, or as a source of convenient analogies, e.g. wave/particle duality *versus* mind/brain duality. A certain exclusion from such QM-rather-than-Consciousness-itself preference is the theory of orchestrated objective reduction (Orch OR) by Hameroff and Penrose (Hameroff, 1994, 1997, 1998, 2001; Penrose, 1996, 2001; Hameroff and Penrose, 1996).

---

[12] Review of this extensive literature is available in a recent monograph (Litscher, 1998). A 'new generation' of the EM field theories of Consciousness (Pockett, 1999; 2000, 2002; McFadden, 2000, 2002a,b; John, 2001, 2002; Romijn, 2002) have been analyzed in detail (Lipkind, 2005).

1.3. The Irreducible Field-Based Theories of Consciousness

The recently suggested theories of Consciousness based on the irreducible field notion, namely, the "Mental field" by B. Libet (1993, 1994, 1996a,b,c, 1999, 2003), the "Unified conscious field" by J. Searle (2000a,b,c), and the "Morphic resonance" by R. Sheldrake (1981, 1986) have a claim to hit the main Consciousness riddle – the Somato-Mental Gap, i.e. they could be immediately connected with the Hard Problem. The Sheldrake's theory usually is not considered seriously by conventional science, e.g. it is not much referred to in the current discussions under the slogan "Toward the Science of Consciousness"[13].

However, in the existing theories based on the autonomous fields, the field notion looses its specific fundamental character acquiring either esoteric meaning with little if any relation to scientific knowledge ("Morphic Resonance" by Sheldrake), or tautological definition ("Conscious mental field" by Libet), or merely metaphoric description ("Unified conscious field" by Searle). Such vague allegoric use of the great principle leads to its emasculation and devaluation.

The above conclusion does not mean that the autonomous irreducible field principle, in general, is far away from the Consciousness problem. On the contrary, potential advantage of employing such concept for understanding Consciousness looks as attractive possibility to join to the grand theoretical edifice of physics pervaded by the field principle from microcosm to macrocosm. Such perspective is stunning, but the indispensable condition that a postulated field must be *irreducible* means that the urge to join to the Grand Physics is limited only to the physical glossary to be used for a formal description of a hypothesized field, apart from its physical (or metaphysical) nature. The problem could be analyzed in the frame of the "naturalistic dualism" (Chalmers, 1995, p.210) if the *Extra Ingredient* is described by means of the *abstract field* vocabulary, but *hic haeret aqua*. As far as 'irreducible' means 'non-physical', a danger is that a postulated irreducible field would lose, together with physicality, its 'field face'. Then, the urge to join to the Grand Physics may be looked as a kind of mimicry when the authority of the universal scientific principle would mask theoretical void of a suggested 'field' hypothesis.

Therefore, for the Consciousness theorizing, the obligatory condition for an autonomous irreducible field is its subordination to the same formal field postulates, which are common to all the established physical fields. The task is to fill such abstract skeleton with novel ontological flesh.

2. Conclusions on the Existing Field-Based Theories of Consciousness

Thus, the theories of Consciousness based on the physical fields may provide only the reductionist explanation with *no need* for the Extra Ingredient, so that the explaining power of these theories is limited to possible revealing the physical basis of the NCC leaving the Hard Problem unsolved. However, the essential finding concerns the occurrence of the oscillations

[13] Although I, myself, also belong to those who consider the Sheldrake's theory as highly speculative and confusing, the theory has been published, spread, popularised, and discussed until recently amongst different *scientific* circles and in the meetings involving highly world-wide reputable scientists, like Prof. H.-P. Dürr (Director of the Max-Planck Institute for Physics), Prof. A. Goswami (Physics, Oregon University), Prof. R. Abraham (Mathematics, Santa Cruz University), Prof. G. Schwartz (Psychology, Arizona University) and some others belonging to the 'hard science' (Conference "Rupert Sheldrake in Discussion", see Dürr & Gottwald, 1997). Therefore, just formally I have had to include it into consideration.

between the involved *distant* brain regions with *zero time delay* (Desmedt and Tomberg, 1994; John, 2001) that is of crucial importance. Such oscillations synchrony is unexplainable by the classic neuronal theory (discrete synaptic transactions) and may be considered as a strong argument for the involvement of a *field-like* factor.

As to the QM theories, an urge to connect Consciousness with quantum mechanics has been based on a deep intuitive feeling. On one hand, there is the Consciousness enigma which stands at the edge of human capacity for understanding, not to say, explanation: being under the shadow of the frightening Explanatory Gap, the science, which is strongly believed to be physical in its ground[5], looks as impotent to grab the Consciousness into its rigid formalistic framework. On the other hand, there is QM with its strictest formalism which is continually gains further complexity and universality supported by the powerful echelon of combined mathematical-physical intellectual forces racing toward the Theory of Everything whose grounding conclusions also have reached the edges of human cognizing capacities. So, the urge is to unite the Consciousness, which has still unsteady scientific background, with the quantum mechanics, which is strictly formalized while embracing all the conceivable levels of the physical world. Besides, such match-making between the QM and Consciousness has another source noted by Chalmers (1995): "The attractiveness of quantum theories of consciousness may stem from a Law of Minimization of Mystery: consciousness is mysterious and quantum mechanics is mysterious, so maybe the two mysteries have a common source" (Chalmers, 1995, p.207).

The conclusion can be expressed by the following citation related to the question "Why are quantum theorists interested in consciousness?" that is the title of the paper taken for the citation:

"Orthodox quantum theory requires *'something else'*. It is a plausible hypothesis to say that this something is a primitive ingredient of the world – that is, not reducible to other things in physics – and to identify it with consciousness. It includes or perhaps we should say *is* the quale of experience. The model naturally shows us why this consciousness is causally effective. To identify the mechanisms in the brain where the required quantum "measurements" that magnify quantum selections into macroscopic effects take place is a problem for the future" (Squires, 1998, p. 617, the *1st boldface* is mine).

This "something else" is just that "Extra Ingredient" (Chalmers, 1995), which was proclaimed for the Consciousness explanation because "no more account of the physical process will tell us why experience arises", i.e. "the emergence of experience goes beyond what can be derived from the physical theory" (Chalmers, 1995, p.208).

However, the elucidation of the Binding Problem may be closely connected to the solution of the Hard Problem, especially if the "binding" concerns the processes located at the opposite shores of the strait formed by the Somato-Mental Gap. A paradox is that although a high potential power of the EM field concept for explanation of the Binding problem is clearly evident, this is in contrast to the in principle impossibility of any physical field theory to jump over the Gap. However, the experimental evidence (using the EM-based measurements) indicates on the possibility for the involvement of a *field-like* factor, which, hence, has to be irreducible to the physical fundamentals, that encourages analysis of the autonomous field-based theories of Consciousness.

V. The Suggested Theory of Consciousness Based on a Novel Principle

A. The Main Premises

The main premises for elaboration of the suggested here theory of Consciousness are as follows:

1. Any attempt of scientific analysis of the Consciousness problem inescapably barges into the Psycho-Physical Gap permeating the whole Binding Problem – Hard Problem entanglement paralyzing further exploration. The inevitable conclusion is that the established physical laws are insufficient for the Consciousness explanation that is expressed by the notion of the Extra Ingredient irreducible to the established physical fundamentals: mass, space-time, and charge. Consequently, a novel basic principle as a factor describing and explaining all the conscious manifestations is to be postulated. The field concept has a promising potential to be in the role of such a factor.
2. Accordingly, the task is to suggest a hypothetical *biological* expression of the Psycho-Physical Gap, i.e. formulate its biological "embodiment".
3. As an expression of the Extra Ingredient principle, a *postulated* notion of the *Protophenomenal Fundamental* is suggested as a basic factor canonical for *all* the conscious manifestations. This means to postulate – within the brain's neurological "flesh" – that ridge ("*locus*") at which the physical-to-mental conversion takes place.
4. However, if to consider the whole chain of the processes occurring along the anatomical receptor-to-brain way, a "spatial representation" (execution) of the Somato-Mental Gap is associated with its respective "*temporal* occurrence", which is a knot point of the Hard Problem - Binding Problem entanglement. The expression "spatial-temporal localization" of the Somato-Mental Gap illuminates that anatomical "locus" at which the physical-to-mental conversion is hypothetically realized at a particular temporal moment of a conscious act performance. As to the Binding Problem, it relates to the sequence and interactions of all the processes that occur within the material stuff (anatomical pattern) that constitutes the receptor-to-brain anatomical channel and provides the realization of the physical-to-mental conversion. The "temporal realization" includes both the pre-Gap realm (before the jump over the Gap) and the post-Gap realm (after the jump). The solution of the Hard Problem is believed to be associated with finding this innermost "*temporal locus*" of the physical-to-mental conversion (the Gap occurrence/localization).
5. Thus, the necessary step is the actual formulation of the Extra Ingredient as a *Protophenomenal Fundamental* being in the role of a new basic fundamental additional to the existing physical fundamentals – mass, space-time, and charge. Before dwelling on this question, I want to formulate the axiomatic demands, which are accounted as obligatory for any physical fundamental and, hence, must be also obligatory for the Protophenomenal Fundamental.

1. Axiomatic Demands to the Extra Ingredient in the Role of the Protophenomenal Fundamental

An incredible difficulty connected with the Consciousness as a notion is the impossibility to define it. The numerous Consciousness' expressions are either almost *synonymous*, e.g., awareness, experience, mind, mentality; or *metaphoric*, e.g., self, "I", psyche, psi, soul, etc.; or manifesting various Consciousness' attributes, e.g., feeling, cognition, comprehension, intention, volition, intuition, thought, mood, purposefulness, decision-making, hesitation, meditation, imagination, creativity, morality, etc. etc., the content of the list depending on one's fantasy and mood. Consequently, the obligatory condition is that the suggested hypothetical notion of the Protophenomenal Fundamental must, as a singular denominator, cover all the attributes and manifestations of Consciousness (Lipkind, 2009). In this respect, an especial care should be taken against the main danger for any definition – tautological shadow, which might creep into the formulation. Therefore, the Protophenomenal Fundamental to be postulated must meet the conditions kept for the established physical fundamentals – mass, charge, and space-time.

Accordingly, like the established physical fundamentals, the Protophenomenal Extra Ingredient must be:

1. *Elementary* (further unsplittable);
2. *Axiomatic* (further unquestionable);
3. Strictly and unequivocally *defined*;
4. Qualitatively *uniform* (homogeneous);
5. *Measurable.*

The 1st and the 2nd demands do not need any comment.

The 3rd demand, which is obligatory for any axiomatic fundamental, is the most problematic since the Consciousness has already been characterized as escaping any definition.

The 4th demand, which is obvious for any physical fundamental, is stipulated here because of the above-mentioned heterogeneous ("hodgepodge") expressions of Consciousness (Flanagan, 1992, p.213-214). However, a physical fundamental, e.g., *mass*, is *constant* in its axiomatic properties in every physical entity and in any physical processes, let it be the mass of an electron, a water molecule, the Earth, the Sun, a Galaxy, or any living entity in any state. The properties of the mass as a basic fundamental are constant in the course of transient time whereas the properties of Consciousness manifestations are changing and interchanging in the course of time. The Protophenomenal Fundamental to be defined *must* be an axiomatically *uniform* basis for *any* manifestation of the Consciousness hodgepodge.

The 5th demand can be realized at least by the simplest mode of quantitative estimation, such as more – less, higher – lower, quicker – slower, more intensive – less intensive, more complicated – less complicated, etc., if there are accepted standards of comparison. Such a demand is hardly realizable in the case of any manifestation of Consciousness, including those that can vary in intensity, such as pain, and some other qualia ("less hurt – more hurt" might sound, but "high – low redness of red" ... is doubtful).

Thus, consecutive analysis clearly shows that not only Consciousness *as it is*, which in principle cannot be in the role of a basic fundamental, but also *none* its manifestations meets the above obligatory demands for being in the role of the axiomatic fundamental. Consequently, the Extra Ingredient to be defined must be expressed using substantially novel concepts and principles.

The situation is perfectly described by M. Delbrück in respect to a living organism: "A mature physicist, acquainting himself for the first time with problems of biology, is puzzled by the circumstance that there are no 'absolute phenomena' in biology. Everything is time-bound and space-bound. The animal or plant or microorganism he is working with is but a link in an evolutionary chain of changing forms, *none* of which has *any permanent validity*" (Delbrück, 1949, cited from Mayr (1961, p.1502, *boldface* is mine). This utterance concerning the living essence relates in full measure to the mental properties.

B. Definition of Consciousness

The numerous utterances cited in the introduction to the present paper (under sacramental slogan *"Ignoramus et ignorabimus"*) lead to the conclusion that "consciousness can only be defined in terms of itself" (Angell, 1904)" [Bisiash, 1988, p.102], i.e. its definition has been proclaimed as "impossible and unnecessary" (Lipkind, 2003).

Thus, as it has been already stated, neither Consciousness as it is, nor any its attributes and manifestations are compatible with the above obligatory demands to the axiomatic fundamental. Therefore, a central assertion is that a formulation of the Protophenomenal Fundamental means the reduction of Consciousness *not* into any of its hodgepodge of features or manifestations, but to an *elementary axiomatic* notion that still *preserves phenomenal quality*. Accordingly, such a Protophenomenal Fundamental must be the same (axiomatically uniform) for *any* attribute of Consciousness. Then, the task is to project the *plurality* of Consciousness expressions to a *common singular* denominator, which is the expression of the Protophenomenal Fundamental.

The consecutive thread of the suggested approach is, first, to formulate an *abstract* coherent definition of the Protophenomenal Fundamental, and then to transcribe it into physicochemical language. Since a strict definition of the Protophenomenal Fundamental is a crucial step of the whole endeavour, it is worth starting the whole analysis from a closer consideration of the meaning of the word "definition" as a term.

1. Explication of the Meaning of "Definition" as a Mental Operation

Apart from the common demands for conceptual clarity and absence of contradictions, a valid definition must be under the following conditions:

It must not be tautological or synonymous;

It must not be reduced to a mere listing of attributes, properties, and features of an entity to be defined;

It must be valid syntactically, i.e., contain a subject (the *Defined*) and a predicate (the *Defining*);

It should be constructed in accordance with the "minimization principle", i.e., contain only necessary and sufficient conditions.

Therefore, besides the semantics, the syntax – the hierarchy between and amongst the connected notions and their attributes – determines logical sufficiency of the definition conditions[14]. Evidently, the central point of any definition is its syntactic validity, while its content (expression, meaning) determines, first, whether the definition is non-tautological and, second, what its essence is.

In this respect, I consider two kinds of definitions: *formal (taxonomic)* and *naturalistic*. The former determines, firstly, which sort (amongst the already established ones) of things the defined something relates to, and, secondly, what differences are displayed by this something as compared with the other things of the same sort. Essentially, such a taxonomic definition means the determination of the exact place of the defined entity within the currently existing hierarchic system of the considered world. As to the naturalistic definition, it is connected with the intrinsic "nature" of the defined entity, or its "mechanism" if the defined entity is a process or a dynamic state. Usually, the naturalistic definition is a reduction of a certain yet unknown "whole" to its parts (portions, ingredients, constituents, fractions, elements, etc.) – to something that is associated with unquestionable (axiomatic) fundamentals.

As far as Consciousness is concerned, it is a realm with *no taxonomic subjection* to any other realms. Consequently, in view of the present state of the art, it seems that the formal taxonomic definition of Consciousness is not possible in principle, i.e., Consciousness can only be "defined in terms of itself" (Angell 1904, p.1). Therefore, any scientific exploration of Consciousness should be associated only with its *naturalistic* definition.

2. The Abstract ("Pre-Naturalistic") Definition of Protoconsciousness

2.1. The Definition of Life as a Basis for the Consciousness Definition

The Consciousness definition to be advanced is grounded on a radical postulation that a conscious capacity is an *inherent* and inalienable property of *any* living entity. Consequently, the variability of Consciousness' expressions ranges in accordance with the complexity of a living system – from an elementary Protoconsciousness of a primitive living cell to the Consciousness *per se* of a normal human individual. Therefore, the postulated association of any form of Life (any living entity) with the respective form of Consciousness leads to the necessity for definition of Life prior to definition of the Protophenomenal Fundamental.

2.1.1. The Existing Definitions of Life

The attempted definitions of Life are so numerous that the limits of the present paper do not permit their detailed consideration. Usually, Life is functionally characterized in a very detailed form based on a variety of aspects: chemical-physical (Loeb, 1906, 1912; Lotka, 1925), biochemical-genetic (Monod, 1971), biophysical-energetic (Szent-Gyorgi, 1957), evolutionary (Dawkins, 1976; Cairns-Smith, 1982; Lifson, 1987; Wächterhauser, 1988; Eigen, 1992; Dennett, 1995), synergetic (Haken, 1977), self-organizational (Prigogine & Stengers, 1984; Babloyantz, 1986), informational (Kauffman, 1996), thermodynamic (Caplan and Essig, 1983; Elitsur, 1994; Dolev and Elitsur, 2000), cybernetic (Korzeniewski, 2001). In those attempts, Life is characterized (rather than strictly defined) according to those specific aspects.

[14] According to Hofstadter (1985, p.445), "*semantics is an emergent quality of complex syntax*" (original italics).

In the present context, the definitions of Life specifically dealing with the Consciousness problem are of particular interest. However, the majority of the authors, including "New Mysterians" (Chalmers, 1995, 1996), "Liberal Naturalists" (Rosenberg, 1996), devoted materialists (Dennett, 1991, Hardcastle, 1996), and "quantologists" (Hameroff and Penrose, 1996), do not strictly define Life but describe it in terms of a set of various Life properties. There is a mere enumeration of arbitrarily chosen manifestations of Life, for example: "DNA, adaptation, reproduction, *and so on*" (Chalmers, 1995, p.203); "reproduction, development, growth, metabolism, self-repair and immunological self-defense" (Dennett, 1996, p.4), "reproduction, metabolism, self-repair, *and the like*" (Dennett, 1996, p.6), "self-organization, metabolism (energy utilization), adaptive behavior, reproduction, and evolution" (Hameroff, 1997b, p.13). Such a situation when the differences in the Life definitions are so drastic and, especially, the usage of the expressions like "*and so on*" and "*and the like*" is intolerable in physics being in contradiction to the above-described conditions for a valid definition[15].

2.1.2. The Abstract ("Pre-Naturalistic") Definition of Life Based on a "Morphic" Principle

As opposed to the invalid "set-like" kind of Life definitions, I suggest one, which, besides meeting the above-declared five conditions, has some especial characteristics. First, the definition is *abstract* without reference to any functional properties of Life (reproduction, development, growth, adaptation, self-organization, self-repair etc. etc.), which, however, are covered by the suggested abstract definition. Second, it has a potential capacity for becoming a starting point for the further definition of the Protophenomenal Fundamental.

Searching for any factors which could provide an abstract definition of Life, I decided in favor of what I call the *morphic* principle, according to which the species-specific *Form (Morpha)* of a living entity is taken as an Ariadne's clew leading to an appropriate definition. The reason is my conviction, that the *morphology* of a living entity cannot be derived from any its physical constituents, e.g., canonic properties of the involved molecules.

Accordingly, the definition is as follows:

- *Life* (any living system) is an evolving and aging species-specific non-equilibrial (i.e. needing incessant energy influx) self-preserving and self-reproducing *geometrical form*, which is continuously refilled with certain substances coming from environment.

The immediate advantage of this definition, as compared to the above-mentioned 'set-like' ones, is its syntax. Namely, instead of a dull enumeration of various manifestations of Life, the definition has one Defining predicate (*"form"*) around which there are necessary attributes (including the attributive subordinate clause) hierarchically subordinated to the predicate. The definition is neither tautological, nor synonymic: semantically and etymologically, *Life* as *Defined* and *Form* as *Defining* have nothing to do with each other. Therefore, the formal side of the definition seems to be irreproachable.

---

15 Sometimes, definition has the Defining predicate, like that by Hameroff (1997, p.13): "[L]ife is a *process* generally described in terms of its properties and functions including ...". However, here the Defining predicate ("process"), on the one hand, is too common and, on the other hand, is not comprehensive: for example, the triplet genetic code as one of the fundamental characteristics of Life cannot be considered as a "process".

The second special attribute of the definition is that it is purely abstract: *none* of the real Life structural-functional properties is mentioned. At the same time, *all* the particular manifestations of Life such as reproduction, heredity, development, growth, constancy of the internal milieu, adaptation, metabolism, differentiation, etc. are covered by this abstract definition[16]. Such compatibility between the abstract attributes of the definition and the real, specific properties and manifestations of Life is due to the morphic principle: the "Form" as a notion has *geometrical* (abstract) meaning whereas all the properties of Life have a real *structural-functional* appearance described by the *physical-chemical* language. This compatibility demonstrates that the suggested definition is in agreement with the "minimization principle" that demands the employment of only necessary and sufficient conditions expressed by the attributes in the definition, which potentially can cover all the Life manifestations. For example, the abstract expression "*self-preserving and self-reproducing geometrical form*" is realized at the cytological level (mitosis), genetic level (DNA triplet code), ontogenetic level (gene expression during embryological development and cell differentiation), biochemical level (DNA transcription, RNA translation, protein post-translational modifications), physiological level (coordinated functioning of different living systems owing to neurological or hormonal regulations), and ecological level (adaptation to environmental conditions providing individual survival). Hence, it appears that the properties of Life taken at random in the "set-like" definitions by Chalmers, Dennett, and Hameroff are covered by the "self-preserving and self-reproducing" attributes of the *Defining* predicate *"Form"*. In a similar way, all the other attributes of the abstract definition can be interpreted in the language of the real manifestations of Life.

Apart from the formal accuracy, the crucial point of the above definition – the *morphic* principle – is a novel one. In the above-cited sources (Loeb 1906, 1912; Lotka 1925, Szent-Gyorgi 1957, Monod 1971, Dawkins 1976, Haken 1977, Cairns-Smith 1982, Caplan and Essig 1983, Prigogine and Stengers 1984, Babloyantz 1986, Lifson 1987, Wächterhauser 1988, Eigen 1992, 1993; Elitsur 1994; Dennett 1995, Kauffman 1996), various aspects – chemical-physical, biochemical-genetic, biophysical-energetic, evolutionary, synergetic, informational, self-organizational, and thermodynamic – have been regarded as being of the paramount importance. In contrast to that, the morphic principle has an abstract category-like meaning that covers all the above aspects, which are associated with the reductionist mode of explanation, whereas the *Form* (as a geometrical notion) in the suggested definition is that biological entity which cannot be reduced to the physical fundamentals.

Such an assertion, which contradicts the reductionist approach, could be contested. An easy conventional explanation is that the things we call "living organisms" are built out of molecules arranged into larger clusters, so that the geometry of the organism is fully determined by the structural properties of the involved molecules, i.e., in the end, by their physical-chemical properties fully reduced to the physical fundamentals. Such a reductionist mode of analysis, if applied to Life, stands in contrast to the robust fact – the *absence of isomorphic identity* between the geometrical morphology of a living entity and its physical constitution. Generally, the problem of isomorphism (Hofstadter 1985), expressed even as a "paradox of isomorphism" (Revonsuo, 2002), is usually applied to the mind-brain problem.

[16] It is worth taking into account that in the definition some of the attributes to the Defining "Form" concern the notion of Life as related to a living individual (e.g., "evolving and aging"), while the others are related to the general notion of Life as a cosmic phenomenon (e.g., "non-equilibrial").

However, I extend the problem of isomorphism to the phenomenology of Life by emphasizing the fact that the *same* macro-morphology of a living system is associated with quite different histological modes of its realization: different cells continuously and dynamically fill and refill the much more precise morphological "receptacle". This is in full contradiction to the formal definition of isomorphism: "[w]hen two formal systems are *isomorphic*, they have strictly analogous internal structures, so that there is a rigorous one-to-one mapping between the roles in the one and the roles in the other" (Hofstadter, 1985, p.60).

The fact is that any living system demonstrates the *absence* of such one-to-one mapping between the abstract geometrical morphology and real, physical, content that fills this abstract draft. This is exactly conveyed by the utterance made by Achilles arguing with Tortoise – the marvelous personages illustrating D. Hofstadter's views (1985, p.626): "Bizarre! The medium is different, but the abstract phenomenon it supports is the same". In such a situation, the general question is how the evolving species-specific geometrical Form of a living system is made up if *a priori* it is not determined by (not supervenient on) the material stuff constituting (filling) it.

2.2. The Abstract "Pre-Naturalistic" Definition of Protoconsciousness Based on the Morphic Principle

The suggested morphic definition of Life has become a basis for the definition of the postulated Protophenomenal Fundamental. The above-described analysis of the imaginable backward evolution of human ontogenesis has led to the zygote as an initial entity that possesses Protoconsciousness.

2.2.1. The First Auxiliary Definition of Protoconsciousness

The suggested definition of Protoconsciousness is based on the same morphic principle as the above definition of Life, and relates to *any* living entity of *any* origin[17]. The expression "any living entity" means any species-specific *cell* as an elementary living system. As an initial step, I suggest the first auxiliary definition of the Protoconsciousness, which is as follows:

- *Protoconsciousness* is the embodied immanent capacity of any living cell to *feel* its own evolving dynamically fluctuating species-specific *geometrical form.*

I designate this "capacity" as "*Geometrical Feeling*" (or "morphic sensation" as a synonym) that is the sought-for *Protophenomenal Fundamental.*

The most comprehensive concept of the Geometrical feeling relates not only to the Form as an external shape of the cell (its morphological contours[18]): the Geometrical feeling in its deepest meaning transpierces through the whole living cell body, i.e., *each geometrical* (stereometrical) *dot* within the three-dimensional cell body is *felt, sensed,* and *being aware of* by the respective *cell.*

---

[17] That is, belonging to any Kingdom of Life: animal, plant, fungal, bacterial.

[18] The essence of the Form as a geometrical entity is open to any interpretation – from the classic Euclidean geometry up to any of the modern proposals based on hyper-dimensional spaces (Pagels 1985, Kaku 1994, Sirag 1993, 1996) and all the versions of the superstring theories (Barrow 1992, Green 1999, 2003; Callender and Huggett 2001, Weingard 2001).

From the formal point of view, the advantage of this definition is that it is not tautological: the capacity to "feel" only Form and nothing else defines those limitations, which express the difference between mathematical equation and mathematical identity (sameness)[19]. Therefore, the aim of this step is to escape what is the most dangerous for any definition – tautology. In accordance with this definition, the Protoconsciousness is an inalienable attribute of Life, i.e., any cell whenever it is alive feels its own geometry.

Thus, the morphic principle reduces Consciousness expressed in the full extensive spectrum of its manifestations to a single principle that is not further reducible.

However, the deficiency of the definition is that it does not describe the ontological difference between physical and conscious: its Defined component – Protoconsciousness – (the left part of the equation) as well as the Defining component – Capacity to Feel Form – (the right part of the equation) are *phenomenal*, whereas the attribute "*embodied*", which is associated with *physical,* is only declared but *does not* "*work*" in the equation, having an abstract geometrical meaning without any connection with the biological "flesh".

2.2.2. The Second Auxiliary Definition of Protoconsciousness

The second auxiliary definition takes into account one of the attributes of the predicate in the definition of Life ("geometrical form"), namely, that the latter "is continuously refilled with specific substances". Consequently, if the abstract geometrical form is "filled" (moreover, "continuously refilled") with material stuff, there is a *non-congruence* (disharmony) between the geometry (morphology) of this abstract "ideal" form and the real distribution of the physical substrate which "fills" (constitutes) this morphic (geometrical) "receptacle". This "physical substrate" includes, essentially, the whole totality of all the intracellular molecular stuff and processes that are under "usual" biochemical and genetic regularities. Consequently, it is not passive stuffing inside the "ideal form" but a chemically highly active medley of different substances including numerous enzymes involved in a dynamic network of metabolic pathways. Besides, this physical substrate is partly included in various subcellular structures (e.g., the cytoskeleton) that, although "solid", are not rigid but dynamically flexible, being under continual regulatonal and restitutional changes. Therefore, the material substrate that fills the geometrically ideal morphic "receptacle" is in the state of dynamic heterogeneity and turbulence, thus "gushing over" the external and internal "geometrical borders" of the ideal form. Just this dynamic non-congruence between the ideal geometrical form and the real continual spatial distributions of the material substances within this form is currently felt ("experienced") by the cell. Therefore, the above-formulated definition of the Protoconsciousness is exchanged for the corrected version, which is as follows:

- *Protoconsciousness* is the embodied immanent capacity of any living cell to *feel* any spatial *non-congruence* between the cell's evolving species-specific "ideal" geometrical *form*, on one hand, and the real distribution of the material stuff "filling" this form, on the other hand.

[19] A tautological version of such a definition would be as follows: "The protoconsciousness is the immanent capacity of any living system to feel its own *state*" (instead of *Form*).

From the formal point of view, the main advantage of this definition is that here the physical component within the definition starts "working": the Defining part contains both the *phenomenal* component ("capacity to *feel*") and the *physical* component ("distribution of the *material* stuff 'filling' the abstract form").

The essence of this definition is that now the "geometrical feeling" is not the feeling of the geometrical form but the feeling of the non-congruence between this ideal geometrical form and its physical (material) realization. Namely, the living cell "feels" the three-dimensional[20] shape of its own body as soon as the metabolically active (dynamically "boiling") material stuff constituting this body "violates" the geometrical abstract borders of this shape. However, since the absolute coincidence of physical and geometrical is ontologically impossible (only an asymptotic approximation to the never reachable coincidence is to be imagined), the discrepancy (non-congruence) between the two will never cease, and neither will the feeling of this non-congruence by a living cell. Consequently, such "non-congruence" is, essentially, a formalized expression of the *Psycho-Somatic Gap* which, allegorically, is analogous to the discrepancy between the *ideal geometry* and its *physical imitation* (*Abstract Mathematics versus Solid Physics*) that is analogous to the confrontation of the geographical meridian-parallel network drawn upon the Earth-sized globe *versus* the physical relief of the actual Earth. Hence, the Protoconsciousness can be imagined as current awareness by the living cell of the fluctuating, never vanishing, Gap between the ideal geometrical form and its physical realization.

In the light of this idea, the living process can be expressed as a continuous dynamic approximation of the real physical form to its geometrical ideal.

However, the above definition of Protoconsciousness still has an internal ontological deficiency, which is as follows.

The final definition of Protoconsciousness must take into account some "active" features from the hodgepodge of manifestations of Consciousness *per se* (to be defined later), namely, those somehow associated with Free Will, e.g., intention, decision, volition, etc. These qualities should somehow be "reflected" in the suggested definition of the Protophenomenal Fundamental.

2.2.3. The Final Definition of Protoconsciousness and Rudimentary Psychic Act

The state of the geometrical-physical non-congruence in living cell is "non-equilibrial" (according to Gurwitsch's expression, 1954/91), constantly fluctuating in accordance with the dynamics of metabolically active intracellular processes, so that the degree of the non-congruence is respectively fluctuating. Here comes another postulated attribute of the Protoconsciousness connected with the capacity of living cell to *preserve* its own species-specific morphology. Such a capacity is expressed by *smoothing* the non-congruence between the "ideal" geometrical form and the physical stuff, which "materializes" it. This (re)action may be described as follows.

Any living system tends to approximate to the minimal non-congruence between the species-specific geometrical form and its physical realization that is under continual influence of numerous fluctuating metabolic pathways. These fluctuations are "brushed up" by the stability of the abstract geometrical contours of the species-specific frame. Such a brushing-up

---

[20] The number of dimensions is an open question (see footnote *28*). It cannot be excluded that the "living state" is characterized by an especial extra-dimensionality.

process that leads to smoothing the non-congruence is a certain background state of a "non-excited" cell whose "at-rest" Protoconsciousness (Geometrical Feeling of an "undisturbed" living cell) can be analogized with Jamesian "stream of consciousness" (James 1890). However, if there is any external disturbance leading to a certain (reversible) disorganization of the cell's physical structure and, thus, drastically increasing the non-congruence degree, this causes *morphological reaction* of the cell leading to restoration of its "perturbed" physical morphology to the "initial" state, i.e. return to the minimal non-congruence. This means that if any physical factor acts upon the cell's material substrate, which "fills" (constitutes) the cell's geometrical form, such factor upsets the current balance by causing "tension" between the "ideal" species-specific geometry (stereometry) of the cell and the real spatial distribution of the material stuff within this "ideal" form. This tension is "felt" by the cell, which "reacts" morphogenically in order "to smooth" the non-congruence. The restored morphology decreases the degree of non-congruence, thus returning the cell to the background state of the dynamic "rest" (minimal non-congruence). Consequently, the combination: *geometrical feeling – morphogenic reaction* can be considered as a *rudimentary psychic act ("morphological mentality")*.

In accordance with this, the final comprehensive definition of Protoconsciousness is as follows:

- *Protoconsciousness* is the embodied immanent capacity of any living cell to *feel* and immediately *minimize* any spatial *non-congruence* between the cell's evolving species-specific "ideal" geometrical *form* and the actual distribution of the material stuff "filling" (constituting) this form.

An additional advantage of this definition is that, if it is applied to a *neuronal* cell, the combination of both the capacities ("...to *feel and immediately minimize* any spatial non-congruence...") symbolizes two main enigmas associated with the Consciousness *per se*: its *origination within* the physical stuff of the brain and its further *influence upon* the physical stuff of respective neurons, thus initiating an efferent impulse.

2.3. Reflections on the Final Abstract Definition of Protoconsciousness

The above-declared advantage of the final definition of Protoconsciousness may bear a certain doubt, namely, such double-potential capacity – to feel *and* minimize the non-congruence – may look incompatible with the 1st demand on any fundamental – to be elementary (further unsplittable). The expression "to feel and to minimize" seems to include two *different* mechanisms, namely, "to feel" the non-congruence that is immediately clear, and "to minimize" the non-congruence that means to "act", i.e., to "move" physically (on the molecular level) the material stuff, so as to reach the maximal approximation of the physical mould to its geometrical prototype. Hence, it would seem that these two aspects of the suggested concept of Protoconsciousness correspond to different fundamentals, i.e., the postulated Geometrical Feeling corresponds to the common capacity "to feel", whereas the capacity to minimize the non-congruence concerns something like a reaction to the feeling associated somehow with an intention to realize a volitional act. However, by intuitive conviction, I can assert that "pure feeling" does not exist: an essential feeling is immediately associated with its experience. This experience would be expressed by different states: from volition leading to incentive to active behavior up to an *internal* state, that is expressed, either by deep intellectual reflection, or ecstatic delight, or imagination, etc. Therefore, if on the conscious level, a feeling and its *immediate* experience are inseparable, i.e., they are part and parcel of the same phenomenal quality, such a principle must be correct for the level of the postulated Protoconsciousness, too. Consequently, both the qualities – to feel and immediately minimize the non-congruence – relate to a *single* Protophenomenal Fundamental, which looks like the two-faced Jánus.

2.3.1. Analogy of the Defined Protoconsciousness with the Le Chatelier Principle of Chemical Equilibrium

Such a situation may be illustrated by the Le Chatelier's principle concerning conditions of the chemical dynamic equilibrium state in solutions and gases:

> "**Le Chatelier's principle states** that if a system at equilibrium is subjected to a disturbance or stress that changes any of the factors determining the state of equilibrium, the system will react in such a way as *to minimize the effect of the disturbance*" (Mahan and Myers 1987, p.171, *boldface* is mine).

I believe this is not only an illustration but also an expression of a deep ontological analogy. The analogy, however, concerns only the "mode of (re)action": certainly, there is no analogy between the dynamic state of the non-congruence between geometry and physics "experienced" by a living cell and the state of the dynamic chemical equilibrium in solutions. The analogy becomes evident at the stage of reaction, which is realized by the *minimization* of the effect of any disturbance caused either by any factors that change the chemical equilibrium (Le Chatelier's principle), or by any factors that upset the balance of the non-congruence (the Protoconsciousness concept). However, the reaction of the chemical equilibrium system expressed "in such a way as to minimize the effect of the disturbance" proceeds as if automatically, according to the natural law. *"The condition of a system at equilibrium represents a compromise between two opposing tendencies: the drive for molecules to assume the state of lowest energy and the urge toward molecular chaos or maximum entropy"* (Mahan and Myers 1987, p.109, *boldface* is mine). Therefore, although the suggested definition of Protoconsciousness is extravagant enough, the above analogy between the "reactive" part of the Protophenomenal Fundamental (minimization of the "non-congruence") and Le Chatelier's principle associated with the general laws of thermodynamics maintains scientific validity of the suggested definition of Protoconsciousness. The anthropomorphic meaning (with a subtle psychological nuance) of the words "drive" and, especially, "urge" in relation to the molecules in ideal solutions used in the above quotation taken from a dry university chemistry text book, emphasizes even more the analogical kinship between the suggested Protoconsciousness formulation and Le Chatelier's principle.

2.3.2. Inquisition of the Concept of Geometrical Feeling as a Hypothetical New Basic Fundamental

Thus, the task to reduce Consciousness to the axiomatic Extra Ingredient, which meets the formal demands for the physical fundamentals but still preserves its phenomenal quality, seems to be fulfilled. The Geometrical Feeling as the expression of the Extra Ingredient designated as the Protophenomenal Fundamental meets all the requirements put to the established physical fundamentals, namely:

1. It is neither tautological, nor metaphorical, nor allegorical: The Geometrical Feeling is synonymous neither with Consciousness as it is, nor with any of its phenomenological attributes and manifestations.
2. It is *elementary*: The Geometrical Feeling, as that related to a unique species-specific “ideal” geometrical form, is irresolvable (indecomposable) into any components.
3. It is *axiomatic*: The Geometrical Feeling like the established physical fundamentals (Mass, Charge, Time/Space) is postulated to lie in the grounds of Protoconsciousness that is the axiomatic basis for the Consciousness *per se*.
4. It is strictly and unequivocally *defined*. Meanwhile, the definition is abstract ("pre-naturalistic").
5. It is *measurable*: The Geometrical Feeling defined as feeling of the non-congruence between the “ideal” geometrical form and the material substrate filling (constituting) this geometrical receptacle permits quantitative estimation without qualitative leaps, i.e., there may be different *degrees* of the non-congruence (e.g., high – low) to be felt.
6. It is *homogeneous*: the Geometrical Feeling is constant in its essence in any hodgepodge of the Consciousness expressions.

Thus, the concept of Geometrical Feeling, as being *elementary, axiomatic, strictly defined, measurable,* and *uniform*, meets all the declared demands for the physical fundamentals, that giving a reason to consider the Geometrical Feeling as a kind of a “*protoquale*”. Consequently, the Geometrical Feeling (morphic sensation) is suggested for the role of the Protophenomenal Fundamental (alongside the physical fundamentals – Mass, Charge, Time/Space). Another aspect of the abstract definition of the Protophenomenal Fundamental is designated as the “capacity to feel and minimize non-congruence” between the abstract “ideal” geometrical form of a living cell and the material stuff filling this ideal form. The execution of such “minimization” realized by “moving” (influencing) the cell material stuff to obtain the maximal approximation of the fluctuating contours of the physical body of the cell to its “ideal” geometry can be considered as a *rudimentary psychic act* (“morphological mentality”).

However, such a conclusion may look like a dry, too abstract idea not having a real connection with the "ordinary" Consciousness *per se*, whose psychological reality and functional realization is impressively self-evident and intuitively clear. Therefore, the above conclusions are poised in mid-air unless, at least logically, are connected with the Consciousness *per se*. The postulated “minimization of the non-congruence” (a part of the final definition of Protoconsciousness), considered as a “rudimentary psychic act”, has *no analogue* in the glossary of the science of Consciousness. This postulation – designated as a reaction of a living cell to a physical stimulus that disturbs the relative harmony between the abstract geometrical form of a cell and the physical stuff “filling” (constituting) this abstract form – remains on the absolutely abstract level. According to the postulation, such a “reaction” is realized perpetually (continually) on the molecular level, and now the main question concerns the *nature* of that “*force*” which “moves” (influences) the molecular substrate causing it to approximate more closely to the minimal non-congruence. The expression "*the force which moves*" invites the idea of a *field* (Lipkind, 2005). The association of a newly suggested notion of Geometrical Feeling that expresses the Extra Ingredient by means of the scientifically well-established and strictly formalized field principle seems promising.

Thus, the general task is to develop a definition of the Consciousness *per se* based on the formulated concept of Geometrical Feeling, i.e., to come from the fundamental Protoconsciousness, as an inalienable attribute of any form of Life, to the Consciousness *per se* as related to the brain of a normal mature human individual. This means to provide a way to elucidate the "*nature*" of the abstract *geometrical form*, which is the predicate in the above-formulated Life and Protoconsciousness definitions. This, in turn, means that the abstract morphic principle, which could be considered as an epistemological tool, must be expressed in the ontological sense. In other words, the main task is to formulate a *naturalistic* factor that determines the species-specific morphology of a living entity, and for this aim, it is intended to use the concept of field.

3. "Naturalistic" Definition of the Extra Ingredient: From Protoconsciousness to the Consciousness per se

The meaning I am attributing to the word "naturalistic" concerns association of the hypothetical field with the *biological reality*, which does not always fit the laws of the today's physics. As a model of such a hypothetical field, the theory of the cellular vectorial field, formulated by Alexander Gurwitsch (1944, 1947a, 1947b, 1954/1991), was the choice that provided a combination of the strictly formalized irreducible field principle with biological regularities.

3.1. The Theory of the Biological Field by A. Gurwitsch

Gurwitsch was the first to introduce the field principle into biology (Gurwitsch 1912), that was acknowledged in contemporary reviews (Bertalanffy 1933, McDougall 1938, Weiss 1939) as well as in more recent works by the biologists who employed the concept of field in their theoretical considerations (Waddington 1966; Haraway 1976, Sheldrake 1986, Goodwin 1986, Welch 1992, Laszlo 1993, Gilbert et al. 1996).

3.1.1. The Basic Postulates of A. Gurwitsch's Theory of the Vectorial Cellular Field

1. Every living cell is a source of the field generated in the nucleus.
2. The field is of a vectorial nature and the vectors are directed centrifugally from the field source. Thus, as opposed to the attractive field of Newtonian gravitation, the Gurwitschian field is repulsive.
3. The generation of the field is associated with certain processes in the cell nucleus related to transformations of chromatin[21]. Extra-nuclear cytochromatin (e.g. that of the neuronal cells) is also considered as a source of the field.

[21] The Gurwitsch's style belongs to the period preceding the DNA-double-helix revolution. Therefore, the classic "chromatin" (generalized designation of the chromosomal apparatus of the cell nucleus) means nucleoprotein, and the "elementary acts in the chromatin metabolism" are hypothetically connected with interactions between the nucleic acid (DNA) and protein parts of the nucleoprotein.

4. There are *elementary "flashes" of generated field* that are connected with certain elementary acts of the chromatin metabolism, depending specifically on its intensity and, hence, on the intensity of the general cell metabolism. A total number of such flashes per time unit determine the *field intensity*.
5. The elementary flashes of generated field can occur only if these acts proceed within the sphere of influence of the already existing field. Essentially, this is the expression of the succession of processes in living systems, or, in other words, the proclamation of the same principles declared by W. Harvey ("*omne vivum ex ovo*") and L. Pasteur (*denial of a spontaneous generation of life*).
6. The field vectors directed from the nucleus are the resulting values from a total statistical number of the elementary flashes of the field at any moment. Therefore, the field intensity is a dynamic fluctuating parameter subtly reacting to metabolic changes.
7. The elementary field is *spatially anisotropic* and this is the main postulate. Its meaning is that the isodynamic surface at which all the vectors are equal *is not spherical* but *ellipsoidal*. The anisotropy of the ellipsoid can be expressed as a particular ratio between its three axes, and this ratio, being species-specific, is considered as an *invariant species-specific field constant*. An infinite number of possible inter-axis ratios cover any number of potentially possible species.
8. The field vector has a certain rate of decrement with increasing distance from the field source. The exact function of the dependence is a matter for empirical examination. In spite of the decrement, the influence of the field is not limited to the cell boundaries.
9. Field vectors influence on the *excited* protein molecules (those that have just got a bit of metabolic energy and are in the excitation state), transforming a portion of the general molecular excitation energy into *directed kinetic energy*, and the direction is determined by the field vector at this spatial point. This is expressed either in the directed movement of the excited protein molecules along the direction of the vector or in specifically directed deformations of the protein molecules, if they are anchored to any structures. This means that under living conditions the field "works" *against* the chaotic movement (agitation) of the protein molecules.
10. The intensity of the field at a certain cell point (the length of the vector at this point) determines what proportion of the molecular excitation energy is transformed into directed kinetic energy. The ratio E_d/E_t, in which E_d is the directed kinetic energy and E_t is the total molecular excitation energy, expresses this proportion. The intensity of the field does not depend on the amount of chromatin, but depends on its *turnover*.
11. The vectors from separate field sources are composed geometrically, and the resulting vector at the point of composition will determine the direction of the kinetic constituent of the full molecular excitation energy. Therefore, all the parts of a living system that consists of a number of cells and, hence, the corresponding number of field sources, are under an *integral (actual) field* that results from the total geometrical composition of all the vectors originating from all the sources (nuclei). Evidently, in such a composition, both the field intensity (being a function of both the metabolic activity and the distance of the point of composition from the field sources, which are nuclei of the involved cells) and the field anisotropy (relation of

the point of composition to the orientations of the axes of the nuclei of the involved cells) contribute to the value (length) of the resulting vector.

12. The last postulate is, essentially, the inference from the above postulates, especially the 11th one concerning the notion of the integral ("actual") field. The geometrical configuration of the integral actual field formed by all the energetically excited protein molecules of the intracellular substrate (the object of the field influence), determines the dynamic spatial configuration of the protein molecular continuum. Such dynamic associations of the energetically excited protein molecules maintained by continuous metabolic energy influx are called "*unbalanced (non-equilibrial) molecular constellations*" (Gurwitsch, 1954/91). These constellations provide dynamic conditions for steric (spatial) facilitation or hindrance for certain reactions. These conditions, therefore, are due not to the canonic chemical properties of the molecules – members of the constellation, but to the specificity of the spatial configuration of the constellation, which is determined by the geometrical configuration of the integral actual field at any considered locus. The unbalanced molecular constellations are considered as a "working substrate" of most biological manifestations.

Although the question of the nature of the biological field is not especially touched upon in the above postulates, two principal comments should be added.

Gurwitsch's biological field cannot be reduced to any known physical field: it is an immanent property only of living objects. According to the postulate 5, the elementary flash of the biological field is induced only by the *existing field*, so that the field is *successive* and *cannot originate de novo.* This is the strictest expression of the vitalistic principle.

Gurwitsch's biological field is *not energetic* which means that no special energy is focused in the field source. The field vector just transforms a portion of the metabolic energy accumulated in the excited protein molecules into directed kinetic energy, moving or deforming the molecules. The energy is not supplied by the field at its point of action, but the field vector as if *harnesses* the *local energy accumulated* at *this point*[22].

Thus, the Gurwitschian field is neither tautological, nor metaphorical; it responds to all the demands for any physical field. The explanatory capacity of the field theory was tested by Gurwitsch using different levels of the biological organization: molecular (metabolism), cellular (mitosis, differentiation and histogenesis) and organismic (morphogenesis, neuromuscular system, brain cortex structure and functioning). The mode of the field action expressed on the morphological level is defined as subjection of equipotential elements (cells) to integral morphogenic field, causing the *spatial* orientation and/or movement of the cells. On the molecular level, the field action is expressed as *vectorization* imposed on the molecules' chaotic movement.

The aim of the further analysis is to employ the Gurwitsch's theory for formulating a definition of Consciousness *per se*. The first step in this way is to express the above-described

[22] Recently, the idea of harnessing of existing energy sources was analyzed in connection with the assertion about violation of physical laws (namely, the second law of thermodynamics and the principle of conservation of momentum) as a result of realization of a volitional act (Wilson, 1999; Lipkind, 2006). Such energy harnessing can be allegorized by the already mentioned idea of the Maxwell's demon whose mode of "functioning" was recently analyzed (Leff and Rex, 1990).

final comprehensive definition of Protoconsciousness in the light of the A. Gurwitsch's theory of the vectorial cellular field.

4. Definition of Consciousness in the Light of the Gurwitschian Field

A new definition of the Protophenomenal Fundamental, like in the case of its previous, "pre-naturalistic" definition, has to be prefaced with a new definition of Life, made in the light of the Gurwitschian field.

4.1. Definition of Life in the Light of the Gurwitschian Field

A new definition of Life, which I call "naturalistic", is as follows:

- *Life* (any living system) is an evolving and aging species-specific non-equilibrial self-preserving and self-reproducing *geometrical form,* which is continuously refilled with certain substances coming from environment and whose spatial configuration is determined by the species-specific anisotropy of the *Gurwitschian field.*

Formally, in contrast to the previous, "pre-naturalistic", definition, in the new definition the concrete geometry of the "abstract geometrical form" is specified by Gurwitsch's field theory postulation. To my conviction, such specification makes the new definition "naturalistic". The reason for such assertion is that the postulates of Gurwitsch's field theory, being deeply rooted in biological reality, comprehensively cover the whole Life phenomenology. However, the new definition, like the "pre-naturalistic" one, retains the advantage of remaining on the abstract level without reference to any specific Life properties. At the same time, the association with the Gurwitschian field concretizes the abstract attributes of the definition, which acquire an explanatory power. For example, the attribute "species-specific" is clarified (detailed, explained) by Gurwitsch's postulate concerning the field anisotropy (postulate #7). Similarly, all the other attributes depend directly on the respective Gurwitsch's field theory postulates, which cover all the essential Life manifestations.

4.2. "Naturalistic" Definition of the Protoconsciousness

In a similar way, the previously formulated definition of Protoconsciousness based on the postulated "geometrical feeling" and "morphogenic mental reaction" has been re-analyzed in the light of Gurwitsch's field theory, and the *naturalistic* definition of the Protoconsciousness has been formulated:

- *Protoconsciousness* is the embodied immanent capacity of any living cell to *feel* and immediately *minimize* any spatial *non-congruence* between the abstract geometrical configuration of the cell's *Gurwitschian field* and the real distribution of the cellular material *stuff* exposed to the field influence.

The difference of this definition from the previously given "pre-naturalistic" final definition is that instead of the "species-specific 'ideal' geometrical form", there is the "geometrical configuration of the cell's Gurwitschian field". Like in the case of the new Life definition, the "naturalistic" meaning of the new Protoconsciousness definition is due to Gurwitsch's field theory postulates covering all the Life properties and manifestations.

However, the non-congruence in the new definition may be imagined differently as compared to the pre-naturalistic definition. The "physical" side of the non-congruence – the material stuff – is the same in both the definitions. As to the "geometrical" side of the non-congruence, in the new definition the abstract *species-specific "ideal" geometrical form* is exchanged for a field, which implies a certain network of *lines of force* determined both by the field species-specific spatial *anisotropy* (postulate #7) and the spatial configuration of the resulted integral actual field (postulate #11). Besides, the "minimization" acquires another meaning. In the previous definition the capacity "to feel and minimize" was considered as a *single* (*unsplittable*) fundamental quality, so that the minimization was realized "automatically" by analogy with Le Chatelier's principle. In the framework of the new definition, the minimization is more concretized. Namely, according to the field postulation (postulate # 9), the material entities sensitive to the Gurwitschian field, are the *excited* protein molecules (those which have just absorbed a portion of metabolic energy), and there is transformation of a portion of the molecule general excitation energy into the *directed kinetic energy*, while the direction of the movement is determined by the field vector at the respective spatial point. Such vector-directed movements of the excited protein molecules along the field lines of force will minimize the non-congruence returning to (in the case of a rough disturbance) or preserving (in the case of the background micro-disturbances) maximally reachable[23] harmony between the cell's field tension and intracellular distribution of the material substrate. The postulated transformation of a portion of the protein molecule's general excitation energy into directed kinetic energy means that under "normal" living conditions, the field "works" *against* the chaotic movement (agitation) of the protein molecules.

4.3. Definition of the Primordial Consciousness

It has been deduced that the initial point of an individual psychic development relates to the zygote. Further analysis is connected with the moment of the first cleavage of the zygote into two blastomeres (embryonic daughter cells), each one being a source of the Gurwitschian field. The great game of the individual development begins, being immanently associated with (led by) the mutually developing *integral* field. The integral field of such a two-cells' system (which from this moment *is* an *embryo*) results from geometrical vectorial composition of the fields[24] of both the blastomeres. The spatial non-congruence between a new geometrical configuration of the integral Gurwitschian field and the material stuff filling (building, constituting, realizing) this abstract configuration is *felt* by this initial embryo whose physical essence (body) "strives" for approximation to its dynamically predetermined "geometrical ideal". This approximation is realized by the corresponding growth of the embryo's physical body that is put into effect by metabolically mediated increasing of the amount of material stuff coming from the environmental medium and assimilated by the embryo's body. However, it is not just growth as expressed in increases of volume and mass (like swelling) – this is a morphological process, i.e., physical embodiment of the abstract geometrical form that is determined by the "lines of force" of the developing integral embryonic field. The next cleavage causes the same non-congruence and the same morphogenic reaction. Such continuous approximation of the physical "content" (body) of the developing embryo to its (her/his!) developing integral field can be considered as the embryo's "actions" – a continuous series of the embryo's "acts" ("deeds") that can be described as an expression of the embryo's "behavior". The latter can be analyzed by means of the same non-congruence notion: now the non-congruence is between the configuration of the Gurwitschian field, on the one hand, and the distribution of the physical substrate within the field-influenced space, on the other hand. This means that both the factors as if located on the opposite sides of a Gap between the geometrical abstract entity and the respective material

[23] A full 100% spatial correspondence between the field configuration and the current distribution of the protein molecules cannot be achieved in principle, similarly to the absence of the absolute coincidence between any physical mould and its geometrical prototype.

[24] The postulate #11.

substrate. The embryo's reaction to the non-congruence, "aiming" to smooth it, is realized via the field-caused vectorization of the molecular movements. Anyhow, the non-congruence is "felt" because of the same Geometrical Feeling that was described in the abstract definition of Protoconsciousness.

Thus, the embryonic development can be imagined as the embryo's behavior. According to Driesch's views, the embryo's entelechy "experiences the current state" and directs morphogenesis purposefully (Driesch, 1915), whereas according to Gurwitsch's views, the developing embryo "knows his[25] actual field" (Gurwitsch, 1954/ 1991, p.280) and "acts morphogenically" toward smoothing geometrical disharmonies (tensions) arising from either "normal" environmental fluctuations, or extravagant interventions, such as Driesch's experiments on "harmonic regulations" (Driesch, 1891). The above difference in views is clearly expressed in the degree of tautology: The Drieschian "experience of the current state", which is vague and indistinct, is confronted by the much more definite Gurwitschian "embryo's knowledge of his own integral field". My suggestion goes further in that direction: the embryo "knows" not "his own integral field", but feels non-congruence between the geometrical configuration of her/his developing abstract integral Gurwitschian field and the actual form of her/his physical body, and strives to minimize the non-congruence. The word "strive" has, on the one hand, a clearly qualitative meaning close to intention, aiming, even yearning, but, on the other hand, this "striving" may be imagined as returning immediately (almost automatically) to the initial state like tumbler or tilting doll (Russian "matryoshka").

Consequently, the Primordial Consciousness concerning the developing embryo at early ("pre-cerebral") stages (before the embryonic cells' differentiation) is defined as follows:

- *Primordial Consciousness* is the embodied immanent capacity of an early non-differentiated ("pre-cerebral") embryo to *feel* and immediately *minimize* (smooth) any spatial *non-congruence* between the geometrical configuration of the embryo's evolving *integral Gurwitschian field* and the actual spatial distribution of the embryo's material *stuff*.

Thus, the rudimentary psychic phenomenology during the "pre-cerebral" stages of the embryo development includes the embryo's *experience* of the continuously changing non-congruence between the momentary configuration of the embryo's integral field and the actual distribution of the molecular stuff, on the one hand, and, on the other hand, morphogenic *action* directed to minimize (to smooth) the non-congruence. Consequently, the early embryogenesis can be imagined as a chain of the embryo's actions, each one being associated with the embryo's *act of choice* among different possibilities. Such embryonic "free will" becomes evident in the case of the experimental interference that was clearly demonstrated in the experiments by H. Driesch ("harmonic regulations"), which led to formulation of the notions of *equipotentiality* and *equifinality*, that means the development of the *same* final species-specific form (equifinality) from equipotential elements (cells) by quite *different* ways (Driesch, 1891, 1908). In the glossary of the modern theory of supervenience (Horgan, 1982; Kim, 1984, 1987, 1999), the equifinality can be expressed as the *absence* of

[25] The embryo's "his" is not chauvinistic. According to Russian grammar, the word "embryo" is masculine. The usage of "its" in relation to a *human* embryo would be considered as degradation of a human embryo equating it to an animal.

the *isomorphic identity* during the morphological development: the *same* final species-specific form displayed on the morphological level can be realized via quite *different* processes occuring on the cellular and molecular levels.

4.4. Transformation of the "Morphological Mentality" into Instinctive/Reflexive Neurophysiology

A turning point of the individual development is connected with cell differentiation as a specification of the functional roles of the cells in the whole organism; it is associated with drastic and irreversible (normally) changes, namely:

1. The decrease and then the loss of the capacity for the Drieschian "harmonic regulations".
2. Morphological "parcellation" of the embryo, leading to his/her geometrical complication. As a result, the embryo loses its "compact" form (morphological "wholeness") and its various parts (e.g., limbs) become morphologically separated. Accordingly, just because of the purely geometrical premises, the integral field of the whole embryo vanishes and is replaced by separated fields (according to the areas of separated organs) that geometrically do not interact with each other.
3. All the above morphological processes advance in parallel with differentiation of specific neural cells and rapid development of the nervous system (neural network) with its central part, the brain, which becomes the main integrating factor for the coherent functioning of the organism as a whole.
4. As the morphological development approaches completion, the morphogenically (geometrically) coordinated cell movements give place to physical (muscular) movements of spatially separated parts of the shaped body[26].
5. Starting from this point, the early "*conscious*" *whole embryo* that feels (is aware of) the non-congruence between the geometrical configuration of the embryo's integral field and the embryo's whole body turns into a *highly sophisticated automaton* that integrates, via the neural network, all the multiple parts of the still developing, morphologically complicated embryo's body into a coherently functioning whole. Thus, the ancient "simple" slowly reacting morphological *mental activity* of the early *non-differentiated* embryo turns into the *neural* (highly sophisticated and quickly reacting instinctive/reflexive) *behavior* of the swiftly complicating (via further differentiation) maturing organism. Now, the embryo's "behavior" is a result of its neurophysiologic activity that includes input and processing of all the signals coming from the developing receptors. The chain of the embryo's actions is now incredibly more effective but such instinctive-reflective activity, in some respect, resembles zombie-like behavior.

4.5. Brain as Geometrical Continuum: Conversion of the Primordial Consciousness into Consciousness *per se*

The development of the brain includes two remarkable features: together with the development of cytoarchitectonics and a highly complicated network of neuronal

[26] In fact, further (much slower) morphological development proceeds during post-embryological period up to full maturation.

interconnections, the brain develops as a *morphologically compact geometrical whole*. In view of Gurwitsch's field theory, this leads to the formation of the brain integral field – a kind of field geometrical continuum. As a result, a remarkable combination is established: unsurpassed complexity of the integral field geometrical configuration is combined with incredible sophistication of the neural cytological infrastructure. There are all the anatomical premises for the manifestations of mental phenomenology in its highly developed form. This stage of the development means "resurrection" of the vanished initial embryonic consciousness, but now this is not that "ancient" rudimentary embryo's "knowledge" ("experience") of the non-congruence between the embryo's integral field configuration and the embryo's material body, characterized by slow "morphological action" directed to the closest approximation between the two. At this stage, the integral field of the whole embryo is replaced by the integral field of the embryonic brain, which is exposed to the current stream of diverse signals (impulses) coming from the developing external and internal somatic receptors via a highly sophisticated neural network becoming more and more sophisticated. All these impulses are conducted in neurophysiologic manner, i.e., the physical energy causing specific excitation of the respective receptors is transformed into the neural impulses (described by biophysical and biochemical terms) that transfer the initial signals from the receptors towards the corresponding areas of the developing brain. The current waves of such a stream flow into (are engulfed by) the Gurwitschian integral field of the brain, thus causing dynamic "disharmonies" in the brain field configuration. These disharmonies are due to the same *non-congruence* – this time between the ideal geometry (stereometry) of the whole-brain integral field continuum (internal "lines of force"), on one hand, and the real (disturbed by incoming impulses) distribution of the structured physical stuff filling (imitating) this geometrical frame, on the other hand. In contrast to the case of the Primordial Consciousness, in which the non-congruence relates to the embryo as a whole, in the case of the Consciousness *per se*, the non-congruence caused by the *physiologically conducted* physical signals, relates not to the whole embryo but to the geometrical continuum of the embryo's whole brain cortex.

4.6. The Final Definition of the Consciousness *per se*

Thus, the definition of the Consciousness *per se*, is as follows:

- *Consciousness per se* is the capacity of a mature human *brain* to *feel* any *non-congruence* between the geometrical configuration of the integral Gurwitschian *field* of the brain and the current fluctuations and disturbances of the spatial distribution of the brain material *stuff*.

The important elucidation to be added to this definition relates to the expression *"current fluctuations and disturbances of the spatial distribution of the brain material stuff"*. The detailed version sounds as follows: *"current fluctuations and disturbances of the spatial distribution of the brain material stuff that are caused by current stream of afferent neural impulses generated in the sense organs (including all the totality of the external, internal, and proprioceptive receptors) and conducted into the brain geometrical field continuum"*. This addition could be syntactically inserted into the complete definition but that would make it too cumbersome. Therefore, it seems to be better arranged in the above form of a supplementary (auxiliary) elucidation.

The aforementioned non-congruence is the same discrepancy between the ideal abstract geometrical form expressed by the Gurwitschian field of the brain and its material realization, i.e., "filling" ("stuffing") the geometrical frame with the physical "content". The expression "*ideal* geometrical form of the brain" is not so easily comprehended, since the rough macroform of brain, although having its characteristic image, is not an example of a strict geometrical appearance. However, the internal cytoarchitectonics of the brain cortex has remarkable geometrical characteristics that served as the basis for Gurwitsch's theorizing. Gurwitsch emphasized strict *geometrical* features of the cortex cytoarchitectonics (which are absent in the histological organization of the subcortical centers), namely: the specific structure and *configuration* of the cells that dominate in a given area (pyramid cells, for example), the *lamellar* character of their spatial arrangement and, especially, the strictly *parallel* orientation of the cell axes (Gurwitsch, 1954/1991)[27]. Just these geometrical features of the brain are meant in the respective part of the above definition of the Consciousness *per se*, namely, "*the geometrical configuration of the integral Gurwitschian field of the brain"*.

As to the "stuffing" of the abstract geometrical frame with the physical "content" as a factor causing the non-congruence, evidently, the absolute congruence between *physical* and abstract *geometrical* cannot be achieved in principle. Therefore, the maximal conformity, i.e., the practical (not absolute) congruence between the geometrical ideal form and its physical realization, is accomplished with the achievement of minimal fluctuations, e.g., the inevitable weak agitation of the molecules that fill the field-determined geometrical framework. Since, according to the suggested postulation, such non-congruence is felt and experienced, the unattainable least minimal non-congruence would mean on the psychic level something like *nirvana.* Against a background of such quasi-congruence, any physical perturbations within the "molecular substrate" disturb the physical-*versus*-geometrical conformity and cause the non-congruence. The current dynamic stream of the afferent neuronal firing originates from various receptors, which are excited by diverse physical stimuli that cause specific sensations, e.g. mechanical (tactile, equilibrial [balancing], proprioceptive), chemical (olfactory, gustatory), acoustic (auditory), and photonic (visual). The specificity of the conducted impulses is determined by the corresponding physical characteristics of the initial stimuli. Since the current dynamic stream of the impulses causing the non-congruence proceeds from the whole totality of the receptors as they are bombarded ceaselessly by all the possible stimuli coming from the external world as well as from the own body, the experience of this integral non-congruence unequivocally reflects the current state of the world as-perceived.

Thus, the causal chain of the psychical loop is as follows:

1) *Activation* (excitement) of the receptors by physical stimuli; 2) *Conduction* of this excitement, via the neuronal network, into the brain; 3) *Disturbance* by the neuronal firing of the molecular substrate of the respective brain part(s), that upsetting *conformity* between the field-determined internal geometrical frame and the molecular substrate within this frame, i.e. increasing the non-congruence; 4) *Feeling* (experience) of the dynamically changing non-congruence, which reflects in every detail a coherent and comprehensive picture of the external world (including the own body) as-perceived by the individual; 5) *Reaction* to such dynamic non-congruence, which is initiated at the micro-level of the interactions between the

[27] The geometrical beauty of the brain cortex was demonstrated histologically by Ramon y Cajal in his classic "Textura" (1899-1904) and has been excellently expressed in Fine Arts by Herms Romijn (see cover illustration in JCS (2001, v. 8, No. 9-10).

Gurwitschian field and the molecular substrate and can be expressed at the phenomenological macro-level through the enormously wide spectrum of all the possible psycho-physiological manifestations ranging from emotional explosions up to deep intellectual reflection with Jamesian stream of Consciousness as a background.

VI. Discussion

The present article looks as grounded on the dualist philosophical ideology that is not welcomed in the modern natural science based on the dominating reductionist doctrine. Just the declared here "irreducibility" of Consciousness to the physical fundamentals immediately ranks the suggested theory with the dualist province. Lately, however, the truly reductionist approach became to be considered as deficient for a comprehensive explanation of the Consciousness problem, and the necessity for counter-intuitive ideas has been proclaimed (Shear, 1995; Chalmers, 1995). Regrettably, the dualism is often associated with the epistemological agnosticism and at times even looks as sliding into the esoteric abyss. The utterance by Du Bois-Reymond (cited in the article's epigraph) that "it is in no way intelligible how consciousness might emerge from the coexistence of carbon, hydrogen, nitrogen, oxygen etc. atoms in ... brain", may be considered as such a combination of dualism and agnosticism.

However, if to omit the overall materialism-dualism dispute, the contemporary philosophical-scientific basis of the reductionist approach does not offer any possibility for the reduction of Consciousness to the established physical fundamentals: even materialists began speculating that, although such a reduction is beyond any doubt, it is probable to occur on a somewhat higher level. Anyhow, the only reasonable reductionism's hope rests upon a strong belief into future new discoveries enabling respective scientific solutions supporting the reductionist credo. Such belief is a kind of a faith.

According to the opposite view, instead of passive waiting for future discoveries, active searching for possible solutions of the problems seemingly unresolved within the reductionist approach are to be initiated, and a practical attempt inevitably leads to the analysis of the *irreducibility* of the Consciousness to the established physical fundamentals. The reaction to such a conclusion can conduct to two possibilities: the first one means falling into the esoteric precipice, while the second one means searching for novel principles permitting formulation of *additional axiomatics*, which, however, must not be in contradiction with the established physical laws. The suggested here approach rested on the second alternative has led to the formulation of the concept of *Geometrical Feeling*, which is such a new fundamental additional to but not contradicting the existing physical fundamentals.

In this respect, the most intriguing riddles concern the anatomical *"localization"* and *temporal occurrence* of the Consciousness' initiation and maintenance which look as peculiar specific manifestations of the widely discussed Psycho-Physical Gap together with the mysterious *spatial-temporal* organization of the conscious phenomenology. The word "localization" in connection to Consciousness, which is the very expression of a non-localized holistic essence, bears perplexity: *where* within the brain does the realization of this physical-conscious transformation occur? This question is indispensably bound to the meaning of the Psycho-Physical Gap, in particular, the Gap's "localization", the respective question being as follows: At what spatial-temporal point/moment within the functioning brain does a current, basically *physical*, neural impulse acquire the "conscious hint"? A more specific question is at which stage along the whole way of the Consciousness' physical-biological-mental realization, namely, *receptor* irritation – neural *impulse* processing – conscious *experience,* the physical-conscious transformation occurs. Theoretically, the magical "switch" could occur at any stage of the above triad: either just after the receptor irritation, or during the further neurological way toward the brain, or at the transfer of the neural impulse into the brain, or during the impulse advancing within the brain, e.g. by the impulse's arriving into any specific brain region. Although the latter possibility would look as the most probable, the problem of the spatial-temporal localization of the physical-conscious transformation remains undetermined. Again and again, any suggestion is broken by the firm fact that there is *no locus* in the cortex, which could be called either *"master map"* (Treisman, 1986), or *multi-modal association areas* (Damasio, 1989), or *central cortical "information exchange"* (Hardcastle, 1994).

An especial point concerns the psychic indeterminism, which is closely interrelated with the problem of the Free Will and mental causation. Namely, the non-physical mind overrides the physical causality: the psycho-physical acts, closely associated with internal awareness and usually terminated by active (efferent) responses, are realized and memorized in contradiction with the physical laws. This leads to the necessity for a theoretical explanation accepting (or at least considering) the situation that the Free Will is realized by the deliberate choice between the existing alternatives. This means that not all the individual's actions are determined by or predictable from the antecedent factors (causes), i.e. a respective theory must hold that not every psychic event has a "natural" (physical) cause. Namely, those psychic acts, which are realized not as responses to any external stimuli but as reactions to certain *internal* feelings and experiences, may look as "spontaneous", i.e. not connected to any particular external stimuli.

However, such "spontaneous" acts may be considered as responses to temporally long chains of such internal feelings and experiences (sometimes, together with thinking), which are to be based on certain hidden but definitely existing causal factors. Then, any such seemingly spontaneous psychic act is a result of the individual's logical analysis based on deterministic grounds.

CONCLUSIONS

The declared irreducibility of the suggested field factor, which is explored for the definition of the postulated Protoconsciousness together with a novel definition of Life, means *combining* the above-defined immaterial Protophenomenal Fundamental with the established physical fundamentals (mass, charge, time/space), that would mean a *new paradigm* including a *non-physical* fundamental.

Such a situation cannot be accepted as a "working tool" in the contemporary science due to the epistemological and ontological premises deeply ingrained into the reductionist principles of the modern scientific exploration. In particular, the search for the "*naturalistic*" meaning of the *irreducible non-physical* Extra Ingredient may induce an ironical attitude, since the existing *physical* laws have been exhaustively formulated and thus exist as a "natural" basis of the modern science, so that namely irreducibility to these laws looks "unnatural" (*meta*physical). As to the mysterious enigmas like the Psycho-Physical Gap, Free Will, mental causation, etc., there is a general confidence that they eventually will become explainable by means of new scientific (physical, in the end) regularities that for sure are to be discovered in the not-too-far-distant future. Such ephemeral hope for future discoveries of new laws, which must be, in principle, *supra-physical* (in relation to the *existing* physical laws) is a kind of a faith leading to mental inertness, i.e. a passive expectation of revolutionary discoveries to occur. A logical alternative to such situation is to accept (or at least to examine) the vitalistic principle, if the traditional dislike of this sacramental term could be overcome. Such examination means an immediate inclusion of active research instead of the passive expectation of future discoveries.

Thus, in the frame of the above-formulated Life and Consciousness definitions, the reduction of Consciousness to Life has been achieved. Postulation of the "Geometrical Feeling" as an inalienable attribute of any living system and (although hypothetical) derivation of the Consciousness *per se* from the primordial 'morphological mind' is in agreement with the ontogenetic and phylogenetic reality. The main advantage of this reduction is transfer through the Brain-to-Mind Explanatory Gap, or, correctly, moving this Gap up to the common basis of the initial unquestionable fundamentals. In this light, the Hard Problem of Consciousness loses its 'hardness', i.e. disappears.

However, if as a result of this reduction the Hard Problem of Consciousness is exchanged for the Hard Problem of Life, the advantage of such reduction may be doubtful. In order to analyze the situation, both the Hard Problems should be confronted by the same Chalmers' stylistic mode: the "Why of Consciousness" (Hardcastle, 1996) should be compared with the "Why of Life". The "Why of Consciousness" means why the information processing does not go on "in the dark", free of any inner feel (Chalmers, 1995). In this respect, the question "Why of Life" loses any sense since it has no analogical connection with the concept of unconscious Zombie perfectly functioning "in the dark". The Life is always alive – there is neither "Living Gadget", nor "Mechanical Life". The general difference between both the Hard Problems is that the Consciousness arises from the living human brain while any form of Life comes only from the respective species of Life. The everyday reality shows that the human Consciousness ontogenetically originates within a developing, proceeding from zygote, individual, i.e. the Consciousness originates from the Life, while any new (progeny) Life originates only from the pre-existing (ancestral) Life *without interruption*, i.e. without any Explanatory or time Gap. The inference from the suggested theory is that there is *no* Hard Problem of Life and, as far as the Consciousness is reducible to the Life, there is *no* Hard Problem of Consciousness.

References

Alfinito, E. and Vitiello, G. (2000). Formation and life-time of memory domains in the dissipative quantum model of brain. *International Journal of Modern Physics, B, 14*, 853-868.

Arecchi, F. T. (2003). Chaotic neuron dynamics, synchronization, and feature binding: quantum aspects. *Mind and Matter, 1*, 15-43.

Allport, A. (1988). What concept of consciousness? In A. Marcel and E. Bisiach (Eds.), *Consciousness in contemporary science.* Oxford, UK: Oxford University Press, pp. 159-182.

Angell J.R. (1904). *Psychology: an introductory study of the structure and functions of human consciousness.* New York: Holt.

Antoniou, I. (1992). Is consciousness decidable? In *Science and Consciousness, Proceedings of the 2nd International Symposium*. Athens, (January 3-7, 1992), pp. 69-91.

Baars, B. J. (1988). *A cognitive theory of consciousness*. Cambridge, UK: Cambridge University Press.

Babloyantz, A. (1986). *Molecules, dynamics, and life*. New York: John Wiley and Sons.

Barber B. (1961). Resistance by scientists to scientific discovery. *Science, 134,* 596-602.

Barrow J. D. (1992). *Theories of everything*. New York: Fawcett-Columbine.

Beeson, M. (1985). *Foundations of constructive mathematics*. Berlin: Springer, E. Bishop.

Bergmann, P. (1979). Unitary field theories. *Physics Today*, *32*, 44-51.

Bertalanffy, *L. von (1933). Modern theories of development: An introduction to theoretical biology*. Oxford, UK: Oxford University Press.

Bieri P. (1995). Why is consciousness puzzling? In. T. Metzinger (Ed.), *Conscious experience*. Schoeningh, Imprint Academic, pp. 45-60.

Bierman, D. J. 2003. Does consciousness collapse the wave-packet? *Mind and Matter*, *1*, 45-57.

Bisiach E. (1988). The (haunted) brain and consciousness. In A. Marcel and E. Bisiach (Eds.), *Consciousness in contemporary science*. Oxford, UK: Oxford University Press, pp. 101-120.

Bohr, N. (1958). *Atomic physics and human knowledge*. New York:Wiley and Sons.

Cairns-Smith, A. G. (1982). G*enetic takeover and the mineral origin of life*. Cambridge, UK: Cambridge University Press.

Callender, C. and Huggett, N. (2001). *Physics meets philosophy at the Planck scale*. Cambridge,UK: Cambridge University Press.

Caplan, S. R. and Essig, A. (1983). *Bioenergetics and linear nonequilibrium thermodynamics*. Cambridge, MA: Harvard University Press.

Chalmers, D. J. (1994). Concluding remarks on the Conference "Toward a scientific basis for consciousness". In Report by Jane Clark. *Journal of Consciousness Studies*, *1*, 152-154.

Chalmers, D. J. (1995). Facing up to the problem of consciousness. *Journal of Consciousness Studies, 2*, 200-219.

Chalmers, D. J. (1996). *Conscious mind: In search of a fundamental theory*. New York: Oxford, UK: Oxford University Press.

Chomsky N. (1988). *Language and problems of knowledge*. Cambridge MA: MIT Press.

Clark A. (1992). *Sensory qualities*. Oxford, UK: Oxford University Press.

Clifford W. K. (1874). Lectures and essays. In L. Stephen and Pollock (Eds.) *Lectures and essays by William Kingdon Clifford*. London: Macmillan, pp. 31-70.

Crick, F. H. C. (1994a). *The astonishing hypothesis - The scientific search for the soul*. London: Simon and Schuster.

Crick F. H. C. (1994b). Interview with J. Clark at the launch of Crick's new book "The astonishing hypothesis". *Journal of Consciousness Studies, 1*, 10-17.

Crick, F. H. C. and Koch, C. (1990). Toward a neurobiological theory of consciousness. *Seminars in the Neurosciences*, *2*, 263-275.

Crick, F. H. C. and Koch, C. (1998). Feature article: Consciousness and neuroscience. *Cerebral Cortex*, *8*, 97-107.

Damasio, A. R. (1989). The brain binds entities and events by multiregional activation from convergence zones. *Neural Computation, 1,* 123-132.

Davidson, D. (1970). Mental events. In L. Foster and J. Swanson (Eds.), *Experience and theory*. Amherst, MA: University of Massachusetts Press, pp. 79-101.

Dawkins, R. (1976). *The selfish gene*. Oxford, UK: Oxford University Press [New edition, 1989].

Dawkins, R. (1982). The extended phenotype: The gene as the unit of selection. Oxford, UK: W.J. Freeman and Co./ Oxford University Press.
Delbrück, M. (1949). Sesquicentennial celebration proceedings: Part II. A physicist looks at biology. *Transactions of the Connecticut Academy of Arts and Sciences, 38*, 173-190.
Dennett, D. C. (1991). *Consciousness explained.* Harmondsworth, UK: Little-Brown and Co.; [2nd edition, 1993, Penguin Books]
Dennett, D. C. (1995). *Darwin's dangerous idea.* New York: Simon and Schuster.
Dennett, D. C. (1996). Facing backwards on the problem of consciousness. *Journal of Consciousness Studies, 3,* 4-6.
De Quincey, C. (2000). Conceiving the 'inconceivable? Fishing for consciousness with a net of miracles. *Journal of Consciousness Studies, 7*, 67-81.
Desmedt, J. D. and Tomberg, C. (1994). Transient phase-locking of 40 Hz electrical oscillations in prefrontal and parietal human cortex reflects the process of conscious somatic perception. *Neuroscience Letters*, *168,*126-129.
Dolev, S. and Elitsur, A. C. (2000). Biology and thermodynamics: seemingly opposite phenomena in search of a unified paradigm. *The Einstein Quaterly: Journal of Biology and Medicine*, *15*, 24-33.
Driesch, H. (1891). Entwicklungsmechanische Studien. 1. Der Wert der beiden ersten Fürchungszellen in der Echinodermenentwicklung Experimentelle Erzeugung von Theil- und Doppelbildungen. *Zeitschrift für Zoology, 53*, 160-78. (In German).
Driesch, H. (1908). *Science and philosophy of the organism.* London: Adam and Charles Black.
Driesch, H. (1915). Vitalism: Its history and system. [Russian Edition, Edited and authorized translation from German into Russian by A.G. Gurwitsch]. Moscow: Nauka. (In Russian).
Du Bois-Reymond E. (1872): Über die Grenzen des Naturerkennens [Quoted from "Vorträge über Philosophie und Gesellschaft", 1974, Hamburg, Meiner] (In German).
Dürr, H. P. and Gottwald, F. T. (2001). Rupert Sheldrake in der Diskussion. München: Scherz.
Edelman, G. M. (1989). *The remembered present: A biological theory of consciousness*. New York: Basic Books.
Edelman, G. M. (1993). Morphology and mind: Is it possible to construct a perception machine? *Frontier Perspectives*, *3*, 7-12.
Eigen, M. (1992). *Steps towards life - A perspective on evolution.* Oxford, UK: Oxford University Press.
Elitzur, A. C. (1994). Let there be life: Thermodynamic reflections on biogenesis and evolution. *Journal of Theoretical Biology*, *164*, 429-459.
Engel, A. K. (2000). Temporal binding and neural correlates of consciousness. *Consciousness and Cognition, 9*, S26.
Engel, A. K., Konig, P., Kreiter, A.K. and Singer, W. (1991). Interhemispheric synchronization of oscillatory neuronal responses in cat visual cortex. *Science, 252*, 1177- 1179.
Flanagan, O. (1992). *Consciousness reconsidered.* Cambridge, MA: MIT Press.
Flohr, H. (1992). Qualia and brain processes. In A. Beckermann, H. Flohr, and J. Kim (Eds.), *Emergence or reduction? Prospects for nonreductive physicalism.* Berlin: De Gruyter, pp. 220-240.
Frith, C., Perry, R. and Lumer, E. (1999). The neural correlates of conscious experience: an experimental framework. *Trends in Cognitive Sciences*, *3*, 105-114.

Galambos, R., Makeig, S. and Talmachoff, A. (1981). A 40 Hz auditory potential recorded from the human scalp. *Proceedings of National Academy of Sciences of USA, 78,* 2643-2647.

Gilbert, S. F., Opitz, J. M. and Raff, R. A. (1996). Resynthesizing evolutionary and developmental biology. *Developmental Biology, 173*, 357-372.

Gödel, K. (1931). Über formal unentscheidbare Satze der Principia Mathematica und verwandter System. *Monatshefte für Mathematik und Physik, 38*, 173-198 (In German).

Goodwin B. C. (1986). Is biology an historical science? In S. Rose and L. Appignanese (Eds.), *Science and beyond.* Oxford, UK: Blackwell, pp. 47-60.

Gray, C. M., King, P., Engel, A. K. and Singer, W. (1989). Oscillatory responses in cat visual cortex exhibit inter-columnar synchronization which reflects global stimulus properties. *Nature, 338*, 334-337.

Greene, B. (1999). *The elegant universe: Superstrings, hidden dimensions, and the quest for the ultimate theory*. New York: Norton.

Greene, B. (2003). The future of string theory (a conversation with Brian Geene). *Scientific American, 289*, 48-53.

Griffin, D. R. (1997). Panexperientialist physicalism and the mind-body problem. *Journal of Consciousness Studies, 4*, 248-268.

Gurwitsch, A. G. (1912). Die Vererbung als Verwirklichungsvorgang. *Biologische Zentralblatt, 32*, 458-486. (In German).

Gurwitsch, A. G. (1914). Der Vererbungsmechanismus der Form. *Roux' Archiv für die Entwicklungsmechanik, 39*, 516-577. (In German).

Gurwitsch, A. G. (1915). On practical vitalism. *The American Naturalist, 49*, 763.

Gurwitsch, A. G. (1922). Über den Begriff des embryonalen Feldes. *Roux' Archiv für die Entwicklungsmechanik, 51*, 383-415. (In German).

Gurwitsch, A. G. (1927). Weiterbildung und Verallgemeinerung des Feldbegriffes. *Roux' Archiv für die Entwicklungsmechanik, 112*, 433-454 (Festschrift für H. Driesch). [In German].

Gurwitsch, A. G. (1929). Der Begriff der Äquipotentialität in seiner Anwendung auf physiologische Probleme. *Roux' Archiv für Entwicklungsmechanik, 116*, 20-35. (In German).

Gurwitsch, A. G. (1930). *Die histologischen Grundlagen der Biologie*. Jena: G. Fischer Verlag, (In German).

Gurwitsch, A. G. (1944). *The theory of the biological field.* Moscow: Sovetskaya Nauka. (In Russian).

Gurwitsch, A. G. (1947a). Une theorie du champ biologique cellulaire. *Bibliotheca Biotheoretica, ser. D, 11*, 1-149, Leiden. (In French).

Gurwitsch, A. G. (1947b). The concept of "whole" in the light of the cell field theory. In A.G. Gurwitsch (Ed.), *Collection of works on mitogenesis and the theory of the biological field.* USSR Moscow: Academy of Medical Sciences Publishing House, Moscow, pp. 141-147. (In Russian).

Gurwitsch, A. G. (1954). *Principles of analytical biology and the theory of cellular fields.* Moscow, Manuscript (In Russian).

Gurwitsch, A. G. (1991). *Principles of analytical biology and the theory of cellular fields.* Moscow: Nauka. (In Russian).

Hagan S. and Hirafuji M. (2001). Constraints on an emergent formulation of conscious mental states. *Journal of Consciousness Studies, 8*, 99-121.

Haken, N. (1977). *Synergetics: An introduction*. Heidelberg, Germany: Springer.

Hameroff, S. (1994). Quantum coherence in microtubules: A neural basis for an emergent consciousness? *Journal of Consciousness Studies, 1*, 91-118.

Hameroff, S. (1997a). Quantum computing in microtubules – An intra-neural correlate of consciousness? *Japanese Bulletin of Cognitive Science, 4*, 67-92.

Hameroff, S. (1997b). Quantum vitalism: Advances. *The Journal of Mind-Body Health, 13*, 13-22.

Hameroff, S. (1998). Funda-Mentality: is the conscious mind subtly linked to a basic level of the universe? *Trends in Cognitive Sciences*, *2*, 119-124.

Hameroff, S. (2001). Consciousness, the brain, and space-time geometry. In P. C. Marijuan (Ed.), *Cajal and consciousness: scientific approaches to consciousness on the centennial of Ramón y Cajal's textura.* New York: Annals of the New York Academy of Science, 929, pp.74-104.

Hameroff, S. and Penrose, R. (1996). Conscious events as orchestrated space-time selections. *Journal of Consciousness Studies*, *3*, 36-53.

Haraway, D. J. (1976). *Crystals, fabrics, and fields: Metaphors of organicism in twentieth-century developmental biology*. New Haven, CT: Yale University Press.

Hardcastle, V. G. (1994). Psychology's binding problem and possible neurobiological solutions. *Journal of Consciousness Studies, 1*, 66-90.

Hardcastle, V. G. (1996). The why of consciousness: a non-issue for materialists. *Journal of Consciousness Studies*, *3*, 7-13.

Hardin, C. L. (1992). Physiology, phenomenology, and Spinoza's true colors. In A. Beckermann, H. Flohr and J. Kim (Eds.), *Emergence or reduction? Prospects for nonreductive physicalism.* Berlin: De Gruyter, pp. 201-219.

Hare, R. M. (1984). Supervenience. *Proceedings of the Aristotelian Society, suppl.*, *58*, 1-16.

Hempel, C. G. (1980). Comments on Goodman's ways of worldmaking. *Synthese*, *45*, 193-194.

Hofstadter, D. R. (1979). *Gödel, Escher, Bach: an eternal golden braid.* New York: Basic Books.

Hofstadter, D. R. (1982). Prelude … ant fugue. In D. Hofstadter and D. Dennett (Eds.), *The mind's I: Fantasies and reflections on self and soul.* Harmondsworth, UK: Penguin Books, pp. 149-201.

Hofstadter, D. R. (1985). *Metamagical themas: Questing for the essence of mind and matter.* New York: Basic Books.

Horgan, T. (1978). Supervenient bridge laws. *Philosophy of Science, 45,* 227-249.

Horgan, T. (1982). Supervenience and microphysics. *Pacific Philosophical Quarterly, 63,* 29-43.

Horgan, T. (1984). Supervenience and cosmic hermeneutics. *Southern Journal of Philosophy, 22*, 19-38.

Horgan, T. (1993). From supervenience to superdupervenience: Meeting the demands of a material world. *Mind, 102*, 555-586.

Hubel, D. H. and Livingstone, M. S. (1987). Segregation of form, color, and stereopsis in primate area 18. *The Journal of Neuroscience, 7*, 3378-3415.

Hughes, J. R. and John, E. R. (1999). Conventional and quantitative electroencephalography in psychiatry. *Journal of Neuropsychiatry and Clinical Neuroscience, 11*, 190-208.

Humphrey, N. (1992). *A history of the mind.* New York: Simon and Schuster.

Hunt H.T. (2001): Some perils of quantum consciousness: epistemological panexperientialism and the emergence-submergence of consciousness. *Journal of Consciousness Studies*, *8*, 35-45.

Huxley, T. H. (1866). *Lessons in elementary physiology.* (Quoted by N. Humphrey, 1992).

Jackendoff, R. (1987). *Consciousness and the computational mind*. Cambridge, MA: MIT Press.

James, W. (1890). *The principles of psychology.* New York: Henry Holt and Co. [Republished in 1950, New York, Dover Books]
Jibu, M., Hagan, S., Hameroff, S., Pribram, K. and Yasue, K. (1994). Quantum optical coherence in cytoskeleton microtubules: implications for brain function. *BioSystems, 32*, 195-209.
Jibu, M., Pribram, K. and Yasue, K. (1996). From conscious experience to memory storage and retrieval: The role of quantum brain dynamics and boson condensation of evanescent protons. *International Journal of Modern Physics B, 10*, 1735-1754.
John, E. R. (1968). Observation learning in cats. *Science, 159*, 1489-1491.
John, E. R. (1972). Switchboard versus statistical theories of learning and memory. *Science, 177,* 850-864.
John, E. R. (2001). A field theory of consciousness. *Consciousness and Cognition, 10*, 18 4-213.
John, E. R. (2002). The neurophysics of consciousness. *Brain Research Reviews, 39*, 1-28.
John, E. R. and Schwartz, E. (1978). The neurophysiology of information processing and cognition. *Annual Review in Psychology, 29*, 1-29.
Kaku, M. (1994). *Hyperspace.* New York: Oxford University Press.
Kauffman, S. (1996). *At home in the universe: The search for laws of self-organization and complexity*. New York: Viking.
Kim, J. (1978). Supervenience and nomological incommensurables. *American Philosophical Quarterly, 15*, 149-156.
Kim, J. (1984). Concepts of supervenience. *Philosophy and Phenomenological Research, 65*, 153-176.
Kim, J. (1987). "Strong" and "global" supervenience revisited. *Philosophy and Phenomenological Research, 68*, 315-326.
Kim, J. (1990). Concepts of supervenience. *Metaphilosophy, 21*, 1-27.
Kim, J. (1993). *Supervenience and mind: Selected philosophical essays.* Cambridge, UK: Cambridge University Press.
Kim, J. (1998). *Mind in a physical world: An essay on the mind-body problem and mental causation.* Cambridge, MA, MIT Press.
Kim, J. (1999). Supervenient properties and micro-based properties: A reply to Noordhof. *Proceedings of the Aristotelian Society, 99*, 115-117.
Korzeniewski, B. (2001). Cybernetic formulation of the definition of life. *Journal of Theoretical Biology, 209*, 275-286.
Laszlo, E. (1993). *The creative cosmos: A unified science of matter, life and mind.* Edinburgh, UK: Floris Books.
Leff, H. S. and Rex, A. F. (1990). *Maxwell's demon: Entropy, information, computing.* Princeton, NJ: Princeton University Press.
Levine, J. (1983). Materialism and qualia: the explanatory gap. *Pacific Philosophical Quarterly, 64*, 354-361.
Libet, B. (1993). *Neurophysiology of consciousness.* Boston, MA: Birkhäuser.
Libet, B. (1994). A testable field theory of mind-brain interaction. *Journal of Consciousness Studies, 1*, 119-126.
Libet, B. (1996a). Conscious mind as a field. *Journal of Theoretical Biology, 178*, 223-224.
Libet, B. (1996b). Solutions to the hard problem of consciousness. *Journal of Consciousness Studies, 3*, 33-35.
Libet, B. (1996c). Neural processes in the production of conscious experience. In M. Velmans (Ed.), *The Science of consciousness.* London: Routledge, pp. 96-117.
Libet B. (1999). Do we have free will? *Journal of Consciousness Studies, 6*, 47-57.

Libet, B. (2003). Can conscious experience affect brain activity? *Journal of Consciousness Studies, 10*, 24-28.

Lifson, S. (1987). Chemical selection, diversity, teleonomy and the second law of thermodynamics: Reflections of Eigen's theory of self-organization of matter. *Biophysical Chemistry*, *26*, 303-311.

Lipkind, M. (1998): The concepts of coherence and "binding problem" as applied to life and consciousness realms: Critical consideration with positive alternative. In J. J. Chang, J. Fisch and F.-A. Popp (Eds.), *Biophotons.* Dordrecht: Kluwer Academic Publishers, pp. 359-373.

Lipkind, M. (2003). Definition of consciousness: Impossible and unnecessary? In F.-A. Popp and L. Beloussov (Eds.), *Integrative biophysics*. Dordrecht: Kluwer Academic Publishers, pp. 467-503.

Lipkind, M. (2005). The field concept in current models of consciousness: A tool for solving the Hard Problem? *Mind and Matter*, *3*, 29-85.

Lipkind, M. (2007). Formulation of Protophenomenal Fundamental and Rudimentary Psychic Act: Vitalistic Theory of Consciousness. In S. Turrini (Ed.), *Consciousness and Learning Research.* New York, Nova Science Publishers, Inc., pp. 1-45.

Lipkind, M. (2007). Free will and violation of physical laws: A new concept of volition based on A. Gurwitsch's field theory. In L. V. Beloussov, V. Voeikov and V. N. Martynyuk (Eds.), *Biophotonics and coherent systems in biology.* New York: Springer Science US, pp. 235-277.

Lipkind, M. (2008). Enigma of "The great encephalization": Explanation by means of irreducible field principle. *Advanced Science Letters, 1*, 199-211.

Lipkind, M. (2009). The hard problem and naturalistic meaning of the extra ingredient. In A. Batthyani and A. Elitzur (Eds.), *Irreducibly Conscious. Selected Papers on Consciousness.* Heidelberg, Germany: Universitätsverlag Winter, pp. 207-301.

Lisman J. (1998). What makes the brain's tickers tick. *Nature, 394*, 132-133.

Livingstone, M. S. and Hubel, D. H. (1987). Psychophysical evidence for separate channels for the perception of form, color, movement, and depth. *The Journal of Neuroscience, 7*, 3416-3468.

Loeb, J. (1906). *The dynamics of living matter*. New York: Columbia University Press.

Loeb, J. (1912). *The mechanistic conception of life*. Chicago, IL: University Chicago Press.

Lotka, A. (1925). *Elements of physical biology.* Baltimore: Williams and Wilkins. [Republished in 1956 as *Elements of Mathematical Biology,* New York, Dover].

Mahler, G. (2004). The partitioned quantum universe: Entanglement and emergence of functionality. *Mind and Matter, 2*, 67-89.

Marshall, I. N. (1989). Consciousness and Bose-Einstein condensates. *New Ideas in Psychology*, *7*, 73-83.

Mayr, E. (1961). Cause and effect in biology. *Science, 134*, 1501-1506.

McDougall, W. (1938). *The riddle of life: A survey of theories*. London: Methuen.

McFadden, J. (2000). *Quantum evolution*. London: Harper Collins.

McFadden, J. (2002a). Synchronous firing and its influence on the brain's electromagnetic field: Evidence for an electromagnetic theory of consciousness. *Journal of Consciousness Studies, 9, No. 4*, 23-50.

McFadden, J. (2002b). The conscious electromagnetic information field theory: The hard problem made easy? *Journal of Consciousness Studies, 9*, *No. 8*, 45-60.

McGinn C. (1991). *The problem of consciousness: Toward a solution*. Oxford, UK: Blackwell.

Metzinger T. (1995). The problem of consciousness. In T. Metzinger (ed.), *Conscious experience.* Paderborn, Ferdinand Schöning, Imprint Academic, pp. 3-43.

Metzinger, T. (2000). *Neural correlates of consciousness: Empirical and conceptual questions.* Cambridge, MA: Bradford Book, MIT Press.

Mölle, M., Marshall, L., Wolf, B., Fehm, H. L. and Bohm E. (1999). EEG complexity and performance measures of creative thinking. *Journal of Psychophysiology, 36*, 95-104.

Monod, J. (1971). *Chance and necessity: An essay on the natural philosophy of modern biology.* New York: Knopf.

Montero, B. (2003). Varieties of causal closure. In S. Walter and H.-D. Heckmann (Eds.), *Physicalism and mental causation: The metaphysics of mind and action.* Exeter, UK, Imprint Academic, pp. 173-187.

Moore, G. E. (1922). *Philosophical studies.* London: Routledge and Kegan Paul.

Näätänen, R., Ilmoniemi, R. J., and Alho, K. (1994). Magnetoencephalography in studies of human cognitive brain function. *Trends in Neuroscience, 17*, 389-395.

Nagel, E. and Newman, J. (1958). *Gödel's proof.* London, UK: Routledge and Kegan Paul.

Pagels, H. R. (1985). *Perfect symmetry.* New York: Simon and Schuster.

Penrose, R. (1989). *The emperor's new mind.* Oxford, UK: Oxford University Press.

Penrose, R. (1994). *Shadows of the mind.* Oxford, UK: Oxford University Press.

Penrose, R. (1996). Beyond the doubting of a shadow. *Psyche, 2*, 89-129.

Penrose R. (1997). On understanding understanding. *International Studies on the Philosophy of Science, 11*, 7-20.

Penrose, R. (2000). Wave function collapse as a real gravitational effect. In A. Fokas, A. Grigouryan, T. Kibble, and B. Zegarlinski (Eds), *Mathematical physics 2000.* London, UK: Imperial College.

Penrose R. (2001). Consciousness, the brain, and space-time geometry: an addendum. Some new developments on the Orch OR model for consciousness. In P.C. Marijuan (Ed.), Cajal and consciousness: Scientific approaches to consciousness on the centennial of ramón y cajal's textura. *Annals of the New York Academy of Science, 929*, 74-104.

Pessa, E. and Vitiello, G. (2003). Quantum noise, entanglement and chaos in the quantum field theory of mind/brain states. *Mind and Matter, 1*, 59-79.

Pockett, S. (1999). Anaesthesia and the electrophysiology of auditory consciousness. *Consciousness and Cognition, 8*, 45-61.

Pockett, S. (2000). *The nature of consciousness: An hypothesis.* Lincoln NE: Universe Ltd.

Pockett, S. (2002). Difficulties with the electromagnetic field theory of consciousness. *Journal of Consciousness Studies, 9*, 51-56.

Prigogine, I and Stengers, I. (1984). *Order out of chaos.* New York: Bantam.

Primas, H. (2003). Time – entanglement between mind and matter. *Mind and Matter, 1* 81-119.

Raichle, M. E. (1998). The correlates of consciousness: an analysis of cognitive skill learning. *Philosophical Transactions of Royal Society, London B, 353*, 1889-1901.

Ramachandran, V. S. (1990). Visual perception in people and machines. In A. Balke and T. Trosciankopp (Eds.), *AI and the eye.* New York: John Wiley and Sons, pp. 21-77.

Ramachandran, V. S. and Anstis, S. M. (1986). The perception of apparent motion. *Scientific American, 254*, 101-109.

Ramón y Cajal, S. (1899-1904). *Textura del sistema nervioso del hombre y de los vertebrados.* Madrid: N. Moya. (In Spanish) [1995, Histology of the Nervous System of Man and the Vertebrates] New York, Oxford University Press (Translated by L. and N. Swanson).

Revonsuo, A. (2000). Prospects for a scientific research program on consciousness. In T. Metzinger (Ed.), *Neural correlates of consciousness.* Cambridge, MA: Bradford Book, MIT Press, pp. 57-75.

Ribary, U., Ioannides, A. A., Singh, K. D., Hasson, R., Bolton, J. P., Lado, F., Mogilner, A., and Llinas, R. (1991). Magnetic field tomography of coherent thalamocortical 40 Hz oscillations in humans. *Proceedings of the National Academy of Sciences of US, 88,* 1037- 1041.

Ricciardi, L.M. and Umezava, H. (1967). Brain physics and many-body problems. *Kybernetik*, *4*, 44-48.

Robinson W. S. (1996). The hardness of the hard problem. *Journal of Consciousness Studies, 3*, 14-25.

Romijn, H. (2002). Are virtual photons the elementary carriers of consciousness? *Journal of Consciousness Studies*, *9,* 61-81.

Rosenberg, G. H. (1996). Rethinking nature: hard problem within the hard problem. *Journal of Consciousness Studies*, *3*, 76-88.

Rudd A. J. (2000). Phenomenal judgment and mental causation. *Journal of Consciousness Studies, 7*, 53-66.

Schacter, D.L., Buckner, R.L., and Koutstraal, W. (1998). Memory, consciousness and neuroimaging. *Philosophical Transactions of the Royal society, London B, 353,*1861-1878.

Seager W. (1995). Consciousness, information and panpsychism. *Journal of Consciousness Studies, 2,* 272-288.

Searle, J. (2000a). Consciousness and free action. In *Toward a Science of Consciousness, Tucson 2000, Consciousness Research Abstracts*, Tucson AZ, April 10-15, 2000, p. 70, Abstract 123, PL9.

Searle, J. (2000b). Consciousness. *Annual Revue of Neuroscience, 23,* 557-578.

Searle, J. (2000c). Consciousness, free action, and the brain. *Journal of Consciousness Studies, 7*, 3-22.

Shear J. (1995). Explaining consciousness – the 'Hard Problem', Editor's introduction. *Journal of Consciousness Studies, 2,* 194-199.

Sheldrake, R. (1981). *A new science of life: the hypothesis of formative causation*. Los Angeles, CA: Tarcher.

Silberstein M. (2001). Converging on emergence: Consciousness, causation and explanation. *Journal of Consciousness Studies, 8,* 61-98.

Sirag, S.-P. (1993). Consciousness: A hyperspace view. In J. Mishlove (Ed.), *Roots of consciousness.* Tulsa: Council Oak, pp. 327-365 [Appendix].

Sirag, S.-P. (1996). A mathematical strategy for a theory of consciousness. In S. R. Hameroff, A. W. Kaczniak, and A. C. Scott (Eds.), *Toward a Science of Cconsciousness.* Cambridge, MA: MIT Press, pp. 579-588.

Squires, E. J. (1998). Why are quantum theorists interested in consciousness? In S. R. Hameroff, A W. Kaczniak, and A. C. Scott (Eds.), *Toward a Science of Consciousness, Tucson II.* Cambridge, MA:. MIT Press, pp. 609-618.

Stapp, H. P. (1993). *Mind, matter, and quantum mechanics.* Berlin, Germany: Springer Verlag.

Stapp, H. P. (2001). Quantum theory and the role of mind in nature. *Foundations of Physics*, *31,* 1465-1499.

Stuart, C. I. J., Takehashi, Y., and Umezava, H. (1978). On the stability and non-local properties of memory. *Journal of Theoretical Biology, 71*, 605-618.

Stuart, C. I. J., Takehashi, Y., and Umezava, H. (1979). Mixed system brain dynamics: neural memory as a macroscopic ordered state. *Foundations of Physics, 9*, 301-327.

Sutherland, N. S. (1989). Consciousness. In N.S. Sutherland (Ed.), *The Macmillan dictionary of psychology.* London: Macmillan, p. 90.

Szent-Gyorgi, A. (1957). *Bioenergetics*. New York: Academic Press.

Tallon-Baudry, C. (2000). Oscillatory synchrony as a signature for the unity of visual experience in humans. *Consciousness and Cognition, 9*, 25-26.

Tononi, G., Srinivasan, R., Russell, D. and Edelman, G. (1998). Investigating neural correlates of conscious perception by frequency-tagged neuromagnetic responses. *Proceedings of National Academy of Sciences USA, 95*, 3198-3203.

Treisman. A. (1986). Features and objects in visual processing. *Scientific American, 254*, 114-125.

Tyndall, J. (1879). *Fragments of science: A series of detached essays, addresses and reviews.* London: Longmans.

Van Gulick, R. (1995). What would count as explaining consciousness? In T.Metzinger (Ed.), *Conscious experience.* Paderborn, Ferdinand Schoening, Imprint Academic, pp. 61-79.

Van Gulick, R. (2000). Inward and upward: Reflection, interspection, and self-awareness. *Philosophical Topics, 28*: 275-305.

Van Gulick, R. (2001). Reduction, emergence and other recent options on the mind/body problem: A philosophical review. *Journal of Consciousness Studies, 8,* 1-34.

Varela F.J. (1996). Neurophenomenology: A methodological remedy for the hard problem. *Journal of Consciousness Studies, 3*, 330-349.

Vitiello, G. (1995). Dissipation and memory capacity in the quantum brain model. *International Journal of Modern Physics B, 9,* 973-989.

Von der Malsburg C. (1987). Synaptic plasticity as basis of brain organization. In J.-P. Changeux and M. Konishi (Eds.), *The neural and molecular bases of learning*. New York: John Wiley and Sons, pp. 411-432.

Wächterhauser, G. (1988). Before enzymes and templates: theory of surface metabolism. *Microbiological Reviews, 52,* 452-484.

Waddington, C. H. (1966). Fields and gradients. In M. Locke (Ed.), *Major problems in developmental biology.* New York and London: Academic Press, pp. 105-124.

Weingard R.(2001). A philosopher looks at string theory. In C. Callender and N. Huggett (Eds.), *Physics meets philosophy at the Planck scale*. Cambridge, UK: Cambridge University Press, pp. 138-151.

Weiss, P. (1939). *Principles of development.* New York: Holt, Rinehart and Winston.

Welch, G. R. (1992). An analogical 'field' construct in cellular biophysics: history and present status. *Progress in Biophysics and Molecular Biology, 57*, 71-128.

Wilkes, K. V. (1988). –, yishi, duh, um, and consciousness. In A. Marcel and E. Bisiach (Eds.), *Consciousness in contemporary science.* Oxford, UK: Oxford University Press, pp. 17-41.

Wilson, D. L. (1999). Mind-brain interaction and violation of physical laws. *Journal of Consciousness Studies, 6*, 185-200.

Zeki, S. (1992). The visual image in mind and brain. *Scientific American 254*, 68-77.

Zeki, S. (2000). The disunity of consciousness. *Consciousness and Cognition, 9*, S30.

Zohar, D. (1996). Consciousness and Bose-Einstein condensates. In S. Hameroff, A. Kaszniak and A. Scott (Eds.), *Toward a science of consciousness (The 1st Tucson Discussions and Debates).* Cambridge, MA: Bradford Book, MIT Press, pp. 439-450.

INDEX

B

C

D

E

F

G

H

I

J

K

L

M

N

O

P

Q

R

S

T

U

V

W

Y

Z